Lecture Notes in Computer Scie

Commenced Publication in 1973
Founding and Former Series Editors:
Gerhard Goos, Juris Hartmanis, and Jan van Leeuwen

Steven D. Galbraith (Ed.)

Cryptography and Coding

11th IMA International Conference
Cirencester, UK, December 18-20, 2007
Proceedings

 Springer

Volume Editor

Steven D. Galbraith
Royal Holloway University of London
Mathematics Department
Egham, Surrey, TW20 0EX, UK
E-mail: steven.galbraith@rhul.ac.uk

Library of Congress Control Number: 2007940956

CR Subject Classification (1998): E.3-4, G.2.1, C.2, J.1

LNCS Sublibrary: SL 4 – Security and Cryptology

ISSN   0302-9743
ISBN-10   3-540-77271-5 Springer Berlin Heidelberg New York
ISBN-13   978-3-540-77271-2 Springer Berlin Heidelberg New York

Springer is a part of Springer Science+Business Media

springer.com

© Springer-Verlag Berlin Heidelberg 2007
Printed in Germany

Typesetting: Camera-ready by author, data conversion by Scientific Publishing Services, Chennai, India
Printed on acid-free paper   SPIN: 12203093   06/3180   5 4 3 2 1 0

Preface

The 11th IMA Conference on Cryptography and Coding was held at the Royal Agricultural College, Cirencester, UK during December 18–20, 2007. As usual, the venue provided a relaxed and convivial atmosphere for attendees to enjoy the conference programme and discuss current and future research ideas.

The programme comprised three invited talks and 22 contributed papers. The invited speakers were Jonathan Katz (University of Maryland, USA), Patrick Solé (Ecole Polytechnique de l'Université de Nice-Sophia Antipolis, France) and Whit Diffie (Sun Microsystems, USA). Special thanks are due to these speakers. Two of the invited speakers provided papers, included in this volume, which highlight the connections between cryptography, coding theory and discrete mathematics.

The contributed talks were selected from 48 submissions. The accepted papers cover a range of topics in mathematics and computer science, including symmetric and public key cryptography, Boolean functions, sequences, efficient implementation and side-channel analysis.

I would like to thank all the people who helped with the conference programme and organization. First, I thank the Steering Committee for their guidance on the general format of the conference and for suggestions of members of the Programme Committee. I also heartily thank the Programme Committee and the sub-reviewers listed on the following pages for their thoroughness during the review process. Each paper was reviewed by at least three people. There was significant online discussion about a number of papers.

The submission and review process was greatly simplified by the ichair software developed by Thomas Baignères and Matthieu Finiasz. Thanks also to Jon Hart for running the submissions Web server and Sriram Srinivasan for designing and maintaining the conference Web page.

Thanks go to the authors of all submitted papers. I also thank the authors of accepted papers for revising their papers according to referee suggestions and returning latex source files in good time. The revised versions were not checked by the Programme Committee so authors bear full responsibility for their contents. I thank the staff at Springer for their help with producing the proceedings.

I thank Hewlett-Packard and Vodafone for their sponsorship of this event.

Finally, I wish to thank the conference staff of the Institute for Mathematics and its Applications, especially Lucy Nye and Sammi Lauesen, for their help with running the conference and handling the finances.

October 2007   Steven Galbraith

Cryptography and Coding 2007

Royal Agricultural College, Cirencester, UK
December 18–20, 2007

Sponsored by
The Institute of Mathematics and its Applications

in cooperation with
Hewlett-Packard Laboratories and *Vodafone Ltd.*

Programme Chair

Steven Galbraith — Royal Holloway University of London

Steering Committee

Bahram Honary — Lancaster University
Chris Mitchell — Royal Holloway
Kenny Paterson — Royal Holloway
Fred Piper — Royal Holloway
Nigel Smart — University of Bristol
Mike Walker — Vodafone Ltd. and Royal Holloway

Programme Committee

Steve Babbage — Vodafone Group Services Ltd.
Nigel Boston — South Carolina/Wisconsin
Pascale Charpin — INRIA Rocquencourt
Liqun Chen — Hewlett-Packard
Carlos Cid — Royal Holloway
YoungJu Choie — Postech, Korea
Arjen Lenstra — EPFL, Lausanne
Alexander May — Ruhr Universität Bochum
Gary McGuire — University College Dublin
Alfred Menezes — University of Waterloo
David Naccache — Ecole Normale Supérieure
Matthew Parker — University of Bergen
Matt Robshaw — France Telecom
Ana Sălăgean — Loughborough University
Berry Schoenmakers — Technical University Eindhoven
Michael Scott — Dublin City University
Amin Shokrollahi — EPFL, Lausanne
Nigel Smart — University of Bristol
Frederik Vercauteren — K. U. Leuven
Gilles Zemor — Université Bordeaux

External Reviewers

Joppe Bos	Christophe De Cannière	Anne Canteaut
Claude Carlet	Mahdi Cheraghchi	Alex Dent
Fabien Galand	Philippe Guillot	Darrel Hankerson
Marcelo Kaihara	Cedric Lauradoux	Lorenz Minder
Marine Minier	Stephane Manuel	Ashley Montanaro
Sean Murphy	Dag Arne Osvik	Gregory Neven
Dan Page	Kenny Paterson	Benny Pinkas
Louis Salvail	Amin Shokrollahi	Andrey Sidorenko
Patrick Solé	Martijn Stam	Søren Steffen Thomsen
Eran Tromer	José Villegas	

Table of Contents

Side Channels

Linear Complexity

Public Key Encryption

Curves

RSA Implementation

Signatures II

Efficient Cryptographic Protocols Based on the Hardness of Learning Parity with Noise

Jonathan Katz[*]

Dept. of Computer Science
University of Maryland
jkatz@cs.umd.edu

Abstract. The problem of *learning parity with noise* (The LPN problem), which can be re-cast as the problem of decoding a random linear code, has attracted attention recently as a possible tool for developing highly-efficient cryptographic primitives suitable for resource-constrained devices such as RFID tags. This article surveys recent work aimed at designing efficient authentication protocols based on the conjectured hardness of this problem.

1 Introduction

1.1 The LPN Problem

Fix a binary vector (i.e., a bit-string) $\mathbf{s}$ of length k. Given a sequence of randomly-chosen binary vectors $\mathbf{a}_1, \ldots, \mathbf{a}_\ell$ along with the values of their inner-product $z_i = \langle \mathbf{s}, \mathbf{a}_i \rangle$ with $\mathbf{s}$, it is a simple matter to reconstruct $\mathbf{s}$ in its entirety as soon as ℓ is slightly larger than k. (All that is needed is to wait until the set $\{\mathbf{a}_i\}$ contains k linearly-independent vectors.) In the presence of *noise*, however, where each bit z_i is flipped (independently) with probability ε, determining $\mathbf{s}$ becomes much more difficult. We refer to the problem of learning $\mathbf{s}$ in this latter case as the problem of learning parity with noise, or *the LPN problem*.

Formally, let Ber_ε be the Bernoulli distribution with parameter $\varepsilon \in (0, \frac{1}{2})$ (so if $\nu \sim \mathsf{Ber}_\varepsilon$ then $\Pr[\nu = 1] = \varepsilon$ and $\Pr[\nu = 0] = 1 - \varepsilon$), and let $A_{\mathbf{s},\varepsilon}$ be the distribution defined by:

$$\left\{ \mathbf{a} \leftarrow \{0,1\}^k; \nu \leftarrow \mathsf{Ber}_\varepsilon : (\mathbf{a}, \langle \mathbf{s}, \mathbf{a} \rangle \oplus \nu) \right\}.$$

Let $A_{\mathbf{s},\varepsilon}$ also denote an oracle which outputs (independent) samples according to this distribution. Algorithm M is said to (t, q, δ)-*solve the* LPN_ε *problem* if

$$\Pr\left[\mathbf{s} \leftarrow \{0,1\}^k : M^{A_{\mathbf{s},\varepsilon}}(1^k) = \mathbf{s}\right] \geq \delta,$$

and furthermore M runs in time at most t and makes at most q queries to its oracle. (This formulation of the LPN problem follows [18]; an alternative but

---

[*] Supported in part by NSF CyberTrust grant #0627306 and NSF CAREER award #0447075.

essentially equivalent formulation allows M to output any $\mathbf{s}$ satisfying at least a $(1-\varepsilon)$ fraction of the equations returned by $A_{\mathbf{s},\varepsilon}$.) In asymptotic terms, the LPN_ε problem is "hard" if every probabilistic polynomial-time algorithm M solves the LPN_ε problem with only negligible probability (where the algorithm's running time and success probability are functions of k).

Note that ε is usually taken to be a fixed constant independent of k, as we will assume here. The value of ε to use depends on a number of tradeoffs and design decisions: although, roughly speaking, the LPN_ε problem becomes "harder" as ε increases, a larger value of ε also affects the error rate (for honest parties) in schemes based on the LPN problem; this will becomes more clear in the sections that follow. For concreteness, the reader can think of $\varepsilon \approx \frac{1}{8}$.

The hardness of the LPN_ε problem, for any constant $\varepsilon \in (0, \frac{1}{2})$, has been studied in many previous works. It can be formulated also as the problem of decoding a random linear code, and is known to be $\mathcal{NP}$-complete [2] as well as hard to approximate within a factor better than 2 (where the optimization problem is phrased as finding an $\mathbf{s}$ satisfying the most equations) [12]. These worst-case hardness results are complemented by numerous studies of the average-case hardness of the problem [3,4,6,20,13,14,24]. Currently, the best algorithms for solving the LPN_ε problem [4,9,23] require $t, q = 2^{\Theta(k/\log k)}$ to achieve $\delta = \mathcal{O}(1)$. We refer the reader to [23] for additional heuristic improvements, as well as a tabulation of the time required to solve the LPN_ε problem (for various settings of the parameters) using the best-known algorithm.

The LPN problem can be generalized to fields other than $\mathbb{F}_2$ (or even other algebraic structures such as rings), and these generalizations have interesting cryptographic consequences also [24]. Such extensions will not be discussed here.

1.2   Cryptographic Applications of the LPN Problem

It is not too difficult to see that hardness of the LPN_ε problem implies the existence of a one-way function. More interesting is that such hardness would imply *efficient* and *direct* constructions of pseudorandom generators [3,24]; see Lemma 1 for an indication of the basic underlying ideas. Furthermore, generating an instance of the distribution $A_{\mathbf{s},\varepsilon}$ is extremely "cheap", requiring only k bit-wise "AND" operations and $k-1$ "XOR" operations.[1] Finally, as mentioned earlier, the best-known algorithms for solving the LPN_ε problem are only slightly sub-exponential in the length k of the hidden vector. Taken together, these observations suggest the possibility of using the LPN_ε problem to construct efficient cryptographic primitives and protocols, as first suggested in [3].

Actually, if the LPN_ε problem is indeed "hard" enough, there is the potential of using it to construct *extremely* efficient cryptographic primitives, suitable either for implementation by humans (using pencil-and-paper) [13,14] or for implementation on low-cost radio-frequency identification (RFID) tags [17] or sensor nodes. Focusing on the case of RFID tags, Juels and Weis [17,25] estimate

---

[1] This assumes that generating the appropriate random coins is "free", which may not be a reasonable assumption in practice.

that current RFID tags contain, in the best case, ≈ 2000 gate equivalents that can be dedicated to performing security functions; even optimized block cipher implementations may require many more gates than this (see [8,1] for current state-of-the-art).

1.3   Efficient Authentication Based on the LPN Problem

In the remainder of this work, we survey recent work directed toward developing authentication protocols based on the LPN problem; these protocols have been suggested as suitable for the secure identification of RFID tags. All protocols we will consider are intended for the shared-key (i.e., symmetric-key) setting, and provide unidirectional authentication only; typically, this would permit an RFID tag, acting as a *prover*, to authenticate itself to a tag reader acting as a *verifier*. We begin with a brief outline of the history of the developments, and defer all technical details to the sections that follow.

The first protocol we will present — following [17], we will refer to it as the *HB protocol* — was introduced by Hopper and Blum [13,14] and provides security against a passive (eavesdropping) adversary. Juels and Weis [17,25] were the first to rigorously prove security of the HB protocol, and to suggest its use for RFID authentication. (Hopper and Blum proposed it as a way to authenticate humans using pencil-and-paper only.) Juels and Weis also proposed a second protocol, called HB^+, that could be proven secure against an active attacker who can impersonate the tag reader to an RFID tag. In each case, Juels and Weis focus on a single, "basic authentication step" of the protocol and prove that a computationally-bounded adversary cannot succeed in impersonating a tag in this case with probability noticeably better than $1/2$; that is, a single iteration of the protocol has *soundness error* $1/2$. The implicit assumption is that repeating these "basic authentication steps" sufficiently-many times yields a protocol with negligible soundness error, though this intuition was not formally proven by Juels and Weis.

Two papers of my own [18,19] (along with Ji-Sun Shin and Adam Smith) provide a simpler and improved analysis of the HB and HB^+ protocols. Besides giving what is arguably a cleaner framework for analyzing the security of these protocols, the proofs in these works also yield the following concrete improvements: (1) they show that the HB^+ protocol remains secure under arbitrary concurrent executions of the protocol; this, in particular, means that the HB^+ protocol can be *parallelized* so as to run in 3 rounds (regardless of the desired soundness error); (2) the proofs explicitly incorporate the dependence of the soundness error on the number of iterations of a "basic authentication step"; and (3) the proofs deal with the inherent error probability in even honest executions of the protocol. (The reader is referred to [18] for further detailed discussion of these points.) The initial work [18] was limited to the case of $\varepsilon < 1/4$; subsequent work [19] extended these results to the case of arbitrary $\varepsilon < 1/2$.

In work tangential to the above, Gilbert et al. [10] show that the HB^+ protocol is not secure against a man-in-the-middle attack, in the sense that a man-in-the-middle attacker is able to reconstruct the entire secret key of the RFID tag after sufficiently-many interactions. (The reader is additionally referred to the work of Wool et al. [21,22], for an illuminating discussion on the feasibility of man-in-the-middle attacks in RFID systems.) This has motivated numerous proposals (e.g., [5]) of HB-variants that are claimed to be secure against the specific attack of Gilbert et al., but I am not aware of any HB-variant that is provably-secure against *all* man-in-the-middle attacks. To my mind, the existence of a man-in-the-middle attack on the HB$^+$ protocol shows that the protocol must be used with care, but does not rule out its usefulness; specifically, I view the aim of the line of research considered here to be the development of protocols which are *exceptionally* efficient while still guaranteeing *some* useful level of (provable) security. The possibility of man-in-the-middle attacks does *not* mean that it is useless to explore the security of authentication protocols in weaker attack models. Furthermore, as a practical matter, Juels and Weis [17, Appendix A] note that the man-in-the-middle attack of [10] does not apply in a *detection-based* system where numerous failed authentication attempts immediately raise an alarm. Nevertheless, the design of an HB-variant with provable security against man-in-the-middle attacks remains an interesting open problem.

1.4   Overview of This Paper

The remainder of this paper is devoted to a description of the HB and HB$^+$ protocols, as well as the technical proofs of security for these protocols (adapted from [18,19]). It is not the goal of this paper to replace [18,19]; instead, the main motivation is to give a high-level treatment of the proofs with a focus on those aspects that might be of greatest interest to coding-theorists. Proof steps that are "technical" but otherwise uninteresting will be glossed over, and some proofs are omitted entirely. The interested reader can find full details of all proofs in [18,19].

2   Definitions and Preliminaries

We have already formally defined the LPN problem in the Introduction. Here, we state and prove the main technical lemma on which we will rely. We also define notion(s) of security for identification; these are standard, but some complications arise due to the fact that the HB/HB$^+$ protocols do not have perfect completeness.

2.1   A Technical Lemma

In this section we prove a key technical lemma due to Regev [24, Sect. 4] (though without the explicit dependence on the parameters given below, which is taken from [18]): hardness of the LPN$_\varepsilon$ problem implies "pseudorandomness" of $A_{\mathbf{s},\varepsilon}$. Specifically, let U_{k+1} denote the uniform distribution on $(k+1)$-bit strings. The following lemma shows that oracle access to $A_{\mathbf{s},\varepsilon}$ (for randomly-chosen $\mathbf{s}$) is indistinguishable from oracle access to U_{k+1}.

Lemma 1. *Say there exists an algorithm D making q oracle queries, running in time t, and such that*

$$\left| \Pr\left[\mathbf{s} \leftarrow \{0,1\}^k : D^{A_{\mathbf{s},\varepsilon}}(1^k) = 1 \right] - \Pr\left[D^{U_{k+1}}(1^k) = 1 \right] \right| \geq \delta.$$

Then there exists an algorithm M making $q' = O\left(q \cdot \delta^{-2} \log k\right)$ oracle queries, running in time $t' = O\left(t \cdot k\delta^{-2} \log k\right)$, and such that

$$\Pr\left[\mathbf{s} \leftarrow \{0,1\}^k : M^{A_{\mathbf{s},\varepsilon}}(1^k) = \mathbf{s} \right] \geq \delta/4.$$

(Various tradeoffs are possible between the number of queries/running time of M and its success probability in solving LPN_ε; see [24, Sect. 4]. We do not discuss these here.)

Proof (Sketch). Algorithm $M^{A_{\mathbf{s},\varepsilon}}(1^k)$ proceeds as follows:

1. Fix random coins for D.
2. Estimate the probability that D outputs 1 when it interacts with oracle U_{k+1}. Call this estimate p.
3. For $i \in [k]$ do:
   (a) Estimate the probability that D outputs 1 when it interacts with an oracle implementing the following distribution:

   $$\mathsf{hyb}_i \overset{\mathrm{def}}{=} \left\{ \mathbf{a} \leftarrow \{0,1\}^k; c \leftarrow \{0,1\}; \nu \leftarrow \mathsf{Ber}_\varepsilon : (\mathbf{a} \oplus (c \cdot \mathbf{e}_i), \langle \mathbf{s}, \mathbf{a} \rangle \oplus \nu) \right\},$$

   where $\mathbf{e}_i$ is the vector with 1 at position i and 0s elsewhere. Note that M can generate this distribution using its own access to oracle $A_{\mathbf{s},\varepsilon}$. Call the estimate obtained in this step p_i.
   (b) If $|p_i - p| \geq \delta/4$ set $s_i' = 0$; else set $s_i' = 1$.
4. Output $\mathbf{s}' = (s_1', \ldots, s_k')$.

Let us analyze the behavior of M. First, note that with "high" probability over choice of $\mathbf{s}$ and random coins for D it holds that

$$\left| \Pr\left[D^{A_{\mathbf{s},\varepsilon}}(1^k) = 1 \right] - \Pr\left[D^{U_{k+1}}(1^k) = 1 \right] \right| \geq \delta/2, \tag{1}$$

where the probabilities are now taken only over the answers D receives from its oracle. We restrict our attention to $\mathbf{s}, \omega$ for which Eq. (1) holds and show that in this case M outputs $\mathbf{s}' = \mathbf{s}$ with probability at least $1/2$. The lemma follows.

Setting the accuracy of our estimations appropriately, we can ensure that

$$\left| \Pr\left[D^{U_{k+1}}(1^k; \omega) = 1 \right] - p \right| \leq \delta/16 \tag{2}$$

except with probability at most $O(1/k)$. Now focus on a particular iteration i of steps 3(a) and 3(b). We may once again ensure that

$$\left| \Pr\left[D^{\mathsf{hyb}_i}(1^k; \omega) = 1 \right] - p_i \right| \leq \delta/16 \tag{3}$$

except with probability at most $O(1/k)$. Applying a union bound (and setting parameters appropriately) we see that with probability at least $1/2$ both Eqs. (2)

and (3) hold (the latter for all $i \in [k]$), and so we assume this to be the case for the rest of the proof.

An easy observation is that if $s_i = 0$ then $\mathsf{hyb}_i = A_{\mathbf{s},\varepsilon}$, while if $s_i = 1$ then $\mathsf{hyb}_i = U_{k+1}$. It follows that if $s_i = 0$ then

$$\left| \Pr\left[D^{\mathsf{hyb}_i}(1^k; \omega) = 1 \right] - \Pr\left[D^{U_{k+1}}(1^k; \omega) = 1 \right] \right| \geq \delta/2$$

(by Eq. (1)), and so $|p_i - p| \geq \frac{\delta}{2} - 2 \cdot \frac{\delta}{16} = \frac{3\delta}{8}$ (by Eqs. (2) and (3)) and $s_i' = 0 = s_i$. When $s_i = 1$ then

$$\Pr\left[D^{\mathsf{hyb}_i}(1^k; \omega) = 1 \right] = \Pr\left[D^{U_{k+1}}(1^k; \omega) = 1 \right],$$

and so $|p_i - p| \leq 2 \cdot \frac{\delta}{16} = \frac{\delta}{8}$ (again using Eqs. (2) and (3)) and $s_i' = 1 = s_i$. Since this holds for all $i \in [k]$, we conclude that $\mathbf{s}' = \mathbf{s}$.   □

2.2 Overview of the HB/HB$^+$ Protocols, and Security Definitions

The HB and HB$^+$ protocols as analyzed here consist of n *parallel* iterations of a "basic authentication step." (As remarked in the Introduction, the fact that these iterations can be run in parallel follows from the proofs we will give here.) We describe the basic authentication step for the HB protocol, and defer a discussion of the HB$^+$ protocol to Section 3.2. In the HB protocol, a tag $\mathcal{T}$ and a reader $\mathcal{R}$ share a random secret key $\mathbf{s} \in \{0,1\}^k$; a basic authentication step consists of the reader sending a random challenge $\mathbf{a} \in \{0,1\}^k$ to the tag, which replies with $z = \langle \mathbf{s}, \mathbf{a} \rangle \oplus \nu$ for $\nu \sim \mathsf{Ber}_\varepsilon$. The reader can then verify whether the response z of the tag satisfies $z \stackrel{?}{=} \langle \mathbf{s}, \mathbf{a} \rangle$; we say the iteration is *successful* if this is the case. See Figure 1.

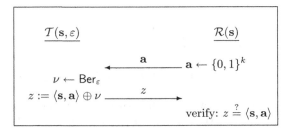

Fig. 1. The basic authentication step of the HB protocol

Even for an honest tag a basic iteration is unsuccessful with probability ε. For this reason, a reader accepts upon completion of all n iterations of the basic authentication step as long as at most $\approx \varepsilon \cdot n$ of these iterations were unsuccessful. More precisely, let $\mathsf{u} = \mathsf{u}(k)$ be such that $\varepsilon \cdot n \leq \mathsf{u}$; then the reader accepts as long as the number of unsuccessful iterations is at most[2] u. (Overall,

---

[2] As suggested in [18], an improvement in practice is to also fix a *lower bound* l and accept iff the number of unsuccessful iterations is in the range $[\mathsf{l}, \mathsf{u}]$. Setting $\mathsf{l} = 0$ (as we do here) makes no difference in an asymptotic sense.

then, the entire HB protocol is parameterized by ε, u, and n.) Since εn is the expected number of unsuccessful iterations for an honest tag, the completeness error ε_c (i.e., the probability that an honest tag is rejected) can be calculated via a Chernoff bound. In particular, we have that for any positive constant δ, setting $\mathsf{u} = (1 + \delta)\varepsilon n$ suffices to achieve ε_c negligible in n.

Observe that by sending random answers in each of the n iterations, an adversary trying to impersonate a valid tag succeeds with probability

$$\delta^*_{\varepsilon,\mathsf{u},n} \stackrel{\text{def}}{=} 2^{-n} \cdot \sum_{i=0}^{\mathsf{u}} \binom{n}{i};$$

that is, $\delta^*_{\varepsilon,\mathsf{u},n}$ is the *best* possible soundness error we can hope to achieve for the given setting of the parameters. Asymptotically, as long as $\mathsf{u} \leq (1 - \delta) \cdot n/2$ for positive constant δ, the success of this trivial attack will be negligible in n. (This can again be analyzed using a Chernoff bound.) Our definitions of security will be expressed in terms of the adversary's ability to do better than $\delta^*_{\varepsilon,\mathsf{u},n}$.

Let $\mathcal{T}^{\mathsf{HB}}_{\mathbf{s},\varepsilon,n}$ denote the tag algorithm in the HB protocol when the tag holds secret key $\mathbf{s}$ (note that the tag algorithm is independent of u), and let $\mathcal{R}^{\mathsf{HB}}_{\mathbf{s},\varepsilon,\mathsf{u},n}$ similarly denote the algorithm run by the tag reader. We denote a complete execution of the HB protocol between a party $\hat{\mathcal{T}}$ and the reader $\mathcal{R}$ by $\left\langle \hat{\mathcal{T}}, \mathcal{R}^{\mathsf{HB}}_{\mathbf{s},\varepsilon,\mathsf{u},n} \right\rangle$ and say this equals 1 iff the reader accepts.

For a passive attack on the HB protocol, we imagine an adversary $\mathcal{A}$ running in two stages: in the first stage the adversary obtains q transcripts[3] of (honest) executions of the protocol by interacting with an oracle $\mathsf{trans}^{\mathsf{HB}}_{\mathbf{s},\varepsilon,n}$ (this models eavesdropping); in the second stage, the adversary interacts with the reader and tries to impersonate the tag. We define the adversary's advantage as

$$\mathsf{Adv}^{passive}_{\mathcal{A},\mathsf{HB}}(\varepsilon, \mathsf{u}, n) \stackrel{\text{def}}{=} \Pr\left[\mathbf{s} \leftarrow \{0,1\}^k; \mathcal{A}^{\mathsf{trans}^{\mathsf{HB}}_{\mathbf{s},\varepsilon,n}}(1^k) : \left\langle \mathcal{A}, \mathcal{R}^{\mathsf{HB}}_{\mathbf{s},\varepsilon,\mathsf{u},n} \right\rangle = 1\right] - \delta^*_{\varepsilon,\mathsf{u},n}.$$

As we will describe in Section 3.2, the HB^+ protocol uses two keys $\mathbf{s}_1, \mathbf{s}_2$. We let $\mathcal{T}^{\mathsf{HB}^+}_{\mathbf{s}_1,\mathbf{s}_2,\varepsilon,n}$ denote the tag algorithm in this case, and let $\mathcal{R}^{\mathsf{HB}^+}_{\mathbf{s}_1,\mathbf{s}_2,\varepsilon,\mathsf{u},n}$ denote the algorithm run by the tag reader. For the case of an active attack on the HB^+ protocol, we again imagine an adversary running in two stages: in the first stage the adversary interacts at most q times with the honest tag algorithm (with concurrent executions allowed), while in the second stage the adversary interacts only with the reader. The adversary's advantage in this case is

$$\mathsf{Adv}^{active}_{\mathcal{A},\mathsf{HB}^+}(\varepsilon, \mathsf{u}, n)$$

$$\stackrel{\text{def}}{=} \Pr\left[\mathbf{s}_1, \mathbf{s}_2 \leftarrow \{0,1\}^k; \mathcal{A}^{\mathcal{T}^{\mathsf{HB}^+}_{\mathbf{s}_1,\mathbf{s}_2,\varepsilon,n}}(1^k) : \left\langle \mathcal{A}, \mathcal{R}^{\mathsf{HB}^+}_{\mathbf{s}_1,\mathbf{s}_2,\varepsilon,\mathsf{u},n} \right\rangle = 1\right] - \delta^*_{\varepsilon,\mathsf{u},n}.$$

---

[3] Following [13,14,17], a transcript comprises only the messages exchanged between the parties and does not include the reader's decision of whether or not to accept. If the adversary is given this additional information, the adversary's advantage may increase by (at most) an additive factor of $q \cdot \varepsilon_c$. Note further that, since u can be set so that the reader accepts the honest tag with all but negligible probability, this has no effect as far as asymptotic security is concerned.

We remark that in both the HB and HB$^+$ protocols, the tag reader's actions are independent of the secret key(s) it holds except for its final decision whether or not to accept. So, allowing the adversary to interact with the reader multiple times (even concurrently) does not give the adversary much additional advantage (other than the fact that, as usual, the probability that the adversary succeeds in at least one impersonation attempt scales linearly with the number of attempts).

3   Proofs of Security for the HB and HB$^+$ Protocols

3.1   Security of the HB Protocol Against Passive Attacks

We first prove security assuming $\varepsilon < 1/4$, and then show how the argument can be extended for the case of $\varepsilon < 1/2$.

Theorem 1. *Say there exists an adversary $\mathcal{A}$ eavesdropping on at most q executions of the HB protocol, running in time t, and achieving $\mathsf{Adv}_{\mathcal{A},\mathsf{HB}}^{passive}(\varepsilon, \mathsf{u}, n) \geq \delta$. Then there exists an algorithm D making $(q + 1) \cdot n$ oracle queries, running in time $O(t)$, and such that*

$$\left| \Pr \left[\mathbf{s} \leftarrow \{0,1\}^k : D^{A_{\mathbf{s},\varepsilon}}(1^k) = 1 \right] - \Pr \left[D^{U_{k+1}}(1^k) = 1 \right] \right|$$

$$\geq \delta + \delta_{\varepsilon,\mathsf{u},n}^* - \varepsilon_c - 2^{-n} \cdot \sum_{i=0}^{2\mathsf{u}} \binom{n}{i}.$$

Asymptotically, for any $\varepsilon < \frac{1}{4}$ and $n = \Theta(k)$ all terms of the above expression (other than δ) are negligible for appropriate choice of u. Thus, assuming the hardness of the LPN_ε problem for $\varepsilon < 1/4$ and appropriate choice of n, u, the HB protocol is secure against a passive adversary.

Proof. D, given access to an oracle returning $(k + 1)$-bit strings $(\mathbf{a}, z)$, proceeds as follows:

1. D runs the first phase of $\mathcal{A}$. Each time $\mathcal{A}$ requests to view a transcript of the protocol, D obtains n samples $\{(\mathbf{a}_i, z_i)\}_{i=1}^n$ from its oracle and returns these to $\mathcal{A}$.
2. For $\mathcal{A}$'s second phase, D again obtains n samples $\{(\bar{\mathbf{a}}_i, \bar{z}_i)\}_{i=1}^n$ from its oracle. D then sends the challenge $(\bar{\mathbf{a}}_1, \ldots, \bar{\mathbf{a}}_n)$ to $\mathcal{A}$ and receives in return a response $Z' = (z_1', \ldots, z_n')$.
3. D outputs 1 iff $\bar{Z} = (\bar{z}_1, \ldots, \bar{z}_n)$ and Z' differ in at most $2\mathsf{u}$ entries.

When D's oracle is U_{k+1}, it is clear that D outputs 1 with probability exactly $2^{-n} \cdot \sum_{i=0}^{2\mathsf{u}} \binom{n}{i}$ since $\bar{Z}$ is in this case uniformly distributed and independent of everything else. On the other hand, when D's oracle is $A_{\mathbf{s},\varepsilon}$ then the transcripts D provides to $\mathcal{A}$ during the first phase of $\mathcal{A}$'s execution are distributed identically to real transcripts in an execution of the HB protocol. Let $Z^* \stackrel{\text{def}}{=} (\langle \mathbf{s}, \bar{\mathbf{a}}_1 \rangle, \ldots, \langle \mathbf{s}, \bar{\mathbf{a}}_n \rangle)$ be the vector of "correct" answers to the challenge $(\bar{\mathbf{a}}_1, \ldots, \bar{\mathbf{a}}_n)$ sent by D in the second phase. Then with probability at least

$\delta + \delta^*_{\varepsilon,l,u,n}$ it holds that Z' and Z^* differ in at most u entries (since $\mathcal{A}$ successfully impersonates the tag with this probability). Also, since $\bar{Z}$ is distributed exactly as the answers of an honest tag, $\bar{Z}$ and Z^* differ in at most u positions except with probability at most ε_c. It follows that with probability at least $\delta + \delta^*_{\varepsilon,u,n} - \varepsilon_c$ the vectors Z' and $\bar{Z}$ differ in at most $2u$ entries, and so D outputs 1 with at least this probability.

When $\varepsilon \geq 1/4$ then $2u \geq 2\varepsilon \cdot n \geq n/2$ and so $2^{-n} \cdot \sum_{i=0}^{2u} \binom{n}{i} \geq 1/2$; thus, the above theorem does not guarantee meaningful security in this case and a different analysis is needed.

Theorem 2. *For $n = \Theta(k)$ and appropriate setting of u, the HB protocol is (asymptotically) secure for any $\varepsilon < 1/2$.*

Proof. Let $u = \varepsilon^+ n$, where ε^+ is a constant satisfying $\varepsilon < \varepsilon^+ < \frac{1}{2}$. Set ε^{++} to be a constant satisfying $\varepsilon^+ - 2\varepsilon^+\varepsilon + \varepsilon < \varepsilon^{++} < \frac{1}{2}$. The reduction D we use here is similar to that used in the previous proof, except that D now outputs 1 only if $\bar{Z}$ and Z' differ in at most $u' \stackrel{\text{def}}{=} \varepsilon^{++} \cdot n$ entries.

When D's oracle is U_{k+1}, it is once again clear that D outputs 1 with probability $2^{-n} \cdot \sum_{i=0}^{u'} \binom{n}{i}$. Since $u' < n/2$, this quantity is negligible in k.

When D's oracle is $A_{\mathbf{s},\varepsilon}$ then the transcripts D provides to $\mathcal{A}$ during the first phase of $\mathcal{A}$'s execution are distributed identically to real transcripts in an execution of the HB protocol. Letting $Z^* \stackrel{\text{def}}{=} (\langle \mathbf{s}, \bar{\mathbf{a}}_1 \rangle, \ldots, \langle \mathbf{s}, \bar{\mathbf{a}}_n \rangle)$ be the vector of correct answers to the challenge $(\bar{\mathbf{a}}_1, \ldots, \bar{\mathbf{a}}_n)$ sent by D in the second phase, it follows that with probability δ (i.e., the impersonation probability of $\mathcal{A}$) the vector of responses Z' given by $\mathcal{A}$ differs from Z^* in at most u entries. We show below that, conditioned on this event, Z' and $\bar{Z}$ differ in at most u' entries with all but negligible probability. Thus, D outputs 1 in this case with probability negligibly close to δ. We conclude from Lemma 1 that δ must be negligible.

Let $\mathbf{wt}(Z)$ denote the *weight* of a vector Z; i.e., the number of entries of Z equal to 1. Note that the distance between two binary vectors Z_1, Z_2 is equal to $\mathbf{wt}(Z_1 \oplus Z_2)$. It remains to show that, conditioned on $\mathbf{wt}(Z' \oplus Z^*) \leq u$, we have $\mathbf{wt}(Z' \oplus \bar{Z}) \leq u'$ with all but negligible probability.

Write $Z' = Z^* \oplus \mathbf{w}$ for some vector $\mathbf{w}$ of weight at most $u = \varepsilon^+ n$. The vector $\bar{Z}$ is selected by the following process: choose an error vector $\mathbf{e}$ by setting each position of $\mathbf{e}$ (independently) to 1 with probability ε, and then set $\bar{Z} = Z^* \oplus \mathbf{e}$. We see that the probability that $\bar{Z}$ differs from Z' in at most u' entries is equal to the probability that

$$\mathbf{wt}(Z' \oplus \bar{Z}) = \mathbf{wt}(\mathbf{w} \oplus \mathbf{e}) \leq u'.$$

It is easy to see that this probability is minimized when $\mathbf{wt}(\mathbf{w}) = u$, and so we assume this to be the case. The random variable $\mathbf{wt}(\mathbf{w} \oplus \mathbf{e})$ can be written as a sum of n indicator random variables, one for each position of the vector $\mathbf{w} \oplus \mathbf{e}$. The expectation of $\mathbf{wt}(\mathbf{w} \oplus \mathbf{e})$ is

$$u \cdot (1 - \varepsilon) + (n - u) \cdot \varepsilon = (\varepsilon^+ - 2\varepsilon^+\varepsilon + \varepsilon) \cdot n.$$

Since ε^{++} is a constant strictly larger than $(\varepsilon^+ - 2\varepsilon^+\varepsilon + \varepsilon)$, a Chernoff bound then implies that $\mathbf{wt}(\mathbf{w} \oplus \mathbf{e}) \leq \varepsilon^{++}n$ with all but negligible probability.

3.2 Security of the HB$^+$ Protocol Against Active Attacks

The HB protocol is insecure against an active attack, as an adversary can repeatedly query the tag with the same challenge $(\mathbf{a}_1, \ldots, \mathbf{a}_n)$ and thereby determine with high probability the correct values of $\langle \mathbf{s}, \mathbf{a}_1 \rangle, \ldots, \langle \mathbf{s}, \mathbf{a}_n \rangle$ (after which solving for $\mathbf{s}$ is easy). To combat such an attack, Juels and Weis [17] propose to modify the HB protocol by having the tag and reader share *two* (independent) keys $\mathbf{s}_1, \mathbf{s}_2 \in \{0,1\}^k$. A basic authentication step now consists of three rounds: first the tag sends a random "blinding factor" $\mathbf{b} \in \{0,1\}^k$; the reader replies with a random challenge $\mathbf{a} \in \{0,1\}^k$ as before; and finally the tag replies with $z = \langle \mathbf{s}_1, \mathbf{b} \rangle \oplus \langle \mathbf{s}_2, \mathbf{a} \rangle \oplus \nu$ for $\nu \sim \mathsf{Ber}_\varepsilon$. As in the HB protocol, the tag reader can then verify whether the response z of the tag satisfies $z \overset{?}{=} \langle \mathbf{s}_1, \mathbf{b} \rangle \oplus \langle \mathbf{s}_2, \mathbf{a} \rangle$, and we again say the iteration is *successful* if this is the case. See Figure 2.

Fig. 2. The basic authentication step of the HB$^+$ protocol

The actual HB$^+$ protocol consists of n parallel iterations of the basic authentication step (and so the entire protocol requires only three rounds). The protocol also depends upon parameters u as in the case of the HB protocol, and the values ε_c and $\delta^*_{\varepsilon,u,n}$ are defined exactly as there.

The following result shows that the HB$^+$ protocol is secure against *active* attacks, assuming the hardness of the LPN$_\varepsilon$ problem. An important point is that the proof does not require any rewinding of the adversary in simulating the first phase of the attack, and the proof therefore holds even when various executions of the HB$^+$ protocol are run in parallel (or even concurrently).

We refer to [18] for a simpler proof when $\varepsilon < 1/4$, and present here only the more complicated proof (from [19]) that deals with any $\varepsilon < 1/2$. This, more complicated proof relies on known bounds on the size of constant-weight codes having certain minimum distance [15,16,11].

Theorem 3. *Let $\varepsilon < 1/2$ and assume the LPN$_\varepsilon$ problem is hard. Let $n = \Theta(k)$ and* u $= \varepsilon^+n$ *where ε^+ is a constant satisfying $\varepsilon < \varepsilon^+ < 1/2$. Then the HB$^+$*

protocol with this setting of the parameters has negligible completeness error, and is (asymptotically) secure against active attacks.

Proof. A standard Chernoff bound shows that the completeness error is negligible for the given setting of the parameters. For any probabilistic polynomial-time (PPT) adversary $\mathcal{A}$ attacking the HB^+ protocol, we construct a PPT adversary D attempting to distinguish whether it is given oracle access to $A_{\mathbf{s},\varepsilon}$ or to U_{k+1} (as in Lemma 1). Relating the advantage of D to the advantage of $\mathcal{A}$ gives the stated result.

D, given access to an oracle returning $(k+1)$-bit strings $(\mathbf{b}, \bar{z})$, proceeds as follows:

1. D chooses $\mathbf{s}_2 \in \{0,1\}^k$ uniformly at random. Then, it runs the first phase of $\mathcal{A}$. To simulate a basic authentication step, D does the following: it obtains a sample $(\mathbf{b}, \bar{z})$ from its oracle and sends $\mathbf{b}$ as the initial message. $\mathcal{A}$ replies with a challenge $\mathbf{a}$, and then D responds with $z = \bar{z} \oplus \langle \mathbf{s}_2, \mathbf{a} \rangle$.

2. When $\mathcal{A}$ is ready for the second phase of its attack, $\mathcal{A}$ sends an initial message $\mathbf{b}_1, \ldots, \mathbf{b}_n$. In response, D chooses random $\mathbf{a}_1^1, \ldots, \mathbf{a}_n^1 \in \{0,1\}^k$, sends these challenges to $\mathcal{A}$, and records $\mathcal{A}$'s response $z_1^1, \ldots, z_n^1$. Then D rewinds $\mathcal{A}$, chooses random $\mathbf{a}_1^2, \ldots, \mathbf{a}_n^2 \in \{0,1\}^k$, sends these to $\mathcal{A}$, and records $\mathcal{A}$'s response $z_1^2, \ldots, z_n^2$.

3. Let $z_i^\oplus := z_i^1 \oplus z_i^2$ and set $Z^\oplus \overset{\text{def}}{=} (z_1^\oplus, \ldots, z_n^\oplus)$. Let $\hat{\mathbf{a}}_i = \mathbf{a}_i^1 \oplus \mathbf{a}_i^2$ and $\hat{z}_i = \langle \mathbf{s}_2, \hat{\mathbf{a}}_i \rangle$, and set $\hat{Z} \overset{\text{def}}{=} (\hat{z}_1, \ldots, \hat{z}_n)$. D outputs 1 iff $Z^\oplus$ and $\hat{Z}$ differ in fewer than $u' = \varepsilon^{++} \cdot n$ entries, for some constant $\varepsilon^{++} < \frac{1}{2}$ to be fixed later.

Let us analyze the behavior of D:

Case 1: Say D's oracle is U_{k+1}. In step 1, since $\bar{z}$ is uniformly distributed and independent of everything else, the answers z that D returns to $\mathcal{A}$ are uniformly distributed and independent of everything else. It follows that $\mathcal{A}$'s view is independent of the secret $\mathbf{s}_2$ chosen by D.

The $\{\hat{\mathbf{a}}_i\}_{i=1}^n$ are uniformly and independently distributed, and so except with probability $\frac{2^n}{2^k}$ they are linearly independent and non-zero (via a standard argument; see [18]). Assuming this to be the case, $\hat{Z}$ is uniformly distributed over $\{0,1\}^n$ from the point of view of $\mathcal{A}$. But then the probability that $Z^\oplus$ and $\hat{Z}$ differ in fewer than u' entries is at most $2^{-n} \cdot \sum_{i=0}^{\lfloor u' \rfloor} \binom{n}{i}$. Since $u'/2$ is a constant strictly less than $\frac{1}{2}$, we conclude that D outputs 1 in this case with negligible probability $\frac{2^n}{2^k} + 2^{-n} \cdot \sum_{i=0}^{\lfloor u' \rfloor} \binom{n}{i}$.

Case 2: Say D's oracle is $A_{\mathbf{s}_1, \varepsilon}$ for randomly-chosen $\mathbf{s}_1$. In this case, D provides a perfect simulation for the first phase of $\mathcal{A}$. Let ω denote all the randomness used to simulate the first phase of $\mathcal{A}$, which includes the keys $\mathbf{s}_1, \mathbf{s}_2$, the randomness of $\mathcal{A}$, and the randomness used in responding to $\mathcal{A}$'s queries. For a fixed ω, let δ_ω denote the probability (over random challenges $\mathbf{a}_1, \ldots, \mathbf{a}_n$ sent by the tag reader) that $\mathcal{A}$ successfully impersonates the tag in the second phase. The probability that $\mathcal{A}$ successfully responds to both sets of queries $\mathbf{a}_1^1, \ldots, \mathbf{a}_n^1$ and $\mathbf{a}_1^2, \ldots, \mathbf{a}_n^2$

sent by D is δ_ω^2. The overall probability that $\mathcal{A}$ successfully responds to both sets of queries is thus

$$\mathsf{Exp}_\omega\left(\delta_\omega^2\right) \geq \left(\mathsf{Exp}_\omega(\delta_\omega)\right)^2 = \delta_{\mathcal{A}}^2,$$

using Jensen's inequality.

We show below that conditioned on both challenges being answered successfully (and for appropriate choice of ε^{++}), $Z^\oplus$ differs from $\hat{Z}$ in fewer than u' entries with *constant* probability. Putting everything together, we conclude that D outputs 1 in this case with probability $\Omega(\delta_{\mathcal{A}}^2)$. It follows from Lemma 1 that $\delta_{\mathcal{A}}$ must be negligible.

We now prove the above claim regarding the probability that $Z^\oplus$ differs from $\hat{Z}$ in fewer than u' entries. Set $\frac{1}{2} > \varepsilon^{++} > \frac{1}{2} \cdot (1 - (1 - 2\varepsilon^+)^2)$. Fixing all randomness used in the first phase (as above) induces a function $f_{\mathcal{A}}$ from queries $\mathbf{a}_1, \ldots, \mathbf{a}_n$ (with each $\mathbf{a}_i \in \{0,1\}^k$) to vectors $(z_1, \ldots, z_n)$ (with each $z_i \in \{0,1\}$) given by the response function of $\mathcal{A}$ in the second phase. Define the function f_{correct} that returns the "correct" answers for a particular query; i.e.,

$$f_{\text{correct}}(\mathbf{a}_1, \ldots, \mathbf{a}_n) \stackrel{\text{def}}{=} (\langle \mathbf{s}_1, \mathbf{b}_1 \rangle \oplus \langle \mathbf{s}_2, \mathbf{a}_1 \rangle, \ldots, \langle \mathbf{s}_1, \mathbf{b}_n \rangle \oplus \langle \mathbf{s}_2, \mathbf{a}_n \rangle)$$

(recall that $\mathbf{b}_1, \ldots, \mathbf{b}_n$ are the vectors sent by $\mathcal{A}$ in the first round). Define

$$\Delta(\mathbf{a}_1, \ldots, \mathbf{a}_n) \stackrel{\text{def}}{=} f_{\mathcal{A}}(\mathbf{a}_1, \ldots, \mathbf{a}_n) \oplus f_{\text{correct}}(\mathbf{a}_1, \ldots, \mathbf{a}_n),$$

and say a query $\mathbf{a}_1, \ldots, \mathbf{a}_n$ is *good* if[4] $\mathbf{wt}(\Delta(\mathbf{a}_1, \ldots, \mathbf{a}_n)) \leq \mathsf{u}$. That is, a query $\mathbf{a}_1, \ldots, \mathbf{a}_n$ is good if $\mathcal{A}$'s response is within distance u of the "correct" response; i.e., $\mathcal{A}$ successfully impersonates the tag in response to such a query.

Let $\mathcal{D}$ denote the distribution over $\Delta(\mathbf{a}_1, \ldots, \mathbf{a}_n)$ induced by a uniform choice of a good query $\mathbf{a}_1, \ldots, \mathbf{a}_n$ (we assume at least one good query exists since we are only interested in analyzing this case). Note that, by definition of a good query, each vector in the support of $\mathcal{D}$ has weight at most u. Our goal is to show that with constant probability over Δ^1, Δ^2 generated according to $\mathcal{D}$, we have $\mathbf{wt}(\Delta^1 \oplus \Delta^2) < \mathsf{u}'$. We remark that this claim does not involve any assumptions regarding the probability that a randomly-chosen query is good.

To see how this maps on to the reduction being analyzed above, note that conditioning on the event that $\mathcal{A}$ successfully responds to queries $\mathbf{a}_1^1, \ldots, \mathbf{a}_n^1$ and $\mathbf{a}_1^2, \ldots, \mathbf{a}_n^2$ is equivalent to choosing these two queries uniformly from the set of good queries. Setting $\Delta^1 \stackrel{\text{def}}{=} \Delta(\mathbf{a}_1^1, \ldots, \mathbf{a}_n^1)$ and Δ^2 analogously, we have

$$\begin{aligned}
\Delta^1 &\oplus \Delta^2 \\
&= f_{\mathcal{A}}(\mathbf{a}_1^1, \ldots, \mathbf{a}_n^1) \oplus f_{\text{correct}}(\mathbf{a}_1^1, \ldots, \mathbf{a}_n^1) \oplus f_{\mathcal{A}}(\mathbf{a}_1^2, \ldots, \mathbf{a}_n^2) \oplus f_{\text{correct}}(\mathbf{a}_1^2, \ldots, \mathbf{a}_n^2) \\
&= Z^\oplus \oplus f_{\text{correct}}(\mathbf{a}_1^1, \ldots, \mathbf{a}_n^1) \oplus f_{\text{correct}}(\mathbf{a}_1^2, \ldots, \mathbf{a}_n^2).
\end{aligned}$$

---

[4] As in the proof of the previous theorem, the weight $\mathbf{wt}(Z)$ of a vector Z is the number of its entries equal to 1.

D cannot compute $f_{\text{correct}}(\mathbf{a}_1^1, \ldots, \mathbf{a}_n^1)$ or $f_{\text{correct}}(\mathbf{a}_1^2, \ldots, \mathbf{a}_n^2)$ since it does not know $\mathbf{s}_1$. However, it *can* compute

$$
\begin{aligned}
f_{\text{correct}}(\mathbf{a}_1^1, \ldots, \mathbf{a}_n^1) &\oplus f_{\text{correct}}(\mathbf{a}_1^2, \ldots, \mathbf{a}_n^2) \\
&= (\langle \mathbf{s}_1, \mathbf{b}_1 \rangle \oplus \langle \mathbf{s}_2, \mathbf{a}_1^1 \rangle, \ldots, \langle \mathbf{s}_1, \mathbf{b}_n \rangle \oplus \langle \mathbf{s}_2, \mathbf{a}_n^1 \rangle) \\
&\quad + (\langle \mathbf{s}_1, \mathbf{b}_1 \rangle \oplus \langle \mathbf{s}_2, \mathbf{a}_1^2 \rangle, \ldots, \langle \mathbf{s}_1, \mathbf{b}_n \rangle \oplus \langle \mathbf{s}_2, \mathbf{a}_n^2 \rangle) \\
&= (\langle \mathbf{s}_2, \mathbf{a}_1^1 \rangle \oplus \langle \mathbf{s}_2, \mathbf{a}_1^2 \rangle, \ldots, \langle \mathbf{s}_2, \mathbf{a}_n^1 \rangle \oplus \langle \mathbf{s}_2, \mathbf{a}_n^2 \rangle) \\
&= (\langle \mathbf{s}_2, (\mathbf{a}_1^1 \oplus \mathbf{a}_1^2) \rangle, \ldots, \langle \mathbf{s}_2, (\mathbf{a}_n^1 \oplus \mathbf{a}_n^2) \rangle) \;=\; \hat{Z}.
\end{aligned}
$$

We thus see that $Z^\oplus$ and $\hat{Z}$ differ in fewer than u' entries exactly when Δ^1 and Δ^2 differ in fewer than $u' = \varepsilon^{++}n$ entries. It is the latter probability that we now analyze.

Let δ be a (positive) constant such that $\varepsilon^{++} = \frac{1}{2} \cdot (1 - \delta)$. Let $\gamma \overset{\text{def}}{=} 1 - 2\varepsilon^+$, and note that by our choice of ε^{++} we have $\delta < \gamma^2$. Set

$$
c \overset{\text{def}}{=} \frac{1 - \delta}{\gamma^2 - \delta} + 1.
$$

We show that for two vectors Δ^1, Δ^2 chosen independently according to distribution $\mathcal{D}$, we have $\mathbf{wt}(\Delta^1 \oplus \Delta^2) < \varepsilon^{++}n$ with (constant) probability at least $\frac{1}{c^2}$. Assume not. So

$$
\Pr[\Delta^1, \Delta^2 \leftarrow \mathcal{D} : \mathbf{wt}(\Delta^1 \oplus \Delta^2) < \varepsilon^{++}n] < \frac{1}{c^2}.
$$

But then, by a union bound,

$$
\Pr[\Delta^1, \ldots, \Delta^c \leftarrow \mathcal{D} : \exists i \neq j \text{ s.t. } \mathbf{wt}(\Delta^1 \oplus \Delta^2) < \varepsilon^{++}n] < \frac{1}{2}.
$$

In particular, there exist c vectors $\Delta^1, \ldots, \Delta^c$ in the support of $\mathcal{D}$ whose pairwise distances are all at least $\varepsilon^{++}n = \frac{1}{2} \cdot (1 - \delta)n$. Furthermore, each Δ^i has weight at most $u = \frac{1}{2} \cdot (1 - \gamma)n$ since it lies in the support of $\mathcal{D}$. However, the Johnson bound [15,16] (our notation was chosen to be consistent with the formulation in [11, Theorem 1]), which gives bounds on the size of constant-weight codes of certain minimum distance, shows that no such set $\{\Delta^i\}_{i=1}^c$ exists.

4   Conclusions and Open Questions

There are a number of interesting questions regarding the HB and HB$^+$ protocols themselves. First, it would be nice to improve the concrete security reductions obtained here, or to propose new protocols with tighter security reductions. It would also be interesting to obtain simpler proofs even for $\varepsilon \geq 1/4$ (there seems to be no inherent reason why the analysis should become more complex in that case). As discussed in the Introduction, it would also be very interesting to see an efficient protocol based on the LPN problem that is provably resistant to man-in-the-middle attacks.

More generally, it remains to be seen whether the LPN problem can be used to construct efficient cryptographic protocols for other tasks. Some hope is provided by Regev's result [24] showing that public-key encryption can be based on a generalization of the LPN problem.

Acknowledgments

I would like to thank Steven Galbraith for inviting me to contribute this survey to the 11th IMA International Conference on Cryptography and Coding Theory.

References

1. Bogdanov, A., Knudsen, L., Leander, G., Paar, C., Poschmann, A., Robshaw, M., Seurin, Y., Vikkelsoe, C.: PRESENT: An Ultra-Lightweight Block Cipher. In: CHES 2007. Workshop on Cryptographic Hardware and Embedded Systems (to appear, 2007)
2. Berlekamp, E.R., McEliece, R.J., van Tilborg, H.C.A.: On the Inherent Intractability of Certain Coding Problems. IEEE Trans. Info. Theory 24, 384–386 (1978)
3. Blum, A., Furst, M., Kearns, M., Lipton, R.: Cryptographic Primitives Based on Hard Learning Problems. 773, 278–291 (1994)
4. Blum, A., Kalai, A., Wasserman, H.: Noise-Tolerant Learning, the Parity Problem, and the Statistical Query Model. J. ACM 50(4), 506–519 (2003)
5. Bringer, J., Chabanne, H., Dottax, E.: HB^{++}: A Lightweight Authentication Protocol Secure against Some Attacks, http://eprint.iacr.org/2005/440
6. Chabaud, F.: On the Security of Some Cryptosystems Based on Error-Correcting Codes. In: De Santis, A. (ed.) EUROCRYPT 1994. LNCS, vol. 950, pp. 131–139. Springer, Heidelberg (1995)
7. Diffie, W., Hellman, M.: New Directions in Cryptography. IEEE Trans. Info. Theory 22(6), 644–654 (1976)
8. Feldhofer, M., Dominikus, S., Wolkerstorfer, J.: Strong Authentication for RFID Systems using the AES Algorithm. In: Joye, M., Quisquater, J.-J. (eds.) CHES 2004. LNCS, vol. 3156, pp. 357–370. Springer, Heidelberg (2004)
9. Fossorier, M.P.C., Mihaljević, M.J., Imai, H., Cui, Y., Matsuura, K.: A Novel Algorithm for Solving the LPN Problem and its Application to Security Evaluation of the HB Protocol for RFID Authentication, http://eprint.iacr.org
10. Gilbert, H., Robshaw, M., Silbert, H.: An Active Attack against HB^{+}: A Provably Secure Lightweight Authentication Protocol, http://eprint.iacr.org/2005/237
11. Guruswami, V., Sudan, M.: Extensions to the Johnson Bound (unpublished manuscript 2001), http://citeseer.ist.psu.edu/guruswami01extensions.html
12. Håstad, J.: Optimal Inapproximability Results. J. ACM 48(4), 798–859 (2001)
13. Hopper, N., Blum, M.: A Secure Human-Computer Authentication Scheme. Technical Report CMU-CS-00-139, Carnegie Mellon University (2000)
14. Hopper, N., Blum, M.: Secure Human Identification Protocols. In: Boyd, C. (ed.) ASIACRYPT 2001. LNCS, vol. 2248, pp. 52–66. Springer, Heidelberg (2001)
15. Johnson, S.M.: Upper Bound for Error-Correcting Codes. IEEE Trans. Info. Theory 8, 203–207 (1962)
16. Johnson, S.M.: Improved Asympotic Bounds for Error-Correcting Codes. IEEE Trans. Info. Theory 9, 198–205 (1963)

17. Juels, A., Weis, S.: Authenticating Pervasive Devices with Human Protocols, vol. 3621, pp. 293–308 (2005),
    http://www.rsasecurity.com/rsalabs/staff/bios/ajuels/publications/pdfs/lpn.pdf
18. Katz, J., Shin, J.-S.: Parallel and Concurrent Security of the HB and HB$^+$ Protocols. In: Vaudenay, S. (ed.) EUROCRYPT 2006. LNCS, vol. 4004, Springer, Heidelberg (2006)
19. Katz, J., Smith, A.: Analyzing the HB and HB$^+$ Protocols in the Large Error Case, Available at http://eprint.iacr.org/2006/326
20. Kearns, M.: Efficient Noise-Tolerant Learning from Statistical Queries. J. ACM 45(6), 983–1006 (1998)
21. Kfir, Z., Wool, A.: Picking Virtual Pockets using Relay Attacks on Contactless Smartcard Systems, http://eprint.iacr.org/2005/052
22. Kirschenbaum, I., Wool, A.: How to Build a Low-Cost, Extended-Range RFID Skimmer, http://eprint.iacr.org/2006/054
23. Levieil, E., Fouque, P.-A.: An Improved LPN Algorithm. In: De Prisco, R., Yung, M. (eds.) SCN 2006. LNCS, vol. 4116, pp. 348–359. Springer, Heidelberg (2006)
24. Regev, O.: On Lattices, Learning with Errors, Random Linear Codes, and Cryptography. In: 37th ACM Symposium on Theory of Computing, pp. 84–93. ACM, New York (2005)
25. Weis, S.: New Foundations for Efficient Authentication, Commutative Cryptography, and Private Disjointness Testing. Ph.D. thesis, MIT (2006)

Galois Rings
and
Pseudo-random Sequences

Patrick Solé[1] and Dmitrii Zinoviev[1,2]

[1] CNRS-I3S, Les Algorithmes, Euclide B, 2000, route des Lucioles, 06 903 Sophia
Antipolis, France
sole@unice.fr, zinoviev@essi.fr
[2] Institute for Problems of Information Transmission, Russian Academy of Sciences,
Bol'shoi Karetnyi, 19, GSP-4, Moscow, 127994, Russia
dzinov@iitp.ru

Abstract. We survey our constructions of pseudo random sequences
(binary, $\mathbb{Z}_8$, $\mathbb{Z}_{2^l}$,...) from Galois rings. Techniques include a local Weil
bound for character sums, and several kinds of Fourier transform. Applications range from cryptography (boolean functions, key generation),
to communications (multi-code CDMA), to signal processing (PAPR reduction).

Keywords: Correlation, Galois Rings, MSB, CDMA, OFDM, PAPR.

1 Introduction

Since the seminal work of Claude Shannon in the 40's the algebraic structure of
coding alphabets was a finite field. However, there was a push towards finite rings
due to modulation requirements, a 4− PSK modulation being more powerful
than an antipodal modulation, for instance [1]. This led researchers in the 70's
to consider cyclic codes over $\mathbb{Z}/M\mathbb{Z}$, with a special attention for M a power of
a prime [2,19]. In the 90's cyclic codes over rings received a lot of attention due
to many pure applications: the solution of the Kerdock Preparata riddle [8], and
a new construction of the Leech lattice [3], (both for $M = 4$), to name but a
few. In this golden dawn, new and powerful mathematical techniques arose, the
most profound being probably a $p−$adic analogue of the Weil bound on character
sums [10], and families of low correlation sequences (based on weighted degree
trace codes) tailored for that bound [20].

In the present decade, new engineering applications emerged that required
that machinery, and are significantly different, but not unrelated to the classical
use in Spread Spectrum and CDMA systems illustrated in [11,7], or to take a
more recent example the 3GPP standard (see www.3gpp.org). In particular an
important quantity for the safe processing of electronic equipment is the **Peak
to Average Power Ratio** (PAPR), signals with high dynamic range being
potentially harmful to operational amplifiers. Two models of signalling sequences

S.D. Galbraith (Eds.): Cryptography and Coding 2007, LNCS 4887, pp. 16–33, 2007.

both due to Paterson and co-workers lead to somewhat different mathematical formulation. In the first scenario, motivated by OFDM applications, and based on DFT calculation, we are to control some hybrid character sums (i.e. involving not only additive but also multiplicative characters) in relation to many valued sequences [17]. In the second approach, motivated by multi-code CDMA, we are to control the **nonlinearity** of binary sequences of length a power of 2 [16]. This notion of nonlinearity coincides with the one familiar from boolean functions theory. In both models sequences live in families and are supposed to form a code either spherical in the first case or binary in the second. Both papers [17,16] already use weighted degree trace codes and our contribution is mostly of extension from $\mathbb{Z}_4$ to $\mathbb{Z}_{2^\ell}$ for $\ell > 2$ [22], or to non primitive periods [24]. In general, the use of larger rings provides extra flexibility in the design choice; the sequences families constructed in that way having a larger size but a somewhat larger PAPR. For a book on PAPR reduction see [13]. For recent results on codes for PAPR reduction see [18]. For books on Galois rings see [25,26].

A distinct, older motivation to study pseudo random sequences over rings is the use of the **Most Significant Bit** (MSB) to obtain highly nonlinear binary sequences, mostly for cryptographic purposes (key generation in stream ciphers). This is due to Dai and her school [5,6]. In [21] we improve some bounds of autocorrelation and imbalance of such sequences in [6].

The material is organized as follows. Section 2 contains definitions and notation on Galois rings. Section 3 collects the bounds we need on characters sums. In particular the bound on Fourier coefficients of the most significant bit (MSB) function that was obtained in [12]. Section 4 defines the polynomials over Galois rings needed to define the trace codes. Section 5 considers the binary sequence constructed from sequences over $\mathbb{Z}_8$ that was introduced in [12]. Section 6 considers short binary sequences of [22], motivated by [16]. Section 7 contains the binary sequences constructed by taking the MSB of 2^l−ary sequences [21].Section 8 contains the PAPR reduction sequences over large rings of [24]. Section 9 considers quadriphase sequences of maximal period of [23].

2   Preliminaries

Let $R = GR(2^l, m)$ denote the Galois ring of characteristic 2^l. It is the unique Galois extension of degree m of $\mathbb{Z}_{2^l}$, with 2^{lm} elements.

$$R = GR(2^l, m) = \mathbb{Z}_{2^l}[X]/(h(X)).$$

where $h(X)$ is a basic irreducible polynomial of degree m. Let ξ be an element in $GR(2^l, m)$ that generates the Teichmüller set $\mathcal{T}$ of $GR(2^l, m)$ which reduces to $\mathbb{F}_{2^m}$ modulo 2. Specifically, let $\mathcal{T}=\{0, 1, \xi, \xi^2, \ldots, \xi^{2^m-2}\}$ and $\mathcal{T}^*=\{1, \xi, \xi^2, \ldots, \xi^{2^m-2}\}$. We use the *convention* that $\xi^\infty = 0$.

The 2-adic expansion of $x \in GR(2^l, m)$ is given by

$$x = x_0 + 2x_1 + \cdots + 2^{l-1}x_{l-1},$$

where $x_0, x_1, \ldots, x_{l-1} \in \mathcal{T}$. The Frobenius operator F is defined for such an x as

$$F(x_0 + 2x_1 + \cdots + 2^{l-1}x_{l-1}) = x_0^2 + 2x_1^2 + \cdots + 2^{l-1}x_{l-1}^2,$$

and the trace Tr, from $GR(2^l, m)$ down to $\mathbb{Z}_{2^l}$, as

$$\mathrm{Tr} := \sum_{j=0}^{m-1} F^j.$$

We also define another trace tr from $\mathbb{F}_{2^m}$ down to $\mathbb{F}_2$ as

$$\mathrm{tr}(x) := \sum_{j=0}^{m-1} x^{2^j}.$$

Throughout this note, we let $n = 2^m$ and $R^* = R \backslash 2R$.
Let $\mathrm{MSB} : \mathbb{Z}_{2^l}^n \to \mathbb{Z}_2^n$ be the most-significant-bit map, i.e.

$$\mathrm{MSB}(y) = y_{l-1}, \text{ where } y = y_0 + 2y_1 + \ldots + 2^{l-1}y_{l-1} \in \mathbb{Z}_{2^l},$$

is its 2-adic expansion.

3   DFT and the Local Weil Bound

We assume henceforth in the whole paper that $l \geq 3$. Let l be a positive integer and $\omega = e^{2\pi i/2^l}$ be a primitive 2^l-th root of 1 in $\mathbb{C}$. Let ψ_k be the additive character of $\mathbb{Z}_{2^l}$ such that

$$\psi_k(x) = \omega^{kx}.$$

Let $\mu : \mathbb{Z}_{2^l} \to \{\pm 1\}$ be the be the mapping $\mu(t) = (-1)^c$, where c is the most significant bit of $t \in \mathbb{Z}_{2^l}$, i.e. it maps $0, 1, \ldots, 2^{l-1} - 1$ to $+1$ and $2^{l-1}, 2^{l-1} + 1, \ldots, 2^l - 1$ to -1. Our goal is to express this map as a linear combination of characters. Recall the Fourier transformation formula on $\mathbb{Z}_{2^l}$:

$$\mu = \sum_{j=0}^{2^l-1} \mu_j \psi_j, \text{ where } \mu_j = \frac{1}{2^l} \sum_{x=0}^{2^l-1} \mu(x)\psi_j(-x). \tag{1}$$

Combining Lemma 4.1 and Corollary 7.4 of [12], we obtain

Lemma 1. *Let $q = 8$. For the constants $\mu_j = (1 + \zeta^{-j} + \zeta^{-2j} + \zeta^{-3j})/4$ $j = 1, 3, 5, 7$ we have*

$$\mu = \mu_1\psi_1 + \mu_3\psi_3 + \mu_5\psi_5 + \mu_7\psi_7,$$

and $\mu_j = 0$, for even j. Furthermore

$$(|\mu_1| + |\mu_3| + |\mu_5| + |\mu_7|)^2 = 2 + \sqrt{2}.$$

Let $q = 2^l$ where $l \geq 4$. Then

$$\sum_{j=0}^{q-1} |\mu_j| < \frac{2}{\pi} \ln(q) + 1. \tag{2}$$

For all $\beta \neq 0$ in the ring $R = GR(2^l, m)$, we denote by Ψ_β the character

$$\Psi_\beta : R \to \mathbb{C}^*, x \mapsto \omega^{\text{Tr}(\beta x)}.$$

Note that for the defined above ψ_k and Ψ_β, we have:

$$\psi_k(\text{Tr}(\beta x)) = \Psi_{\beta k}(x). \tag{3}$$

Let $f(X)$ denote a polynomial in $R[X]$ and let

$$f(X) = F_0(X) + 2F_1(X) + \ldots + 2^{l-1}F_{l-1}(X)$$

denote its 2-adic expansion. Let d_i be the degree in x of F_i. Let Ψ be an arbitrary additive character of R, and set D_f to be the *weighted degree* of f, defined as

$$D_f = \max\{d_0 2^{l-1}, d_1 2^{l-2}, \ldots, d_{l-1}\}.$$

With the above notation, we have (under mild technical conditions) the bound

$$\left| \sum_{x \in T} \Psi(f(x)) \right| \leq (D_f - 1)2^{m/2}. \tag{4}$$

See [10] for details.

4 Polynomials over the Galois Ring $GR(2^l, m)$

Recall that $R = GR(2^l, m)$. A polynomial

$$f(X) = \sum_{j=0}^{d} c_j x^j \in R[X]$$

is called **canonical** if $c_j = 0$ for all even j.

Given an integer $D \geq 4$, define

$$S_D = \{f(X) \in R[X] \mid D_f \leq D, f \text{ is canonical}\},$$

where D_f is the weighted degree of f. Observe that S_D is an $GR(2^l, m)-$module.

Recall [21, Lemma 4.1]. For a weaker condition on D see [18, Theorem 6.13].

Lemma 2. *For any integer $D \geq 4$, we have:*

$$|S_D| = 2^{(D - \lfloor D/2^l \rfloor)m},$$

where $\lfloor x \rfloor$ is the largest integer $\leq x$.

Recall the following property of the weighted degree [22, Lemma 3.1]

Lemma 3. *Let $f(X) \in R[X]$ and $\alpha \in R^* = R \backslash 2R$ is a unit of R and let $g(X) = f(\alpha X) \in R[X]$. Then*

$$D_g = D_f,$$

where D_f, D_g are respectively the weighted degrees of the polynomials $f(X)$ and $g(X)$.

5 Binary Sequences over Z_8

In this section we consider the case of $GR(8, m)$ and will consider the periodic sequences $c_0, c_1, \ldots$ of period $2^m - 1$. Let $\alpha \in R^*$, then define

$$c_t = \mathrm{MSB}(\mathrm{Tr}(\alpha \xi^t)), \tag{5}$$

where $t = 0, \ldots, n - 2$ and $n = 2^m$. This sequence was introduced in [12].

We now have the following results on, respectively, the imbalance and the crosscorrelation function of the binary sequence $(c_t)_{t \in \mathbb{N}}$.

Theorem 1. *With notation as above, we have*

$$\left| \sum_{t=0}^{n-2} (-1)^{c_t} \right| \le \sqrt{2 + \sqrt{2}} (3\sqrt{2^m} + 1).$$

Proof. By definition of ψ_j and Ψ_α, (where $0 \ne \alpha \in R$), for any $0 \le j \le 7$, we have:

$$\psi_j(\mathrm{Tr}(\alpha x)) = \Psi_{\alpha j}(x).$$

As we have $c_t = \mathrm{MSB}(\mathrm{Tr}(\alpha \xi^t))$, and by (1), we obtain that $(-1)^{c_t}$ is equal to:

$$\mu(\mathrm{Tr}(\alpha \xi^t)) = \sum_{j=0}^{7} \mu_j \psi_j(\mathrm{Tr}(\alpha \xi^t)) = \sum_{j=0}^{7} \mu_j \Psi_{\alpha j}(\xi^t). \tag{6}$$

Changing the order of summation, we obtain that:

$$\sum_{t=0}^{n-2} (-1)^{c_t} = \sum_{j=0}^{7} \mu_j \sum_{t=0}^{n-2} \Psi_{\alpha j}(\xi^t). \tag{7}$$

Inequality (4) implies that:

$$\left| \sum_{x \in T^*} \Psi_\lambda(x) \right| \le 3\sqrt{2^m} + 1, \tag{8}$$

for all $\lambda \in GR(8, m), \lambda \ne 0$. Thus, the absolute value of the Right Hand Side of (7) can be estimated from above by:

$$(3\sqrt{2^m} + 1) \sum_{j=0}^{7} |\mu_j|.$$

Applying Lemma 1 the Theorem follows.   □

Theorem 2. *With notation as above, and for all phase shifts* $\tau, 0 < \tau < 2^m - 1$, *let*

$$\Theta(\tau) = \sum_{t \in \mathcal{I}} (-1)^{c_t} (-1)^{c'_{t+\tau}},$$

where $c_t = \text{MSB}(\text{Tr}(\alpha \xi^t))$ and $c'_t = \text{MSB}(\text{Tr}(\alpha' \xi^t))$ are two elements such that for any pair $0 \le j_1, j_2 \le 7$ the following condition holds

$$j_1 \alpha + j_2 \alpha' \xi^\tau \ne 0.$$

We then have the bound

$$|\Theta(\tau)| \le 3(2 + \sqrt{2})\sqrt{2^m}.$$

Proof. As we have $c_t = \text{MSB}(\text{Tr}(\alpha \xi^t))$ and $c'_t = \text{MSB}(\text{Tr}(\alpha' \xi^t))$, where t ranges between 0 and $n-2$ and by (1), applying (6) and changing the order of summation, we obtain that:

$$\Theta(\tau) = \sum_{j_1=0}^{7} \sum_{j_2=0}^{7} \mu_{j_1} \mu_{j_2} \sum_{t=0}^{n-2} \Psi_{j_1}(\alpha \xi^t) \Psi_{j_2}(\alpha' \xi^{t+\tau}). \tag{9}$$

By definition of Ψ, we have:

$$\Psi_{j_1}(\alpha \xi^t) \Psi_{j_2}(\alpha' \xi^{t+\tau}) = \Psi_\beta(\xi^t),$$

where $\beta = j_1 \alpha + j_2 \alpha' \xi^\tau \ne 0$. Applying (4), we obtain:

$$\left| \sum_{t=0}^{n-2} \Psi_{j_1}(\alpha \xi^t) \Psi_{j_2}(\alpha' \xi^{t+\tau}) \right| = \left| \sum_{t=0}^{n-2} \Psi_\beta(\xi^t) \right| \le 3\sqrt{2^m} + 1. \tag{10}$$

Applying Lemma 1, we obtain

$$\sum_{j_1=0}^{7} \sum_{j_2=0}^{7} |\mu_{j_1} \mu_{j_2}| = \left(\sum_{j=0}^{2^l-1} |\mu_j| \right)^2 = 2 + \sqrt{2}. \tag{11}$$

Combining it with (10) the result follows. $\qquad\square$

6   Binary Sequences of Period $2^m - 1$

In this section we will consider periodic sequences $c_0, c_1, \ldots$ of period $n-1$. For any integer $D \ge 4$, let $f \in S_D$ and set

$$c_t = \text{MSB}(\text{Tr}(f(\xi^t))), \tag{12}$$

where $t = 0, \ldots, n-2$.

We now have the following results on, respectively, the imbalance and the crosscorrelation function of the binary sequence defined by (12).

Theorem 3. *With notations as above, we have:*

$$\left| \sum_{t=0}^{n-2} (-1)^{c_t} \right| \le (2l \ln(2)/\pi + 1)(D-1)\sqrt{n},$$

where $c_t = \text{MSB}(\text{Tr}(f(\xi^t)))$.

Proof. As we have $c_t = \text{MSB}(\text{Tr}(f(\xi^t)))$, and by (1), we obtain that $(-1)^{c_t}$ is equal to:

$$\mu(\text{Tr}(f(\xi^t))) = \sum_{j=0}^{2^l-1} \mu_j \psi_j(\text{Tr}(f(\xi^t))) = \sum_{j=0}^{2^l-1} \mu_j \Psi_j(f(\xi^t)).$$

Changing the order of summation, we obtain that:

$$\sum_{t=0}^{n-2} (-1)^{c_t} = \sum_{j=0}^{2^l-1} \mu_j \sum_{x \in \mathcal{T}^*} \Psi_j(f(x)). \tag{13}$$

Applying (4), the absolute value of the Right Hand Side of (13) can be estimated from above by:

$$((D_f - 1)\sqrt{2^m} + 1) \sum_{j=0}^{2^l-1} |\mu_j|. \tag{14}$$

Applying Lemma 1 the sum (14) can be estimated from above by:

$$(2l\ln(2)/\pi + 1)(D_f - 1)\sqrt{2^m}.$$

The result follows. □

Theorem 4. *With notation as above, and for all phase shifts τ, in the range $0 < \tau < 2^m - 1$, let $(n = 2^m)$*

$$\Theta(\tau) = \sum_{t=0}^{n-2} (-1)^{c_t}(-1)^{c'_{t+\tau}},$$

where $c_t = \text{MSB}(\text{Tr}(f_1(\xi^t)))$ and $c'_t = \text{MSB}(\text{Tr}(f_2(\xi^t)))$. We then have the bound $(l \geq 4)$:

$$|\Theta(\tau)| \leq \left(\frac{2l}{\pi}\ln(2) + 1\right)^2 [1 + (D-1)\sqrt{2^m}],$$

where $D = max\{D_{f_1}, D_{f_2}\}$ and for any pair $0 \leq j_1, j_2 \leq 2^l - 1$ the following condition holds

$$j_1 f_1(x) + j_2 f_2(x\xi^\tau) \neq 0.$$

Proof. As we have $c_t = \text{MSB}(\text{Tr}(f_1(\xi^t)))$ and $c'_t = \text{MSB}(\text{Tr}(f_2(\xi^t)))$, where t ranges between 0 and $n-2$ and by (1), we obtain that $(-1)^{c_t}$ is equal to:

$$\mu(\text{Tr}(f_1(\xi^t))) = \sum_{j=0}^{2^l-1} \mu_j \psi_j(\text{Tr}(f_1(\xi^t))) = \sum_{j=0}^{2^l-1} \mu_j \Psi_j(f_1(\xi^t)).$$

Changing the order of summation, we obtain that:

$$\Theta(\tau) = \sum_{j_1=0}^{2^l-1} \sum_{j_2=0}^{2^l-1} \mu_{j_1} \mu_{j_2} \sum_{t=0}^{n-2} \Psi_{j_1}(f_1(\xi^t))\Psi_{j_2}(f_2(\xi^{t+\tau})). \tag{15}$$

By definition of Ψ, we have:

$$\Psi_{j_1}(f_1(\xi^t))\Psi_{j_2}(f_2(\xi^{t+\tau})) \;=\; \Psi(g(\xi^t)),$$

where $g(X) = j_1 f_1(X) + j_2 f_2(X\xi^\tau)$. Note that if $f(X) \in S_D$ then $f(X\xi^\tau) \in S_D$ since, by Lemma 3 the change of variable $X \to X\xi^\tau$ does not increase the weighted degree. Moreover S_D is an R-linear space. Thus the polynomial $g(X)$ belongs to S_D along with f_1 and f_2. Applying (4), we obtain:

$$\left| \sum_{t=0}^{n-2} \Psi_{j_1}(f_1(\xi^t))\Psi_{j_2}(f_2(\xi^{t+\tau})) \right| \;=\; \left| \sum_{x \in T^*} \Psi(g(x)) \right| \;\le\; 1 + (D-1)\sqrt{2^m}. \quad (16)$$

Applying Lemma 1, we obtain

$$\sum_{j_1=0}^{2^l-1}\sum_{j_2=0}^{2^l-1} |\mu_{j_1}\mu_{j_2}| \;=\; \left(\sum_{j=0}^{2^l-1} |\mu_j| \right)^2 \;\le\; \left(\frac{2l\ln(2)}{\pi} + 1 \right)^2. \quad (17)$$

Combining (16) with (17) the result follows.   □

7   Binary Codes and Sequences of Maximal Length

Let $\gamma = \xi(1+2\lambda) \in R$, where $\xi \in T$ and $\lambda \in R^*$. Assume $1+2\lambda$ is of order 2^{l-1}. Since ξ is of order $2^m - 1$ then γ is an element of order $N = 2^{l-1}(2^m - 1)$.

In this section we consider the periodic sequences $c_0, c_1, \ldots$ of period N. Let $\alpha \in R^*$, then define

$$c_t = \mathrm{MSB}(\mathrm{Tr}(\alpha\gamma^t)), \quad (18)$$

where $t = 0, \ldots, N-1$. This sequence was introduced and studied in [21].

We now have the following results on, respectively, the imbalance and the crosscorrelation function of the binary sequence $(c_t)_{t\in\mathbb{N}}$, (18) under the MSB map.

First, we need the following technical lemma:

Lemma 4. *Let* $\gamma = \xi(1 + 2\lambda) \in R$, *where* $\xi \in T^*$, $\lambda \in R^* = R\backslash 2R$, *and* $1 + 2\lambda \in R$ *is an element of order* 2^{l-1}. *Set* $N = 2^{l-1}(2^m - 1)$. *Then, for any* $0 \ne \beta \in R$, *we have:*

$$\sum_{j=0}^{N-1} \Psi_\beta(\gamma^j) \;=\; \sum_{j=0}^{2^{l-1}-1} \left(\sum_{x\in T^*} \Psi_{\beta(1+2\lambda)^j}(x) \right).$$

Proof. Since $(2^{l-1}, 2^m - 1) = 1$, as j ranges over $\{0, 1, \ldots, N - 1\}$, the set of ordered pairs

$$\{(j \,(\mathrm{mod}\, 2^{l-1}),\, j \,(\mathrm{mod}\, 2^m - 1))\}$$

runs over all pairs (j_1, j_2), where $j_1 \in \{0, 1, \ldots, 2^{l-1}-1\}$ and $j_2 \in \{0, 1, \ldots, 2^m - 2\}$. Thus the set

$$\{\gamma^t; t = 0, 1, \ldots, N - 1\} \;=\; \{\xi^t(1 + 2\lambda)^t; t = 0, 1, \ldots, N - 1\}.$$

is equal to the cartesian product of sets:

$$\{\xi^{t_1}; t_1 = 0, 1, \ldots, 2^m - 2\} \times \{(1 + 2\lambda)^{t_2}; t = 0, 1, \ldots, 2^{l-1} - 1\}. \tag{19}$$

From (19) we obtain:

$$\sum_{j=0}^{N-1} \Psi_\beta(\gamma^j) = \sum_{j_1=0}^{2^{l-1}-1} \sum_{j_2=0}^{2^m-2} \Psi_\beta((1 + 2\lambda)^{j_1} \xi^{j_2}),$$

whereas observing that

$$\Psi_\beta((1 + 2\lambda)^{j_1} \xi^{j_2}) = \Psi_{\beta'}(\xi^{j_2}),$$

where $\beta' = \beta(1 + 2\lambda)^{j_1}$, yields:

$$\sum_{j_1=0}^{2^{l-1}-1} \sum_{j_2=0}^{2^m-2} \Psi_{\beta(1+2\lambda)^{j_1}}(\xi^{j_2}).$$

The Lemma follows.   □

Theorem 5. *With notation as above, we have:*

$$\left| \sum_{t=0}^{N-1} (-1)^{c_t} \right| \le (2l \ln(2)/\pi + 1) 2^{l-1} [(2^{l-1} - 1)\sqrt{2^m} + 1].$$

Proof. Recall that $\gamma = \xi(1 + 2\lambda) \in R$, where $\xi \in T^*$ is a generator of the Teichmüller set and $\lambda \in R^*$. By definition of ψ_j and Ψ_α, (where $0 \neq q \in R$), for any $0 \le j \le 2^l - 1$, we have:

$$\psi_j(\mathrm{Tr}(\alpha x)) = \Psi_{\alpha j}(x).$$

As we have $c_t = \mathrm{MSB}(\mathrm{Tr}(\alpha \gamma^t))$, and by (1), we obtain that $(-1)^{c_t}$ is equal to:

$$\mu(\mathrm{Tr}(\alpha \gamma^t)) = \sum_{j=0}^{2^l-1} \mu_j \psi_j(\mathrm{Tr}(\alpha \gamma^t)) = \sum_{j=0}^{2^l-1} \mu_j \Psi_{\alpha j}(\gamma^t).$$

Changing the order of summation, we obtain that:

$$\sum_{t=0}^{N-1} (-1)^{c_t} = \sum_{j=0}^{2^l-1} \mu_j \sum_{t=0}^{N-1} \Psi_{\alpha j}(\gamma^t). \tag{20}$$

Applying Lemma 4, we have:

$$\sum_{t=0}^{N-1} \Psi_{\alpha j}(\gamma^t) = \sum_{t=0}^{2^{l-1}-1} \left(\sum_{x \in T^*} \Psi_{\alpha j(1+2\lambda)^t}(x) \right). \tag{21}$$

Applying Lemma 3.1 of [21] to each of the 2^{l-1} sums over x, we obtain that:

$$\left| \sum_{t=0}^{N-1} \Psi_{\alpha j}(\gamma^t) \right| \le 2^{l-1}[(2^{l-1}-1)\sqrt{2^m}+1].$$

Thus, the absolute value of the Right Hand Side of (20) can be estimated from above by:

$$2^{l-1}[(2^{l-1}-1)\sqrt{2^m}+1] \sum_{j=0}^{2^l-1} |\mu_j|. \tag{22}$$

Recall Corollary 7.4 of [12] which states that for $l \ge 4$ the following estimate holds:

$$\sum_{j=0}^{2^l-1} |\mu_j| < \frac{2l\ln(2)}{\pi} + 1.$$

Thus (22) can be estimated from above by:

$$(2l\ln(2)/\pi + 1)2^{l-1}[(2^{l-1}-1)\sqrt{2^m}+1].$$

The Lemma follows.   □

We now proceed to bound the crosscorrelation.

Theorem 6. *With notation as above, and for all phase shifts $\tau, 0 < \tau < 2^m - 1$, let*

$$\Theta(\tau) = \sum_{t=0}^{N-1} (-1)^{c_t}(-1)^{c'_{t+\tau}},$$

where $c_t = MSB(\mathrm{Tr}(\alpha\gamma^t))$ and $c'_t = MSB(\mathrm{Tr}(\alpha'\gamma^t))$ are such that for any pair $0 \le j_1, j_2 \le 2^l - 1$ the following condition holds

$$j_1\alpha + j_2\alpha'\xi^\tau \ne 0.$$

We then have (for $l \ge 4$) the bound :

$$|\Theta(\tau)| \le \left(\frac{2l}{\pi} \ln(2) + 1 \right)^2 2^{l-1}[(2^{l-1}-1)\sqrt{2^m}+1] = C'_l\sqrt{2^m}.$$

Here C'_l is a constant in l of order $l^2 2^{2l}$, i.e.

$$C'_l = (\ln(2)/\pi)^2(l^2 2^{2l} + o(1)).$$

Proof. Again let $\gamma = \xi(1+2\lambda) \in R$, where $\xi \in T$ is a generator of the Teichmüller set and $\lambda \in R^*$. As we have $c_t = MSB(\mathrm{Tr}(\alpha\gamma^t))$, and by (1), we obtain that $(-1)^{c_t}$ is equal to:

$$\mu(\mathrm{Tr}(\alpha\gamma^t)) = \sum_{j=0}^{2^l-1} \mu_j \psi_j(\mathrm{Tr}(\alpha\gamma^t)) = \sum_{j=0}^{2^l-1} \mu_j \Psi_{\alpha j}(\gamma^t).$$

Changing the order of summation, we obtain that:

$$\Theta(\tau) = \sum_{j_1=0}^{2^l-1} \sum_{j_2=0}^{2^l-1} \mu_{j_1}\mu_{j_2} \sum_{t=0}^{N-1} \Psi_\beta(\gamma^t).$$

Here $\beta = j_1\alpha + j_2\alpha'\gamma^\tau \neq 0$. Applying Corollary 7.4 of [12] (for $l \geq 4$), we have:

$$\sum_{j_1=0}^{2^l-1} \sum_{j_2=0}^{2^l-1} |\mu_{j_1}\mu_{j_2}| = \left(\sum_{j=0}^{2^l-1} |\mu_j|\right)^2 \leq \left(\frac{2l}{\pi}\ln(2)+1\right)^2. \tag{23}$$

Applying Lemma 4, we have:

$$\sum_{j=0}^{N-1} \Psi_\beta(\gamma^j) = \sum_{j=0}^{2^{l-1}-1} \left(\sum_{x \in \mathcal{T}} \Psi_{\beta(1+2\lambda)^j}(x)\right),$$

where $0 \neq \beta(1+2\lambda)^j \in R$ so that each sum over x can be estimated using Lemma 3.1 of [21]. Thus, we have:

$$\left|\sum_{j=0}^{N-1} \Psi_\beta(\gamma^j)\right| \leq 2^{l-1}[(2^{l-1}-1)\sqrt{2^m}+1]. \tag{24}$$

Combining (23) with (24) the Lemma follows.   □

8   Non Binary Codes of Maximal Length

Let $\gamma = \xi(1+2\lambda) \in R$, where $\xi \in \mathcal{T}$ and $\lambda \in R^*$. Assume $1+2\lambda$ is of order 2^{l-1}. Since ξ is of order $2^m - 1$ then γ is an element of order $N = 2^{l-1}(2^m - 1)$.

Following [6, Lemma 2], we define the code of length N :

$$S_{l,m,D} = \{(\mathrm{Tr}(f(\gamma^t)))_{t=0}^{N-1} \mid f \in S_D\}. \tag{25}$$

This sequence was introduced and studied in [24].

We prepare the upper bound on the PAPR of codewords by a result on a character sum.

Theorem 7. *Let* $\gamma = \xi(1+2\lambda) \in R$, *where* $\xi \in \mathcal{T}^*$, $\lambda \in R^* = R\backslash 2R$, *and* $1+2\lambda \in R$ *is an element of order* 2^{l-1}. *Set* $N = 2^{l-1}(2^m-1)$. *Then for any* $j \in [0, 2^{l-1}-1]$, *we have*

$$\left|\sum_{k=0}^{N-1} \Psi(f(\gamma^k))e^{2\pi ikj/N}\right| \leq 2^{l-1}[D_f\sqrt{2^m}+1]. \tag{26}$$

Proof. Since $(2^{l-1}, 2^m - 1) = 1$, as j ranges over $\{0, 1, \ldots, N-1\}$, the set of pairs

$$\{(j \,(\mathrm{mod}\, 2^{l-1}),\ j \,(\mathrm{mod}\, 2^m - 1))\}$$

runs over all pairs (j_1, j_2), where $j_1 \in \{0, 1, \ldots, 2^{l-1}\}$ and $j_2 \in \{0, 1, \ldots, 2^m - 1\}$. Thus the set

$$\{\gamma^k; k = 0, 1, \ldots, N-1\} \ = \ \{\xi^k(1+2\lambda)^k; k = 0, 1, \ldots, N-1\}.$$

is equal to the direct product of sets:

$$\{\xi^{k_1}; t_1 = 0, 1, \ldots, 2^m - 1\} \times \{(1+2\lambda)^{k_2}; t = 0, 1, \ldots, 2^{l-1} - 1\}, \tag{27}$$

where

$$k \equiv k_1 (\mathrm{mod}\, 2^m - 1), \ k \equiv k_2 (\mathrm{mod}\, 2^{l-1}).$$

By the CRT there exist integers c_1, c_2 such that for all $k = 0, 1, \ldots, N-1$ we can write

$$k = 2^{l-1} c_1 k_1 + (2^m - 1) c_2 k_2$$

Consequently since $\gamma = \xi(1+2\lambda)$, where $\xi^{2^m - 1} = 1$ and $(1+2\lambda)^{2^{l-1}} = 1$, the sum of the left hand side of (26) is equal to:

$$\sum_{k_2=0}^{2^{l-1}-1} \sum_{k_1=0}^{2^m - 2} \Psi(f(\xi^{k_1}(1+2\lambda)^{k_2}))e^{2\pi i k j/N}. \tag{28}$$

Set $f_{k_2}(x) = f(x(1+2\lambda)^{k_2})$. Then, expressing k as a function of k_1 and k_2, the above sum is equal to:

$$\sum_{k_2=0}^{2^{l-1}-1} \sum_{k_1=0}^{2^m - 2} \Psi(f_{k_2}(\xi^{k_1}))e^{2\pi i c_1 k_1 j/(2^m - 1)} e^{2\pi i c_2 k_2 j/2^{l-1}}.$$

Using Lemma 3 the weighted degree of f_{k_2} is bounded by D_f. Applying (4) to each of the 2^{l-1} inner sums yields:

$$\left| \sum_{t=0}^{2^m - 2} \Psi(f_{k_2}(\xi^t))e^{2\pi i c_1 k_1 j/(2^m - 1)} \right| \le D_f \sqrt{2^m} + 1.$$

Thus, the absolute value of (28) can be estimated from above by

$$2^{l-1}[D_f \sqrt{2^m} + 1].$$

The result follows. $\square$

This character sum estimate translates immediately in terms of PAPR.

Corollary 1. *For every* $c \in S_{l,m,D}$ *the PAPR is at most*

$$\frac{2^{l-1}}{2^m - 1}(1 + D\sqrt{2^m})^2 \left(\frac{2}{\pi} \log(2N) + 2\right)^2,$$

where log *stands for the natural logarithm.*

We prepare the bound on the minimum Euclidean distance of the code $S_{l,m,D}$ by a correlation approach.

Theorem 8. *With notation as above, and for all phase shifts τ, in the range $0 < \tau < N$, let*

$$\Theta(\tau) = \sum_{t=0}^{N-1} \omega^{c_t - c'_{t+\tau}},$$

where $c_t = \mathrm{Tr}(f_1(\gamma^t))$ and $c'_t = \mathrm{Tr}(f_2(\gamma^t))$. We then have the bound ($l \geq 4$):

$$|\Theta(\tau)| \leq 2^{l-1}[(D_f - 1)\sqrt{2^m} + 1].$$

Proof. As we have $c_t = \mathrm{Tr}(f_1(\gamma^t))$ and $c'_t = \mathrm{Tr}(f_2(\gamma^t))$, where t ranges between 0 and $N - 1$ and $\gamma = \xi(1 + 2\lambda)$ is of order $N = 2^{l-1}(2^m - 1)$.

We obtain that:

$$\Theta(\tau) = \sum_{t=0}^{N-1} \Psi(f_1(\gamma^t) - f_2(\gamma^{t+\tau})). \tag{29}$$

By definition of Ψ, we have:

$$\Psi(f_1(\gamma^t) - f_2(\gamma^{t+\tau})) = \Psi(f_3(\gamma^t)),$$

where $f_3(x) = f_1(x) - f_2(x\gamma^\tau)$. Note that if $f(x) \in S_D$ then by Lemma 3 $f(x\gamma^\tau) \in S_D$ since the change of variables $x \to x\gamma^\tau$ does not increase the weighted degree. Moreover S_D is an R-linear space. Thus the polynomial $f_3(x)$ belongs to S_D along with f_1 and f_2. Further, as t ranges between 0 and $N - 1$, the set $\{\gamma^t; t = 0, 1, \ldots, N - 1\}$ is equal to the product

$$\{\xi^{k_1}; t_1 = 0, 1, \ldots, 2^m - 1\} \times \{(1 + 2\lambda)^{k_2}; t = 0, 1, \ldots, 2^{l-1} - 1\}.$$

Thus, in (29), the sum over t is equal to

$$\sum_{k_2=0}^{2^{l-1}-1} \sum_{k_1=0}^{2^m-2} \Psi(f_3(\xi^{k_1}(1 + 2\lambda)^{k_2})) = \sum_{k_2=0}^{2^{l-1}-1} \sum_{k_1=0}^{2^m-2} \Psi(g_{k_2}(\xi^{k_1})), \tag{30}$$

where $g_{k_2}(x) = f_3(x(1 + 2\lambda)^{k_2})$ and for any k_2, $g_{k_2} \in S_D$. Applying (4) to each of the 2^{l-1} sums:

$$\left| \sum_{t=0}^{2^m-2} \Psi(f_{k_2}(\xi^t)) \right| \leq (D_f - 1)\sqrt{2^m} + 1.$$

Thus, the absolute value of (30) can be estimated above by

$$2^{l-1}[(D_f - 1)\sqrt{2^m} + 1]. \tag{31}$$

The result follows.   □

We are now in a position to estimate the minimum Euclidean distance d_E of our code $S_{l,m,D}$.

Corollary 2. *For all integers $l \geq 4$ and $m \geq 3$ we have*

$$d_E \geq 2^l(2^m - 2 - D\sqrt{2^m}).$$

9   Biphase Sequences

In this section, we construct biphase sequences using the composition of the Carlet map and inverse Gray map (see [23] for details).

Let $a = a_0 + 2a_1 + 4a_2 \in \mathbb{Z}_8$ be 2-adic expansion of a. Define two maps Z_1, Z_2 from $\mathbb{Z}_8$ to $\mathbb{Z}_4$ and a map Z from $\mathbb{Z}_8$ to $\mathbb{Z}_4 \times \mathbb{Z}_4$:

$$Z(a) = (Z_1(a), Z_2(a)) = (a_1 + 2(a_0 + a_2), a_1 + 2a_2),$$

where $a_0, a_1, a_2 \in \mathbb{Z}_2$.

For any integer $k \geq 1$, extend this map (also denoted by Z) to the following map from $\mathbb{Z}_8^k$ to $\mathbb{Z}_4^k \times \mathbb{Z}_4^k$:

Definition 1. *Let* $\mathbf{c} = (c_1, c_2, \ldots, c_k) \in \mathbb{Z}_8^k$ *and* $Z(c_j) = (a_j, b_j) \in \mathbb{Z}_4 \times \mathbb{Z}_4$ *where* $a_j = Z_1(c_j)$ *and* $b_j = Z_2(c_j)$ *for* $j = 1, 2, \ldots, k$. *Define* Z *as follows:*

$$Z(\mathbf{c}) = (\mathbf{a}, \mathbf{b}), \tag{32}$$

where $\mathbf{a} = (a_1, a_2, \ldots, a_k) \in \mathbb{Z}_4^k$ *and* $\mathbf{b} = (b_1, b_2, \ldots, b_k) \in \mathbb{Z}_4^k$ *defined above.*

Define the map MSBZ: $\mathbb{Z}_8 \to \mathbb{Z}_2 \times \mathbb{Z}_2$ via

$$\mathrm{MSBZ}(u) = (\mathrm{MSB}(Z_1(u)), \mathrm{MSB}(Z_2(u))),$$

in other words

$$\mathrm{MSBZ}(a_0 + 2a_1 + 4a_2) = (a_0 + a_2, a_2),$$

obtained by taking the most significant bit in each component of Z (which are elements of $\mathbb{Z}_4$).

In this section we consider the periodic sequences $c_0, c_1, \ldots$ of period $2(n-1)$, where $n = 2^m$. Let $\beta = 5\xi$. Since $5^2 = 1$ in $\mathbb{Z}_8$, we have that

$$\beta^{n-1} = (5\xi)^{n-1} = 5,$$

and thus $\beta^{2(n-1)} = 1$.

Any two polynomials $f(X), g(X) \in R[X]$ are considered equivalent if $g(X) = f(aX\alpha)$ for some $a \in \mathbb{Z}_8^*$ and $\alpha \in \mathcal{T}^*$. For any positive integer $D \geq 4$, let $f \in S_D$ modulo the equivalence relation, and set

$$c_t = \begin{cases} \mathrm{MSB}(Z_1(\mathrm{Tr}(f(\beta^t)))), & \text{if } 0 \leq t < n-1 \\ \mathrm{MSB}(Z_2(\mathrm{Tr}(f(\beta^t)))), & \text{if } n-1 \leq t < 2(n-1). \end{cases}$$

The same DFT technique as above can be applied to the functions $(-1)^{MSB(Z_1(u))}$, $(-1)^{MSB(Z_2(u))}$, which map the element $a_0 + 2a_1 + 4a_2$ of $\mathbb{Z}_8$ into the respectively $(-1)^{a_0+a_2}$ and $(-1)^{a_2}$. Then the following holds

Lemma 5. *Let* $\omega = e^{2\pi i/8}$ *be a primitive 8th root of* 1, *and set set* $\psi_k(u) = \omega^{ku}$, *where* k *is an integer. Then we have that*

$$(-1)^{MSB(Z_1(u))} = \mu_1\psi_1(u) + \mu_3\psi_3(u) + \mu_5\psi_5(u) + \mu_7\psi_7(u),$$

$$(-1)^{MSB(Z_2(u))} = \mu_5\psi_1(u) + \mu_7\psi_3(u) + \mu_1\psi_5(u) + \mu_3\psi_7(u),$$

where

$$\mu_1 = \frac{1}{4}(1 + \omega - \omega^2 + \omega^3), \qquad \mu_3 = \frac{1}{4}(1 + \omega + \omega^2 + \omega^3),$$

$$\mu_5 = \frac{1}{4}(1 - \omega - \omega^2 - \omega^3), \qquad \mu_7 = \frac{1}{4}(1 - \omega + \omega^2 - \omega^3);$$

and moreover

$$(|\mu_1| + |\mu_3| + |\mu_5| + |\mu_7|)^2 = 2 + \sqrt{2}.$$

Proof. Recall the Fourier transformation formula on the additive group of $\mathbb{Z}_8$. Let μ be an arbitrary function from $\mathbb{Z}_8$ to the complex numbers. Then for any $u \in \mathbb{Z}_8$ we have

$$\mu(u) = \sum_{j=0}^{7} \mu_j\psi_j(u), \quad \text{where} \quad \mu_j = \frac{1}{8}\sum_{x=0}^{7}\mu(x)\psi_j(-x).$$

In particular, when $\mu(u) = (-1)^{MSB(Z_1(u))}$ and $x = a_0 + 2a_1 + 4a_2$ we have

$$\mu_j = \frac{1}{8}\sum_{a_0,a_1,a_2}(-1)^{a_0+a_2}\omega^{-j(a_0+2a_1+4a_2)}.$$

Once simplified, we obtain

$$\mu_j = \frac{1}{8}(1 - \omega^{-j})(1 + i^{-j})(1 + (-1)^{1-j}).$$

Substituting $j = 0, 1, \ldots, 7$ and simplifying, the result follows (in particular $\mu_j = 0$, $j = 0, 2, 4, 6$). The derivation for $(-1)^{MSB(Z_2(u))}$ is analogous and omitted. $\square$

We now proceed to bound the imbalance.

Theorem 9. *With notations as above, we have that*

$$\left|\sum_{t=0}^{n-2}((-1)^{MSB(Z_1(u_t))} + (-1)^{MSB(Z_2(u_t))})\right| \le 2(2 + \sqrt{2})^{1/2}(D-1)\sqrt{2^m},$$

where $u_t = \mathrm{Tr}(f(\beta^t))$

Proof. Applying Lemma 5, and changing the order of summation, we obtain that

$$\sum_{t=0}^{n-2}((-1)^{MSB(Z_1(u_t))} + (-1)^{MSB(Z_2(u_t))})$$

is equal to

$$\sum_{t=0}^{n-2}\Big((\mu_1+\mu_5)(\Psi_1(f(\beta^t))+\Psi_5(f(\beta^t))) + (\mu_3+\mu_7)(\Psi_3(f(\beta^t))+\Psi_7(f(\beta^t)))\Big), \quad (33)$$

where $\Psi_k(u) = \psi_k(\mathrm{Tr}(u)) = \psi(k\mathrm{Tr}(u))$. Let $\Psi = \Psi_1$. Then the sum (33) equals to

$$\sum_{t=0}^{n-2}\Big((\mu_1+\mu_5)(\Psi(f(\beta^t))+\Psi(5f(\beta^t)))+(\mu_3+\mu_7)(\Psi(3f(\beta^t))+\Psi(7f(\beta^t)))\Big). \quad (34)$$

Recall that

$$f(\beta^t) \;=\; \begin{cases} f(\xi^t), & \text{if } t \text{ is even} \\ 5f(\xi^t), & \text{if } t \text{ is odd} \end{cases}$$

Thus, since $5^2 \equiv 1 \pmod 8$, we have

$$\sum_{t=0}^{n-2}(\Psi(f(\beta^t))+\Psi(5f(\beta^t))) = \sum_{t=0}^{n-2}(\Psi(f(\xi^t))+\Psi(5f(\xi^t))).$$

Since $3 \times 5 \equiv 7 \pmod 8$ and $7 \times 5 \equiv 3 \pmod 8$, we have

$$\sum_{t=0}^{n-2}(\Psi(3f(\beta^t))+\Psi(7f(\beta^t))) = \sum_{t=0}^{n-2}(\Psi(3f(\xi^t))+\Psi(7f(\xi^t))).$$

Recalling that

$$\left| \sum_{t=0}^{n-2}\Psi_j(\mathrm{Tr}(f(\beta^t))) \right| \;\le\; (D-1)\sqrt{2^m} \quad (35)$$

and combining it with (5) the result follows.   □

We now proceed to bound the crosscorrelation.

Theorem 10. *With notations as above, and for all phase shifts τ in the range* $0 < \tau < 2^m - 1$, *let*

$$\Theta(\tau) = \sum_{t=0}^{n-2}((-1)^{\mathrm{MSB}(Z_1(u_t))-\mathrm{MSB}(Z_1(v_{t+\tau}))} + (-1)^{\mathrm{MSB}(Z_2(u_t))-\mathrm{MSB}(Z_2(v_{t+\tau}))}),$$

where $u_t = \mathrm{Tr}(f_1(\beta^t))$ *and* $v_{t+\tau} = \mathrm{Tr}(f_2(\beta^{t+\tau}))$. *We then have the bound*

$$|\Theta(\tau)| \le 2(2+\sqrt{2})(D-1)\sqrt{2^m}.$$

Proof. For $i = 1, 2$, define

$$\Theta_i(\tau) = \sum_{j=0}^{n-2}((-1)^{\mathrm{MSB}(Z_1(u_j))+\mathrm{MSB}(Z_i(v_{j+\tau}))}.$$

Consider the contribution from $\Theta_1(\tau)$. Applying Lemma 5 to $(-1)^{\mathrm{MSB}(Z_1(u_t))+\mathrm{MSB}(Z_1(v_{t+\tau}))}$, we obtain

$$\Theta_1(\tau) = \sum_{j_1=0}^{7}\sum_{j_2=0}^{7}\mu_{j_1}\mu_{j_2}\sum_{t=0}^{n-2}\Psi_{j_1}(f_1(\beta^t))\Psi_{j_2}(f_2(\beta^{t+\tau})).$$

By definition of Ψ, we have:

$$\Psi_{j_1}(f_1(\beta^t))\Psi_{j_2}(f_2(\beta^{t+\tau})) = \Psi(g(\beta^t)),$$

where $g(X) = j_1 f_1(X) + j_2 f_2(X\xi^\tau)$ or $g(X) = j_1 f_1(X) + j_2 5 f_2(X\xi^\tau)$.

Note that if $f(X) \in S_D$ then $f(X\beta^\tau) \in S_D$ since, by Lemma 3 the change of variable $X \to X\beta^\tau$ does not increase the weighted degree. Moreover S_D is an R-linear space. Thus the polynomial $g(X)$ belongs to S_D along with f_1 and f_2. Following the ideas of Theorem 9, we reduce to the sums

$$\left|\sum_{t=0}^{n-2} \Psi_{j_1}(f_1(\xi^t))\Psi_{j_2}(f_2(\xi^{t+\tau}))\right| = \left|\sum_{t=0}^{n-2} \Psi(g(\xi^t))\right| \le (D-1)\sqrt{2^m}. \tag{36}$$

Applying Lemma 5, we obtain

$$\sum_{j_1=0}^{7}\sum_{j_2=0}^{7} |\mu_{j_1}\mu_{j_2}| = \left(\sum_{j=0}^{7} |\mu_j|\right)^2 = 2 + \sqrt{2}. \tag{37}$$

Combining it with (36) the result follows. The constant 2 in the Theorem comes from the contributions of Z_1 and Z_2. □

Acknowledgement. We thank Steven Galbraith, Kenny Paterson, Ana Salagean, and Kai-Uwe Schmidt for helpful remarks that greatly improved the presentation of this material.

References

1. Berlekamp, E.R.: Algebraic coding theory. McGraw-Hill, New York (1968)
2. Blake, I.F.: Codes over integer residue rings. Information and Control 29(4), 295–300 (1975)
3. Bonnecaze, A., Solé, P., Calderbank, A.R.: Quaternary quadratic residue codes and unimodular lattices. IEEE Trans. Inform. Theory IT-41(2), 366–377 (1995)
4. Carlet, C., Charpin, P., Zinoviev, V., etal.: Codes, bent functions and permutations suitable for DES-like cryptosystems. Designs Codes and Cryptography 15, 125–156 (1998)
5. Dai, Z.-D.: Binary sequences derived from ML-sequences over rings I: period and minimal polynomial. J. Cryptology 5, 193–507 (1992)
6. Fan, S., Han, W.: Random properties of the highest level sequences of primitive Sequences over $\mathbb{Z}_{2^e}$. IEEE Trans. Inform. Theory IT-49, 1553–1557 (2003)
7. Fan, P., Darnell, M.: Sequence Design for Communications Applications. John Wiley and sons Inc, Chichester (1996)
8. Hammons Jr., A.R., Kumar, P.V., Calderbank, A.R., Sloane, N.J.A., Solé, P.: The $\mathbb{Z}_4$-linearity of Kerdock, Preparata, Goethals and related codes. IEEE Trans. Inform. Theory IT-40, 301–319 (1994)
9. Helleseth, T., Kumar, P.V.: Sequences with low Correlation. In: Pless, V.S., Huffman, W.C. (eds.) Handbook of Coding theory, North Holland, vol. II, pp. 1765–1853 (1998)

10. Kumar, P.V., Helleseth, T., Calderbank, A.R.: An upper bound for Weil exponential sums over Galois rings and applications. IEEE Trans. Inform. Theory IT-41, 456–468 (1995)
11. Lam, A.W.: Theory and applications of spread spectrum systems. IEEE Press, Los Alamitos (1994)
12. Lahtonen, J., Ling, S., Solé, P., Zinoviev, D.: $\mathbb{Z}_8$-Kerdock codes and pseudo-random binary sequences. Journal of Complexity 20(2-3), 318–330 (2004)
13. Litsyn, S.: Peak Power Control in Multicarrier Communications, Cambridge (2007)
14. MacWilliams, F.J., Sloane, N.J.A.: The Theory of Error-Correcting Codes, North-Holland (1977)
15. McDonald, B.R.: Finite Rings with Identity. Marcel Dekker, New York (1974)
16. Paterson, K.G.: On Codes with Low Peak-to-Average Power Ratio for Multi-Code CDMA. IEEE Trans. on Inform. Theory IT-50, 550–559 (2004)
17. Paterson, K.G., Tarokh, V.: On the existence and construction of good codes with low peak-to-average power ratios. IEEE Trans. on Inform. Theory IT-46, 1974–1987 (2000)
18. Schmidt, K.-U.: PhD thesis, On Spectrally Bounded Codes for Multicarrier Communications, Vogt Verlag, Dresden, Germany (2007),
    http://www.ifn.et.tu-dresden.de/~schmidtk/#publications
19. Shankar, P.: On BCH codes over arbitrary integer rings. IEEE Trans. Inform. Theory IT-25, 480–483 (1979)
20. Shanbhag, A., Kumar, P.V., Helleseth, T.: Improved Binary Codes and Sequence Families from $\mathbb{Z}_4$-Linear Codes. IEEE Trans. on Inform. Theory IT-42, 1582–1586 (1996)
21. Solé, P., Zinoviev, D.: The Most Significant Bit of Maximum Length Sequences Over $\mathbb{Z}_{2^l}$: Autocorrelation and Imbalance. IEEE Transactions on Information Theory 50, 1844–1846 (2004)
22. Solé, P., Zinoviev, D.: Low Correlation, High Nonlinearity Sequences for multi-code CDMA. IEEE Transactions on Information Theory 52, 5158–5163 (2006)
23. Solé, P., Zinoviev, D.: Quaternary Codes and Biphase Sequences from $\mathbb{Z}_8$-Codes. Problems of Information Transmission 40(2), 147–158 (2004)
24. Solé, P., Zinoviev, D.: Weighted degree trace codes for PAPR reduction. In: Helleseth, T., Sarwate, D., Song, H.-Y., Yang, K. (eds.) SETA 2004. LNCS, vol. 3486, pp. 406–413. Springer, Heidelberg (2005)
25. Wan, Z.X.: Quaternary Codes. World Scientific, Singapore (1997)
26. Wan, Z.X.: Lectures on Finite Fields and Galois Rings. World Scientific, Singapore (2003)

Finding Invalid Signatures in Pairing-Based Batches*

Laurie Law[1] and Brian J. Matt[2,**]

[1] National Security Agency, Fort Meade, MD 20755, USA
lelaw@orion.ncsc.mil
[2] JHU Applied Physics Laboratory Laurel, MD, 21102, USA
brian.matt@jhuapl.edu

Abstract. This paper describes efficient methods for finding invalid digital signatures after a batch verification has failed. We present an improvement to the basic binary "divide-and-conquer" method, which can identify an invalid signature in half the time. We also present new, efficient methods for finding invalid signatures in some pairing-based batches with low numbers of invalid signatures. We specify these methods for the Cha-Cheon signature scheme of [5]. These new methods offer significant speedups for Cha-Cheon batches as well as other pairing-based signature schemes.

Keywords: Pairing-based signatures, ID-based signatures, Batch verification.

1 Introduction

When a large number of digital signatures need to be verified, it is sometimes possible to save time by verifying many signatures together. This process is known as *batch verification*. If the batch verification fails, then it is often necessary to identify the invalid ("bad") signature(s) that caused the batch to fail. This process can be time consuming, so batch sizes are typically chosen to be small enough so that most batches will be expected to pass. "Divide-and-conquer" techniques have been proposed for identifying invalid signatures in bad batches. These methods are faster than verifying each signature individually, requiring only $O(\log_2 N)$ verifications, where N is the number of signatures in the batch [11].

---

* The views and conclusions contained in this presentation are those of the authors and should not be interpreted as representing the official policies, either expressed or implied, of the National Security Agency, the Army Research Laboratory, or the U. S. Government.
** Dr. Matt's participation was through collaborative participation in the Communications and Networks Consortium sponsored by the U. S. Army Research Laboratory under the Collaborative Technology Alliance Program, Cooperative Agreement DAAD-19-01-2-0011. The U. S. Government is authorized to reproduce and distribute reprints for Government purposes notwithstanding any copyright notation thereon.

S.D. Galbraith (Eds.): Cryptography and Coding 2007, LNCS 4887, pp. 34–53, 2007.
© Springer-Verlag Berlin Heidelberg 2007

Several digital signature schemes have recently been proposed that are based on bilinear pairings [2,5,6]. This is because the mathematical properties of pairings can be used to design signatures with improved features such as a shorter length or the ability to derive the public verification key from the signer's identity (i.e., *identity-based signatures*). Identity-based signatures can reduce the number of bits that need to be transmitted because the signer does not need to transmit his public key or certificate to the verifier. Some identity-based signatures also have short signature lengths, making them ideal for bandwidth-constrained environments. However, verification of these signatures often involves bilinear pairing operations, which are relatively expensive to compute. While it may be feasible to verify a small number of pairing-based signatures, the number of verifications required to find the invalid signatures in a large batch can be prohibitive, even with divide-and-conquer methods. Therefore, finding faster methods for identifying invalid signatures in bad batches is very important for pairing-based signature schemes.

All pairing-based schemes discussed in this paper are assumed to use bilinear pairings on an elliptic curve E, defined over F_q, where q is a large prime. G_1 and G_2 are distinct subgroups of prime order r on this curve, where G_1 is a subset of the points on E with coordinates in F_q and G_2 is a subset of the points on E with coordinates in F_{q^d}, for a small integer d. The pairing e is a map from $G_1 \times G_2$ into F_{q^d}.

In this paper, we present an improvement to the divide-and-conquer method for finding invalid signatures in a bad batch. We also present new, efficient methods for finding invalid signatures in some pairing-based batches. We specify these methods for the Cha-Cheon signature scheme of [5]. However, these methods are applicable to other pairing-based schemes with similar form, such as BLS short signatures [2] when all signatures in the batch are applied to the same message or signed by the same signer.

2   Background

2.1   Batch Verification

Batch verification of digital signatures was introduced by Naccache *et al.* [10] to verify modified DSA signatures. Yen and Laih [17] applied batch verification to a variant of the Schnorr signature and improved its performance by removing a requirement on small exponents that are generated by the verifier. A similiar test, called the "small exponents test", was one of three batch verification techniques for modular exponentiation given in [1]. The small exponents test has been applied to several signatures schemes including BLS short signatures [2,4].

In the small exponents test, the verifier chooses a new random value (a "small exponent") for each signature in the batch. Each random value is mathematically combined with its corresponding signature and then all the "randomized" signatures are combined into two values that should be equal if and only if all of the signatures in the batch are valid. The random values prevent an attacker from inserting invalid signatures that may cancel each other out, resulting in a batch that appears valid.

Small exponents test:

Input: a security parameter l, a generator g of the group G of prime order q, and $(x_1, y_1), (x_2, y_2), \ldots, (x_n, y_n)$ with $x_i \in Z_q$ and $y_i \in G$.
Check: That $g^{x_i} = y_i$ for all i, $1 \leq i \leq n$.

1. Choose n random integers $r_1, \ldots, r_n$ in the range $[0, 2^l - 1]$.
2. Compute $x = \sum_{i=1}^{n} x_i r_i$ and $y = \prod_{i=1}^{n} y_i^{r_i}$.
3. If $g^x = y$ then accept; else reject.

The probability that this test accepts a batch containing invalid signatures is at most 2^{-l} [1]. The order of G must be prime to prevent a weakness described in [3]. Observe that we can set $r_1 = 1$ without any impact on the security of this test.

2.2   Cha-Cheon Signature Scheme

Cha and Cheon [5] proposed an identity-based signature scheme that can be constructed with bilinear pairings. We describe their scheme using pairings that map $G_1 \times G_2$ into F_{q^d} as defined in Section 1. $H(m, U)$ is a cryptographic hash that maps a bit string m and a point $U \in G_1$ to an integer between 1 and r.

1. *Setup phase.* The system manager selects an order r point $T \in G_2$ and randomly selects an integer s in the range $[1, r-1]$. The manager computes $S = sT$. The public system parameters are T and S, which are made available to anyone that will need to verify signatures. The master secret key is s, which is known to the system manager only.
2. *Extract phase.* Each user is given a key pair. The user's public key, Q, is a point in G_1 that is derived from the user's identity using a public algorithm. The user's private key, $C = sQ$ is computed by the system manager and given to the user through a secure channel.
3. *Signing.* To sign a message m, the signer randomly generates an integer t in the range $[1, r-1]$ and outputs a signature (U, V) where
   $U = tQ$
   $V = (t + H(m, U))C$
4. *Verification.* To verify a signature (U, V) of message m, the verifier derives the signer's public key Q from the purported signer's identity and computes $h = H(m, U)$. If $e(U + hQ, S) = e(V, T)$ then the signature is accepted. Otherwise, the signature is rejected.

2.3   Fast Batch Verification for the Cha-Cheon Signature Scheme

Cheon *et al.* [6] proposed a batch signature scheme for a variant of the Cha-Cheon scheme. Their scheme is partially aggregate, meaning that a portion of each signature can be combined into into a single short string. Aggregate signatures are shorter than batch signatures, but when a batch verification fails they do not provide enough information to allow identification of the bad signatures. Cheon *et al.* claimed that batch verification is not secure for Cha-Cheon and demonstrated an attack. However, their attack is for a particular batch verification scheme and they did not consider the batch verification methods of [1].

By applying the small exponents test to the Cha-Cheon signature we can obtain an efficient batch verification method that can support verification of multiple signatures by distinct signers on distinct messages. The verifier obtains N messages m_k, for $k = 1$ to N, and the signature (U_k, V_k) and signer's identity for each message. The verifier derives each public key Q_k from the signer's identity. The verifier sets $r_1 = 1$ and generates random values r_k from $[0, 2^l - 1]$, for $k = 2$ to N.

The batch is valid if

$$e\left(\sum_{k=1}^{N} r_k\,(U_k + H(m_k, U_k) \cdot Q_k), S\right) = e\left(\sum_{k=1}^{N} r_k V_k, T\right). \tag{1}$$

If this equality does not hold, then at least one signature in the batch is not valid.

This batch verification requires only two pairings (and some other relatively inexpensive operations). This is much more efficient than the aggregate scheme of [6], which requires $N + 1$ pairings.

2.4   Divide-and-Conquer Methods

Although many papers have been written on efficient methods for batch verification, there has been less attention devoted to finding the invalid signatures after a batch verification fails. One method for identifying invalid signatures was applied to batches of RSA signatures [9]. It was shown in [15] that this approach is not secure for RSA signatures. This method is similar to an independently developed method we present in Section 4.1 for the special case of identifying a single bad signature. Pastuszak *et al.* [11] investigated the "divide-and-conquer" method, in which the set of signatures in a failed batch is repeatedly split into z smaller sub-batches to verify. We refer to the case where $z = 2$ as "Simple Binary Search". They found that Simple Binary Search is more efficient than naively testing each signature individually if there are fewer than $N/8$ invalid signatures in the batch.

The following recursive algorithm describes the Simple Binary Search on a batch of N messages and their corresponding digital signatures (called "message / signature pairs"). On the initial call to Algorithm $SBS(X, V)$, X contains the entire batch of N signatures and $V \leftarrow true$.

Note: $verify(X, V)$ is a function of a list $X = ((m_1, s_1), \dots, (m_n, s_n))$ of n message / signature pairs and a flag V. If $V = true$ then the list is verified, and the function returns $true$ if the verification passes and $false$ otherwise. If $V = false$, then no verification is performed and the function returns $false$.

Algorithm $SBS(X, V)$.   (Simple Binary Search)

Input: A list X of n message / signature pairs, and a flag V.

Output: A list of all invalid signatures in the batch.

1. If $n = 1$, then call $verify((m_1, s_1), V)$. If the verification passes ($true$), then return. Otherwise ($false$), output (m_1, s_1) and return.
2. If $n \neq 1$, then call $verify(X, V)$. If verification passes ($true$), then return. Otherwise ($false$), go to Step 3.
3. Divide the batch X into 2 sub-batches, $left(X)$ with $\lceil n/2 \rceil$ signatures, and $right(X)$ with $\lfloor n/2 \rfloor$ signatures. Call $SBS(left(X), true)$. If $SBS(left(X), true)$ or a descendant sub-instance of SBS finds an invalid signature, then call $SBS(right(X), true)$ otherwise call $SBS(right(X), false)$.

Algorithm SBS can be illustrated with a perfect binary tree. The tree's root represents all N message / signature pairs. Each left child represents half of the message / signature pairs from its parent node, and each right child represents the other half. The depth of the tree is $k + 1$, so there are N leaves, each representing a single message / signature pair. A leaf is called "invalid" if its corresponding signature is invalid. Note that if the left child of an invalid node is valid, then its sibling is not verified because it must be invalid.

Costs for Simple Binary Search. If there is only one invalid signature in the batch, there will be at most 2 verifications for each non-root level of the tree. Therefore, as shown in [11], the maximum number of verifications to find a single invalid signature after a batch has failed is $2\lceil \log_2 N \rceil$.

If there are $0 < w < N/2$ invalid signatures in a batch of N signatures, the worst-case cost is $2(2^{\lceil \log_2 w \rceil} - 1 + w(\lceil \log_2 N \rceil - \lceil \log_2 w \rceil))$ batch verifications when the tree is perfectly balanced. If $w > N/2$ then the worst-case cost is the same as the cost for $w = N/2$.

Pastuszak *et al.* [11] also observed that if the verifier knows in advance that the batch contains exactly one invalid signature, then the number of verifications can be reduced to $\lceil \log_2 N \rceil$. In the following section, we will show how to modify Simple Binary Search to find a single invalid signature with $\lceil \log_2 N \rceil$ verifications without advance knowledge of the number of invalid signatures. This modified method also reduces the worst-case cost when there are multiple invalid signatures.

3   Binary Quick Search

Batch verification tests typically compare two quantities, X and Y, and the batch is accepted as valid if they are equal. In many batch signature methods [1,2,3,6,17,18] as well as the batch Cha-Cheon method of Section 2.3, an equivalent test is to compute $A = XY^{-1}$ and accept the batch as valid if $A = 1$. In most of these signature schemes, XY^{-1} can be computed at least as quickly as computing X and Y individually. When this equivalent form of verification is used, the "A" values can be used to eliminate some of the verifications required by divide-and-conquer methods.

For a batch of N signatures, A can be written as the product $A_1 A_2 \ldots A_N$. Let $\bar{A} = A_1 A_2 \ldots A_j$ for a sub-batch of $j < N$ signatures labeled 1 through j. Suppose that $A \neq 1$ and $\bar{A} \neq 1$. Since $A_i = 1$ if and only if signature i is valid, we can instantly determine whether or not the remaining signatures $j+1$ through N are a valid batch. If $A = \bar{A}$ then $A_{j+1} A_{j+2} \ldots A_N = 1$, so signatures $j+1$ through N can be accepted as valid. Otherwise, the set of signatures $j+1$ through N must contain at least one invalid signature.

We can modify the Simple Binary Search method of the previous section to obtain the *Binary Quick Search* method as follows. The initial verification and verification of all $left(X)$ sub-batches are replaced by a computation of $\bar{A}$ and checking whether $\bar{A} = 1$. Verification of all $right(X)$ sub-batches are replaced by a comparison of the A values for its left sibling and its parent. This means that the expensive verification will *never* be needed for right sub-batches. When required, the A value for invalid right sub-batches can be efficiently computed from the A values of its left sibling and its parent.

The Binary Quick Search can be applied to pairing-based batch verification schemes that are equivalent to accepting the batch as valid if $\alpha_0 = 1$, where

$$\alpha_0 = e \left(\sum_{k=1}^{N} B_k, P \right) e \left(\sum_{k=1}^{N} D_k, R \right). \tag{2}$$

For example, the Cha-Cheon Batch verification method of Section 2.2, equation (1), is equivalent to this test, where

$B_k = r_k \left(U_k + H(m_k, U_k) Q_k \right),$
$D_k = r_k V_k,$
$P = S,$ and
$R = -T.$

Binary Quick's worst-case cost (beyond the initial batch verification) is half that of Simple Binary Search. If there are $0 < w < N/2$ invalid signatures in a batch of N signatures, the worst-case cost for this new method is

$$2^{\lceil \log_2 w \rceil} - 1 + w(\lceil \log_2 N \rceil - \lceil \log_2 w \rceil)$$

batch verifications when the tree is perfectly balanced. If $w > N/2$ then the cost is the same as the cost for $w = N/2$.

4   Finding Invalid Signatures in Pairing-Based Batches

The Binary Quick Search method will find invalid signatures in pairing-based batches using no more than approximately $w \log_2 N$ verifications. While faster than previously known methods, Binary Quick Search's lower bound is expensive for large N when the verifications require pairing computations. In this section we propose two alternative methods for finding invalid signatures in pairing-based batch signature schemes such as the Cha-Cheon verification presented above, when the verification can be written as in equation (2).

These new methods make use of the following observation. We can insert integer multipliers $m_{j,k}$ into equation (2) to obtain

$$\alpha_j = e\left(\sum_{k=1}^{N} m_{j,k} B_k, P\right) e\left(\sum_{k=1}^{N} m_{j,k} D_k, R\right)$$

where $m_{0,k} = 1$ for all $k = 1$ to N.

Let I be the set of indices of invalid signatures, and let

$$X_k = e(B_k, P)\, e(D_k, R).$$

Using the properties of bilinear pairings, we have

$$\alpha_j = \prod_{k=1}^{N} X_k^{m_{j,k}}.$$

However, $X_k = 1$ if the kth signature is valid, so we have

$$\alpha_j = \prod_{k \in I} X_k^{m_{j,k}}.$$

We can identify the set I of invalid signatures by looking for relationships between the α_j's. For example, if the kth signature is the only invalid signature in the batch, then we have $\alpha_1 = \alpha_0^{m_{1,k}}$.

Since we are working in groups of prime order, we are free to choose the values for the scalars $m_{j,k}$ to maximize the efficiency of our methods. Note that these scalars can have lengths much smaller than the security parameter l used in generating the randomizers r_k in the initial batch verification.

4.1   Exponentiation Method

We begin with the same initial batch verification as in the Binary Quick Search method of Section 3, given by equation (2). If $\alpha_0 = 1$ then all signatures in the batch are accepted as valid. Otherwise, the following algorithm will identify the bad signatures in the batch:

Input: A bad batch of N message / signature pairs to be verified, along with the identities of each signer.
Output: A list of the invalid signatures.

1. Perform an computation similar to the batch computation of equation (2), but first multiply each B_k and D_k by the signature identifier k, ($k = 1$ through N):

$$\alpha_1 = e\left(\sum_{k=1}^{N} k B_k, P\right) e\left(\sum_{k=1}^{N} k D_k, R\right).$$

Search for an i, $1 \leq i \leq N$, such that $\alpha_1 = \alpha_0^i$.
If such an i is found, then output "the ith signature is invalid" and exit. If no match is found, then there are at least 2 bad signatures in the batch. Go to the next step.

2. Compute

$$\alpha_2 = e \left(\sum_{k=1}^{N} k \, (kB_k), P \right) e \left(\sum_{k=1}^{N} k \, (kD_k), R \right).$$

Search for an (i, j), $1 \leq i \leq N$, $1 \leq j \leq N$, $i < j$ such that $\alpha_2 = \alpha_1^{i+j} \alpha_0^{-ij}$.
If an (i, j) pair is found, then output "the ith and j^{th} signatures are invalid" and exit. If no match is found, then there are at least 3 bad signatures in the batch. Set $w \leftarrow 3$ and go to the next step.

3. Compute

$$\alpha_w = e \left(\sum_{k=1}^{N} k \, (k^{w-1} B_k), P \right) e \left(\sum_{k=1}^{N} k \, (k^{w-1} D_k), R \right). \tag{3}$$

For all w-subsets of signatures $x_1, \ldots, x_w$, $x_1 < x_2 < \ldots < x_w$
Check that

$$\alpha_w = \prod_{t=1}^{w} (\alpha_{w-t})^{(-1)^{t-1} p_t} \tag{4}$$

where p_t is the tth elementary symmetric polynomial in $x_1, \ldots, x_w$.
Equivalently, we can search for $p_1, \ldots, p_w$ that will solve equation (4). Once we know $p_1, \ldots, p_w$, it is easy to solve for the signature identifiers $x_1, \ldots, x_w$. If a match is found, then output "signatures $x_1, \ldots,$ and x_w are invalid" and exit. If no match is found, then there are at least $w+1$ bad signatures in the batch. Set $w \leftarrow w + 1$ and repeat Step 3, or stop and switch to a different method.

Cost. We will assume that the number of invalid signatures, w, is small. We need to compute each α_i, for $i = 1$ to w, and to then solve equation (4).

To compute each α_i in (3), we first need to compute the quantities $\sum_{k=1}^{N} (k^i B_k)$
and $\sum_{k=1}^{N} (k^i D_k)$. Solinas [14] observed that these two quantities can be computed for $i = 1$ to w with only $2w(N-1)$ elliptic curve additions (see Appendix B). Each α_i computation will also require 2 pairings and 1 multiplication in F_{q^d}.

For $w \geq 2$, each step of this algorithm will require the inverse of α_{w-2} (other inverses can be saved from previous steps), so there will be $w - 1$ inverse computations in F_{q^d}.

Finally, we need to solve equation (4) for $x_1, \ldots, x_w$, where $1 \leq x_i \leq N$. If $w = 1$, then this equation reduces to the discrete logarithm problem in F_{q^d}, where the exponents are from an interval of length N. This can be solved using square-root methods such as Shanks' baby-step giant-step method [12] with only $2\sqrt{N}$ multiplications in F_{q^d}.

For $w \geq 2$, we can solve equation (4) for $x_1, \ldots, x_w$ with $\frac{8}{(w-1)!} N^{w-1} +$ $O(N^{w-2})$ multiplications in F_{q^d} (see Appendix A).

The approximate upper bound for the total cost to find w invalid signatures in a bad batch of N signatures using the exponentiation method is $2w$ pairings, $2w(N-1)$ elliptic curve additions, $w-1$ inverses in F_{q^d}, and the number of multiplications in F_{q^d} is $2\sqrt{N}$ for $w = 1$ or $\frac{8}{(w-1)!} N^{w-1}$ for $w \geq 2$.

4.2   Exponentiation with Sectors Method

The exponentiation method of the previous section works efficiently when there is a single invalid signature, but the number of combinations of potentially invalid signatures grows quickly as the number of bad signatures increases, even for just a few invalid signatures. We can slow this growth by dividing the N signatures in a failed batch into Z sectors of $T \approx \frac{N}{Z}$ signatures each and modifying the Exponentiation method of the previous section to identify bad sectors instead of bad signatures. Once these bad sectors have been identified, we use a variant of the Exponentiation Method to identify the bad signatures from within the bad sectors. We call this the Exponentiation with Sectors Method.

We begin with the same initial batch verification of the entire batch as in the Binary Quick Search method of Section 3, given by equation (2). If the batch fails, we use the following algorithm to identify the bad signature(s).

Input: A bad batch of N message / signature pairs to be verified, along with the identities of each signer.
Output: A list of the invalid signatures.
Stage 1. Divide the N signatures into Z sectors of $T \approx \frac{N}{Z}$ signatures each. Label the sectors s_1 through s_Z. This stage will identify the bad sectors.

1. Compute

$$\beta_1 = e\left(\sum_{j=1}^{Z} j\left(\sum_{k \in s_j} B_k\right), P\right) e\left(\sum_{j=1}^{Z} j\left(\sum_{k \in s_j} D_k\right), R\right).$$

Search for an i, $1 \leq i \leq Z$, such that $\beta_1 = \alpha_0^i$. If such an i is found, then Sector i is the only bad sector. Set $v \leftarrow 1$ and go to Step 1 of *Stage 2*. If no match is found, then there are at least 2 bad sectors in the batch. Set $v \leftarrow 2$ and go to the next step.

2. Compute

$$\beta_v = e\left(\sum_{j=1}^{Z} j\left(j^{v-1} \sum_{k \in s_j} B_k\right), P\right) e\left(\sum_{j=1}^{Z} j\left(j^{v-1} \sum_{k \in s_j} D_k\right), R\right). \quad (5)$$

Search for a v-subset of sectors $s_1, \ldots, s_v$, $s_1 < s_2 < \ldots < s_v$ such that

$$\beta_v = \prod_{t=1}^{v} (\beta_{v-t})^{(-1)^{t-1} p_t} \quad (6)$$

where $\beta_0 = \alpha_0$ and p_t is the tth elementary symmetric polynomial in $s_1, \ldots, s_v$. If a v-subset is found, then go to Step 1 of *Stage 2*. If no match is found, then there are at least $v + 1$ bad sectors. Set $v \leftarrow v + 1$ and repeat Step 2, or stop and switch to a different method.

Stage 2. Now v bad sectors of T signatures each have been identified. Therefore, there are at least v invalid signatures in the batch. Set $w \leftarrow v$. *Stage 2* will identify these invalid signatures.

1. Label the signatures in the bad sectors from 1 to vT. Compute γ_i for $i = 1$ to w as follows:

$$\gamma_i = e \left(\sum_{k=1}^{vT} k^i B_k, P \right) e \left(\sum_{k=1}^{vT} k^i D_k, R \right). \tag{7}$$

   Compute the inverses of α_0 (if w is even) and $\gamma_{w-2}, \gamma_{w-4}, \ldots$ Search for a w-subset of signatures $x_1, \ldots, x_w$, $x_1 < x_2 < \ldots < x_w$ such that

$$\gamma_w = \prod_{t=1}^{w} (\gamma_{w-t})^{(-1)^{t-1} p_t} \tag{8}$$

   where $\gamma_0 = \alpha_0$ and p_t is the tth elementary symmetric polynomial in $x_1, \ldots, x_w$. If a match is found then output the list of w invalid signatures and exit. If no match is found, then there are more that w invalid signatures in the batch. Set $w \leftarrow w + 1$ and go to the next step.
2. Compute γ_w as in equation (7) with $i = w$. Compute the inverses of γ's as needed and search for a w-subset of signatures $x_1, \ldots, x_w$, $x_1 < x_2 < \ldots < x_w$ to solve equation (8). If a match is found then output the list of w invalid signatures and exit. If no match is found, then there are more than w invalid signatures in the batch. Set $w \leftarrow w + 1$ and repeat this step.

Cost. To simplify the cost computation, we will assume that there are w bad sectors and exactly 1 invalid signature per sector ($w = v$), which is the worst case. This is a reasonable assumption because batch verification is most useful when most batches are valid. When a batch does fail, the number of invalid signatures is expected to be very small (unless the batch was intentionally flooded with bad signatures). In most cases, there is unlikely to be more than one bad signature in any given sector.

The *Stage 1* cost to identify w bad sectors in a batch with Z Sectors will be the same as the cost for the exponentiation method of Section 4.1 to identify w bad signatures in a batch of Z signatures. This cost is $2w$ pairings, $2w(Z - 1)$ elliptic curve additions, $w - 1$ inverses in F_{q^d} and the number of multiplications in F_{q^d} is $2\sqrt{Z}$ for $w = 1$ or $\frac{8}{(w-1)!} Z^{w-1}$ for $w \geq 2$.

In *Stage 2*, Step 1, we need to identify w invalid signatures in w distinct bad sectors of $T = \frac{N}{Z}$ signatures each. (We are assuming $w = v$ so we will be done after Step 1.) There are wT signatures in the w bad sectors. The costs for equation (7) are $2w(wT - 1)$ elliptic curve additions to compute $\sum_{k=1}^{vT} k^i B_k$ and $\sum_{k=1}^{vT} k^i D_k$ for $i = 1$ to w (see Appendix B), and $2w$ pairings to compute the γ_i's.

For equation (8), we will need to compute $\lfloor \frac{w}{2} \rfloor$ inverses in F_{q^d}.

Finally, we need to solve equation (8) for $x_1, \ldots, x_w$, where $1 \leq x_i \leq wT$. If $w = 1$, then this equation reduces to the discrete logarithm problem in F_{q^d}, where the exponents are from an interval of length T. This can be solved using square-root methods such as Shanks' baby-step giant-step method [12] with only $2\sqrt{T}$ multiplications in F_{q^d}.

For $w \geq 2$, we can solve equation (8) for $x_1, \ldots, x_w$ with $\frac{8}{(w-1)!}(wT)^{w-1}$ multiplications in F_{q^d} (see Appendix A).

The approximate upper bound for the total cost for both stages of the Exponentiation with Sectors method to identify w bad signatures in a batch of N signatures (assuming exactly one bad signature per sector) is $4w$ pairings, $2w(Z + \frac{wN}{Z} - 2)$ elliptic curve additions, $\lfloor 1.5w \rfloor - 1$ inverses in F_{q^d} and the number of multiplications in F_{q^d} is: $2\left(Z^{1/2} + \left(\frac{N}{Z}\right)^{1/2}\right)$ for $w = 1$ and $\frac{8}{(w-1)!}\left(Z^{w-1} + \left(\frac{wN}{Z}\right)^{w-1}\right)$ for $w \geq 2$.

To minimize the cost, we need to select the number of sectors, Z, to balance the work in *Stage 1* and *Stage 2*. We can minimize the work for the most likely case, $w = 1$, by choosing $Z = \sqrt{N}$ sectors. In this case the total cost is $4w$ pairings, $2w((w + 1)\sqrt{N} - 2)$ elliptic curve additions, $\lfloor 1.5w \rfloor - 1$ inverses in F_{q^d} and the number of multiplications in F_{q^d} is $4(N)^{1/4}$ for $w = 1$ and $\frac{8(w^{w-1}+1)}{(w-1)!}N^{(w-1)/2}$ for $w \geq 2$.

5   Performance

Table 1 summarizes the approximate upper bound for the number of operations to identify a small number of invalid signatures in a bad batch for each of the three new methods presented in this paper, and compares them to the cost of Simple Binary Search and to the cost of verifying each signature individually. For each batch method, we have omitted the cost of the initial verification of the entire batch. Note that other computational techniques exist that may lower these costs in some cases, such as techniques for computing products of pairings [8].

As shown in Table 1, Binary Quick Search is twice as fast as the worst case for Simple Binary Search. The Exponentiation and Sector methods use fewer pairings than Binary Quick Search, but the number of multiplications in $F(q^d)$ for these methods increases proportionally to a power of the batch size. For very large batch sizes, Binary Quick Search will always be the best method. However, for reasonably sized batches of mostly valid signatures, the Exponentiation or Sector method will often be the most efficient.

Table 1. Number of Operations

Method	Pairings	Inverses in F_{q^d}	EC additions	Multiplications in F_{q^d}
N individual	$2N^\star$	0	0	0
Simple Binary	$4w \log_2 N$	0	0	0
Binary Quick	$2w \log_2 N$	0	0	0
Exponentiation				
$\quad w = 1$	2	0	$2(N-1)$	$2\sqrt{N}$
$\quad w \geq 2$	$2w$	$w-1$	$2w(N-1)$	$\frac{8}{(w-1)!}N^{w-1}$
$\sqrt{N}$ Sectors				
$\quad w = 1$	4	0	$4(\sqrt{N}-1)$	$4N^{1/4}$
$\quad w \geq 2$	$4w$	$\lfloor 1.5(w) \rfloor - 1$	$2w((w+1)\sqrt{N}-2)$	$\frac{8(w^{w-1}+1)}{(w-1)!}N^{(w-1)/2}$

Table 2. Number of multiplies in F_q, where r and q are 160-bit values and $d = 6$

w	Method	Batch Size				
		16	64	256	1024	4096
	N individual	$3.18 \cdot 10^{05}$	$1.27 \cdot 10^{06}$	$5.08 \cdot 10^{06}$	$2.03 \cdot 10^{07}$	$8.13 \cdot 10^{07}$
0	Initial Verify	$6.29 \cdot 10^{04}$	$1.99 \cdot 10^{05}$	$7.45 \cdot 10^{05}$	$2.93 \cdot 10^{06}$	$1.17 \cdot 10^{07}$
1	Simple Binary	$1.46 \cdot 10^{05}$	$2.19 \cdot 10^{05}$	$2.92 \cdot 10^{05}$	$3.65 \cdot 10^{05}$	$4.38 \cdot 10^{05}$
	Binary Quick	$7.30 \cdot 10^{04}$	$1.09 \cdot 10^{05}$	$1.46 \cdot 10^{05}$	$1.82 \cdot 10^{05}$	$2.19 \cdot 10^{05}$
	Exponent.	$\langle 1.87 \cdot 10^{04} \rangle$	$\langle 1.99 \cdot 10^{04} \rangle$	$\langle 2.43 \cdot 10^{04} \rangle$	$4.17 \cdot 10^{04}$	$1.10 \cdot 10^{05}$
	$\sqrt{N}$ Sectors	$3.67 \cdot 10^{04}$	$3.69 \cdot 10^{04}$	$3.73 \cdot 10^{04}$	$\langle 3.82 \cdot 10^{04} \rangle$	$\langle 3.97 \cdot 10^{04} \rangle$
2	Simple Binary	$2.55 \cdot 10^{05}$	$4.01 \cdot 10^{05}$	$5.47 \cdot 10^{05}$	$6.93 \cdot 10^{05}$	$8.39 \cdot 10^{05}$
	Binary Quick	$1.28 \cdot 10^{05}$	$2.01 \cdot 10^{05}$	$2.74 \cdot 10^{05}$	$3.47 \cdot 10^{05}$	$4.20 \cdot 10^{05}$
	Exponent.	$\langle 3.91 \cdot 10^{04} \rangle$	$\langle 4.70 \cdot 10^{04} \rangle$	$\langle 7.85 \cdot 10^{04} \rangle$	$2.04 \cdot 10^{05}$	$7.08 \cdot 10^{05}$
	$\sqrt{N}$ Sectors	$7.48 \cdot 10^{04}$	$7.67 \cdot 10^{04}$	$8.07 \cdot 10^{04}$	$\langle 8.85 \cdot 10^{04} \rangle$	$\langle 1.04 \cdot 10^{05} \rangle$
3	Simple Binary	$3.28 \cdot 10^{05}$	$5.47 \cdot 10^{05}$	$7.66 \cdot 10^{05}$	$9.85 \cdot 10^{05}$	$1.20 \cdot 10^{06}$
	Binary Quick	$1.64 \cdot 10^{05}$	$2.74 \cdot 10^{05}$	$3.83 \cdot 10^{05}$	$\langle 4.92 \cdot 10^{05} \rangle$	$\langle 6.02 \cdot 10^{05} \rangle$
	Exponent.	$\langle 7.12 \cdot 10^{04} \rangle$	$3.05 \cdot 10^{05}$	$4.00 \cdot 10^{06}$	$6.30 \cdot 10^{07}$	$1.01 \cdot 10^{09}$
	$\sqrt{N}$ Sectors	$1.20 \cdot 10^{05}$	$\langle 1.50 \cdot 10^{05} \rangle$	$\langle 2.67 \cdot 10^{05} \rangle$	$7.32 \cdot 10^{05}$	$2.58 \cdot 10^{06}$
4	Simple Binary	$4.01 \cdot 10^{05}$	$6.93 \cdot 10^{05}$	$9.85 \cdot 10^{05}$	$1.28 \cdot 10^{06}$	$1.57 \cdot 10^{06}$
	Binary Quick	$2.01 \cdot 10^{05}$	$\langle 3.47 \cdot 10^{05} \rangle$	$\langle 4.92 \cdot 10^{05} \rangle$	$\langle 6.38 \cdot 10^{05} \rangle$	$\langle 7.84 \cdot 10^{05} \rangle$
	Exponent.	$\langle 1.56 \cdot 10^{05} \rangle$	$5.32 \cdot 10^{06}$	$3.36 \cdot 10^{08}$	$2.15 \cdot 10^{10}$	$1.37 \cdot 10^{12}$
	$\sqrt{N}$ Sectors	$2.30 \cdot 10^{05}$	$8.14 \cdot 10^{05}$	$5.48 \cdot 10^{06}$	$4.28 \cdot 10^{07}$	$3.41 \cdot 10^{08}$

$\star$ Plus N point multiplications by a scalar the size of the group order r.

Tables 2 and 3 compare methods for finding invalid signatures in bad batches for Cases A and C of [7]. In Case A, the group order r is a 160-bit value, the elliptic curve E is defined over F_q, where q is a 160-bit value, and the embedding degree $d = 6$. In Case C, the group order r is a 256-bit value, q is a 256-bit value, and the embedding degree $d = 12$. All costs are given in terms of the number of multiplications (m) in F_q using the following estimates from [7]:

- For Case A, 1 pairing $= 9120m$, 1 multiplication in $F_{q^6} = 15m$, 1 inverse in $F_{q^6} = 44m$ (assuming 1 inverse in $F_q = 10m$), 1 elliptic curve addition $= 11m$, and an elliptic point multiplication by a 160-bit value is $614m$ and by an 80-bit value is $827m$. The cost of a pair of elliptic point multiplications by 160-bit and 80-bit integers simultaneously is $2017m$, using the Joint Sparse Form [13].
- For Case C, 1 pairing $= 43,703m$, 1 multiplication in $F_{q^{12}} = 45m$, 1 inverse in $F_{q^{12}} = 104m$, 1 elliptic curve addition $= 11m$, and an elliptic point multiplication by a 256-bit value is $2535m$ and by an 128-bit value is $1299m$. The cost of a pair of elliptic point multiplications by 256-bit and 128-bit values simultaneously is $3225m$, using the Joint Sparse Form.

Table 3. Number of multiplies in F_q, where r and q are 256-bit values and $d = 12$

w	Method	Batch Size 16	64	256	1024	4096
	N individual	$1.44 \cdot 10^{06}$	$5.76 \cdot 10^{06}$	$2.30 \cdot 10^{07}$	$9.21 \cdot 10^{07}$	$3.68 \cdot 10^{08}$
0	Initial Verify	$1.58 \cdot 10^{05}$	$3.76 \cdot 10^{05}$	$1.24 \cdot 10^{06}$	$4.72 \cdot 10^{06}$	$1.86 \cdot 10^{07}$
1	Simple Binary	$6.99 \cdot 10^{05}$	$1.05 \cdot 10^{06}$	$1.40 \cdot 10^{06}$	$1.75 \cdot 10^{06}$	$2.10 \cdot 10^{06}$
	Binary Quick	$3.50 \cdot 10^{05}$	$5.24 \cdot 10^{05}$	$6.99 \cdot 10^{05}$	$8.74 \cdot 10^{05}$	$1.05 \cdot 10^{06}$
	Exponent.	$\langle 8.81 \cdot 10^{04} \rangle$	$\langle 8.95 \cdot 10^{04} \rangle$	$\langle 9.45 \cdot 10^{04} \rangle$	$\langle 1.13 \cdot 10^{05} \rangle$	$1.83 \cdot 10^{05}$
	$\sqrt{N}$ Sectors	$1.75 \cdot 10^{05}$	$1.76 \cdot 10^{05}$	$1.76 \cdot 10^{05}$	$1.77 \cdot 10^{05}$	$\langle 1.79 \cdot 10^{05} \rangle$
2	Simple Binary	$1.22 \cdot 10^{06}$	$1.92 \cdot 10^{06}$	$2.62 \cdot 10^{06}$	$3.22 \cdot 10^{06}$	$4.02 \cdot 10^{06}$
	Binary Quick	$6.12 \cdot 10^{05}$	$9.61 \cdot 10^{05}$	$1.31 \cdot 10^{06}$	$1.66 \cdot 10^{06}$	$2.01 \cdot 10^{06}$
	Exponent.	$\langle 1.81 \cdot 10^{05} \rangle$	$\langle 2.01 \cdot 10^{05} \rangle$	$\langle 2.78 \cdot 10^{05} \rangle$	$5.89 \cdot 10^{05}$	$1.83 \cdot 10^{06}$
	$\sqrt{N}$ Sectors	$3.54 \cdot 10^{05}$	$3.59 \cdot 10^{05}$	$3.69 \cdot 10^{05}$	$\langle 3.88 \cdot 10^{05} \rangle$	$\langle 4.27 \cdot 10^{05} \rangle$
3	Simple Binary	$1.57 \cdot 10^{06}$	$2.62 \cdot 10^{06}$	$3.67 \cdot 10^{06}$	$4.72 \cdot 10^{06}$	$5.77 \cdot 10^{06}$
	Binary Quick	$7.87 \cdot 10^{05}$	$1.31 \cdot 10^{06}$	$1.84 \cdot 10^{06}$	$\langle 2.36 \cdot 10^{06} \rangle$	$\langle 2.88 \cdot 10^{06} \rangle$
	Exponent.	$\langle 3.09 \cdot 10^{05} \rangle$	$1.00 \cdot 10^{06}$	$1.21 \cdot 10^{07}$	$1.89 \cdot 10^{08}$	$3.02 \cdot 10^{09}$
	$\sqrt{N}$ Sectors	$5.54 \cdot 10^{05}$	$\langle 6.41 \cdot 10^{05} \rangle$	$\langle 9.89 \cdot 10^{05} \rangle$	$2.38 \cdot 10^{06}$	$7.91 \cdot 10^{06}$
4	Simple Binary	$1.92 \cdot 10^{06}$	$3.32 \cdot 10^{06}$	$4.72 \cdot 10^{06}$	$6.12 \cdot 10^{06}$	$7.52 \cdot 10^{06}$
	Binary Quick	$9.61 \cdot 10^{05}$	$\langle 1.66 \cdot 10^{06} \rangle$	$\langle 2.36 \cdot 10^{06} \rangle$	$\langle 3.06 \cdot 10^{06} \rangle$	$\langle 3.76 \cdot 10^{06} \rangle$
	Exponent.	$\langle 5.97 \cdot 10^{05} \rangle$	$1.61 \cdot 10^{07}$	$1.01 \cdot 10^{09}$	$6.44 \cdot 10^{10}$	$4.12 \cdot 10^{12}$
	$\sqrt{N}$ Sectors	$9.50 \cdot 10^{05}$	$2.70 \cdot 10^{06}$	$1.67 \cdot 10^{07}$	$1.29 \cdot 10^{08}$	$1.02 \cdot 10^{09}$

To indicate the best method for each batch size and number of invalid signatures, the table entry with the lowest cost is given in brackets.

The results in Table 2 for Case A show that the Exponentiation method is the most efficient method when there are one or two invalid signatures in a batch of up to 256 signatures, with costs that are as much as 83% lower than the cost of Binary Quick Search and 92% lower than Simple Binary Search. For batches of 1024 to 4096 signatures with one or two invalid signatures, the Sector method with $\sqrt{N}$ sectors is the most efficient, with costs that are up to 82% lower cost than Binary Quick Search and 91% lower than Simple Binary Search. If the batch size is large enough, Binary Quick Search will be the most efficient method, and the size at which this happens is smaller if there are more invalid signatures in the batch. However, when there are 4 invalid signatures, the Exponentiation method is still 22% faster than Binary Quick Search, and 61% faster than Simple Binary, for batches of 16 signatures.

The results in Table 3 for Case C show that the benefit of using the Exponentiation and Sector methods can be even more significant at higher security levels. For example, the cost of the Exponentiation or Sector methods are up to 87% less than the cost of Binary Quick Search, and up to 94% less than the cost of Simple Binary Search, for batches with up to 4096 signatures in which one or two signatures are invalid.

6   Conclusion

We have presented two new methods for identifying invalid signatures in pairing-based, batch signature schemes with low numbers of invalid signatures, and have analyzed their performance. These methods are applicable to batch verification schemes employing small exponent techniques such as the Cha-Cheon scheme described here, or BLS [2] short signatures when all signatures in the batch are applied to the same message or signed by the same signer. These new methods offer significant speedups for such schemes.

We have presented Binary Quick Search, an improvement to previous "divide-and-conquer" methods. This method is better suited for identifying larger numbers of invalid signatures, especially in large batches, than our other methods, and has application to batch signature schemes that do not use pairings, including [1,3,17].

References

1. Bellare, M., Garay, J., Rabin, T.: Fast Batch Verification for Modular Exponentiation and Digital Signatures. In: Nyberg, K. (ed.) EUROCRYPT 1998. LNCS, vol. 1403, pp. 236–250. Springer, Heidelberg (1998)
2. Boneh, D., Lynn, B., Shacham, H.: Short Signatures from the Weil Pairing. In: Boyd, C. (ed.) ASIACRYPT 2001. LNCS, vol. 2248, pp. 514–532. Springer, Heidelberg (2001)

3. Boyd, C., Pavlovski, C.: Attacking and Repairing Batch Verification Schemes. In: Okamoto, T. (ed.) ASIACRYPT 2000. LNCS, vol. 1976, pp. 58–71. Springer, Heidelberg (2000)
4. Camenisch, J., Hohenberger, S., Pedersen, M.: Batch Verification of Short Signatures. In: EUROCRYPT 2007. LNCS, vol. 4515, pp. 246–263. Springer, Heidelberg (2007), See also Cryptology ePrint Archive, Report 2007/172 (2007), http://eprint.iacr.org/2007/172
5. Cha, J., Cheon, J.: An Identity-Based Signature from Gap Diffie-Hellman Groups. In: Desmedt, Y.G. (ed.) PKC 2003. LNCS, vol. 2567, pp. 18–30. Springer, Heidelberg (2002)
6. Cheon, J., Kim, Y., Yoon, H.: A New ID-based Signature with Batch Verification, Cryptology ePrint Archive, Report 2004/131 (2004), http://eprint.iacr.org/2004/131
7. Granger, R., Page, D., Smart, N.P.: High Security Pairing-Based Cryptography Revisited. In: Hess, F., Pauli, S., Pohst, M. (eds.) ANTS VII. LNCS, vol. 4076, pp. 480–494. Springer, Heidelberg (2006)
8. Granger, R., Smart, N.P.: On Computing Products of Pairings, Cryptology ePrint Archive, Report 2006/172 (2006), http://eprint.iacr.org/2006/172
9. Lee, S., Cho, S., Choi, J., Cho, Y.: Efficient Identification of Bad Signatures in RSA-Type Batch Signature. IEICE Transactions on Fundamentals of Electronics, Communications and Computer Sciences E89-A(1), 74–80 (2006)
10. Naccache, D., M'Raihi, D., Vaudenay, S., Raphaeli, D.: Can D.S.A. be improved? Complexity Trade-offs with the Digital Signature Standard. In: De Santis, A. (ed.) EUROCRYPT 1994. LNCS, vol. 950, pp. 77–85. Springer, Heidelberg (1995)
11. Pastuszak, J., Michalek, D., Pieprzyk, J., Seberry, J.: Identification of Bad Signatures in Batches. In: Imai, H., Zheng, Y. (eds.) PKC 2000. LNCS, vol. 1751, pp. 28–45. Springer, Heidelberg (2000)
12. Shanks, D.: Class Number, a Theory of Factorization and Genera. Proc. Symp. Pure Math. 20, 415–440 (1969) (AMS 1971)
13. Solinas, J.: Low-Weight Binary Representations for Pairs of Integers, Technical Report CORR 2001-41, Centre for Applied Cryptographic Research (2001)
14. Solinas, J.: Personal communication
15. Stanek, M.: Attacking LCCC Batch Verification of RSA Signatures, Cryptology ePrint Archive, Report 2006/111 (2006), http://eprint.iacr.org/2006/111
16. Sury, B., Wang, T., Zhao, F.: Identities Involving Reciprocals of Binomial Coefficients. Journal of Integer Sequences 7, Article 04.2.8 (2004)
17. Yen, S., Laih, C.: Improved Digital Signature Suitable for Batch Verification. IEEE Transactions on Computers 44(7), 957–959 (1995)
18. Yoon, H., Cheon, J.H., Kim, Y.: Batch verifications with ID-based signatures. In: Park, C., Chee, S. (eds.) ICISC 2004. LNCS, vol. 3506, pp. 223–248. Springer, Heidelberg (2005)

A   Solving Equations (4), (6) and (8) Using the Factor Method

In equation (4) of Section 4.1, and equations (6) and (8) of Section 4.2, we need to solve an equation of the form

$$B = \prod_{t=1}^{w} (A_t)^{p_t} \tag{9}$$

where p_t is the tth symmetric polynomial in $x_1, x_2, \ldots, x_w$, $A_t \in F_{q^d}$ and

$$1 \le x_i \le M.$$

For equation (4): $B = \alpha_w$, $M \leftarrow N$ and $A_t = (\alpha_{w-t})^{(-1)^{t-1}}$ for $t = 1$ to w.
For equation (6): $B = \beta_w$, $M \leftarrow Z$ and $A_t = (\beta_{w-t})^{(-1)^{t-1}}$ for $t = 1$ to w.
For equation (8): $B = \gamma_w$, $M \leftarrow \frac{wN}{Z}$ and $A_t = (\gamma_{w-t})^{(-1)^{t-1}}$ for $t = 1$ to w.

A.1 $w = 2$: Factor Method

When $w = 2$, equation (9) reduces to

$$B = A_1^{(x_1 + x_2)} A_2^{(x_1 x_2)}. \tag{10}$$

If we raise both sides of equation (10) to the 4th power and observe that $4x_1 x_2 = (x_2 + x_1)^2 - (x_2 - x_1)^2$, equation (10) becomes

$$B^4 (A_1^{-4})^{(x_2 + x_1)} (A_2^{-1})^{(x_2 + x_1)^2} = (A_2^{-1})^{(x_2 - x_1)^2}.$$

This can be written as

$$\delta_0 (\delta_1)^s (\delta_2)^{(s)^2} = (\delta_2)^{(m)^2} \tag{11}$$

where

$$\delta_0 = B^4, \; \delta_1 = A_1^{-4}, \; \delta_2 = A_2^{-1}$$

$$s = x_2 + x_1, \; m = x_2 - x_1$$

$$1 \le s \le 2M - 1$$

$$1 \le m \le M - 1.$$

To solve equation (11), the $M - 1$ possible values for the right hand side of the equation, $RHS_{(i)}$, are computed and stored in a search tree. Then the $2M - 1$ values for the left side, $LHS_{(i)}$, are computed and compared, each in $\log M$ time, with the stored values. If a match is found, the values x_1 and x_2 can be easily computed. To compute the values for the right hand side

1. Set $RHS_1 = \delta_2$ and compute $\delta_2^{(2)}$.
2. For $2 \le i < M$, compute
   (a) $(\delta_2)^{(2i-1)} = (\delta_2)^{(2i-3)} (\delta_2)^{(2)}$ and
   (b) $RHS_{(i)} = (\delta_2)^{(i)^2} = (\delta_2)^{(i-1)^2} (\delta_2)^{(2i-1)}$.

Computing the values for the left hand side of equation (11) is performed in two stages. In the first stage, the values $(\delta_2)^{(i)^2}$, which have already been computed for the right hand side, are re-used to reduce the number of multiplications.

Stage 1:

1. Compute $\zeta_3 = \delta_0\,\delta_1^{(3)}$, lookup $\delta_2^{(3)^2}$ and compute $LHS_3 = \zeta_3\,\delta_2^{(3)^2}$.
2. For $4 \le i < M$, compute
   (a) $\zeta_{(i)} = \zeta_{(i-1)}\,\delta_1$,
   (b) lookup $\delta_2^{(i)^2}$ and compute $LHS_{(i)} = \zeta_{(i)}\,\delta_2^{(i)^2}$.

Stage 2:

1. For $M \le i < 2M$ compute
   (a) $\zeta_{(i)} = \zeta_{(i-1)}\,\delta_1$,
   (b) $(\delta_2)^{(2i-1)} = (\delta_2)^{(2i-3)}\,(\delta_2)^{(2)}$,
   (c) $(\delta_2)^{(i)^2} = (\delta_2)^{(i-1)^2}\,(\delta_2)^{(2i-1)}$ and
   (d) $LHS_{(i)} = \zeta_{(i)}\,(\delta_2)^{(i)^2}$.

Cost.

Computing the values for the right hand side requires $2M$ multiplications. Computing the values for the left hand side requires $2M$ multiplications for Stage 1 and $4M$ for Stage 2. The total cost of using the Factor method for $w = 2$ is no more than $8M$ multiplications in F_{q^d}.

A.2 $w \ge 3$: Iterative Use of the Factor Method

When $w \ge 3$, equation (9) can be solved by reducing the problem to the $w = 2$ case for each of the $\binom{M-2}{w-2}$ possible combinations of the values of $x_3, \ldots, x_w$. We can express p_t, the tth elementary symmetric polynomial in $x_1, \ldots, x_w$ as

$$p_t = (x_1 x_2)u_{t-2} + (x_1 + x_2)u_{t-1} + u_t.$$

If $0 \le i \le w - 2$ then u_i is the ith elementary symmetric polynomial in $x_3, \ldots, x_w$ and $u_i = 0$ otherwise. Equation (9) becomes

$$B = \prod_{t=1}^{w} (A_t)^{((x_1 x_2)\,u_{t-2} + (x_1 + x_2)\,u_{t-1} + u_t)}.$$

Rewriting this equation in the form of equation (11)

$$B^4 \underbrace{\prod_{t=1}^{w-2} (A_t)^{-4u_t}}_{\delta_0} \left(\underbrace{\prod_{t=1}^{w-1} (A_t)^{-4u_{t-1}}}_{\delta_1}\right)^{(s)} \left(\underbrace{\prod_{t=2}^{w} (A_t)^{-u_{t-2}}}_{\delta_2}\right)^{(s)^2} = \left(\underbrace{\prod_{t=2}^{w} (A_t)^{-u_{t-2}}}_{\delta_2}\right)^{(m)^2}.$$

The Factor method can be used with the values δ_0, δ_1 and δ_2 as shown above, with $1 \le s \le 2x_3 - 3$ and $1 \le m \le x_3 - 1$.

Cost.

The Factor method is used a maximum of $\binom{M-2}{w-2}$ times to test equation (9). Therefore, for $w \geq 2$, the cost of all calls to the Factor method is bounded by

$$\binom{M-2}{w-2} 8M < \frac{8}{(w-2)!} M^{w-1}$$

multiplications in F_{q^d}.

We can establish a tighter upper bound if we consider the fact that x_1 and x_2 must lie in the range $[1, x_3 - 1]$ instead of $[1, M]$. The number of times the Factor method is called for any given value of x_3 is no more than $\binom{M-x_3}{w-3}$. Therefore, excluding the cost of computing the δ's, the cost of the using the Factor method for $w \geq 3$ is

$$8 \sum_{x_3=3}^{M-(w-3)} \binom{M-x_3}{w-3} (x_3 - 1)$$

$$\leq \frac{8}{(w-3)!} \sum_{x_3=3}^{M-w+3} x (M - x)^{w-3} \qquad (12)$$

multiplications in F_{q^d}.

The most significant term of (12) can be written as

$$\frac{8}{(w-3)!} M^{w-1} \left(\sum_{j=1}^{w-2} (-1)^{j-1} \binom{w-3}{j-1} \left(\frac{1}{j+1} \right) \right).$$

However, by an identity given in [16], we have

$$\sum_{j=1}^{w-2} (-1)^{j-1} \binom{w-3}{j-1} \left(\frac{1}{j+1} \right) = \frac{1}{(w-1)(w-2)}.$$

Therefore, the most significant term of (12) becomes

$$\frac{8}{(w-1)!} M^{w-1}. \qquad (13)$$

Cost of computing the δ's. We also need to compute the different δ_0, δ_1 and δ_2 required for each use of the Factor method.

$$\delta_0 = B^4 \prod_{t=1}^{w-2} (A_t^{-4})^{u_t}, \quad \delta_1 = \prod_{t=0}^{w-2} (A_{t+1}^{-4})^{u_t}, \quad \delta_2 = \prod_{t=0}^{w-2} A_{t+2}^{u_t}.$$

To compute δ_0 in the following algorithm, all possible values of each $\left(A_t^{-4}\right)^{u_t}$ are computed and combined to produce the value of δ_0 for each possible combination of the values $x_3, \ldots, x_w$.

Next, all possible values of $\left(A_t^{-4}\right)^{u_t}$ are computed for each t. Since the maximum value of each u_t is less than $\binom{w-2}{t} M^t$, we can compute all possible values of $\left(A_t^{-4}\right)^{u_t}$ in $\binom{w-2}{t} M^t$ multiplications.

Finally, we compute the δ's. We can compute all possible values of δ_0 by computing all possible values for $\eta_1 = B^4 (A_1^{-4})^{u_1}$ and then computing all possible values for $\eta_t = B^4 \prod_{z=1}^{t} (A_z^{-4})^{u_z}$, for $2 \le t \le w - 2$, from η_{t-1} and $(A_t^{-4})^{u_t}$.

Since there are less than $\binom{M}{w-2}$ possible values of $x_3, \ldots, x_w$, we can compute $\delta_0 = \eta_{w-2}$ with $(w-3)\binom{M}{w-2}$ multiplications.

The method computes δ_1 and δ_2 in a similar fashion. The total number of multiplications required to compute the δ's is bounded by

$$3 \left(\sum_{t=1}^{w-2} \binom{w-2}{t} M^t + (w-3) \binom{M}{w-2} \right) < 3\,(M+1)^{w-2} + \frac{3\,(w-3)}{(w-2)!} M^{w-2}. \quad (14)$$

The total cost to iteratively use the Factor Method is the sum of equations (13) and (14), but this cost is dominated by equation (13). Therefore, the approximate number of multiplications in F_{q^d} to solve equation (9) is

$$\frac{8}{(w-1)!} M^{w-1} \quad (15)$$

for $w \ge 2$.

B Faster Method for Computing Sums of the Form $\sum_{k=1}^{N} (k^t P_k)$

Let P_k, for $k = 1$ to N, be points on an elliptic curve. We want to compute

$$SUM_t = \sum_{k=1}^{N} \left(k^t P_k\right) \quad (16)$$

for $t = 1$ to w. Computing these values in the obvious way would require $w(N-1)$ elliptic scalar multiplications by a $\log_2 N$ bit-integer and $w(N-1)$ elliptic curve additions. We show a method due to Solinas [14] to compute these w quantities with only $w(N-1)$ elliptic curve additions.

First, we compute

$$U_t = \sum_{k=1}^{N} \binom{t+k-1}{t} P_k$$

for $t = 1$ to w. These w values can be computed from the P_k's as follows:

```
V_k ← P_k for 1 ≤ k ≤ N
For t = 0 to w
        For k from N − 1 downto 1 do
              V_k ← V_k + V_{k+1}
        Next k
        U_t ← V_1
Next t
```

The cost of this algorithm is $w(N-1)$ elliptic curve additions. (We don't count the cost for the first time through the loop because the computations for $t = 0$ can be computed during the initial batch verification.)

We can now compute SUM_t (equation (16)) from $U_1, \ldots, U_t$:

$$SUM_t = \sum_{k=1}^{t} (-1)^{t-k}(k!)s_{t,k}U_k \tag{17}$$

where the $s_{t,k}$'s are the *Stirling numbers of the second kind*. We can compute SUM_t, for $t = 1$ to w, with only $O(w^2)$ operations:

```
SUM_1 ← U_1
s_0 ← 0
s_1 ← 1
For i = 2 to w
        s_i ← 0
Next i
For t = 2 to w
        SUM_t ← 0
        For k from t downto 1 do
              s_k ← s_{k-1} + k s_k
              SUM_t ← k ((-1)^{t-k} s_k U_k + SUM_t)
        Next k
        Return SUM_t
Next t
```

For large N and small w, we can ignore the $O(w^2)$ operations. Therefore, the total cost of computing SUM_t, for $t = 1$ to w, is approximately $w(N-1)$ elliptic curve additions.

How to Forge a Time-Stamp
Which Adobe's Acrobat Accepts

Tetsuya Izu, Takeshi Shimoyama, and Masahiko Takenaka

FUJITSU LABORATORIES Ltd.
4-1-1, Kamikodanaka, Nakahara-ku,
Kawasaki, 211-8588, Japan
{izu,shimo,takenaka}@labs.fujitsu.com

Abstract. This paper shows how to forge a time-stamp which the latest version of Adobe's Acrobat and Acrobat Reader accept improperly. The target signature algorithm is RSASSA-PKCS1-v1.5 with a 1024-bit public composite and the public key $e = 3$, and our construction is based on Bleichenbacher's forgery attack presented in CRYPTO 2006. Since the original attack is not able to forge with these parameters, we used an extended attack described in this paper. Numerical examples of the forged signatures and times-stamp are also provided.

Keywords: Bleichenbacher's forgery attack, RSASSA-PKCS-v1.5, time-stamp forgery, Acrobat, Acrobat Reader.

1 Introduction

In the rump session of CRYPTO 2006, held on August 2006, Bleichenbacher presented a new forgery attack [6] against the signature scheme RSASSA-PKCS1-v1.5 (PKCS#1v1.5 for short) defined in PKCS#1 [13] and RFC 3447 [16], a cryptographic standard developed and maintained by RSA Laboratories [13]. The attack allows an adversary to forge a valid signature on an (almost) arbitrary message in a very simple way, if an implementation of the signature scheme is loose, namely, a format check in the verification is not adequate. In fact, several implementations of PKCS#1v1.5 including OpenSSL, Firefox2 and Sun's JRE (Java Runtime Environment) library had this vulnerability. In response to Bleichenbacher's attack, US-CERT published a vulnerability note on September 2006 [7], and these implementations resist the attack now.

Since Bleichenbacher's presentation was limited to the case when the bit-length of the public composite n (denoted by $|n|$) is 3072 and the public exponent e is 3, applicability to other parameters was unclear. Though Tews showed the applicability of the extended forgery attack when $|n| = 1024$ and $e = 3$ [19], other cases such as $e = 17, 65537$ has not been discussed yet.

In this paper, we analyze Bleichenbacher's forgery attack and show applicable composite sizes for given exponents. Then we propose the extended attack with assuming the same implementational error, which is a generalization of the

S.D. Galbraith (Eds.): Cryptography and Coding 2007, LNCS 4887, pp. 54–72, 2007.

original attack and Tew's extended attack. For fixed n and e, the success probability of the proposed attack is $2^{(|n|-15)/e-353}$ in the random oracle model. When $|n| = 1024$ and $e = 3$, the proposed attack succeeds the forgery with probability $2^{-16.6}$ which coincides the Tew's experiment [19]. Note that the preliminary version of the proposed attack was published in [9].

In the proposed attack, the success probability of a forgery on the chosen message is $2^{(|n|-15)/e-353}$. However, when the message is 'redundant', namely, it includes supplementary data (such as a name of used tool, a name of author, date) besides a body of the message and whose size is large enough, the adversary can forge on the chosen message by changing the redundant part. As an example of this scenario, we show how to forge a time-stamp defined in RFC 3161 [4] (in which 64-bit nonce space is available) with our proposed attack. Finally and most importantly, we check some verification client programs whether they accept forged time-stamps by (1) the proposed attack and (2) a variant of Bleichenbacher's attack by Oiwa et al. [11]., As a result, a forged time-stamp by the proposed attack embedded in a PDF (Portable Document Format) file was improperly accepted by the latest version of Adobe's Acrobat and Acrobat Reader [1].

The rest of this paper is organized as follows: in section 2, Bleichenbacher's forgery attack against PKCS#1v1.5 and analytic results are described. Then the extended attack is proposed in section 3, and its application to the time-stamp scheme is described in section 4. Some numerical examples of forged signatures are in the appendix.

2   Bleichenbacher's Attack

This section describes Bleichenbacher's forgery attack [6] against RSASSA-PKCS1-v1_5 (PKCS#1v1.5 for short) with the loose implementation. Let n be an RSA composite whose size is denoted by $|n|$ (in bit). In the followings, a variable in the `typewriter font` denotes an octet string and a variable in the Roman font denotes an integer. Two variables in the same letter correspond to each other, namely, `A` is an octet representation of an integer A, vice versa.

2.1   RSASSA-PKCS1-v1_5

Let us introduce the signature scheme RSASSA-PKCS1-v1_5 defined in PKCS#1 [13]. For a given public composite n (a product of two large primes with same size) such that $|n|$ is a multiple of 8, a message m (to be signed) is encoded to an integer M, an integer representation of an octet string `M` defined by

$$\texttt{M} = \texttt{00}||\texttt{01}||\texttt{PS}||\texttt{00}||\texttt{T}||\texttt{H}, \quad \text{(PKCS\#1v1.5 message format)}$$

where `PS` is an octet string with `ff` such that $|\texttt{M}| = |n|$ (and $|\texttt{PS}| \geq 64$), `T` is an identifier of the signature scheme and the hash function (Table 1), and `H` is an

---
[1] The authors have already submitted the vulnerability report to a governmental organization.

Table 1. Identifiers of the algorithm and the hash function [13]

Hash Function	Length (bit)	Octet String
MD2	144	3020300c06082a864886f70d020205000410
MD5	144	3020300c06082a864886f70d020505000410
SHA-1	120	3021300906052b0e03021a05000414 ($= \mathrm{T_{SHA1}}$)
SHA-256	152	3031300d060960864801650304020105000420
SHA-384	152	3041300d060960864801650304020205000430
SHA-512	152	3051300d060960864801650304020305000440

octet representation of the hash value $H(m)$. Then, a signature s is generated by $s = M^d \bmod n$ for the signer's secret integer d.

On input the original message m, its signature s and the signer's public exponent e, a verifier obtains an octet string M' representing an integer $M' = s^e \bmod n$ and checks whether it satisfies the format

$$\mathsf{M}' = \mathsf{00}||\mathsf{01}||\mathsf{PS}||\mathsf{00}||\mathsf{T}||\mathsf{H}'.$$

Then the verifier obtains a value H', an integer representation of the octet string H', and compares whether $H' = H(m)$. If this equation holds, the signature is accepted by the verifier.

In the implementation level, a part of the format check is sometimes inadequate by some reasons. For example, when an octet string

$$\mathsf{00}||\mathsf{01}||\mathsf{PS}||\mathsf{00}||\mathsf{T}||\mathsf{H}'||\mathsf{garbage}.$$

(a garbage data is followed) is obtained by a verifier as a decoded message, it should be rejected because it is in the illegal format. However, some implementations accept the string because they do not check the number of ff and they stop the scan at the end of H' (namely, they do not notice the existence of the garbage). Such loose implementation is the target of Bleichenbacher's forgery attack described in the next subsection.

2.2   Outline of Bleichenbacher's Attack

Next, let us introduce Bleichenbacher's forgery attack [6] against PKCS#1 v1.5. Here we assume that the hash function SHA-1 and parameters $|n| = 3072$ and $e = 3$ are used [8]. In the attack, an adversary chooses a message $\bar{m}$ with arbitrary bit-length such that

$$a = 2^{288} - (T \times 2^{160} + H(\bar{m}))$$

is divisible by 3, where T is an integer representation of the octet string $\mathrm{T_{SHA1}}$ (as in Table 1). Note that such $\bar{m}$ can be obtained by generating $\bar{m}$ randomly (3 trials are required on average). The adversary also computes two integers

$$g = a^2/3 \times 2^{1087} - a^3/27 \times 2^{102}, \quad \bar{s} = 2^{1019} - a/3 \times 2^{34}.$$

Observe that

$$\bar{s}^e = (2^{1019} - a/3 \times 2^{34})^3$$
$$= 2^{3057} - a \times 2^{2072} + a^2/3 \times 2^{1087} - a^3/27 \times 2^{102}$$
$$= 2^{3057} - 2^{2360} + T \times 2^{2232} + H(\bar{m}) \times 2^{2072} + g$$
$$= (2^{985} - 2^{288} + T \times 2^{160} + H(\bar{m})) \times 2^{2072} + g.$$

Since an integer $2^{985} - 2^{288} + T \times 2^{160} + H(\bar{m})$ corresponds to an octet string $00||01||\mathtt{ff}...\mathtt{ff}||00||\mathtt{T}||\mathtt{H}'$ (the number of $\mathtt{ff}$ is different from that of the original PS), $\bar{s}$ is a forged signature on the message $\bar{m}$, if an implementation of the verification ignores the number of $\mathtt{ff}$ and the garbage g. In the forgery, the adversary only requires to compute $H(\bar{m})$, a and $\bar{s}$. This is why the attack is called "the pencil and paper attack" [6]. Note that the adversary does not use modulus computations and thus integers n, d are not required in the forgery.

A numerical example of Bleichenbacher's forgery attack with a 3072-bit composite and $e = 3$ is shown in Table 7 in appendix A.

2.3 Analysis

This subsection analyzes Bleichenbacher's forgery attack with general parameters. Only SHA-1 is considered in the following, however, similar attacks and analysis can be easily obtained for other hash functions. For simplicity, we consider the public composite n with arbitrary length (rather than a multiple of 8).

Firstly, we consider the case with general n but $e = 3$. Since the padding $00||\mathtt{T}_{\mathrm{SHA1}}$ is 128-bit and the hash value is 160-bit, we use the same a as in the original attack, namely $a = 2^{288} - (T \times 2^{160} + H(\bar{m}))$ such that $3|a$. Let

$$\bar{s}(\alpha, \beta) = 2^\alpha - a/3 \times 2^\beta,$$

be a forged signature. Then, we have

$$\bar{s}(\alpha, \beta)^3 = 2^{3\alpha} - a \times 2^{2\alpha+\beta} + g(\alpha, \beta)$$

for the garbage $g(\alpha, \beta) = a^2/3 \times 2^{\alpha+2\beta} - a^3/27 \times 2^{3\beta}$. Since $\bar{s}(\alpha, \beta)^3$ should be in the PKCS#1v1.5 format, we have $3\alpha = |n| - 15$, namely, $\alpha = (|n| - 15)/3$ and $|n|$ should be divisible by 3. On the other hand, since the garbage should be smaller than $2^{2\alpha+\beta}$, we have $2\alpha+\beta > 576+\alpha+2\beta-\log_2 3$, namely, $\beta < |n|/3-581+\log_2 3$. By substituting $\beta \geq 0$ in this inequality, we have a condition on n that

$$|n| > 1743 - 3\log_2 3 = 1738.24....$$

Consequently, Bleichenbacher's attack with $e = 3$ is applicable to the case with $|n| \geq 1739$ with $|n|$ is divisible by 3. More precisely, $|n|$ can be parameterized by $|n| = 3k$ for $k \geq 580$ and β is in a form $\beta = 8\ell + 2$ $(0 \leq \ell \leq 55)$ since PS is a repetition of the octet string $\mathtt{ff}$.

Next, let us discuss with general n and e. Similar to the above discussion, we set $\bar{s}(\alpha, \beta) = 2^\alpha - a/e \times 2^\beta$ for $a = 2^{288} - (T \times 2^{160} + H(\bar{m}))$ such that $e|a$ and $\alpha = (|n| - 15)/e$. Then, we have

$$\bar{s}(\alpha, \beta)^e = 2^{e\alpha} - a \times 2^{(e-1)\alpha+\beta} + g(\alpha, \beta)$$

for the garbage $g(\alpha, \beta) = a^2(e-1)/(2e) \times 2^{(e-2)\alpha+2\beta} + \ldots$. By the same discussion, we have conditions on n that

$$|n| > 576e + 15 - e\log_2\left(\frac{2e}{e-1}\right)$$

and $|n| - 15$ is divisible by e. Also, we have $0 \le \beta < |n|/3 - 581 + \log_2 3$ and $\beta \equiv 2 \pmod 8$ on β. Especially, we have $|n| = 17k + 15$ $(k \ge 575)$ for $e = 17$, and $|n| = 65537k + 15$ $(k \ge 1061)$ for $e = 65537$. Consequently, Bleichenbacher's attack for general e is far from feasible. Even if $e = 3$, Bleichenbacher's attack cannot be applicable to 1024-bit (since 1024 is smaller than 1739) or 2048-bit composites (since $n - 15 = 2033$ is not divisible by 3, 17, 65537).

2.4   Oiwa et al.'s Variant

Recently, Oiwa et al. proposed a variant of Bleichenbacher's attack [11]. In the message format 00||01||PS||00||T||H, T||H can be described by {{OID, PF}, H} in ASN.1 language, where {, ..., } denotes the enumerate type, OID is the hash object ID and PF is the parameter field. In PKCS#1, PF is defined as NULL. When PF is replaced by non-null data, though the message format is not accepted in PKCS#1, it is acceptable by an ASN.1 parser. An idea of a variant attack by Oiwa et al. [11] is to insert the garbage into the parameter field rather than to the end of the message format. If message format is checked by generic ASN.1 parser, the forgery will be successful. In fact, they actually forged a signature and found the vulnerability in GNUTLS ver 1.4.3 and earlier (though they are resistant to Bleichenbacher's attack).

By the same analysis, it is easily shown that Oiwa et al.'s variant has the same ability to Bleichenbacher's attack. Moreover, the same extension proposed in the next section can be possible.

3   Extending Bleichenbacher's Attack

The security of PKCS#1v1.5 relies on the hardness of factoring n and computing the e-th root mod n. A key idea of Bleichenbacher's forgery is to set the forged signature $\bar{s}$ in the special form so that upper bits of $\bar{s}^e$ are in the PKCS#1v1.5 message format by using the garbage g. In this scenario, an adversary computes the e-th power only, however, because of the speciality of the forged signature, the public composites should be large as described in the previous section. In this section, we extend Bleichenbacher's attack by using computers rather than pencils and papers. Our strategy is to obtain a forged signature in non-special forms. To do so, for a given hash value $H(\bar{m})$, we search $\bar{s}^e$ such that the e-th root over integer exists, by computing the e-th root over real numbers (note that the e-th root computation over real number is easy with computers).

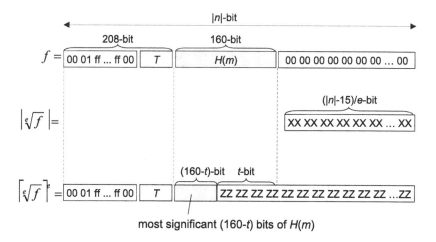

Fig. 1. Outline of the proposed forgery attack

3.1 Description of Proposed Forgery Attack

Let $\bar{m}$ be a message and $H(\bar{m})$ be its hash value by SHA-1. For given bit-length of the composite $|n| \geq 369$, define f as a function of $\bar{m}$ by

$$f = 2^{|n|-15} + 15 \times (2^{|n|-23} + \cdots + 2^{|n|-79}) + T \times 2^{|n|-208} + H(\bar{m}) \times 2^{|n|-368}$$
$$= (2^{192} - 2^{128} + T) \times 2^{|n|-208} + H(\bar{m}) \times 2^{|n|-368}$$

where T denotes an integer representation of the octet string $\mathrm{T_{SHA1}}$ (Table 1). Note that the integer $2^{192} - 2^{128} + T$ in the above equation represents an octet string $00\|01\|\mathtt{ffffffffffffffff}\|00\|T$. Next, compute the e-th root $\sqrt[e]{f}$ as a real number and its ceiling $\lceil \sqrt[e]{f} \rceil$. If the difference $g = \lceil \sqrt[e]{f} \rceil^e - f$ is smaller than $2^{|n|-368}$, the forgery succeeds, since the forged signature $\bar{s} = \sqrt[e]{f+g} = \lceil \sqrt[e]{f} \rceil$ is valid on the message $\bar{m}$ with the garbage g. If g is not small, change the message $\bar{m}$ until the forgery succeeds.

Let us analyze the proposed forgery attack with general n and e. In the failure case, some least significant bits of $H(\bar{m})$ in f, say t bits, differs from the corresponding part in $\lceil \sqrt[e]{f} \rceil^e$ (in other words, these t bits coincide in the successful case), where t is a parameter determined by n and e (see Figure 1). In the forgery, f is $(|n| - 15)$-bit and the integer part of $\sqrt[e]{f}$ is $(|n| - 15)/e$-bit, and the uncontrolled part of $\lceil \sqrt[e]{f} \rceil^e$ is $(e - 1)(|n| - 15)/e$-bit. Thus we have a condition $|n| - 208 - (160 - t) > (e - 1)(|n| - 15)/e$, namely,

$$|n| > (353 - t)e + 15. \tag{1}$$

Here we implicitly used the random oracle assumption. Especially, when $|n| = 1024$ and $e = 3$, this condition implies that $t > 50/3 \approx 16.6$. That is, in order to forge a signature with 1024-bit composites, the proposed forgery attack succeeds

Table 2. A forged signature by the extended forgery attack (1024-bit, $e = 3$)

$\bar{m}$	"00002e36"
$H(\bar{m})$	701f0dd6 f28a0bab 4b647db8 ddcbde40 1f810d4e
f	0001ffff ffffffff ffff0030 21300906 052b0e03 021a0500 0414<u>701f 0dd6f28a</u>
	<u>0bab4b64 7db8ddcb de401f81 0d4e</u>0000 00000000 00000000 00000000 00000000
	00000000 00000000 00000000 00000000 00000000 00000000 00000000 00000000
	00000000 00000000 00000000 00000000 00000000 00000000 00000000 00000000
$\bar{s}$	00000000 00000000 00000000 00000000 00000000 00000000 00000000 00000000
	00000000 00000000 00000000 00000000 00000000 00000000 00000000 00000000
	00000000 00000000 00000000 00000000 00000000 0001428a 2f98d728 ae220823
	1fc5cff6 ac440735 9b078378 24846240 4cebfc71 5690f34c 7119d1da 99227fd0
$\bar{s}^e$	0001ffff ffffffff ffff0030 21300906 052b0e03 021a0500 0414<u>701f 0dd6f28a</u>
	<u>0bab4b64 7db8ddcb de401f81 0d4e</u> 06dd 391b3fd4 ace323ee de903694 dd78887f
	5f8a73e0 5ea698ae 72a6bdfa cb7c359e 1f78cbee 96939eea 4d9b8f3e 47aebae3
	90f4fe61 73ef7535 80c4cb88 edd95623 84b7e5ed ccc19fa3 ca64c0a2 a37e5000

with probability $2^{-16.6}$, namely, $2^{16.6}$ messages are required to forge a signature, which is feasible in practice. Note that the proposed attack is a generalization of the extension by Tews [19] in which $275992 \approx 2^{18.1}$ messages are required in the experiment.

In the above construction, the number of the octet ff in f was fixed to 8 (this is minimum for the forgery). Similar construction is possible with more octets than 8, but requires larger composites instead.

Numerical Example. As an example of the proposed forgery attack, a forged signature on the message $\bar{m} = $ "00002e36" (as a binary data with big endian) with $|n| = 1024$ and $e = 3$ is shown in Table 2, where underlined octets correspond to the hashed value $H(\bar{m})$ and masked octets correspond to the garbage g. Here, the messages were incrementally generated (as integers) from "00000000", and $0x00002e36 = 262144 = 2^{13.53}$ messages were generated until the forgery succeeds.

3.2 Special Cases

Let us consider two special cases of the proposed forgery attack, namely $t = 0$ or $t = 160$ cases.

When we set $t = 0$, the forgery attack always succeeds. In this case, the condition (1) implies

$$|n| > 353e + 15. \qquad (2)$$

Even when $e = 3$, this condition implies that $|n| > 1074$ which is beyond 1024. Also, we have $|n| > 6017$ for $e = 17$ and $|n| > 23134577$ for $e = 65537$. Since this case only uses the garbage space, it allows a forgery on arbitrary chosen messages with smaller composites than the original attack. Especially, this attack does not require a condition on the target message $\bar{m}$ and the attack always succeeds. As a numerical example, a forged signature on the message $\bar{m}$ ("pkcs-1v2-1.doc" [14]) with $|n| = 1152$ and $e = 3$ in Table 8 in appendix B, which succeed a forgery for $e = 3$. Note that Bleichenbacher's original attack cannot forge for 1152-bit composites nor the exponent $e = 3$.

On the other hand, when we set $t = 160$, the attack becomes most powerful but the adversary can not control the hash value at all. In this case, the condition (1) implies

$$|n| > 193e + 15 \tag{3}$$

which is obtained by substituting $t = 160$ into the condition (1). Consequently, we have $|n| > 595$ for $e = 3$, $|n| > 3297$ for $e = 17$ and $|n| > 12648657$ for $e = 65537$. However, since the adversary can not control the hash value, the success probability (in the sense that the adversary obtains the target message $\bar{m}$) is 2^{-160} which is beyond feasible. Another forged signature on the hash value

$$H = \texttt{7fa66ee7 e5cc4a9f bd6e13a8 11d298c2 6b9b3302}$$

with $|n| = 4096$ and $e = 17$ in Table 9 in appendix B, which succeed a forgery for $e = 17$, however, the success probability is 2^{-113}. Note that Bleichenbacher's original attack cannot forge for 1024-bit composites nor the exponent $e = 17$.

Table 3. A comparison of forgery attacks

	Bleichenbacher's Attack	Proposed Attack		
		$t = 0$	General t	$t = 160$
$M(e)$	$576e + 15 - e\log_2\left(\frac{2e}{e-1}\right)$	$353e + 15$	$(353 - t)e + 15$	$193e + 15$
	$\lvert n\rvert - 15$ is divisible by e			
$M(3)$	1740	1075	$1075 - 3t$	595
$M(17)$	9790	6017	$6017 - 17t$	3297
$M(65537)$	37683790	23134577	$23134577 - 65537t$	12648657
Success Probability	1	1	2^{-t}	2^{-160}

3.3   Comparison

A comparison between the original and proposed attacks are shown in Table 3, where $M(e)$ denotes the minimum bit-length of the composites to which the attack succeeds with a general exponent e. Since exponents $e = 3$, 17, 65537 are widely used, corresponding values $M(3)$, $M(17)$, $M(65537)$ are also included in the comparison. As in the table, the proposed attack with $t = 160$ forges with smallest composites. Especially, it only forges for $|n| = 1024$ (with $e = 3$).

4   Time-Stamp Forgery

In the previous section, we proposed a forgery attack against PKCS#1v1.5 based on Bleichenbacher's attack. When $|n| = 1024$ and $e = 3$, the success probability

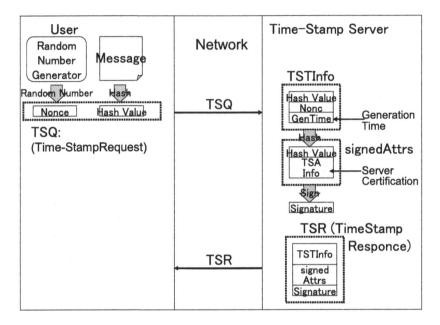

Fig. 2. Outline of the time-stamp protocol

of a forgery on a chosen message is $2^{-16.6}$. However, when the message is 're-dundant', namely, it includes supplementary data besides a body of the message whose space is larger than $2^{16.6}$, the adversary can always forge a signature on the chosen message by the proposed attack. In order to clarify this scenario, we forge time-stamps by the proposed attack and (extended) Oiwa et al.'s attack, since a time-stamp contains a 64-bit nonce. We also checked some time-stamp verification applications whether they accept the forged time-stamp. As a result, we would like to report that the latest version of Adobe's Acrobat and Acrobat Reader improperly accept the forged time-stamp (generated by the proposed attack).

4.1 Time-Stamp Protocol

The time-stamp is a cryptographic scheme which assures the existence of a document at a certain time. We briefly explain the time-stamp protocol defined in RFC 3161 [4], which is a target of our forgery attack. When a user requires a time-stamp on his document m, he sends its hash value $H(m)$ with a 64-bit random value (as a nonce) to a time-stamp server. The server firstly creates a file TSTInfo which includes the given hash value $H(m)$ (messageImprint), the given random value (nonce) and the time information (genTime). Then the server computes signedAttrs which includes a hash value of TSTInfo as messageDigest. Finally, the time-stamp server generates a signature signature on signedAttrs with his secret key and publishes it as a time-stamp on the document m and the random value. By verifying the time-stamp, a verifier confirms the existence of the document at the time described in genTime. An outline of the time-stamp protocol is shown in Figure 2.

In the time-stamp protocol [4], PKCS#1v1.5 with SHA-1 can be used as a signature scheme. The center column in Table 4 shows an example of a time-stamp issued by a trial time-stamp server maintained by Seiko Instruments Inc. [17] at 2006-09-21 10:29:27.64 on the document m ("pkcs-1v2-1.pdf" [15]) based on PKCS#1v1.5 ($|n| = 1024$ and $e = 3$) with SHA-1. TSTInfo and signerInfos are in PKCS#1v1.5 format. A verification result ($s^e \bmod n$) is also included in the table, where underlined octets correspond to a hash value of signerInfos.

4.2 Applying Proposed Attack to Time-Stamp Protocol

Since the time-stamp is a kind of signature scheme, we can apply the proposed forgery attack. A difference is that since a message has a 64-bit nonce, an adversary can fix the message in the forgery attack: the adversary only has to change the nonce until the forgery succeeds. Thus, the adversary can forge a chosen message if 64-bit space is enough.

In the attack, an adversary firstly fixes the time to be stamped and sets genTime. Then, on a chosen message, he randomly generates a 64-bit nonce and computes messageDigest from chosen genTime and nonce. If he can forge a signature on messageDigest, the time-stamp forgery is finished. When $|n| = 1024$ and $e = 3$, it is estimated that about $2^{16.6}$ messageDigest, namely, $2^{16.6}$ nonce are required to forge a time-stamp. Since the nonce space is 64-bit, the adversary always succeeds a forgery.

An example of a forged time-stamp is shown in the right-most column in Table 4, where changed data (from a valid time-stamp) are only described. Here underlined octets correspond to a hash value of signerInfos and masked octets correspond to the garbage. If a verification implementation is loose, the forged time-stamp will be accepted, namely, it is confirmed that the document existed at 0001-01-01 00:00:00.0. Note that in this forgery, only genTime is changed from a valid time-stamp for simplicity. The adversary can change the message, if required.

In the above description, we only dealt with the proposed attack which is based on Bleichenbacher's attack. By the same approach, Oiwa et al.'s forgery attack can be extended and applied to the time-stamp protocol with the same ability (but the different implementation assumptions).

4.3 Checking Verification Implementations

We checked some time-stamp verification client programs whether they accept two forged time-stamps generated by the proposed attack (denoted by TST-P) and by Oiwa et al.'s variant (TST-O).

Table 4. Valid and forged time-stamps (1024-bit, $e = 3$)

	A valid time-stamp	A forged time-stamp
TSTInfo	3081f6	
version	020101	
policy	06090283 388c9b3d 01030230	
	21300906 052b0e03 021a0500	
messageImprint	0414784a 21902486 c00c6dbb	
	a87ef918 7bdf08fc 3f9f	
serialNumber	020357ee 1b	
genTime	18123230 30363039 32313130	18123030 30313031 30313030
	32393237 2e36345a	30303030 2e30305a
	("2006-09-21 10:29:27.64")	("0001-01-01 00:00:00.0")
accuracy	300a0201 00800201 f4810100	
nonce	02082dff 78d3789b 9c97	02080000 00000001 bf67
tsa	a08193a4 81903081 8d310b30	
	09060355 04061302 4a50311f	
	301d0603 55040a13 16536569	
	6b6f2049 6e737472 756d656e	
	74732049 6e632e31 14301206	
	0355040b 130b4368 726f6e6f	
	74727573 74312d30 2b060355	
	040b1324 536f7665 72656967	
	6e205469 6d652054 53205365	
	72766572 20534e3a 39314430	
	30363234 31183016 06035504	
	03130f44 656d6f54 53536f6e	
	53494931 3031	
signerInfos		
signedAttrs	3181f3	
contentInfo	301a0609 2a864886 f70d0109	
	03310d06 0b2a8648 86f70d01	
	09100104 30230609 2a864886	
	f70d0109 043116	
messageDigest	0414feca 1088d59e e69a1553	0414e8e6 4aa7ec9c 2fdb3d22
	172fbc92 a8636195 6335	f7b5682a bcf9afd0 f9c7
eSSSigningCertificate	3081af06 0b2a8648 86f70d01	
	0910020c 31819f30 819c3081	
	99307f04 140bde7d 9e80ee5b	
	4d802804 b4c8382b ac8c4d0c	
	05306730 5aa45830 56310b30	
	09060355 04061302 4a50311f	
	301d0603 55040a13 16536569	
	6b6f2049 6e737472 756d656e	
	74732049 6e632e31 14301206	
	0355040b 130b4368 726f6e6f	
	74727573 74311030 0e060355	
	04031307 44656d6f 43413102	
	0900f6e2 0bb648ca fd8e3016	
	0414342d 517f9a5f 7ebfa4ab	
	4bcaffdf cf1ae903 30d6	

Table 4. (*continued*)

signature (s)	3868bb42 9f71a918 e906f2e0	00000000 00000000 00000000
	674798fd ffeef81d 5942a300	00000000 00000000 00000000
	08c45e1b a8b2a966 3be95650	00000000 00000000 00000000
	d6bb8501 06dea5c7 e373f820	00000000 00000000 00000000
	f538f860 cf06bd78 313e51dc	00000000 00000000 00000000
	070e03b5 f286bace 7dff72af	00000000 00000000 00000000
	88e32a5a 6629ae72 65541ea3	00000000 00000000 00000000
	28014181 b7a6424c aee030d6	0001428a 2f98d728 ae220823
	548636af 20f5d6d4 f43936a2	1fc5cff6 ac440735 9b078378
	f732e728 b1fbdfc6 a4cf5b0e	2484756e 10093903 9319a13e
	3637bcd9 c8aa8f73	eb2d174d 90fe10aa
A verification result	0001ffff ffffffff ffffffff	0001ffff ffffffff ffff0030
(s^e mod n)	ffffffff ffffffff ffffffff	21300906 052b0e03 021a0500
	ffffffff ffffffff ffffffff	0414cb74 56c4991f 270c1044
	ffffffff ffffffff ffffffff	5a7c525b 14836ec0 fe33 6a89
	ffffffff ffffffff ffffffff	ea7f9c08 e357202c 288839f2
	ffffffff ffffffff ffffffff	0ce9cbd3 75925ad5 45f6ebf2
	ffffffff ffffffff ffffffff	99a247b1 f995ae2e 7365203c
	ffffffff ffffffff 00302130	ba83acf1 7ca3964d 1a204b0b
	0906052b 0e03021a 05000414	2be547f6 91771716 55f5e7dd
	26eee9a4 46e35f03 5c1f6857	51c0a4a3 6b06b235 6eb173da
	95fc927d dc158653	482f36d7 5a1db768

Table 5. Verification results of forged time-stamps

Client	Version	TST-P	TST-O
TrustPort		Rejected	Rejected
Chronotrust client program	1.0	Rejected	Rejected
PFU time-stamp client tool	V2.0L30	Rejected	Rejected
Acrobat (Professional / Standard / Reader)	7.x, 8.x	**Accepted**	Rejected
e-timing EVIDENCE Verifier for Acrobat	2.30	(input error)	(input error)

Target verification clients are as follows:

- TrustPort by AEC [2]
- Chronotrust client program by Seiko Instruments Inc. [18]
- Time-stamp client tool by PFU [12]
- Acrobat (Professional, Standard, Reader) by Adobe [1]
- e-timing EVIDENCE Verifier for Acrobat by Amano [3]

These clients are categorized by two groups: a time-stamp in the first group should be held separately from the corresponding message while a time-stamp in the second group should be embedded in the same file as the message.

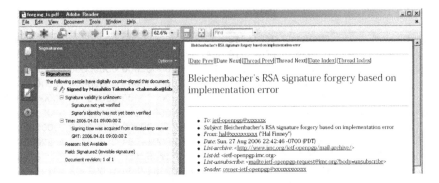

Fig. 3. Verification with the forged time-stamp by Adobe's Acrobat Reader 8.00 (1)

Verification results of above clients are summarized in Table 5. As in the table, only Acrobat improperly accepts the forged time-stamp TST-P. More details are described in the next subsection. Note that the output by e-timing EVIDENCE Verifier for Acrobat are "input error", (not "rejected"), since the time-stamp was not generated by Amano's time-stamp server. Thus, we couldn't check whether Amano's client has the implementation error or not.

4.4 How to Deceive Adobe's Acrobat

A Portable Document Format (PDF) is a *de facto standard* for describing digital documents developed by Adobe. The format allows a PDF file to embed signatures and time-stamps in the same file. A verifier confirms the signatures and time-stamps by specific tools (application softwares). In the current version of PDF, the time-stamp based on RFC 3161 [4] can be used. Adobe's Acrobat is a widely used application for making digital files in PDF, and Adobe's Acrobat Reader is used for reading PDF files. In addition, Acrobat and Acrobat Reader can verify the signatures and time-stamps embedded in the file.

We are motivated by an observation that Adobe has not published any response to the vulnerability report VU#845620 [7]. A PDF file of the schedule of the rump session of CRYPTO 2006 (held on August 2006) was used as a target message. First, we obtain a time-stamp from a trial-server maintained by Seiko Instruments Inc. [17] based on PKCS#1v1.5 ($|n| = 1024$ and $e = 3$) with SHA-1. Then, we generate a forged time-stamp which asserts that the document existed at 2006-04-01 09:00:00.0. In fact, Adobe's Acrobat Reader 7.09 accepts the forged time-stamp as in Figure 3 and 4. As far as we examined, the same vulnerability is found in Acrobat 7.09 and Acrobat Reader 7.09, 8.00. Note that we have already submitted the vulnerability report to a governmental organization.

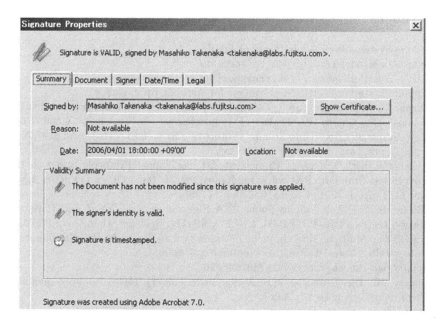

Fig. 4. Verification with the forged time-stamp by Adobe's Acrobat Reader 8.00 (2)

5   Concluding Remarks

This paper analyzes Bleichenbacher's forgery attack against the signature scheme RSASSA-PKCS1-v1.5 (PKCS#1v1.5) with the implementation error, and proposed an extended attack. As an example, we forged a time-stamp with 1024-bit composite and the exponent 3, and showed that the latest version of Adobe's Acrobat and Acrobat Reader accept the forged time-stamp improperly. Resisting the original and proposed forgery attacks is easy and simple: fixing the loose implementation is enough. Avoiding small exponents might be another counter-measure, however, once a certificate with exponent $e = 3$ is issued by a trusted party, one cannot refuse it when the verification is accepted.

As described in this paper, the proposed attack succeeds a forgery on an arbitrary message with redundancy when $|n| = 1024$ and $e = 3$. Though this paper only dealt with the time-stamp, similar forgery is possible on other messages with redundancy. One future work is a signature forgery on a message in PDF or PS files and other is a certificate forgery. Searching loose implementations which accept these forgeries is also required.

Acknowledgements

The authors would like to thank the program committee of the conference and anonymous reviewers on the preliminary version of this paper for their helpful comments and suggestions.

References

1. Adobe Systems Inc., Adobe Acrobat family,
   http://www.adobe.com/products/acrobat/
2. AEC, TrustPort. http://www.trustport.cz/?content=tsa
3. Amano, E-timing EVIDENCE Verifier for Acrobat (in Japanese),
   https://www.e-timing.ne.jp/download/evidence-verifier/formmail.html
4. Adams, C., Chain, P., Pinkas, D., Zuccherato, R.: Internet X.509 Public
   Key Infrastructure: Time-Stamp Protocol (TSP), RFC 3161 (August 2001),
   http://www.ietf.org/rfc/rfc3161.txt
5. Bleichenbacher, D.: Chosen Ciphertext Attacks against Protocols Based on RSA
   Encryption Standard PKCS#1. In: Krawczyk, H. (ed.) CRYPTO 1998. LNCS,
   vol. 1462, pp. 1–12. Springer, Heidelberg (1998)
6. Bleichenbacher, D.: Forging Some RSA Signatures with Pencil and Paper. In:
   Dwork, C. (ed.) CRYPTO 2006. LNCS, vol. 4117, Springer, Heidelberg (2006)
7. US-CERT, Multiple RSA Implementations Fail to Properly Handle Signatures,
   Vulnerability Note VU#845620 (September 5, 2006),
   http://www.kb.cert.org/vuls/id/845620
8. Finney, H.: Bleichenbacher's RSA Signature Forgery Based on Implementation
   Error. e-mail (August 27, 2006),
   http://www.imc.org/ietf-openpgp/mail-archive/msg14307.html
9. Izu, T., Takenaka, M., Shimoyama, T.: Analysis on Bleichenbacher's Forgery Attack. In: WAIS 2007, pp. 1167–1174. IEEE Computer Society, Los Alamitos (2007)
10. NTT Communications, Certificates for the Internal Credit Application CA. (in
    Japanese) http://www.ntt.com/creditca/x509.pdf
11. Oiwa, Y., Kobara, K., Watanabe, H.: A New Variant for an Attack Against RSA
    Signature Verification using Parameter Field. In: EUROPKI 2007 (June 2007)
12. PFU, PFU time-stamp service (in Japanese), http://www.pfu.fujitsu.com/tsa/
13. RSA Laboratories, RSA PKCS #1 v2.1: RSA Cryptography Standard (June 14,
    2002)
14. A digital file of [13] in WORD format,
    ftp://ftp.rsasecurity.com/pub/pkcs/pkcs-1/pkcs-1v2-1.doc
15. A digital file of [13]in PDF format,
    ftp://ftp.rsasecurity.com/pub/pkcs/pkcs-1/pkcs-1v2-1.pdf
16. RSA Laboratories, Public-Key Cryptography Standards (PKCS) #1:
    RSA Cryptography Specifications Version 2.1, RFC 3447 (February 2003)
    http://www.ietf.org/rfc/rfc3447.txt
17. Seiko Instruments Inc., a trial time-stamp service (in Japanese),
    http://www.sii.co.jp/ni/tss/trial/
18. Seiko Instruments Inc., Chronotrust (in Japanese),
    http://www.sii.co.jp/ni/tss/index.html
19. Tews, E.: Real World Exploit for Bleichenbacher's Attack on SSL. e-mail submitted
    to the Cryptography Mailing List (September 14, 2006),
    http://www.imc.org/ietf-openpgp/mail-archive/msg14307.html

A   Numerical Example of Bleichenbacher's Forgery

We show a numerical example of Bleichenbacher's original forgery attack [6]. For obtaining a valid signature, a 3072-bit composite n and a secret key d were generated by OpenSSL as in Table 6. We used the hash function SHA-1 and a public key $e = 3$ and chose a digital file "pkcs-1v2-1.doc" [14] as a message m (whose corresponding a is divisible by 3, fortunately).

A valid signature s on the message m, and a forged signature $\bar{s}$ on the same message m are shown in Table 7, where underlined octets correspond to the hashed value $H(m)$ and masked octets correspond to the garbage g. Comparing s^e and $\bar{s}^e$ in Table 7, all octets are same except the number of the octet ff and the garbage. Thus, if an implementation ignores these differences, the forged signature $\bar{s}$ is accepted in the verification. Actually, OpenSSL 0.9.7f accepts the forged signature $\bar{s}$ on the message m.

Table 6. RSA parameters (3072-bit)

n	d9057e4d 2e231c66 f0a35c2c b7eddb75 04b181d6 535b81b3 83eb4765 1d76950d
	76c0c513 9efc0933 16255a5a a958007c 1b698c4c 2641418a dab6419f 8c8cf6a9
	ac799a12 7b0ec916 b5837e9c 0ecb3dc3 9629427c 08b9b076 1014d3fb c2d6d26f
	aade8a49 7aa8b03a 8e0fa396 6f6b54bd 2735a972 85cbaaed 4760ff5c 7c8b4fe2
	3d6c053c 69d0fa64 ef3ec8ad 4fa03c16 9b8e5a68 466f7dbb 1f05f6ec caf9706c
	d524b148 c41ccb67 512bcf40 b6456321 1a420f22 fedeaf1a 44ff940d eeec2117
	9ce14bec 73b5b294 f0723d03 3a810ac3 a98dc56a a9e94eca 798c2033 3fa79eb8
	ea10d25b cca36cc2 b14f4c53 3c42560c aafbb7c6 5d524591 68f8b4e0 99351f23
	8f5fbf52 ee002fb8 240f7323 938207e3 59a17330 b7df56ef e8660f9a 5cc319ce
	d3d93f25 84f5e42a 80f0acdd dec65d4d 629e2250 cbbb06f5 7ceab655 b22216d7
	9120bdc9 216310be 4c3b81ea 92017a0b 8205e92d afb9c402 9b0f4603 2a847f67
	ba0c271e a3c8f60d 5c48f4fe 22e0d3e9 3b72e9ce 1e5191bc 6167decd cde29c89
d	90ae5433 74176844 a06ce81d cff3e7a3 5876568e e23d0122 57f22f98 be4f0e08
	f9d5d8b7 bf52b0cc b96e3c3c 70e555a8 12465d88 1980d65c 91ced66a 5db34f1b
	c8511161 a75f30b9 ce57a9bd 5f32292d 0ec62c52 b07bcaf9 600de2a7 d739e19f
	c73f06db a71b2027 095fc264 4a478dd3 6f791ba1 ae87c748 da40aa3d a85cdfec
	28f2ae28 468b5198 9f7f3073 8a6ad2b9 bd09919a d99fa927 6a03f9f3 31fba048
	8e187630 82bddcef 8b728a2b 242e4216 11815f6c a9e9ca11 83550d5e 9f48160e
	83a1eb18 fd99ce23 eb14095c 333f0375 747bec29 cbe110e8 4aee7d3b 98b0e20a
	53586ce9 319c9857 50fd3c8f 7cc6613f 773748a6 9aa5550c fb691771 f5921b52
	dedacd6c cbcee703 1663f656 2019fcd7 2fcd66f5 2f6b5d86 f148f420 5eed94b1
	170937ea 8cd536d2 932435c3 adb9d529 98ab1613 8a24f1e2 c9b1c7ad cd57713c
	c08f4ccf d8a2dc47 681ef8c0 3fe709d9 52dd12ca edbba76c 21629613 fe8e0343
	193e73a5 26533256 aedda14e 6517c092 52a66013 4a2acb98 2c5e2ec1 9fdd7cab
e	03

Table 7. Valid and forged signatures (3072-bit, $e = 3$)

m	"pkcs-1v2-1.doc" [14]
$H(m)$	f7497dac 551ec010 2f0da8f1 bc8cad52 f93476c3
s	8e33fd97 65de866e 6af1c2ee 0beea1fc 26f7207c 3c9881ef f37876a0 6332d88c
	526f8102 93d21d6e 392c248a 1d2b0d6f 2f8ade54 29420bdb 78bd384c 7ef5a52f
	2249759e 1edef3f3 88f5d67f c53e8e68 f3dcb403 59716aca 1c3d911d 73fb031d
	8cb7b0d3 c3b4a378 02ad1ad5 595859e9 1bd61f51 95e7c275 cc0bfe93 96aee5d2
	69474578 7f8b2488 95fd7676 d1dbd964 50cf6ad6 10869c65 aa1520df 508a4376
	354b27b5 49677f28 5bcd54e3 b4c3aaa9 1225a955 7e630201 3343b6f8 56de4cbd
	af8e227e 4c755675 71c86627 af4ea910 8ecc1d1f 00331169 597d31b5 2028877c
	3904b4c1 03077f11 fe4cf28a 79e41bf3 473083ce af4039ae aa92ac62 2826fc90
	aef29c49 66bfc99c 01421130 d2b6313d 07031652 1862e9d5 fb3715e7 00fc168b
	abc17ac4 c3b1a83c abe59ab6 34e29539 0c51fafa 685aeeb9 c53aa717 c2cb3960
	eae314b8 ba09ef93 bef18bea 59502641 08e31ffc 569ed6aa b3f145f8 d0e82466
	8d2ca851 e6a279c7 474387ea 3d300923 dbbaa193 a0baf928 2668fa60 469ecc14
$s^e \bmod n$	0001ffff ffffffff ffffffff ffffffff ffffffff ffffffff ffffffff ffffffff
	ffffffff ffffffff ffffffff ffffffff ffffffff ffffffff ffffffff ffffffff
	ffffffff ffffffff ffffffff ffffffff ffffffff ffffffff ffffffff ffffffff
	ffffffff ffffffff ffffffff ffffffff ffffffff ffffffff ffffffff ffffffff
	ffffffff ffffffff ffffffff ffffffff ffffffff ffffffff ffffffff ffffffff
	ffffffff ffffffff ffffffff ffffffff ffffffff ffffffff ffffffff ffffffff
	ffffffff ffffffff ffffffff ffffffff ffffffff ffffffff ffffffff ffffffff
	ffffffff ffffffff ffffffff ffffffff ffffffff ffffffff ffffffff ffffffff
	ffffffff ffffffff ffffffff ffffffff ffffffff ffffffff ffffffff ffffffff
	ffffffff ffffffff ffffffff ffffffff ffffffff ffffffff ffffffff ffffffff
	ffffffff ffffffff ffffffff ffffffff ffffffff ffffffff ffffffff 00302130
	0906052b 0e03021a 05000414 f7497dac 551ec010 2f0da8f1 bc8cad52 f93476c3
$\bar{s}$	00000000 00000000 00000000 00000000 00000000 00000000 00000000 00000000
	00000000 00000000 00000000 00000000 00000000 00000000 00000000 00000000
	00000000 00000000 00000000 00000000 00000000 00000000 00000000 00000000
	00000000 00000000 00000000 00000000 00000000 00000000 00000000 00000000
	00000000 00000000 00000000 00000000 00000000 00000000 00000000 00000000
	00000000 00000000 00000000 00000000 00000000 00000000 00000000 00000000
	00000000 00000000 00000000 00000000 00000000 00000000 00000000 00000000
	00000000 00000000 00000000 00000000 00000000 00000000 00000000 00000000
	07ffffff ffffffff ffffffff ffffffff ffffffff ffffffff ffffffff ffffffff
	ffffffff ffffffff ffffffff ffffffff ffffffff ffffffff ffffffff ffffffff
	ffffffff ffffffff ffffffff ffffffff ffffffff fffffffe aaead6ea b6b2b18e
	bd595822 b1555ac6 9f0ca790 717e556a e9678bec fb663c6e a19b4904 00000000

Table 7. (*continued*)

$\bar{s}^e$	0001ffff ffffffff ffffffff ffffffff ffffffff ffffffff ffffffff ffffffff
	ffffffff ffffffff ffffffff ffffffff ffffffff ffffffff ffffffff ffffffff
	ffffffff ffffffff ffffffff ffffffff ffffffff ffffffff ff003021 30090605
	2b0e0302 1a050004 14<u>f7497d</u> <u>ac551ec0</u> <u>102f0da8</u> <u>f1bc8cad</u> <u>52f93476</u> <u>c3</u> 000000
	00000000 00000000 00000000 00000000 00000000 00000000 00000000 00000000
	00000000 00000000 00000000 00000000 2a9aa11c bb60cb35 cb569ddd 576c2729
	34a1298d 905793b0 24ba9a39 7f041398 a7622310 78e8099f 87faed46 0fbb8f46
	67ace20c a1940f81 bced58bf 9ac3671c a2551f73 4cb80ec1 7fffffff ffffffff
	ffffffff fffffffd a285694c d9347ab7 528d15f9 d0dbf0cc 704f592f da3facc6
	210397ee 5d034b6d 269467e8 329d478c 53a8e99d 80f0732a 05d709d4 00e7ada7
	7ddc41a8 e640296f b2a8eae6 f4888211 591f0578 a07d6ec4 f147f08e ccb06340
	4439cb38 fc8144b0 cb0e382b 65583078 a7e9b040 00000000 00000000 00000000

B   Numerical Example of Proposed Forgery with $t = 0$

As an example of the proposed attack with $t = 0$, we show a forged signature on the message $\bar{m}$ ("pkcs-1v2-1.doc" [14]) with $|n| = 1152$ and $e = 3$ in Table 8. Here, <u>underlined octets</u> correspond to the hashed value $H(\bar{m})$ and masked octets correspond to the garbage g. Note that this case succeeds a forgery for 1152-bit composites while the original attack cannot. Also note that a certificate with $|n| = 1152$ and $e = 3$ is used in practice [10].

Table 8. A forged signature by the extended forgery attack (1152-bit, $e = 3$, $t = 0$)

$\bar{m}$	"pkcs-1v2-1.doc" [14]
$H(\bar{m})$	f7497dac 551ec010 2f0da8f1 bc8cad52 f93476c3
$\bar{s}$	00000000 00000000 00000000 00000000 00000000 00000000 00000000 00000000
	00000000 00000000 00000000 00000000 00000000 00000000 00000000 00000000
	00000000 00000000 00000000 00000000 00000000 00000000 00000000 00000000
	07ffffff ffffffff feaaead6 eab6b2b1 8e848b2b fc6229a1 298029f9 27529629
	bb642126 87226bf8 913ab27d 52295002
$\bar{s}^e$	0001ffff ffffffff ffff0030 21300906 052b0e03 021a0500 0414<u>f749</u> <u>7dac551e</u>
	<u>c0102f0d</u> <u>a8f1bc8c</u> <u>ad52f934</u> <u>76c3</u> 0000 008e30ab d25ce35d 65cd0c25 1fc29df3
	37419efd 4d08694d f3b45d86 42970cbe ef3cb225 c0e88433 552da1d0 dc35aaa1
	73f1189f e0b341fc 56d5c5ea 45db5483 15e79d2a 71b6235a 44891287 00bb02f9
	ffabe940 83af15c8 eabb0c30 2fefc008

C   Numerical Example of Proposed Forgery with $t = 160$

As an example of the proposed attack with $t = 160$, we show a forged signature on the hash value $H(\bar{m})$ with $|n| = 4096$ and $e = 17$ in Table 8. Note that this case succeeds a forgery for $e = 17$ while the original attack cannot. However, the adversary cannot obtain the message $\bar{m}$ in practice.

Table 9. A forged signature by the extended forgery attack (4096-bit, $e = 17$)

$H(\bar{m})$	7fa66ee7 e5cc4a9f bd6e13a8 11d298c2 6b9b3302
$\bar{s}$	00000000 00010aa7 58ccbbf7 7d970c35 9e1c3dc0 f20d32ad 2cf9e18a 463ea7c6 346e7f90
$\bar{s}^e$	0001ffff ffffffff ffff0030 21300906 052b0e03 021a0500 04147fa6 6ee7e5cc 4a9fbd6e 13a811d2 98c26b9b 3302 448a 78e5e262 89a4190f 7d18916a 7aaaf897 feeb1e94 5866a030 208c1f48 2c906901 5f70eb66 97253c87 49790ff7 c175fc06 bddf8bb4 d2ba1cdd c626336a dda2165c dc3f425a 12cc59bc be11883e bbccc73a 0d130b94 83ac2a29 19850778 f066ff4f 374e7a96 f4fb3343 fd397d9c f7a1b8ce 16340da6 f9876f1f cca76cb4 7bfb368b a95a5842 e99c0bfb a2de62cf dbf2c635 c2c268f3 2dc228f7 2f0ebfe2 776dae35 3b82b9d9 474777ed c85eed79 e147fa2b 7500f1d4 23189a7b 9b08abb6 0df908f0 7c1c0fbb 528b3e22 df358b24 8bef05b8 f2449d0b f3fb6dc6 31a809ed 31000210 3df7ae2e 80f3f822 ae5a9f69 2948a2b5 a4529bf0 2b30fc99 1874a25f 28b5de4d 4f9c76cc 419a6848 4536e2fe 2771af8b 989e5fef 1a3aaeea f1694ebf 36e8685c 7f65eff8 b99d956b 676b5a5d f68c4519 330b4b7b 82037bd7 502d7823 4e952ba7 b9662cc2 e4389d00 76e16a47 e3dad8af e7f86e37 f164aa90 b377dfbc 9d5cc1a4 e1a966fe 3902fea5 2526240b 99ecf6b3 ced8e16e 2d085131 e5ca1676 25459ca0 0821ff8e 03cde17d 3509de96 cbe40f6f 97d5dd5b b7c977fa be2be4f5 79abcbf7 7093ad52 c346371b 5b2708fc 8b831412 9a023cfc 6b2ff020 105db3ac ef80a605 e3c1ea94 d0af9790 00000000 00000000

Efficient Computation of the Best Quadratic Approximations of Cubic Boolean Functions*

Nicholas Kolokotronis[1,2], Konstantinos Limniotis[1],
and Nicholas Kalouptsidis[1]

[1] Department of Informatics and Telecommunications
National and Kapodistrian University of Athens
TYPA Buildings, University Campus, 15784 Athens, Greece
{nkolok, klimn, kalou}@di.uoa.gr
[2] Department of Computer Science and Technology
University of Peloponnese
End of Karaiskaki Street, 22100 Tripolis, Greece
nkolok@uop.gr

Abstract. The problem of computing best quadratic approximations of a subset of cubic functions with arbitrary number of variables is treated in this paper. We provide methods for their efficient calculation by means of best affine approximations of quadratic functions, for which formulas for their direct computation, without using Walsh-Hadamard transform, are proved. The notion of second-order nonlinearity is introduced as the minimum distance from all quadratic functions. Cubic functions, in the above subset, with maximum second-order nonlinearity are determined, leading to a new lower bound for the covering radius of the second order Reed-Muller code $\Re(2,n)$. Moreover, a preliminary study of the second-order nonlinearity of known bent functions constructions is also given.

Keywords: Boolean functions; bent functions; covering radius; second order nonlinearity; low-order approximations; Reed-Muller codes.

1 Introduction

Boolean functions play a prominent role in the security of cryptosystems. Their most important cryptographic applications are in the analysis and design of s-boxes for block ciphers, as well as, filter/combining functions for stream ciphers [23]. In general, resistance of cryptosystems to various cryptanalytic attacks depends on the properties of the Boolean functions used. The *nonlinearity* of Boolean functions is one of the most significant cryptographic properties; it is defined as the minimum distance from all affine functions, and indicates the degree to which attacks based on linear cryptanalysis [21], and best affine approximations [9], are prevented. For an even number of variables the maximum possible nonlinearity can only be attained by the so-called *bent functions* [25].

---

* This work was partially supported by a Greece–Britain *Joint Research & Technology Programme*, co-funded by the Greek General Secretariat for Research & Technology (GSRT) and the British Council, under contract GSRT 132 − C.

S.D. Galbraith (Eds.): Cryptography and Coding 2007, LNCS 4887, pp. 73–91, 2007.

The appearance of recent attacks, like algebraic [7] and low order approximation attacks [12,16,18], necessitates the design of Boolean functions that cannot be approximated efficiently by low degree functions. This problem has also been studied in [24], where an algorithm to determine good low order approximations by repetitive Hamming sphere sampling was presented. The computation of the best rth order approximations and rth order nonlinearity are known to be difficult tasks, even for the case $r = 2$ [4]; in particular, the second order nonlinearity is unknown, with the exception of some special cases, or when the number of variables is adequately small.

Motivated by the aforementioned need, in this paper we focus on the efficient computation of best quadratic approximations of a certain class of cubic Boolean functions. This led to generalizing notions that are familiar from best affine approximations, e.g. nonlinearity is extended to *2nd order nonlinearity*, defined as the minimum distance from quadratic functions. Explicit formulas are proved that compute all best affine (resp. quadratic) approximations of the quadratic (resp. cubic) functions, without use of the Walsh-Hadamard transform, by applying the Shannon expansion formula. The results obtained are general, and most importantly, they hold for an arbitrary number of variables; they focus on the theoretical aspects of the subject rather than the algorithmic ones. Cubic functions with maximum (within the subset considered in this paper) second order nonlinearity are determined, yielding a lower bound on the covering radius of $\mathfrak{R}(2, n)$. It is shown that their structure is similar to that of quadratic bent functions. Finally, a preliminary analysis of the second-order nonlinearity for known constructions of bent functions is given, indicating potential weaknesses when parameters are not properly chosen. Since several constructions of cryptographic primitives, based on quadratic and cubic functions, have been proposed in the literature (see e.g. [1] and [11] respectively) due to their efficient implementation, our results are of high cryptographic value when such functions need to be applied.

The paper is organized as follows: Section 2 provides the background and introduces the notation. The properties of best affine approximations of quadratic functions are treated in Section 3, whereas Section 4 studies best quadratic approximations of cubic Boolean functions and determines efficient ways for their computation. Constructions of bent functions are analyzed in Section 5, and Section 6 summarizes our conclusions.

2   Preliminaries

Let $f : \mathbb{F}_2^n \to \mathbb{F}_2$ be a Boolean function, where $\mathbb{F}_2 = \{0, 1\}$ is the binary field. The set of Boolean functions on n variables is denoted by $\mathbb{B}_n$; they are commonly expressed in their *algebraic normal form* (ANF) as

$$f(x_1, \ldots x_n) = \sum_{j \in \mathbb{F}_2^n} a_j \, x_1^{j_1} \cdots x_n^{j_n} \,, \qquad a_j \in \mathbb{F}_2 \tag{1}$$

where $j = (j_1, \ldots, j_n)$ and the sum is performed modulo 2. The *algebraic degree* of function f is defined as $\deg(f) = \max\{\mathrm{wt}(j) : a_j = 1\}$, where $\mathrm{wt}(j)$ is the

Hamming weight of vector j. The terms of degree $k \leq \deg(f)$ that appear in
(1) are called the *kth degree part* of $f \in \mathbb{B}_n$; its truth table is the binary vector
$f = (f(0,0,\ldots,0), f(1,0,\ldots,0),\ldots, f(1,1,\ldots,1))$ of length 2^n, also denoted by
f for simplicity. If $\deg(f) \leq r$, it is well-known that f is codeword of the rth
order binary Reed-Muller code $\mathfrak{R}(r,n)$ [20]. The Boolean function $f \in \mathbb{B}_n$ is
said to be *balanced* if $\mathrm{wt}(f) = 2^{n-1}$, and its *Hamming distance* from $g \in \mathbb{B}_n$ is
defined as $\mathrm{wt}(f+g)$. Any Boolean function admits the following decomposition.

Definition 1. *Let $j_1,\ldots,j_k$ be integers such that $1 \leq j_1 < \cdots < j_k \leq n$ and
$k < n$. Further, let $r = r_1 + 2r_2 + \cdots + 2^{k-1}r_k$ be the binary expansion of the
integer $0 \leq r < 2^k$. The expression*

$$f(x_1,\ldots,x_n) = \sum_{r=0}^{2^k-1} \left(\prod_{i=1}^{k}(x_{j_i} + \bar{r}_i) \right) f_r \tag{2}$$

*where $\bar{r}_i = r_i + 1$ denotes the complement of r_i, and each $f_r \in \mathbb{B}_{n-k}$ does not
depend on $x_{j_1},\ldots,x_{j_k}$, is called the kth order Shannon's expansion formula of
f with respect to the variables $x_{j_1},\ldots,x_{j_k}$ [17].*

The *sub-functions* $f_0,\ldots,f_{2^k-1}$ in (2) are uniquely determined from f by setting
$x_{j_i} = r_i$, $1 \leq i \leq k$. Subsequently, we write $f = f_0 \|_{\mathcal{J}} \cdots \|_{\mathcal{J}} f_{2^k-1}$ instead of (2),
where $\mathcal{J} = \{j_1,\ldots,j_k\}$, and $f = f_0 \|_j f_1$ when $\mathcal{J} = \{j\}$. If $\mathcal{J} = \{n-k+1,\ldots,n\}$,
the truth table of $f(x_1,\ldots,x_n)$ is constructed by *concatenating* the truth tables
of the sub-functions $f_r(x_1,\ldots,x_{n-k})$; this case will be denoted by $f = f_0 \| \cdots \|$
f_{2^k-1} and simply referred to as the *kth order Shannon's expansion formula* of f.

The Walsh-Hadamard transform $\widehat{\chi}_f(a)$ of the Boolean function $f \in \mathbb{B}_n$ at
$a \in \mathbb{F}_2^n$ is the real-valued function given by [20]

$$\widehat{\chi}_f(a) = \sum_{x \in \mathbb{F}_2^n} \chi_f(x)(-1)^{\langle a,x \rangle} = 2^n - 2\,\mathrm{wt}(f + \phi_a) \tag{3}$$

where $\chi_f(x) = (-1)^{f(x)}$ and $\phi_a(x) = \langle a,x \rangle = a_1 x_1 + \cdots + a_n x_n$. Clearly f is
balanced if and only if $\widehat{\chi}_f(0) = 0$. The *nonlinearity* of f is its minimum distance
from all affine functions, and is determined by

$$\mathcal{NL}_f = \min_{v \in \mathfrak{R}(1,n)} \{\mathrm{wt}(f+v)\} = 2^{n-1} - \frac{1}{2} \max_{a \in \mathbb{F}_2^n} |\widehat{\chi}_f(a)|. \tag{4}$$

Any affine function v such that $\mathrm{wt}(f+v) = \mathcal{NL}_f$ is called a *best affine approximation* of f and is denoted by λ_f, whereas the set comprising of the best
affine approximations of f is denoted by $\mathcal{A}_f \subseteq \mathfrak{R}(1,n)$. The definition of the
nonlinearity leads directly to the following well-known result [4].

Lemma 1. *Let $f \in \mathbb{B}_n$ and $v \in \mathfrak{R}(1,n)$. Then, $\lambda \in \mathcal{A}_f$ if and only if $\lambda + v \in
\mathcal{A}_{f+v}$. Further, $|\mathcal{A}_{f+v}| = |\mathcal{A}_f|$, i.e. both sets have the same cardinality.*

An equivalent statement of Lemma 1, which is subsequently used, is that $\lambda_{f+v} =
\lambda_f + v$ for any linear function v. Likewise, the minimum distance between f and

all quadratic functions is called *second-order nonlinearity* of f and is denoted by $\mathcal{N}\mathcal{Q}_f$; it is given by

$$\mathcal{N}\mathcal{Q}_f = \min_{u \in \mathfrak{R}(2,n)} \left\{ \mathrm{wt}(f + u) \right\}. \tag{5}$$

From (4), (5) we clearly obtain that $\mathcal{N}\mathcal{Q}_f \leq \mathcal{N}\mathcal{L}_f$. Likewise, any quadratic function u with the property $\mathrm{wt}(f + u) = \mathcal{N}\mathcal{Q}_f$ is said to be a *best quadratic approximation* of f and is denoted by ξ_f, whereas the set comprising of best quadratic approximations of f is denoted by $\mathcal{Q}_f \subseteq \mathfrak{R}(2,n)$.

3   Best Affine Approximations of Quadratic Functions

Let $f \in \mathbb{B}_n$ be a quadratic function and $x = (x_1, \ldots, x_n)$; then, f can be written as $f = xQx^T + Lx^T + \epsilon$ for some upper triangular binary matrix Q, binary vector L, and a constant $\epsilon \in \mathbb{F}_2$, where xQx^T is the quadratic part of f. It is well-known that (*see* e.g. [20, pp. 434–442]) the rank of the symplectic matrix $Q + Q^T$ equals $2h$, for some $1 \leq h \leq \lfloor n/2 \rfloor$. Then, by Dickson's theorem there exists a nonsingular matrix $R = (r_{i,j})_{i,j=1}^n$ such that under the transformation of variables $g = xR$, function f becomes

$$f = g_0 + \sum_{i=1}^h g_{2i-1}g_{2i}, \qquad \deg(g_0) \leq 1 \text{ and } \deg(g_j) = 1 \tag{6}$$

where $g_0 = \tilde{g} + L(R^{-1})^T g^T + \epsilon$, for some linear function $\tilde{g}$ obtained from the quadratic part of f, and $\{g_1, \ldots, g_{2h}\}$ are linearly independent linear functions (actually, we have $g_j = \sum_{i=1}^n r_{i,j}x_i$). Since h only depends on the quadratic part of function f, it is denoted by h_f; by convention $h_f = 0$ if $f \in \mathfrak{R}(1,n)$. As seen below, the Walsh spectra of f are fully determined by the value of h_f, as well as its nonlinearity.

Theorem 1 ([20]). *Let* $\mathcal{B}_f = \{f + v : v \in \mathfrak{R}(1,n)\}$ *for a fixed quadratic Boolean function* $f \in \mathfrak{R}(2,n)$. *Then* $\mathrm{wt}(f + v)$ *is three-valued*

$$\mathrm{wt}(f + v) \in \left\{ 2^{n-1}, 2^{n-1} \pm 2^{n-h_f-1} \right\}$$

where each value $2^{n-1} \pm 2^{n-h_f-1}$ *occurs* 2^{2h_f} *times, and* 2^{n-1} *occurs the remaining* $2^{n+1} - 2^{2h_f+1}$ *times.*

According to (4), the nonlinearity of any quadratic function $f \in \mathbb{B}_n$ equals $2^{n-1} - 2^{n-h_f-1}$. The following statement allows the direct computation of all best affine approximations of a quadratic function.

Theorem 2. *Let the Boolean function* $f \in \mathfrak{R}(2,n)$ *be given by (6). Then, for* $b = (b_1, \ldots, b_{2h})$, *its best affine approximations are given by*

$$\mathcal{A}_f = \left\{ \lambda_f^b \in \mathfrak{R}(1,n) : \lambda_f^b = g_0 + \sum_{i=1}^{2h} b_i g_i + \sum_{i=1}^h b_{2i-1}b_{2i}, \ b \in \mathbb{F}_2^{2h} \right\}.$$

Proof. First note that the affine Boolean functions λ_f^b are pairwise distinct since from hypothesis we have that $\{g_1, \ldots, g_{2h}\}$ are linearly independent; thus $|\mathcal{A}_f| = 2^{2h}$. Furthermore, for all $b \in \mathbb{F}_2^{2h}$, we have that

$$f + \lambda_f^b = \sum_{i=1}^{h} \left(g_{2i-1}g_{2i} + b_{2i-1}g_{2i-1} + b_{2i}g_{2i} + b_{2i-1}b_{2i}\right)$$

$$= \sum_{i=1}^{h} \left(g_{2i-1} + b_{2i}\right)\left(g_{2i} + b_{2i-1}\right). \tag{7}$$

Since $\{g_1 + b_2, \ldots, g_{2h} + b_{2h-1}\}$ are also linearly independent, we get that for any choice of $b \in \mathbb{F}_2^{2h}$ it holds $\mathrm{wt}(f + \lambda_f^b) = 2^{n-1} - 2^{n-1-h}$ [20], which by Theorem 1 is the minimum distance between f and all affine functions. The fact that the number of best affine approximations of f is 2^{2h} (equal to the number of different λ_f^b constructed here) concludes our proof.   □

Example 1. Let $f \in \mathbb{B}_5$ be the quadratic function given by $f(x_1, \ldots, x_5) = (x_1 + x_3)(x_2 + x_5) + x_2 + x_4$. Its best affine approximations are

$$\lambda_f^0 = x_2 + x_4 \qquad\qquad \lambda_f^1 = x_1 + x_2 + x_3 + x_4$$
$$\lambda_f^2 = x_4 + x_5 \qquad\qquad \lambda_f^3 = x_1 + x_3 + x_4 + x_5 + 1$$

by Theorem 2. Note that one of the solutions is the linear part of f.   □

The complexity of determining $\mathcal{A}_f$ by Theorem 2 is $\mathcal{O}(n^3)$ and rests with finding (6) [19,20]; thus, the above method is more efficient than the application of the fast Walsh transform, whose complexity is $\mathcal{O}(n2^n)$, for computing all the best affine approximations.

Proposition 1. *Let the Boolean functions $f_1, \ldots, f_s \in \mathfrak{R}(2, n)$ be given by (6) and $\bigcup_{i=1}^{s}\{g_{i,1}, \ldots, g_{i,2h_{f_i}}\}$ be linearly independent, $s \geq 2$. Further, let $b_i = (b_{i,1}, \ldots, b_{i,2h_{f_i}})$ be binary vectors of length $2h_{f_i}$, $1 \leq i \leq s$. Then*

$$\lambda_{f_1 + \cdots + f_s}^b = \lambda_{f_1}^{b_1} + \cdots + \lambda_{f_s}^{b_s}, \quad \forall\, b = (b_1, \ldots, b_s) \in \mathbb{F}_2^{2h_{f_1} + \cdots + 2h_{f_s}}. \tag{8}$$

Proof. Clearly, (8) holds for $s = 1$; we will only prove its validity for $s = 2$, since the general case is then established by induction on s. Let us write $f = f_1 + f_2$, and denote h_f, h_{f_i} as h, h_i respectively. By hypothesis we get $f_i = g_{i,0} + \sum_{j=1}^{h_i} g_{i,2j-1}g_{i,2j}$; since $\bigcup_{i=1}^{2}\{g_{i,1}, \ldots, g_{i,2h_i}\}$ are linearly independent we obtain $h = h_1 + h_2$, and from Theorem 2 the best affine approximations of f, for all $b \in \mathbb{F}_2^{2h}$, are given by

$$\lambda_f^b = g_{1,0} + g_{2,0} + \sum_{j=1}^{2h_1} b_j\, g_{1,j} + \sum_{j=1}^{2h_2} b_{2h_1+j}\, g_{2,j} + \sum_{i=1}^{h} b_{2i-1}b_{2i} = \lambda_{f_1}^{c_1} + \lambda_{f_2}^{c_2}$$

where $c_1 = (b_1, \ldots, b_{2h_1}) \in \mathbb{F}_2^{2h_1}$, $c_2 = (b_{2h_1+1}, \ldots, b_{2h_1+2h_2}) \in \mathbb{F}_2^{2h_2}$.   □

4   Best Quadratic Approximations of Cubic Functions

In this section we develop a detailed theoretical framework for efficiently computing the best quadratic approximation of cubic Boolean functions. First, we need to introduce the following classification.

Definition 2. *The Boolean function $f \in \mathfrak{R}(3, n)$ is said to be a class-m function if there exists a set $\mathcal{J} = \{j_1, \ldots, j_m\}$ with $1 \leq j_1 < \cdots < j_m \leq n$ such that each cubic term of $f(Ax + b)$ involves at least one variable with index in $\mathcal{J}$, and $m > 0$ is the smallest integer that is obtained by a suitable invertible affine transformation of f.*

The above definition implies that a cubic function belongs to exactly one class, due to the minimality of m (however $\mathcal{J}$ is not always unique). An important subset of class-m functions are the *separable class-m functions* whose cubic terms involve *exactly one* variable with index in $\mathcal{J}$; all others are referred to as *inseparable*. The equivalence classes for cubic functions of certain weight found in [13,14] are separable. An important property of class-m cubic Boolean functions is given below.

Proposition 2. *All class-m cubic Boolean functions $f \in \mathfrak{R}(3, n)$, with $\mathcal{J} = \{j_1, \ldots, j_m\}$, admit the following properties*

1. *Let $\mathcal{J}' \subset \mathcal{J}$ with cardinality k, $1 \leq k \leq m - 1$. From the decomposition $f = f_0 \|_{\mathcal{J}'} \cdots \|_{\mathcal{J}'} f_{2^k - 1}$ we have that all $f_i \in \mathbb{B}_{n-k}$ are class-$(m - k)$ cubic Boolean functions with the same cubic part;*
2. *Moreover, m is the least integer such that $f = f_0 \|_{\mathcal{J}} \cdots \|_{\mathcal{J}} f_{2^m - 1}$ with $\deg(f_i) < \deg(f) = 3$ for all $0 \leq i < 2^m$.*

Proof. The proof is provided in Appendix A.   □

Proposition 2 leads to an alternative definition of class-m cubic functions; that is, $m > 0$ is the least integer such that a proper choice of m variables leads to a decrease in the degree of sub-functions, if we apply mth order Shannon expansion with respect to these variables. Next we focus on the separable cubic functions, and prove a series of results to determine their best quadratic approximations.

Lemma 2. *Let $f \in \mathfrak{R}(3, n)$ be a separable class-m function, with cubic part $c = \sum_{i=1}^{m} x_{j_i} q_i$, where $q_i \in \mathbb{B}_{n-m}$ is quadratic function not depending on variables with index in $\mathcal{J} = \{j_1, \ldots, j_m\}$. Then, from the decomposition $f = f_0 \|_{\mathcal{J}} \cdots \|_{\mathcal{J}} f_{2^m - 1}$ we get that*

$$f_r = q + \langle r, p \rangle + l_r, \qquad 0 \leq r < 2^m \tag{9}$$

for a quadratic $q \in \mathbb{B}_{n-m}$ and affine $l_r \in \mathbb{B}_{n-m}$ Boolean functions, where $r = (r_1, \ldots, r_m)$ is the binary representation of r, and $p = (q_1, \ldots, q_m)$.

Proof. The Boolean function is written as $f = c + q + l$, where c, q, and l is its cubic, quadratic, and linear part respectively. By hypothesis, f is a class-m cubic Boolean function, and therefore according to Definition 2 we necessarily have that $q_1, \ldots, q_m \neq 0$ and linearly independent. Indeed, let us assume that there exist $a_1, \ldots, a_m \in \mathbb{F}_2$, not all of them being zero, such that $a_1 q_1 + \cdots + a_m q_m = 0$; without loss of generality let $a_m = 1$. Therefore, we have $c = (x_{j_1} + a_1 x_{j_m}) q_1 + \cdots + (x_{j_{m-1}} + a_{m-1} x_{j_m}) q_{m-1}$, and there exists an invertible linear transformation (mapping $x_{j_i} + a_i x_{j_m}$ to x_{j_i}, for $1 \leq i < m$, and all the remaining variables to themselves) such that f become class-$(m-1)$ cubic Boolean function—contradiction. The quadratic and linear parts of f can be similarly written as

$$q = \sum_{i=1}^{m-1} \sum_{k=i+1}^{m} x_{j_i} x_{j_k} \epsilon_{i,k} + \sum_{i=1}^{m} x_{j_i} l_i' + q' \quad \text{and} \quad l = \sum_{i=1}^{m} x_{j_i} \epsilon_i + l' \quad (10)$$

for some quadratic $q' \in \mathbb{B}_{n-m}$ and linear functions $l', l_i' \in \mathbb{B}_{n-m}$ that do not depend on $x_{j_1}, \ldots, x_{j_m}$. Let us next introduce the auxiliary functions

$$g^s = \left(\sum_{i=1}^{s} x_{j_i} q_i \right) + \left(\sum_{i=1}^{s-1} \sum_{k=i+1}^{s} x_{j_i} x_{j_k} \epsilon_{i,k} + \sum_{i=1}^{s} x_{j_i} l_i' + q' \right) + \left(\sum_{i=1}^{s} x_{j_i} \epsilon_i + l' \right)$$

where the parentheses are used to indicate its cubic, quadratic, and linear parts respectively (note that $g^m = f$ whereas $g^0 = q' + l'$), and

$$h_i^s = \sum_{k=1}^{s} x_{j_k} \epsilon_{k,i} + \sum_{k=i+1}^{m} r_k \epsilon_{i,k} , \qquad 0 \leq s < i \leq m$$

where $r_k \in \mathbb{F}_2$. By applying Shannon's expansion formula recursively, we obtain at the first step $f = f_0 \parallel_{j_m} f_1$, where $f_{r_m} = g^{m-1} + r_m(q_m + l_m' + \epsilon_m + h_m^{m-1})$ for $r_m = 0, 1$. Further expansion of these sub-functions gives $f = (f_0 \parallel_{j_{m-1}} f_1) \parallel_{j_m} (f_2 \parallel_{j_{m-1}} f_3)$, where

$$f_r = g^{m-2} + \sum_{i=m-1}^{m} r_i(q_i + l_i' + \epsilon_i + h_i^{m-2}) , \qquad 0 \leq r < 4$$

and $r = r_{m-1} + 2r_m$ is the binary expansion of r. If we continue this way, we get the decomposition $f = f_0 \parallel_{\exists} \cdots \parallel_{\exists} f_{2^m-1}$ after $m-2$ steps, which for all $0 \leq r < 2^m$ leads to

$$f_r = q' + \sum_{i=1}^{m} r_i q_i + \left(l' + \sum_{i=1}^{m} r_i (l_i' + \epsilon_i + \sum_{k=i+1}^{m} r_k \epsilon_{i,k}) \right) \quad (11)$$

and $r = r_1 + 2r_2 + \cdots + 2^{m-1} r_m$ is the binary expansion of r. The claim is proved by noting that the expression inside the parentheses corresponds to l_r in (9), q' corresponds to q, and $\langle r, p \rangle = \sum_{i=1}^{m} r_i q_i$. $\square$

We next introduce a commonly used partial ordering of elements of the vector space $\mathbb{F}_2^n$. For all $a, b \in \mathbb{F}_2^n$ we write $a \preceq b$ if and only if $a_i \leq b_i$ for all $1 \leq i \leq n$.

Lemma 3. *Let $f = f_0 \| \cdots \| f_{2^m-1}$ be a Boolean function in $\mathbb{B}_n$ where $m > 0$ and all sub-functions $f_r \in \mathfrak{R}(2, n-m)$ have the same quadratic part $f_r = q + l_r$, for $0 \leq r < 2^m$. Then, $\deg(f) = 2$ if and only if we have $\sum_{r \preceq c} l_r = \epsilon_c$ for some $\epsilon_c \in \mathbb{F}_2$ if $\mathrm{wt}(c) = 2$ and zero if $\mathrm{wt}(c) \geq 3$, $c \in \mathbb{F}_2^m$.*

Proof. The proof is provided in Appendix B; however, it is seen that it also holds for the case $f = f_0 \|_\mathfrak{J} \cdots \|_\mathfrak{J} f_{2^m-1}$ with $\mathfrak{J} = \{j_1, \ldots, j_m\}$. It should be noted that the family of affine functions $l_r \in \mathbb{B}_{n-m}$ introduced in (11) satisfies the conditions implied by this Lemma. □

Lemma 4. *For all $s \geq 1$ and vectors $a = (a_1, \ldots, a_s) \in \mathbb{Z}^s$ we have that*

$$\sum_{r \in \mathbb{F}_2^s} 2^{\langle r, a \rangle} = \prod_{i=1}^{s} \left(1 + 2^{a_i}\right). \tag{12}$$

Proof. Note that (12) holds for $s = 1$; suppose that it holds for $s = k \geq 1$ and all $a = (a_1, \ldots, a_k) \in \mathbb{Z}^k$. Then, for $s = k+1$ we have from (12) that

$$\sum_{(r,t) \in \mathbb{F}_2^{k+1}} 2^{\langle (r,t),(a,b) \rangle} = \left(1 + 2^b\right) \sum_{r \in \mathbb{F}_2^k} 2^{\langle r, a \rangle} = \left(1 + 2^b\right) \prod_{i=1}^{k} \left(1 + 2^{a_i}\right)$$

by the induction hypothesis, for all $(a, b) = (a_1, \ldots, a_k, b) \in \mathbb{Z}^{k+1}$. □

Subsequently, we present the main result of our paper.

Theorem 3. *With the notation of Lemma 2, assume $f \in \mathbb{B}_n$ is a class-m cubic function, and let $q_i \in \mathbb{B}_{n-m}$ be given by (6). If all linear functions in $\bigcup_{i=1}^{m} \{ g_{i,1}, \ldots, g_{i,2h_{q_i}} \}$ are linearly independent, then the best quadratic approximations of f have one of the following forms*

$$\xi_f^s = \xi_{f,0}^s \|_\mathfrak{J} \cdots \|_\mathfrak{J} \xi_{f,2^m-1}^s, \qquad s \in \mathbb{F}_2^m \tag{13}$$

where $\xi_{f,r}^s = q + \langle s, p \rangle + l_r + \lambda_{\langle r+s, p \rangle}$ for all $r \in \mathbb{F}_2^m$.

Proof. For fixed $s \in \mathbb{F}_2^m$ the sub-functions in (13) have the same quadratic part $q + \langle s, p \rangle$ and yield a quadratic function ξ_f^s by Lemma 3. Indeed, by Proposition 1 we have $\lambda_{\langle r+s, p \rangle} = \sum_{i=1}^{m} (r_i + s_i) \lambda_{q_i}$, whereas (11) leads to

$$\sum_{r \preceq c} \xi_{f,r}^s = 2^{\mathrm{wt}(c)} \left(q + \langle s, p \rangle + l'\right) + 2^{\mathrm{wt}(c)-1} \sum_{i=1}^{m} \left(c_i \left(l_i' + \epsilon_i\right) + a_i \lambda_{q_i}\right)$$

$$+ 2^{\mathrm{wt}(c)-2} \sum_{i=1}^{m-1} \sum_{j=i+1}^{m} c_i c_j \epsilon_{i,j} = \begin{cases} \epsilon_{u,v}, & \text{if } \mathrm{wt}(c) = 2 \ (c_u = c_v = 1) \\ 0, & \text{if } \mathrm{wt}(c) > 2 \end{cases}$$

since $\sum_{r \prec c}(r_i + s_i)\lambda_{q_i}$ equals $2^{\mathrm{wt}(c)}s_i\lambda_{q_i}$ if $c_i = 0$ (in which case $a_i = 2s_i$), and $2^{\mathrm{wt}(c)-1}\lambda_{q_i}$ otherwise (where $a_i = 1$). Thus ξ_f^s are quadratic functions for all $s \in \mathbb{F}_2^m$; their distance from f is given by

$$\mathrm{wt}(f + \xi_f^s) = \sum_{r \in \mathbb{F}_2^m \setminus \{s\}} \mathrm{wt}(\langle r + s, p \rangle + \lambda_{\langle r+s, p \rangle}) = \sum_{r \in \mathbb{F}_2^m \setminus \{0\}} \mathrm{wt}(\langle r, p \rangle + \lambda_{\langle r, p \rangle})$$

which is independent of s, and therefore all ξ_f^s, $s \in \mathbb{F}_2^m$, are equidistant from f. By Theorem 1, the definition of nonlinearity, and $\mathcal{NL}_0 = h_0 = 0$ by convention, we have $1 \leq h_{\langle r, p \rangle} \leq \lfloor (n - m)/2 \rfloor$ for $r \neq 0$ and

$$\mathrm{wt}(f + \xi_f^s) = 2^{n-1} - \sum_{r \in \mathbb{F}_2^m} 2^{n-m-1-h_{\langle r, p \rangle}} = 2^{n-1} - 2^{n-1} \prod_{i=1}^{m} \left(\frac{1}{2} + \frac{1}{2^{h_{q_i}+1}} \right)$$

since $\bigcup_{i=1}^{m} \{ g_{i,1}, \ldots, g_{i,2h_{q_i}} \}$ are linearly independent by hypothesis, and thus $h_{\langle r, p \rangle} = h_{r_1 q_1 + \cdots + r_m q_m} = r_1 h_{q_1} + \cdots + r_m h_{q_m}$ from Proposition 1 (whereas the last equality is derived from Lemma 4). Next, we prove that the distance of any other quadratic function from f is greater than $\mathrm{wt}(f + \xi_f^s)$. Assume there exists $u \in \mathfrak{R}(2, n)$, which does not coincide with ξ_f^s for any $s \in \mathbb{F}_2^m$, and $u = (q' + l'_0) \|_\partial \cdots \|_\partial (q' + l'_{2^m - 1})$ where l'_r satisfy the condition of Lemma 3. Consequently $\tilde{q} = q' + q, q_1, \ldots, q_m$ are linearly independent, otherwise u would be identical to one of ξ_f^s; indeed, if $q' = q + \langle s, p \rangle$ for some $s \in \mathbb{F}_2^m$ then we would have

$$\mathrm{wt}(f + u) = \sum_{r \in \mathbb{F}_2^m \setminus \{s\}} \mathrm{wt}(\langle r + s, p \rangle + l'_r + l_r)$$

$$\geq \sum_{r \in \mathbb{F}_2^m \setminus \{s\}} \mathrm{wt}(\langle r + s, p \rangle + \lambda_{\langle r+s, p \rangle}) = \mathrm{wt}(f + \xi_f^s)$$

where equality holds if and only if we set $l'_r = l_r + \lambda_{\langle r+s, p \rangle}$, for all $r \in \mathbb{F}_2^m$ by Lemma 1. Thus, in order to minimize $\mathrm{wt}(f + u)$ we get $u = \xi_f^s$. Hence we only consider $q' \neq q + \langle s, p \rangle$ for all $s \in \mathbb{F}_2^m$. Then, we likewise find that

$$\mathrm{wt}(f + u) = \sum_{r \in \mathbb{F}_2^m} \mathrm{wt}(\tilde{q} + \langle r, p \rangle + \tilde{l}_r) \geq 2^{n-1} - \sum_{r \in \mathbb{F}_2^m} 2^{n-m-1-h_{\tilde{q}+\langle r, p \rangle}}$$

where $\tilde{l}_r = l'_r + l_r$ and equality holds if and only if $\tilde{l}_r = \lambda_{\tilde{q}+\langle r, p \rangle}$, for all $r \in \mathbb{F}_2^m$ by Lemma 1. The weight of $f + u$ is minimized if $\tilde{q}$ is chosen such that all $h_{\tilde{q}+\langle r, p \rangle}$ take their minimum possible value. The fact $\tilde{q} + \langle r, p \rangle \neq 0$ implies that $h_{\tilde{q}+\langle r, p \rangle} \geq 1$ for all $r \in \mathbb{F}_2^m$; still, not all $h_{\tilde{q}+\langle r, p \rangle}$ can be made simultaneously equal to 1. We write $\tilde{q}$ as $\tilde{q} = \tilde{g}_0 + \sum_{1 \leq j \leq h_{\tilde{q}}} \tilde{g}_{2j-1}\tilde{g}_{2j}$, where the linear part $\tilde{g}_0$ is obtained by applying Dickson's theorem on $\tilde{q}$. Let us define d_i as the number of $\tilde{g}_j$ that are not linearly independent from $g_{i,j}$ of q_i, and e_i as the number of products $\tilde{g}_{2j-1}\tilde{g}_{2j}$ shared by $\tilde{q}, q_i, 1 \leq i \leq m$. It is then seen that $0 \leq d_i \leq 2\min\{h_{\tilde{q}}, h_{q_i}\}$ and $0 \leq e_i \leq \lfloor d_i/2 \rfloor$. Further, it can be proved that $h_{\tilde{q}+q_i} \geq h_{\tilde{q}} + h_{q_i} - d_i$ and the lower bound is always attained[1] if d_i is even and $e_i = d_i/2$. Thus $\mathrm{wt}(f+u)$ can be minimized by letting $\tilde{q}$ have common products $\tilde{g}_{2j-1}\tilde{g}_{2j}$ with $q_1, \ldots, q_m$. By hypothesis $\bigcup_{i=1}^m \{ g_{i,1}, \ldots, g_{i,2h_{q_i}} \}$ are linearly independent, leading to (for all $r \in \mathbb{F}_2^m$)

$$h_{\tilde{q}+\langle r,p \rangle} = h_{\tilde{q}} + \sum_{i=1}^m r_i (h_{q_i} - 2e_i) \quad \text{and} \quad 0 \le \sum_{i=1}^m r_i e_i \le \min \left\{ h_{\tilde{q}}, \sum_{i=1}^m r_i h_{q_i} \right\}.$$

Hence, we see that $\mathrm{wt}(f+u) \ge 2^{n-1} - 2^{n-1-h_{\tilde{q}}} \prod_{i=1}^m (1/2 + 1/2^{h_{q_i}-2e_i+1})$ and we need to examine if there exists a particular choice of parameters $e_1, \ldots, e_m$ such that $\mathrm{wt}(f+u) < \mathrm{wt}(f+\xi_f^s)$, which is equivalent to

$$\mathrm{wt}(f+u) < \mathrm{wt}(f+\xi_f^s) \iff 2^{h_{\tilde{q}}-(e_1+\cdots+e_m)} \prod_{i=1}^m \frac{1+2^{h_{q_i}}}{2^{e_i}+2^{h_{q_i}-e_i}} < 1. \tag{14}$$

Since it holds $0 \le e_i \le \min\{h_{\tilde{q}}, h_{q_i}\}$, all terms $(2^{h_{q_i}}+1)(2^{h_{q_i}-e_i}+2^{e_i})^{-1}$ in (14) are greater than or equal to 1, where equality is attained if either $e_i = 0$ or $e_i = h_{q_i} = \min\{h_{\tilde{q}}, h_{q_i}\}$; the latter case is valid for all $1 \le i \le m$ only if $h_{\tilde{q}} \ge \max\{h_{q_1}, \ldots, h_{q_m}\}$. Moreover, $e_i = 0$ implies that $\tilde{q}$ does not have common products with q_i, whereas $e_i = h_{q_i}$ that $\tilde{q}$ is written as the sum of q_i and another quadratic function. Since by hypothesis $\tilde{q} \ne \langle r,p \rangle$, we either have $0 < e_i < \min\{h_{\tilde{q}}, h_{q_i}\}$ for some $1 \le i \le m$, or that $\tilde{q}$ has a product whose functions do not depend on $\bigcup_{i=1}^m \{ g_{i,1}, \ldots, g_{i,2h_{q_i}} \}$, hence $2^{h_{\tilde{q}}-(e_1+\cdots+e_m)} > 1$, due to $0 < e_1 + \cdots + e_m < \min\{h_{\tilde{q}}, h_{q_1} + \cdots + h_{q_m}\}$. Therefore, in any case we get $\mathrm{wt}(f+u) > \mathrm{wt}(f+\xi_f^s)$.   □

By assuming that all $\bigcup_{i=1}^m \{ g_{i,1}, \ldots, g_{i,2h_{q_i}} \}$ are linearly independent (see Remark 1 below how such a constraint can be relaxed) and $h_{q_i} \ge 1$, for $1 \le i \le m$, the fact $h_{q_1} + \cdots + h_{q_m} = h_{q_1+\cdots+q_m} \le \lfloor (n-m)/2 \rfloor$ leads to $h_{q_i} \le \lfloor (n-3m)/2 \rfloor + 1$. Thus, we have the following result.

Corollary 1. *With the notation of Theorem 3, for any separable class-m cubic function $f \in \mathfrak{R}(3,n)$ we have $m \le \lfloor n/3 \rfloor$ and*

$$\mathfrak{NQ}_f = 2^{n-1} - 2^{n-1} \prod_{i=1}^m \left(\frac{1}{2} + \frac{1}{2^{h_{q_i}+1}} \right) \tag{15}$$

for some $1 \le h_{q_i} \le \lfloor (n-3m)/2 \rfloor + 1$.

Remark 1. The applicability of the results in Theorem 3 and Corollary 1 can be more general than currently stated. In particular, let us consider an $m \times m$ nonsingular matrix $P = (p_{i,j})_{i,j=1}^m$, $Q = (q_1, \ldots, q_m)$, and vector $Q' = (q_1', \ldots, q_m')$

---

[1] From the above, $\tilde{q} + q_i$ has $h_{\tilde{q}} + h_{q_i} - 2e_i$ products (excluding common ones) of the form $g_{2j-1}'g_{2j}'$, where $d_i - 2e_i$ of the terms g_j' are not linearly independent (l.i.); they can be chosen (from the $d_i - 2e_i$ l.i. "parity check" equations) so that they reside in different products, forming u'. Thus $h_{\tilde{q}+q_i} = h_{\tilde{q}} + h_{q_i} - 2e_i - (d_i - 2e_i) + h_{u'}$, where $0 \le h_{u'} \le \lfloor d_i/2 \rfloor - e_i$, leading to $h_{\tilde{q}} + h_{q_i} - d_i \le h_{\tilde{q}+q_i} \le h_{\tilde{q}} + h_{q_i} - e_i - \lceil d_i/2 \rceil$.

such that $Q' = QP$. Then, from the linear independence of $\bigcup_{i=1}^{m} \{ g_{i,1}, \ldots, g_{i,2h_{q_i}} \}$ we get $h_{q'_j} = \sum_{i=1}^{m} p_{i,j} h_{q_i}$, and

$$h_{r_1 q'_1 + \cdots + r_m q'_m} = \sum_{i=1}^{m} \left(\sum_{j=1}^{m} r_j p_{i,j} \right) h_{q_i} = \sum_{i=1}^{m} r'_i h_{q_i} = h_{r'_1 q_1 + \cdots + r'_m q_m}$$

where $r'_i = \sum_{j=1}^{m} r_j p_{i,j} \bmod 2$, for $1 \le i \le m$. Hence, the results obtained in Theorem 3 and Corollary 1 would still hold in this case, if we replace $h_{q'_i}$ with the ith element of vector $Q' P^{-1}$.

In the simple case of the class-1 Boolean functions, Corollary 1 gives that $\mathcal{N}\mathcal{Q}_f = 2^{n-2} - 2^{n-2-h_{q_1}}$ with $1 \le h_{q_1} \le \lfloor (n-1)/2 \rfloor$. Next, we prove that the maximum second-order nonlinearity attained by a separable class-m cubic function grows with $m \le \lfloor n/3 \rfloor$ (see also Table 1 in Appendix C). Hence, class-1 functions constitute the most cryptographically weak class with respect to second-order nonlinearity. The security offered by class-m cubic functions, for large values of m, is also attributed to the difficulty of finding a set $\mathcal{J}$ such that all the 2^m sub-functions in $f = f_0 \|_{\mathcal{J}} \cdots \|_{\mathcal{J}} f_{2^m-1}$ are quadratic (in order to find the best quadratic approximations via the Theorem 3); the complexity of this step grows exponentially with m, for fixed number of variables n.

Theorem 4. *With the notation of Theorem 3, the maximum 2nd-order nonlinearity of a separable class-m cubic function $f \in \mathfrak{R}(3, n)$ grows with m. Further, the maximum 2nd-order nonlinearity of separable class-$\lfloor n/3 \rfloor$ cubic functions is given by $2^{n-1} - \frac{1}{2} 6^{n/3}$.*

Proof. In order to establish the first part we need to determine the class of separable cubic functions with the highest second-order nonlinearity. From Corollary 1 we see that this is maximized if and only if the product $\prod_{i=1}^{m} \left(1/2 + 1/2^{h_{q_i}+1} \right)$ depending on $m, h_{q_1}, \ldots, h_{q_m}$ is minimized. Since each product term is an integer less than 1, the number m of terms must be sufficiently large. However, the constraints on the values taken by each h_{q_i} need also be considered. Let H be the set of distinct integers from $h_{q_1} + 1, \ldots, h_{q_m} + 1$, and suppose $a - r, a + r \in H$ for some $a > r + 1 > 1$. Then, we have the property

$$\left(\frac{1}{2} + \frac{1}{2^{a-r}} \right) \left(\frac{1}{2} + \frac{1}{2^{a+r}} \right) > \cdots > \left(\frac{1}{2} + \frac{1}{2^{a-1}} \right) \left(\frac{1}{2} + \frac{1}{2^{a+1}} \right) > \left(\frac{1}{2} + \frac{1}{2^a} \right)^2$$

from which we derive that $\max_{a \in H} \{a\} - \min_{a \in H} \{a\}$, and the cardinality of H, should be relatively small. Moreover, by noting that the sequence $\{(1/2 + 1/2^a)^i\}_{i \ge 0}$ is purely decreasing for any $a \in H$ (since then $a \ge 2$), we conclude that the highest possible second-order nonlinearity achieved by separable class-m cubic Boolean functions, by Corollary 1, is given by

$$\max_{\text{class-}m} \{\mathcal{N}\mathcal{Q}\} = 2^{n-1} - 2^{n-1} \left(\frac{1}{2} + \frac{1}{2^{a_m+2}} \right)^{b_m} \left(\frac{1}{2} + \frac{1}{2^{a_m+1}} \right)^{m-b_m}$$

$$= 2^{n-1} - 2^{n-1}\left(\frac{2^{a_m+1}+1}{2^{a_m+1}+2}\right)^{b_m}\left(\frac{1}{2}+\frac{1}{2^{a_m+1}}\right)^m \qquad (16)$$

where $a_m = \lfloor(n-m)/2m\rfloor$ and $b_m = \lfloor(n-m)/2\rfloor \bmod m$, as a result of letting b_m functions q_i have $h_{q_i} = a_m + 1$, and the remaining $m - b_m$ have $h_{q_i} = a_m$. It is clear from (16) that for small values of m, the integer a_m is large and therefore the contribution of $(2^{a_m+1}+1)(2^{a_m+1}+2)^{-1} \approx 1$ is negligible (b_m is also small). So, the maximum second-order nonlinearity attained by separable class-m cubic functions grows with $m \leq \lfloor n/3\rfloor$. If $m = \lfloor n/3\rfloor$, then we have $a_m = 1$, $b_m = \lfloor(n \bmod 3)/2\rfloor$, and (16) becomes

$$\mathcal{NQ}_f = 2^{n-1} - 2^{\lceil(n \bmod 3)/2\rceil-1}\left(\tfrac{5}{3}\right)^{\lfloor(n \bmod 3)/2\rfloor}6^{\lfloor n/3\rfloor} = 2^{n-1} - b_n\tfrac{1}{2}6^{n/3}$$

where the term b_n equals 1 if $n \equiv 0 \pmod 3$, $(4/3)^{1/3}$ if $n \equiv 1 \pmod 3$, and $(250/243)^{1/3}$ if $n \equiv 2 \pmod 3$. Thus, $b_n \approx 1$ in all cases.   □

The above result also leads to a lower bound on the covering radius of the 2nd order binary Reed-Muller code $\mathfrak{R}(2, n)$, that is $\rho(2, n) \geq \rho_3(2, n) \geq 2^{n-1} - \tfrac{1}{2}6^{n/3}$ (since we only considered cubic Boolean functions satisfying the conditions of Theorem 3). The lower bound given behaves well with respect to the upper bound $2^{n-1} - \sqrt{15}\,2^{n/2-1}$ that has been proved in [5] (see Fig. 1 in Appendix C, and the analysis therein). The cubic part of any separable class-$\lfloor n/3\rfloor$ Boolean function is equivalent (under some transformation of variables $y = xR$) to the following

$$\sum_{i=1}^{\lfloor\frac{n}{3}\rfloor-1} y_{3i-2}\,y_{3i-1}\,y_{3i} + y_{3\lfloor\frac{n}{3}\rfloor-2}\left(y_{3\lfloor\frac{n}{3}\rfloor-1}\,y_{3\lfloor\frac{n}{3}\rfloor} + a\,y_{3\lfloor\frac{n}{3}\rfloor+1}\,y_{3\lfloor\frac{n}{3}\rfloor+2}\right)$$

where $a = 1$ if $n \equiv 2 \pmod 3$ and zero otherwise; they can be considered as a natural extension of bent functions (they have similar representation and the maximum possible distance from all functions of degree one less). Next we prove a formula for directly computing ξ_f^s from f; comparison of (7), (17) illustrates similarities on the way that best affine and quadratic approximations are obtained in terms of the Boolean function f.

Proposition 3. *With the notation of Theorem 3, the best quadratic approximations ξ_f^s of the separable class-m cubic function f are given by*

$$\xi_f^s = f + \sum_{i=1}^{m}(x_{j_i} + s_i)(q_i + \lambda_{q_i}), \qquad s \in \mathbb{F}_2^m. \qquad (17)$$

Proof. From the proof of Theorem 3 and Definition 1, we have that for all $s \in \mathbb{F}_2^m$ the best quadratic approximation ξ_f^s of the class-m cubic function f, with cubic part $c = \sum_{i=1}^{m} x_{j_i}q_i$, is such that

$$\xi_f^s + f = (\xi_{f,0}^s + f_0) \parallel_\partial \cdots \parallel_\partial (\xi_{f,2^m-1}^s + f_{2^m-1})$$

$$= \sum_{r \in \mathbb{F}_2^m}(x_{j_1} + \bar{r}_1)\cdots(x_{j_m} + \bar{r}_m)\sum_{i=1}^{m}(r_i + s_i)(q_i + \lambda_{q_i})$$

due to the linear independence of the functions in $\bigcup_{i=1}^{m}\{g_{i,1},\ldots,g_{i,2h_{q_i}}\}$ and the fact that we may write $\lambda_{r_1 q_1 + \cdots + r_m q_m} = r_1 \lambda_{q_1} + \cdots + r_m \lambda_{q_m}$, for all $r \in \mathbb{F}_2^m$. By writing the above expression as the sum of those terms for which $r_m = 0$ and those for $r_m = 1$, then simple calculations give

$$\xi_f^s + f = \sum_{r \in \mathbb{F}_2^{m-1}} (x_{j_1} + \bar{r}_1) \cdots (x_{j_{m-1}} + \bar{r}_{m-1}) \sum_{i=1}^{m-1}(r_i + s_i)(q_i + \lambda_{q_i})$$

$$+ (x_{j_m} + s_m)(q_m + \lambda_{q_m}) \sum_{r \in \mathbb{F}_2^{m-1}} (x_{j_1} + \bar{r}_1) \cdots (x_{j_{m-1}} + \bar{r}_{m-1})$$

$$= \sum_{r \in \mathbb{F}_2^{m-1}} (x_{j_1} + \bar{r}_1) \cdots (x_{j_{m-1}} + \bar{r}_{m-1}) \sum_{i=1}^{m-1}(r_i + s_i)(q_i + \lambda_{q_i})$$

$$+ (x_{j_m} + s_m)(q_m + \lambda_{q_m})$$

since $\sum_r (x_{j_1} + \bar{r}_1) \cdots (x_{j_{m-1}} + \bar{r}_{m-1}) = 1$ corresponds to the constant all-one Boolean function. Repeated application of the above steps will lead to (17). □

Remark 2. Given class-m cubic Boolean function $f = \sum_{i=1}^{m} x_{j_i} q_i + q + l$, where q, l are its quadratic and linear parts respectively, and $q_1, \ldots, q_m$ satisfying conditions of Theorem 3, its best quadratic approximations are directly computed by means of Proposition 3 as follows

$$\xi_f^s = \left(q + \sum_{i=1}^{m}(s_i q_i + x_{j_i} \lambda_{q_i}) \right) + \left(l + \sum_{i=1}^{m} s_i \lambda_{q_i} \right), \qquad s \in \mathbb{F}_2^m$$

where the parentheses indicate its quadratic and linear part respectively.

Example 2. Let $f \in \mathbb{B}_8$ be the class-2 Boolean function $f(x_1, \ldots, x_8) = (x_1 + x_3)(x_2 + x_7)(x_3 + x_5) + (x_4 + x_7)(x_5(x_6 + x_8) + (x_7 + x_8)x_8)$. It is seen that it satisfies the conditions of Theorem 3, and therefore its best quadratic approximations (from Proposition 3) are given by

$$\xi_f = s_1 q_1 + s_2 q_2 + (x_1 + x_3 + s_1)\lambda_{q_1} + (x_4 + x_7 + s_2)\lambda_{q_2}, \qquad s_i \in \mathbb{F}_2$$

where $q_1 = (x_2 + x_7)(x_3 + x_5)$ and $q_2 = x_5(x_6 + x_8) + (x_7 + x_8)x_8$. From Section 3 we know that the best affine approximations of q_1, q_2 are

$$\lambda_{q_1} = a_1(x_2 + x_7) + a_2(x_3 + x_5) + a_1 a_2, \qquad a_i \in \mathbb{F}_2,$$
$$\lambda_{q_2} = b_1 x_5 + b_2(x_6 + x_8) + b_3(x_7 + x_8) + b_4 x_8 + b_1 b_2 + b_3 b_4, \qquad b_i \in \mathbb{F}_2.$$

Then, its second-order nonlinearity is equal to $\mathcal{NQ}_f = 2^7 - 2^2 \cdot 3 \cdot 5 = 68$. Note that $\mathcal{NQ}_f$ depends only on the choice of the cubic terms, which is a well-known result [4]. □

5 Implications on Bent Functions Constructions

Many constructions of *bent functions* [25], having maximum nonlinearity $2^{n-1} - 2^{n/2-1}$ for even n, are based on the concatenation of sub-functions with low degree. Next, we analyze the second-order nonlinearity of known bent constructions and demonstrate the applicability of our results.

5.1 Maiorana-McFarland Construction

Let n be a positive integer, $\phi : \mathbb{F}_2^n \to \mathbb{F}_2^n$ be a permutation of the elements of the vector space $\mathbb{F}_2^n$, and $g : \mathbb{F}_2^n \to \mathbb{F}_2$ be an arbitrary Boolean function. Then $f : \mathbb{F}_2^n \times \mathbb{F}_2^n \simeq \mathbb{F}_2^{2n} \to \mathbb{F}_2$, which is given by

$$f(x,y) = \langle x, \phi(y) \rangle + g(y), \qquad x, y \in \mathbb{F}_2^n \tag{18}$$

is a *Maiorana-McFarland function* [8,22]; it is known that all functions of this form are bent (of highest possible degree n if g is properly chosen). Furthermore, f can be considered as the concatenation of 2^n affine sub-functions of $\mathbb{B}_n$ [3]. As shown next, the second-order nonlinearity of cubic Boolean functions obtained by (18) is very low in certain cases.

Proposition 4. *Let $f \in \mathbb{B}_{2n}$ be a cubic function given by (18) and let ϕ be a linear invertible mapping. If g is separable satisfying the condition of Theorem 3, then $\mathcal{NQ}_f \leq \max_{class-\lfloor n/3 \rfloor}\{\mathcal{NQ}\}$.*

Proof. Since ϕ is linear, the expression $\langle x, \phi(y) \rangle$ contains only quadratic terms, leading to $\deg(g) = 3$. By the fact $g \in \mathbb{B}_n$ we get that the maximum second-order nonlinearity that $f \in \mathbb{B}_{2n}$ can attain is $\max_{class-\lfloor n/3 \rfloor}\{\mathcal{NQ}\}$, which is far below $\max_{class-\lfloor 2n/3 \rfloor}\{\mathcal{NQ}\}$. □

The trade-off between nonlinearity and second-order nonlinearity is easily seen even for small values of n; e.g. if $n = 3$ then g becomes a class-1 cubic Boolean function, while the nonlinearity and second-order nonlinearity of f equals 28 and 8 respectively, according to Corollary 1.

5.2 Charpin-Pasalic-Tavernier Construction

The construction of cubic bent functions proposed in [6] is based on the concatenation of two quadratic semi-bent functions $f_b, f_c \in \mathbb{B}_n$ for odd n, vectors b, c satisfying $\mathrm{wt}(b) \not\equiv \mathrm{wt}(c) \pmod 2$, and each function given by

$$f_a(x) = \sum_{i=1}^{(n-1)/2} a_i \, \mathrm{tr}\left(x^{2^i+1}\right), \qquad x \in \mathbb{F}_{2^n} \text{ and } a \in \mathbb{F}_2^{(n-1)/2} \tag{19}$$

where $a = (a_1, \ldots, a_{(n-1)/2})$ and $\mathrm{tr}(x) = x + x^2 + \cdots + x^{2^{n-1}}$ is the trace function mapping elements of finite field $\mathbb{F}_{2^n}$ onto $\mathbb{F}_2$ [19]. It is seen that the Boolean function $f = f_b \parallel f_c \in \mathbb{B}_{n+1}$ is a class-1 cubic function.

Proposition 5. *With the above notation, let f_b, f_c be such that $f_b + f_c$ is a quadratic semi-bent function. Then, the second-order nonlinearity of $f = f_b \parallel f_c$ is $\mathcal{NQ}_f = 2^{n-1} - 2^{(n-1)/2} = \max_{class-1}\{\mathcal{NQ}\}$.*

Proof. From Theorem 3, the Boolean function $\xi_f^0 = f_b \parallel (f_b + \lambda_{f_b + f_c})$ is a best quadratic approximation of f. From Corollary 1, its second-order nonlinearity is $\mathcal{NQ}_f = \mathcal{NL}_{f_b + f_c} = 2^{n-1} - 2^{(n-1)/2}$, since n is odd, $f_b + f_c$ is a semi-bent quadratic function and $h_{f_b + f_c} = (n-1)/2$ [15]. □

A generalization of the above concatenation was presented in [6]; for odd n, if $f, f' \in \mathbb{B}_{n+1}$ are bent such that $f = f_b \parallel f_c$, and $f' = f_d \parallel f_e$, where $\mathrm{wt}(b), \mathrm{wt}(d)$ are odd and $\mathrm{wt}(c), \mathrm{wt}(e)$ are even, then

$$g = f \parallel f' \in \mathbb{B}_{n+2} \quad \text{and} \quad g' = f \parallel f' \parallel (1 + f) \parallel f' \in \mathbb{B}_{n+3} \tag{20}$$

are semi-bent and bent respectively. If we assume that $f_e = f_b + f_c + f_d$, then both g, g' are cubic with $g = f_b + x_{n+1}(f_b + f_c) + x_{n+2}(f_b + f_d)$ and $g' = g + x_{n+3}(1 + x_{n+2})$. As a result, class-2 cubic functions can also be obtained by this construction, having higher second-order nonlinearity.

Proposition 6. *The second-order nonlinearity of any function g, g' given by (20) satisfies $\mathcal{NQ}_g \geq \mathcal{NQ}_f + \mathcal{NQ}_{f'}$ and $\mathcal{NQ}_{g'} \geq 2(\mathcal{NQ}_f + \mathcal{NQ}_{f'})$.*

Proof. The claim follows by noting that, if there exist $\xi_f \in \mathcal{Q}_f, \xi_{f'} \in \mathcal{Q}_{f'}$ with the same quadratic part, then $\xi_g = \xi_f \parallel \xi_{f'} \in \mathcal{Q}_g$. Similar arguments hold for the function g', since $\mathcal{NQ}_{1+f} = \mathcal{NQ}_f$ [4]. □

6   Conclusions

The common characteristic of cubic functions studied in this paper, called *separable*, is that the structure of their highest degree part is undesirable in most cryptographic applications; it allows the efficient computation of their best quadratic approximations, by Theorem 3, but can be the source of other cryptanalytic attacks as well. Among these functions, class-$\lfloor n/3 \rfloor$ are the most cryptographically strong, as they attain higher second-order nonlinearity and their best quadratic approximations are more difficult to find. The analysis of bent constructions raises the need to find Boolean functions of nearly maximum nonlinearity and second-order nonlinearity. Research in progress focuses on extending the above results to functions of higher degree, and identify any trade-offs of second-order nonlinearity with other cryptographic criteria.

Acknowledgment

The authors would like to thank the anonymous reviewers for the helpful comments and suggestions.

References

1. Berbain, C., Billet, O., Canteaut, A., et al.: DECIM - a new stream cipher for hardware applications. eSTREAM, ECRYPT Stream Cipher Project, Report 2005/004 (2005) http://www.ecrypt.eu.org/stream

2. Carlet, C.: Partially-bent functions. Designs, Codes and Cryptography 3, 135–145 (1993)

3. Carlet, C.: On the confusion and diffusion properties of Maiorana-McFarland's and extended Maiorana-McFarland's functions. Journal of Complexity 20, 182–204 (2004)

4. Carlet, C.: Recursive lower bounds on the nonlinearity profile of Boolean functions and their applications. Cryptology ePrint Archive, Report 2006/459 (2006) http://eprint.iacr.org

5. Carlet, C., Mesnager, S.: Improving the upper bounds on the covering radii of binary Reed-Muller codes. IEEE Transactions on Information Theory 53, 162–173 (2007)

6. Charpin, P., Pasalic, E., Tavernier, C.: On bent and semi-bent quadratic boolean functions. IEEE Transactions on Information Theory 51, 4286–4298 (2005)

7. Courtois, N., Meier, W.: Algebraic attacks on stream ciphers with linear feedback. In: Biham, E. (ed.) EUROCRPYT. LNCS, vol. 2656, pp. 345–359. Springer, Heidelberg (2003)

8. Dillon, J.F.: Elementary Hadamard Difference Sets. Ph.D. Thesis, University of Maryland (1974)

9. Ding, C., Xiao, G., Shan, W.: The Stability Theory of Stream Ciphers. LNCS, vol. 561. Springer, Heidelberg (1991)

10. Dobbertin, H.: Construction of bent functions and balanced Boolean functions with high nonlinearity. In: Preneel, B. (ed.) Fast Software Encryption. LNCS, vol. 1008, pp. 61–74. Springer, Heidelberg (1995)

11. Gammel, B., Göttfert, R., Kniffler, O.: The Achterbahn stream cipher. eSTREAM, ECRYPT Stream Cipher Project, Report 2005/002 (2005) http://www.ecrypt.eu.org/stream/

12. Johansson, T., Meier, W., Muller, F.: Cryptanalysis of Achterbahn. In: Robshaw, M. (ed.) FSE 2006. LNCS, vol. 4047, pp. 1–14. Springer, Heidelberg (2006)

13. Kasami, T., Tokura, N.: On the weight structure of Reed-Muller codes. IEEE Transactions on Information Theory 16, 752–759 (1970)

14. Kasami, T., Tokura, N., Azumi, S.: On the weight enumeration of weights less than $2.5d$ of Reed-Muller codes. Information and Control 30, 380–395 (1976)

15. Khoo, K., Gong, G., Stinson, D.: A new characterization of semi-bent and bent functions on finite fields. Designs, Codes and Cryptography 38, 279–295 (2006)

16. Knudsen, L.R., Robshaw, M.J.B.: Non-linear approximations in linear cryptanalysis. In: Maurer, U.M. (ed.) EUROCRYPT 1996. LNCS, vol. 1070, pp. 224–236. Springer, Heidelberg (1996)

17. Kohavi, Z.: Switching and Finite Automata Theory. McGraw-Hill Book Company, New York (1978)

18. Kurosawa, K., Iwata, T., Yoshiwara, T.: New covering radius of Reed-Muller codes for t-resilient functions. IEEE Transactions on Information Theory 50, 468–475 (2004)

19. Lidl, R., Niederreiter, H.: Finite Fields. Encyclopedia of Mathematics and its Applications, 2nd edn., vol. 20. Cambridge University Press, Cambridge (1996)

20. MacWilliams, F.J., Sloane, N.J.A.: The Theory of Error Correcting Codes. North-Holland, Amsterdam (1977)
21. Matsui, M.: Linear cryptanalysis method for DES cipher. In: Helleseth, T. (ed.) EUROCRYPT 1993. LNCS, vol. 765, pp. 386–397. Springer, Heidelberg (1994)
22. McFarland, R.L.: A family of noncyclic difference sets. Journal of Combinatorial Theory, Series A 15, 1–10 (1973)
23. Menezes, A.J., Van Oorschot, P.C., Vanstone, S.A.: Handbook of Applied Cryptography. CRC Press, Boca Raton (1996)
24. Millan, W.L.: Low order approximation of cipher functions. In: Dawson, E.P., Golić, J.D. (eds.) Cryptography: Policy and Algorithms. LNCS, vol. 1029, pp. 144–155. Springer, Heidelberg (1996)
25. Rothaus, O.S.: On bent functions. Journal of Combinatorial Theory, Series A 20, 300–305 (1976)

A   Proof of Proposition 2

Without loss of generality let us assume that $\mathcal{J}'$ is comprised of the last k elements of the set $\mathcal{J}$, with $1 \leq k \leq m - 1$. We proceed by induction on the cardinality k of $\mathcal{J}'$. It is easily seen that Property 1 holds for $k = 1$, since by $f = f_0 \|_{j_m} f_1$ and the hypothesis, we conclude that $\deg(f_0) = 3$ and its cubic part includes the cubic terms involving at least one variable with index in $\mathcal{J} \setminus \mathcal{J}'$; hence, f_0 is a class-$(m - 1)$ cubic Boolean function. Then, we have that $f_1 = f_0 + f_1'$ where $\deg(f_1') < 3$, and thus f_1 has the same cubic part with f_0. Next, assume that Property 1 holds for some k and $\mathcal{J}' = \{j_{m-k+1}, \ldots, j_m\}$, $1 \leq k < m - 1$. The fact that it also holds for $k + 1$ is established by the identity

$$f = f_0 \|_{\mathcal{J}'} \cdots \|_{\mathcal{J}'} f_{2^k-1} = (f_0' \|_{j_{m-k}} f_1') \|_{\mathcal{J}'} \cdots \|_{\mathcal{J}'} (f_{2^{k+1}-2}' \|_{j_{m-k}} f_{2^{k+1}-1}')$$
$$= f_0' \|_{\{j_{m-k}\} \cup \mathcal{J}'} \cdots \|_{\{j_{m-k}\} \cup \mathcal{J}'} f_{2^{k+1}-1}' \tag{21}$$

due to Definitions 1, 2, and the fact that $j_{m-k} < \min \mathcal{J}'$ (note that (21) would still hold, up to a re-ordering of the resulting sub-functions, if this was not true). The sub-functions f_i (resp. f_i') have cubic terms involving at least one variable with index in $\mathcal{J} \setminus \mathcal{J}'$ (resp. $\mathcal{J} \setminus \{j_{m-k}\} \cup \mathcal{J}'$).

In order to prove Property 2 we need only consider (21) for $k = m - 1$. From Property 1 we get that $f = f_0 \|_{\mathcal{J} \setminus \{j_1\}} \cdots \|_{\mathcal{J} \setminus \{j_1\}} f_{2^{m-1}-1}$, where all f_i are class-1 cubic Boolean functions with the same cubic part (that of f_0). From (21) we have that $f_i = f_{2i}' \|_{j_1} f_{2i+1}'$, where both functions f_{2i}', f_{2i+1}' are quadratic.

B   Proof of Lemma 3

By hypothesis and Definition 1, we have $\mathcal{J} = \{n - m + 1, \ldots, n\}$. First we prove by induction on the cardinality m of $\mathcal{J}$ that f can be written as

$$f = \sum_{c \in \mathbb{F}_2^m} \left(\sum_{r \preceq c} f_r \right) x_{n-m+1}^{c_1} \cdots x_n^{c_m} . \tag{22}$$

Indeed, it is seen that (22) holds for $m = 1$ and $m = 2$ since then we find $f = f_0 + x_{n-1}(f_0 + f_1) + x_n(f_0 + f_2) + x_{n-1}x_n(f_0 + f_1 + f_2 + f_3)$. Let us assume that (22) holds for $m = s$ and let $f \in \mathbb{B}_n$ be the concatenation of 2^{s+1} sub-functions. From the identity $f = f_0 \| f_1 \| \cdots \| f_{2^{s+1}-2} \| f_{2^{s+1}-1} = (f_0 + x_{n-s}(f_0 + f_1)) \| \cdots \| (f_{2^{s+1}-2} + x_{n-s}(f_{2^{s+1}-2} + f_{2^{s+1}-1}))$ we see that f is written as the concatenation of 2^s sub-functions. Hence, by the induction hypothesis we obtain

$$f = \sum_{c \in \mathbb{F}_2^s} \left(\sum_{r \preceq c} f_{2r} + x_{n-s} \sum_{r \preceq c} (f_{2r} + f_{1+2r}) \right) x_{n-s+1}^{c_1} \cdots x_n^{c_s}$$

$$= \sum_{b \in \mathbb{F}_2} \sum_{c \in \mathbb{F}_2^s} \left(\sum_{t \leq b} \sum_{r \preceq c} f_{t+2r} \right) x_{n-s}^{b} x_{n-s+1}^{c_1} \cdots x_n^{c_s}$$

which leads to (22) if we define $\tilde{c} = (b, c) \in \mathbb{F}_2^{s+1}$, and $\tilde{r} = (t, r) \in \mathbb{F}_2^{s+1}$. Hence, if $f_r = q + l_r$ then we have $\sum_{r \preceq c} f_r = 2^{\mathrm{wt}(c)} q + \sum_{r \preceq c} l_r = \sum_{r \preceq c} l_r$ for all nonzero c, and the claim is a direct consequence of (22).

C   Bounds on the Second-Order Nonlinearity of Separable Functions

Let us denote by $\ell(n) = 2^{n-1} - \frac{1}{2} 6^{n/3}$ the lower bound which was proved in Theorem 4, and by $u(n) = 2^{n-1} - \sqrt{15} \, 2^{n/2-1}$ the upper bound derived in [5]. Moreover, let $\Theta(u) = \{v : v \sim u\}$ be the set of functions that are asymptotically equivalent to $u(n)$. Then, it can be verified that

$$\tfrac{1}{2} u(n) < \ell(n) < u(n), \qquad \text{for all } n \geq 6$$

which in turn implies that $\ell = \Theta(u)$; this is a standard notation in the literature, although it would be more natural to write $\ell \in \Theta(u)$. In fact, we can prove the much stronger relation

Table 1. The maximum second-order nonlinearity attained by class-1 and class-$\lfloor n/3 \rfloor$ separable cubic Boolean functions in $\mathbb{B}_n$, for $3 \leq n \leq 32$, as computed by Corollary 1

n	class-1	class-$\lfloor n/3 \rfloor$	n	class-1	class-$\lfloor n/3 \rfloor$	n	class-1	class-$\lfloor n/3 \rfloor$
3	1	1	4	2	2	5	6	6
6	12	14	7	28	31	8	56	68
9	120	148	10	240	315	11	496	667
12	992	1400	13	2016	2918	14	4032	6052
15	8128	12496	16	16256	25703	17	32640	52698
18	65280	107744	19	130816	219754	20	261632	447260
21	523776	908608	22	1047552	1842813	23	2096128	3732139
24	4192256	7548800	25	8386560	15251183	26	16773120	30781447
27	33550336	62070016	28	67100672	125061533	29	134209536	251797546
30	268419072	506637824	31	536854528	1018804657	32	1073709056	2047656190

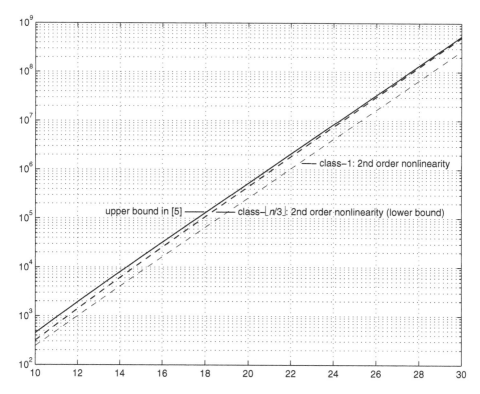

Fig. 1. The lower and upper bounds on the covering radius of $\Re(2, n)$, as determined by Theorem 4 and [5], for $10 \leq n \leq 30$. The second-order nonlinearity of class-1 cubic functions is also depicted.

$$\lim_{n \to +\infty} \frac{\ell(n)}{u(n)} = 1$$

which again states that $\ell(n)$ and $u(n)$ have the same rate of growth, as also depicted in Fig. 1.

On the Walsh Spectrum of a New APN Function

Carl Bracken[*], Eimear Byrne[**], Nadya Markin[**], and Gary McGuire[**]

School of Mathematical Sciences
University College Dublin
Ireland
carlbracken@yahoo.com, ebyrne@ucd.ie, nadyamarkin@gmail.com,
garymcguire@ucd.ie

Abstract. We compute the Walsh spectrum of a new quadratic APN function, $x^3 + \mathrm{Tr}(x^9)$, showing that its Walsh transform is 3-valued for odd n, and is 5-valued for even n. Therefore, the distribution of the values of the Walsh transform of $x^3 + \mathrm{Tr}(x^9)$ is the same as that of the APN Gold functions. Moreover, for odd n the function is AB, which gives an alternative proof of the APN property of the function.

1 Introduction

Let $L = GF(2^n)$ for some positive integer n and let $\mathrm{Tr}(x)$ denote the absolute trace map on L.

A function $f : L \longrightarrow L$ is said to be *almost perfect nonlinear* (APN) on L if the number of solutions in L of the equation

$$f(x + q) + f(x) = p$$

is at most 2, for all $p, q \in L$, $q \neq 0$. Equivalently, f is APN if the set $\{f(x+q) + f(x) : x \in L\}$ has size 2^{n-1} for each $q \in L^*$.

Given a function $F : GF(2)^n \longrightarrow GF(2)^n$, the *Walsh transform* of F at $(\mathbf{a}, \mathbf{b}) \in GF(2)^n \times GF(2)^n$ (cf. [5]) is given by

$$F^W(\mathbf{a}, \mathbf{b}) := \sum_{\mathbf{x} \in GF(2)^n} (-1)^{\langle \mathbf{a}, \mathbf{x} \rangle + \langle \mathbf{b}, F(\mathbf{x}) \rangle}, \tag{1}$$

where $\langle \cdot, \cdot \rangle$ denotes the usual inner product on $GF(2)^n$. For a function $f : L \longrightarrow L$, we may equivalently define the Walsh transform of f at (a, b) by

$$f^W(a, b) := \sum_{x \in L} (-1)^{\mathrm{Tr}(ax + bf(x))}, \tag{2}$$

[*] Research supported by Irish Research Council for Science, Engineering and Technology Postdoctoral Fellowship.
[**] Research supported by the Claude Shannon Institute, Science Foundation Ireland Grant 06/MI/006.

S.D. Galbraith (Eds.): Cryptography and Coding 2007, LNCS 4887, pp. 92–98, 2007.

for each $a, b \in L$. Then $f^W(a, b)$ is the discrete binary Fourier transform of $g(x) = (-1)^{\mathrm{Tr}(bf(x))}$. We define the *Walsh spectrum* of f as the set

$$\Lambda_f = \{f^W(a, b) : a, b \in L, b \neq 0\}.$$

Functions with small Walsh spectra (up to 5 values) have arisen in various contexts and given a map $f : L \longrightarrow L$, it is often of interest to know its Walsh spectrum. For example, it is well known that if n is odd and $(n, d) = 1$ then the functions x^{2^d+1} and $x^{2^{2d}-2^d+1}$ have 3-valued Walsh spectra given by $\{0, \pm 2^{\frac{n+1}{2}}\}$ (see [9,11]). These correspond to the APN Gold functions x^{2^d+1} and the Welch-Kasami functions $x^{2^{2d}-2^d+1}$. Niho has made several conjectures in [12] on the Walsh spectra of many functions. If $f(x)$ has a 3-valued Walsh spectrum (in which case n must be odd and $\Lambda_f = \{0, \pm 2^{\frac{n+1}{2}}\}$) we say that f is *almost bent* (AB) or *maximally nonlinear*.

Both APN and AB functions are used in block ciphers. APN functions were characterized in [13] as the mappings with highest resistance to differential cryptanalysis and are precisely the functions for which the plaintext difference $x + y$ yields the ciphertext difference $f(x) + f(y)$ with probability $1/2^{n-1}$. AB functions offer the best resistance to linear cryptanalysis, having maximal Hamming distance to the space of all affine maps. For a comprehensive survey of APN, AB, and related functions that arise in cryptography and coding theory, see [5].

When n is odd, every AB function on L is also APN [8]. If f is quadratic (so that each of its exponents is of the form $2^i + 2^j$ for some integers i, j) and f is also APN then it is necessarily an AB function [7]. Thus computing the Walsh spectrum of a quadratic function on L for odd n can be used to establish the APN property of a function. On the other hand, for n even, an APN function may have a large Walsh spectrum (more than 5 values), in which case the function could be less resistant to a linear attack.

Carlet-Charpin-Zinoviev (CCZ) equivalence, introduced in [7], is a standard measure to determine whether or not a pair of APN functions are essentially the same. This relation generalizes *extended affine* (EA) equivalence. A pair of CCZ equivalent functions have the same resistance to linear and differential attacks. A family of APN functions is determined to be "new" if they are CCZ inequivalent to any previously known family.

Until recently, all known APN functions had been found to be EA equivalent to one of a short list of monomial functions, namely the Gold, Kasami-Welch, inverse, Welch, Niho and Dobbertin functions. For some time it was conjectured that this list was the complete list of APN functions up to EA equivalence.

In 2006, new examples began to appear in the literature. A sporadic example of a binomial APN function that is not CCZ equivalent to any power mapping was given in [10]. A family of APN binomials on L, where n is divisible by 3 but not 9, was presented in [1] and shown to be EA inequivalent to any monomial function, and CCZ inequivalent to the Gold or Kasami-Welch functions in [2]. A method for constructing new quadratic APN functions from known ones has been outlined in the preprint [3], and has resulted in the discovery of the function

$$f(x) = x^3 + \mathrm{Tr}(x^9),$$

which is APN on L for any n. In the next section, we will compute the Walsh spectrum of this function, showing that it is 3-valued (as expected) for odd n, and 5-valued for even n. This provides another proof of the APN property of this function for odd n, and is a new result for even n.

For the case n even, if $f(x)$ is an APN function and its Walsh transform has values in $\{0, \pm2^{\frac{n}{2}}, \pm2^{\frac{n+2}{2}}\}$, then the distribution of values in the Walsh spectrum is uniquely determined and must therefore be the same as that of the Gold functions x^{2^d+1}, where $(d, n) = 1$. The proof that the distribution is uniquely determined is well known, but we include the argument as Corollary 1.

Finally, we remark that not all quadratic APN functions have the same values as the Gold functions. For odd n, a quadratic APN function has a 3-valued Walsh spectrum, and so the distribution of its values is determined (and is the same as Gold functions). However, for even n, the Walsh spectrum of a quadratic APN function may have more than five values. The following example is due to Dillon. Let u be primitive in $GF(2^6)$. Then

$$g(x) = x^3 + u^{11}x^5 + u^{13}x^9 + x^{17} + u^{11}x^{33} + x^{48}$$

is a quadratic APN function on $GF(2^6)$ whose Walsh transform takes 7 distinct values.

2   New APN Functions

The main result is given by Theorem 1, in which we compute the Walsh spectrum of $x^3 + \mathrm{Tr}(x^9)$. We will do this by obtaining an upper bound on $|f^W(a, b)|$ for $a, b \in L$, $b \neq 0$. This turns out to reduce to the problem of obtaining an upper bound on the size of the kernel of an $\mathbb{F}_2$-linear map $\mathcal{L}_b$ on L.

Theorem 1. *The Walsh spectrum of the function*

$$f(x) = x^3 + \mathrm{Tr}(x^9)$$

is contained in $\{0, \pm2^{\frac{n+1}{2}}\}$ when n is odd, and is contained in $\{0, \pm2^{\frac{n}{2}}, \pm2^{\frac{n+2}{2}}\}$ when n is even.

Proof: Throughout the proof, the notation $x^{1/2}$ and $x^{2^{-1}}$ means $x^{2^{n-1}}$.

By definition, we have

$$f^W(a, b) = \sum_{x \in L} (-1)^{\mathrm{Tr}(ax+b(x^3+\mathrm{Tr}(x^9)))} \tag{3}$$

$$= \sum_{x \in L} (-1)^{\mathrm{Tr}(ax+bx^3+\mathrm{Tr}(b)x^9)}, \tag{4}$$

since

$$\mathrm{Tr}(b\mathrm{Tr}(x^9)) = \mathrm{Tr}(x^9)\mathrm{Tr}(b) = \mathrm{Tr}(x^9\mathrm{Tr}(b)).$$

Computing the square of $f^W(a,b)$ gives

$$|f^W(a,b)|^2 = \sum_{u\in L}\sum_{x\in L}(-1)^{\text{Tr}(ax+bx^3+\text{Tr}(b)x^9+a(x+u)+b(x+u)^3+\text{Tr}(b)(x+u)^9)}$$

$$= \sum_{u\in L}(-1)^{\text{Tr}(au+bu^3+\text{Tr}(b)u^9)}\sum_{x\in L}(-1)^{\text{Tr}(b(x^2u+xu^2)+\text{Tr}(b)(x^8u+xu^8))}.$$

Using the fact that $\text{Tr}(\theta) = \text{Tr}(\theta^2)$ for any $\theta \in L$, we write

$$|f^W(a,b)|^2 = \sum_{u\in L}(-1)^{\text{Tr}(au+bu^3+\text{Tr}(b)u^9)}\sum_{x\in L}(-1)^{(x\mathcal{L}_b(u))},$$

where $\mathcal{L}_b(u) := bu^2 + b^{2^{-1}}u^{2^{-1}} + \text{Tr}(b)(u^{2^3} + u^{2^{-3}})$. Since $\mathcal{L}_b$ is linearized in u, it follows that

$$|f^W(a,b)|^2 = 2^n \sum_{u\in\ker\mathcal{L}_b}(-1)^{\text{Tr}(au+bu^3+\text{Tr}(b)u^9)}. \tag{5}$$

Moreover, writing $g_b(u) := \text{Tr}(au + bu^3 + \text{Tr}(b)u^9)$, we obtain

$$g_b(u + v) = g_b(u) + g_b(v) + \text{Tr}(v\mathcal{L}_b(u)) = g_b(u) + g_b(v),$$

for all $u \in \ker\mathcal{L}_b$. Then g_b is linear on $\ker\mathcal{L}_b$ and (5) is a character sum over a linear space, so we get

$$|f^W(a,b)|^2 = \begin{cases} 2^n|\ker\mathcal{L}_b| & \text{if } g_b(u) = 0 \text{ for all } u \in \ker\mathcal{L}_b \\ 0 & \text{otherwise.} \end{cases} \tag{6}$$

We will show that $|\ker\mathcal{L}_b| \le 4$, and then it will follow from (6) that

$$|f^W(a,b)| \le (2^n.4)^{1/2} = 2^{\frac{n+2}{2}}. \tag{7}$$

For the moment assume that $f^W(a,b)$ is an integer restricted by (6) and (7). If the inequality (7) holds and n is odd we have $|f^W(a,b)| \le 2^{\frac{n+1}{2}}$, and the Walsh spectrum of f must be $\{0, \pm 2^{\frac{n+1}{2}}\}$. If the same inequality holds and n is even, we deduce that f has Walsh spectrum contained in $\{0, \pm 2^{\frac{n}{2}}, \pm 2^{\frac{n+2}{2}}\}$.

We pause to remark that the techniques used up to now are standard in computing an upper bound on the absolute value of the Walsh transform of an arbitrary quadratic function.

Now we begin the proof that $|\ker\mathcal{L}_b| \le 4$. Once we show this, the proof is complete.

If $\text{Tr}(b) = 0$ then $f^W(a,b) = \sum_{x\in L}(-1)^{\text{Tr}(ax+bx^3)}$, by (4). It is well-known that x^3 has Walsh spectrum $\{0, \pm 2^{\frac{n+1}{2}}\}$ for odd n, and $\{0, \pm 2^{\frac{n}{2}}, \pm 2^{\frac{n+2}{2}}\}$ for even n. In fact this is easy to see, since then (5) becomes

$$|f^W(a,b)|^2 = 2^n \sum_{u\in\ker\mathcal{L}_b}(-1)^{\text{Tr}(au+bu^3)},$$

with $\mathcal{L}_b(u) := bu^2 + b^{2^{-1}}u^{2^{-1}}$, which has at most 4 roots in L, so that (7) holds.

We therefore assume that $\text{Tr}(b) = 1$. Applying this hypothesis, (5) becomes

$$|f^W(a,b)|^2 = 2^n \sum_{u \in \ker \mathcal{L}_b} (-1)^{\text{Tr}(au+bu^3+u^9)},$$

with $\mathcal{L}_b(u) = bu^2 + b^{2^{-1}} u^{2^{-1}} + u^{2^3} + u^{2^{-3}}$. Our goal now is to show that $\mathcal{L}_b(u)$ has at most 4 zeroes in L.

We now adopt a trick similar to that used in [9]. Consider the polynomial

$$\Gamma_b(u) := bu^3 + u^9 + u^{9/2} + u^{9/4}.$$

It is straightforward to check that $\Gamma_b(u) + \Gamma_b(u)^{2^{-1}} = u\mathcal{L}_b(u)$. If $\mathcal{L}_b(u) = 0$ then $\Gamma_b(u) \in GF(2)$. We claim that $\Gamma_b(u) = 0$. Suppose, to the contrary that $\Gamma_b(u) = 1$. Then $b = u^6 + u^{3/2} + u^{-3/4} + u^{-3}$ and hence $\text{Tr}(b) = 0$, which contradicts our assumption that $\text{Tr}(b) = 1$.

It follows that, for $u \in L$, $\mathcal{L}_b(u) = 0$ if and only if $\Gamma_b(u) = 0$. Now $\Gamma_b(u) = 0$ if and only if

$$0 = b^4 u^{12} + u^9 + u^{18} + u^{36} = u^9(b^4 u^3 + 1 + u^9 + u^{27})$$

Observe that the set of zeroes in L of Γ_b is a linear space. Fix an arbitrary nonzero $v \in \ker \mathcal{L}_b$. For $u \in \ker \mathcal{L}_b$

$$u(u+v)\Gamma_b(v) + v(u+v)\Gamma_b(u) + uv\Gamma_b(u+v) = 0. \tag{8}$$

This yields

$$\begin{aligned}
0 &= u(u+v)(bv^3 + v^9 + v^{9/2} + v^{9/4}) + v(u+v)(bu^3 + u^9 + u^{9/2} + u^{9/4}) \\
&\quad + uv(b(u+v)^3 + (u+v)^9 + (u+v)^{9/2} + (u+v)^{9/4}), \\
&= u^2(v^{2^2+2^{-1}} + v^{2^{-2}+2}) + v^2(u^{2^2+2^{-1}} + u^{2^{-2}+2}) \\
&\quad + uv(u^{2^2} v^{2^{-1}} + u^{2^{-1}} v^{2^2}) + uv(u^2 v^{2^{-2}} + u^{2^{-2}} v^2),
\end{aligned}$$

which factorizes as

$$(u^4 v + uv^4)(u^{1/2}v + uv^{1/2}) + (u^2 v + uv^2)(u^{1/4}v + uv^{1/4}) = 0 \tag{9}$$

Now perform the substitution $u \to zv$ in (9) and divide by v^2 to obtain

$$v^{9/2}(z + z^4)(z + z^{1/2}) + v^{9/4}(z + z^2)(z + z^{1/4}) = 0. \tag{10}$$

For fixed $v \in \ker \mathcal{L}_b$, any z satisfying $u = zv \in \ker \mathcal{L}_b$ for some u also satisfies (10). Since each z is uniquely determined by u for fixed nonzero v, the number of solutions to (10) gives an upper bound on $|\ker \mathcal{L}_b|$.

Observe that the solution set (in z) of (10) is an $\mathbb{F}_2$-linear space. Let $w = z + z^2$ and rewrite (10) to get

$$\begin{aligned}
0 &= v^{9/2}(w + w^2)w^{1/2} + v^{9/4}w(w^{1/2} + w^{1/4}), \tag{11} \\
&= v^{9/4}(v^{9/4}w^{5/2} + (v^{9/4} + 1)w^{3/2} + w^{5/4}). \tag{12}
\end{aligned}$$

The solution set of (12) is also an $\mathbb{F}_2$-linear space, say $\mathcal{K}$. Since $x \mapsto x^2 + x$ is a 2-1 map on L, there are at most 2 solutions z in L satisfying $w = z^2 + z$.

Now divide (12) by $v^{9/4}w$ to obtain the polynomial

$$G_v(w) := v^{9/4}w^{3/2} + (v^{9/4} + 1)w^{1/2} + w^{1/4},$$

whose set of zeroes is $\mathcal{K}$. Suppose that $r, w, r + w$ are all solutions of (12) in $\mathcal{K}$. We have

$$\begin{aligned} 0 &= G_v(r) + G_v(w) + G_v(r + w), \\ &= v^{9/4}(w^{3/2} + r^{3/2} + (r + w)^{3/2}), \\ &= v^{9/4}(r^{1/2}w + rw^{1/2}), \end{aligned}$$

which gives

$$rw^2 + r^2w = 0. \tag{13}$$

For fixed $r \neq 0$, (13) has only 2 solutions in w since $rw^2 + r^2w$ is quadratic in w. On the other hand, $0, r, w, w + r$ are all solutions of (13). We deduce that either $w = 0$ or $w = r$, so there exist at most 2 distinct members w of $\ker \mathcal{L}_b$ satisfying $w = z^2 + z$.

We deduce that $|\ker \mathcal{L}_b| \leq 4$ and hence that $f^W(a, b)$ is 3-valued for n odd, and is 5-valued for n even.

Corollary 1. *Let $f(x) = x^3 + \mathrm{Tr}(x^9)$ as in Theorem 1. Then the distribution of values in the Walsh spectrum of f is the same as the distribution of the Gold functions.*

Proof: By Theorem 1 we know that the Walsh spectrum is contained in $\{0, \pm 2^{\frac{n+1}{2}}\}$ when n is odd, and $\{0, \pm 2^{\frac{n}{2}}, \pm 2^{\frac{n+2}{2}}\}$ when n is even. In either case the spectrum is at most 5-valued. Let N_V be the number of times the value V is taken by $f^W(a, b)$. For any APN function, it is well known (see [5] for example) that $\sum_V V^j N_V$ is uniquely determined for $j = 0, 1, 2, 3, 4$. These five equations determine the N_V uniquely. □

Acknowledgement. The authors are indebted to the referees for their helpful comments and remarks.

References

1. Budaghyan, L., Carlet, C., Felke, P., Leander, G.: An infinite class of quadratic APN functions which are not equivalent to power mappings. In: ISIT 2006. Proceedings of the International Symposium on Information Theory, Seattle, USA (July 2006)
2. Budaghyan, L., Carlet, C., Leander, G.: A class of quadratic APN binomials inequivalent to power functions (preprint)
3. Budaghyan, L., Carlet, C., Leander, G.: Constructing new APN functions from known ones, Finite Fields and Applications (preprint submitted)

4. Budaghyan, L., Carlet, C., Pott, A.: New constructions of almost bent and almost perfect nonlinear functions. IEEE Transactions on Information Theory 52(3), 1141–1152 (2006)

5. Carlet, C.: Vectorial Boolean functions for Cryptography. In: Hammer, P., Crama, Y. (eds.) Boolean methods and models, Cambridge University Press, Cambridge (to appear)

6. Canteaut, A., Charpin, P., Dobbertin, H.: Weight divisibility of cyclic codes, highly nonlinear functions on $GF(2^m)$ and crosscorrelation of maximum-length sequences. SIAM Journal on Discrete Mathematics 13(1), 105–138 (2000)

7. Carlet, C., Charpin, P., Zinoviev, V.: Codes, bent functions and permutations suitable for DES-like cryptosystems. Designs, Codes and Cryptography 15(2), 125–156 (1998)

8. Chabaud, F., Vaudenay, S.: Links between differential and linear cryptanalysis. In: De Santis, A. (ed.) EUROCRYPT 1994. LNCS, vol. 950, pp. 356–365. Springer, Heidelberg (1995)

9. Dobbertin, H.: Another proof of Kasami's Theorem. Designs, Codes and Cryptography 17, 177–180 (1999)

10. Edel, Y., Kyureghyan, G., Pott, A.: A new APN function which is not equivalent to a power mapping. IEEE Transactions on Information Theory 52(2), 744–747 (2006)

11. Kasami, T.: The weight enumerators for several classes of subcodes of the second order binary Reed-Muller codes. Information and Control 18, 369–394 (1971)

12. Niho, Y.: Multi-valued cross-correlation functions between two maximal linear recursive sequences, Ph.D. thesis, Dept Elec. Eng., University of Southern California (USCEE Rep. 409) (1972)

13. Nyberg, K.: Differentially uniform mappings for cryptography. In: Helleseth, T. (ed.) EUROCRYPT 1993. LNCS, vol. 765, pp. 55–64. Springer, Heidelberg (1994)

Non-linear Cryptanalysis Revisited:
Heuristic Search for Approximations to S-Boxes

Juan M.E. Tapiador[1], John A. Clark[1], and Julio C. Hernandez-Castro[2]

[1] Department of Computer Science, University of York
York YO10 5DD, England, UK
{jet, jac}@cs.york.ac.uk
[2] Department of Computer Science, Carlos III University of Madrid
28911 Leganes (Madrid), Spain
jcesar@inf.uc3m.es

Abstract. Non-linear cryptanalysis is a natural extension to Matsui's linear cryptanalitic techniques in which linear approximations are replaced by non-linear expressions. Non-linear approximations often exhibit greater absolute biases than linear ones, so it would appear that more powerful attacks may be mounted. However, their use presents two main drawbacks. The first is that in the general case no joint approximation can be done for more than one round of a block cipher. Despite this limitation, Knudsen and Robshaw showed that they can be still very useful, for they allow the cryptanalist greater flexibility in mounting a classic linear cryptanalysis. The second problem concerning non-linear functions is how to identify them efficiently, given that the search space is superexponential in the number of variables. As the size of S-boxes (the elements usually approximated) increases, the computational resources available to the cryptanalyst for the search become rapidly insufficient.

In this work, we tackle this last problem by using heuristic search techniques –particularly Simulated Annealing– along with a specific representation strategy that greatly facilitates the identification. We illustrate our approach with the 9×32 S-box of the MARS block cipher. For it, we have found multiple approximations with biases considerably larger (e.g. $151/512$) than the best known linear mask ($84/512$) in reasonable time. Finally, an analysis concerning the search dynamics and its effectiveness is also provided.

1 Introduction

The adoption of the Data Encryption Standard (DES) [26,27] provided an extraordinary stimulus for the development of public cryptology, and particularly for the advancement of modern cryptanalytic methods. In the context of DES analysis, Reeds and Manferdelli [28] introduced in 1984 the idea of "partial linearity", pointing out that a block cipher with such a characteristic may be vulnerable to known- or chosen-plaintext attacks faster than exhaustive key search. Chaum and Evertse subsequently extended the notion of a per round linear factor to that of "sequence of linear factors" [5], proving that DES versions with more than 5 rounds had no partial linearity caused by such a sequence. These concepts were later generalized to that of "linear structure" [8], which

S.D. Galbraith (Eds.): Cryptography and Coding 2007, LNCS 4887, pp. 99–117, 2007.
© Springer-Verlag Berlin Heidelberg 2007

embraced properties such as the complementarity of DES or the existence of bit inde-
pendencies (i.e. some bits of the ciphertext being independent of the values of certain
plaintext and key bits), among others.

In a sense, linear cryptanalysis constituted the natural extension to these efforts. Linear cryptanalysis was introduced by Matsui in [17] as a potential technique to attack
DES. Its applicability was corroborated soon after [18], in what is commonly accepted
as the first –although barely practical at the time– compromise of the cipher.

Several refinements to the basic idea of linear cryptanalysis have attempted to improve the efficiency of the attacks, either in specific circumstances or in all cases. As
soon as 1994, Kaliski and Robshaw proposed an extension based on the use of multiple linear approximations [13]. Harpes, Kramer and Massey [11] presented in 1995 a
generalisation in which linear expressions are replaced by I/O sums. An I/O sum is the
XOR of a balanced binary-valued function of the input and a balanced binary-valued
function of the output.

Beyond these improvements, the most natural idea is to consider whether the linear approximations can be replaced with non-linear approximations. In 1996, Knudsen
and Robshaw introduced the idea of extending Matsui's linear cryptanalytic techniques
to the more general case in which non-linear relations are also considered [15]. To
motivate this approach, they provide a practical example showing that it is feasible to
obtain much more accurate approximations to DES S-boxes by considering non-linear
relations instead of linear ones. In that same work, they identified non-linear approximations to the S-boxes of LOKI91 [2], a DES-like block cipher that operates on 64-bit
blocks and uses a 64-bit key. One of the most remarkable features of LOKI91 is that it
uses four identical S-boxes which map 12 to 8 bits. While the three best linear approximations known to these S-boxes exhibited biases of 88/4096, 108/4096 and 116/4096,
the authors found non-linear relations with biases of 136/4096, 130/4096 and 110/4096.

1.1 Motivation and Related Work

Linear cryptanalysis was proposed to attack block ciphers and mainly applied to Feistel-
like constructions [9]. In this type of design, the overall effect of a round is entirely
linear (and often independent of the key) except for a single component, which is typically implemented using one or more S-boxes. It is precisely in this context wherein
the search for (linear) approximations of these components makes sense.

As S-boxes (especially in the past) were often fairly small, the search space in which
to look for linear approximations was small enough, in many cases allowing an exhaustive search in a reasonable amount of time. As a result, the method for determining the
best linear approximation has not been itself a matter of extensive study (an early exception to this was [19]). However, the situation becomes dramatically different when
considering non-linear approximations: there are 2^{2^n} different Boolean functions of n
variables (recall that only 2^n are linear). Even for a low number of inputs (e.g. $n = 8$)
the search space is astronomically huge, so a brute-force approach will simply not work.

There is, however, a different but very related field that has been tackled quite successfully by applying heuristic search techniques: the design and analysis of cryptographic Boolean functions.

The design of Boolean functions with desirable cryptographic properties (e.g. high non-linearity, low autocorrelation, high algebraic degree, reasonable order of correlation immunity, etc.) has traditionally been a central area of cryptological research. In the latter half of the 1990s, a few works suggested that heuristic search techniques could be applied to efficiently derive good Boolean functions. Millan et al. [20] were the first to show that small changes to the truth table of Boolean functions do not radically alter their non-linearity nor their autocorrelation properties. This provides for an efficient delta-fitness function for local searches. Later works by the same authors (e.g. [21]) demonstrated that Boolean functions that are correlation-immune or satisfying the strict avalanche criterion can be found too with smart hill-climbing.

This approach was subsequently generalized in a number of ways. Millan et al. applied it to the design of bijective [22] and regular [23] S-boxes. Clark and Jacob used Simulated Annealing in [6] to achieve some results hitherto unattained by other means. Some of the results presented in that work constituted also a counter-example to a then existing conjecture on autocorrelation.

1.2   Contribution and Overview

The main purpose of this paper is to show that heuristic methods very similar to that used for designing cryptographic Boolean functions can be applied to find non-linear approximations to S-boxes. In Section 2 we introduce some basic concepts on non-linear cryptanalysis, with particular emphasis in the elements which are essentially different to linear cryptanalysis. We describe our approach and the experimental setup used in Section 3. The most relevant results of the experiments carried out are shown in Section 4, together with an analysis concerning the efficiency and statistical significance of the search. Finally, in Section 5 we make a few concluding remarks and outline some possible extensions to this work.

2   Basic Concepts of Non-linear Cryptanalysis

Consider an n variable Boolean function $f : GF(2^n) \rightarrow GF(2)$. The (binary) truth table (TT) of f is a vector of 2^n elements representing the output of the function for each input. Each Boolean function has a unique representation in the Algebraic Normal Form (ANF) as sum of product terms:

$$f(x_1, \ldots, x_n) = a_0 \oplus a_1 x_1 \oplus a_2 x_2 \oplus \cdots \oplus a_n x_n$$
$$a_{12} x_1 x_2 \oplus \cdots a_{n-1,n} x_{n-1} x_n \qquad (1)$$
$$\oplus \cdots \oplus a_{1,\ldots,n} x_1 x_2 \cdots x_n$$

The order of each product term in the ANF is defined as the number of variables the product term contains. The *algebraic order* of f, denoted $ord(f)$, is the maximum order of the product terms in the ANF for which the coefficient is 1.

There are other widely-known representations of Boolean functions, such as the polarity truth table or the Walsh-Hadamard spectrum, which nonetheless shall not be used in this work.

The *Hamming weight* of a Boolean function f of n variables, denoted by $hw(f)$, is the number of ones in its TT:

$$hw(f) = \sum_{x=0}^{2^n-1} f(x) \tag{2}$$

A function f is said to be *balanced* iff $hw(f) = 2^{n-1}$.

Let $x = (x_1, x_2, \ldots, x_n)$ and $\omega = (\omega_1, \omega_2, \ldots, \omega_n)$ be binary n-tuples, and:

$$\omega \cdot x = \bigoplus_{i=1}^{n} \omega_i x_i \tag{3}$$

their dot product. Then the linear function $L_\omega(x)$ is defined as $L_\omega(x) = \omega \cdot x$. The set of affine functions consists of the set of linear functions and their complements. Note that every linear function is balanced.

A $n \times m$ substitution box (or simply S-box) is a mapping from n input bits to m output bits. It can also be viewed as an ordered set of m Boolean functions of n variables each.

2.1 Non-linear Approximations to S-Boxes

In the following definitions S represents a $n \times m$ S-box, $x = (x_1, \ldots, x_n)$ the input to S, and $y = (y_1, \ldots, y_m)$ the corresponding output. For compatibility with the notation often used in the literature on linear cryptanalysis, we shall write the inner product $\omega \cdot x$ of two binary n-tuples ω and x, as $x[\omega]$. We shall keep this notation even when approximations are not linear, i.e. if f is a non-linear function, then $x[f] = f(x)$. An exception to this is the output produced by S-boxes, which always shall be written as $S(x)$.

Definition 1. *A non-linear approximation for S is a pair*

$$A = \langle \Gamma_x, \Gamma_y \rangle \tag{4}$$

where $\Gamma_x : GF(2^n) \to GF(2)$ and $\Gamma_y : GF(2^m) \to GF(2)$.

It should be clear that, even though the domains of Γ_x and Γ_y are, respectively, $GF(2^n)$ and $GF(2^m)$, in order for them to be cryptanalytically useful they must not depend on all the variables in x and y. This implies that it must be possible to write them as functions in $GF(2^{\hat{n}})$ and $GF(2^{\hat{m}})$, respectively, with $\hat{n} < n$ and $\hat{m} < m$.

Definition 2. *Let $A = \langle \Gamma_x, \Gamma_y \rangle$ be a non-linear approximation and $P_S(A)$ the probability that the relation:*

$$x[\Gamma_x] \oplus S(x)[\Gamma_y] = 0 \tag{5}$$

holds for an S-box S. Let $P_R(A)$ be the probability that the same relation holds for a random $n \times m$ bit permutation R. The deviation of A with respect to S, denoted by $\delta_S(A)$, is given by:

$$\delta_S(A) = P_S(A) - P_R(A) \tag{6}$$

Definition 3. *The absolute value of the deviation is called the bias of A with respect to S:*

$$\epsilon_S(A) = |\delta_S(A)| \tag{7}$$

When the context will be clear enough, the subscript S will be omitted for denoting the deviation and bias.

2.2   Computing the Bias of a Non-linear Approximation

The computation of $P_S(A)$ can be done by exploring the 2^n possible inputs to $S(x)$ and checking how many times expression (5) holds. In cases where n is large, a random sampling over the inputs will provide an estimation of $P_S(A)$.

On the other hand, the computation of $P_R(A)$ depends on the Hamming weight of the functions integrating the approximation. Under the assumption of randomness of x, we have that:

$$Px(0) = \text{Prob}\big(x[\Gamma_x] = 0\big) = 1 - \frac{1}{2^n} hw(\Gamma_x) \tag{8}$$

$$Px(1) = \text{Prob}\big(x[\Gamma_x] = 1\big) = \frac{1}{2^n} hw(\Gamma_x) \tag{9}$$

$$Py(0) = \text{Prob}\big(R(x)[\Gamma_y] = 0\big) = 1 - \frac{1}{2^m} hw(\Gamma_y) \tag{10}$$

$$Py(1) = \text{Prob}\big(R(x)[\Gamma_y] = 1\big) = \frac{1}{2^m} hw(\Gamma_y) \tag{11}$$

Expression (5) can be now computed by simply applying the definition of the sum over GF(2):

$$P_R(A) = Px(0) \cdot Py(1) + Px(1) \cdot Py(0) \tag{12}$$

Substituting and regrouping terms we have:

$$P_R(A) = \frac{1}{2^n} hw(\Gamma_x) + \frac{1}{2^m} hw(\Gamma_y) - \frac{2}{2^{n+m}} hw(\Gamma_x) \cdot hw(\Gamma_y) \tag{13}$$

A particularly interesting observation is that if at least one of the two functions (e.g. Γ_x) is balanced, then:

$$P_R(A) = \frac{1}{2^n} hw(\Gamma_x) + \frac{1}{2^m} hw(\Gamma_y) - \frac{2}{2^{n+m}} hw(\Gamma_x) \cdot hw(\Gamma_y)$$
$$= \frac{1}{2^n} \frac{2^n}{2} + \frac{1}{2^m} hw(\Gamma_y) - \frac{2}{2^{n+m}} \frac{2^n}{2} \cdot hw(\Gamma_y) = \frac{1}{2} \tag{14}$$

The same is applicable in case of Γ_y being balanced. Note that this is precisely the case when one of the two functions is linear. In these cases, calculating the bias is considerably more efficient.

2.3   Joint Approximations, Limitations and Cryptanalytic Utility

In order to make a practical use of a non-linear approximation to an S-box, one needs to extend it into an approximation across the entire round function. This task is usually strongly dependant on the particular structure of the round function used and the way its output is subsequently processed through the rest of operations comprising a round.

In nearly all cases, a common problem is how to relate the input to the S-box with the actual inputs to the round function. The usual operation of a round function consists in first combining certain key bits (or a subkey) with some bits from the input data block, and then apply the S-box to the result. Suppose that d and k are the input data block and key to a round function, respectively, and that both are of the same length. Assume that the round function operates by applying an S-box to $x = d \oplus k$. In case of L being a linear approximation, the dependency on the key can be easily peeled off, since:

$$x[L] = (d \oplus k)[L] = d[L] \oplus k[L] \tag{15}$$

However, it is not generally possible to do this for a non-linear approximation. Furthermore, the actual approximation applied depends on the specific values of d and k. A detailed analysis of this phenomenon along with specific examples of how to approximate a round function can be found in [15] and [25].

Consider now a Feistel cipher [9] where the input data to round i is denoted as C_h^{i-1} and C_l^{i-1} (the high-order and low-order halves of the data block, respectively). The action of the round function will be denoted by $f(C_l^{i-1}, k_i)$, k_i being the subkey corresponding to this round. With this notation, the output from the ith round is written as $C_h^i = C_l^{i-1}$ and $C_l^i = C_h^{i-1} \oplus f(C_l^{i-1}, k_i)$.

If α, β, γ and δ are linear masks specifying a selected subgroup of bits, then a *linear* approximation to a single round can be written as [15]:

$$C_h^{i-1}[\alpha] \oplus C_l^{i-1}[\beta] = C_h^i[\gamma] \oplus C_l^i[\alpha] \oplus k_i[\delta] \tag{16}$$

Rewriting previous expression as:

$$C_l^{i-1}[\beta \oplus \gamma] \oplus k_i[\delta] = C_h^{i-1}[\alpha] \oplus C_l^i[\alpha] = (C_h^{i-1} \oplus C_l^i)[\alpha] \tag{17}$$

we obtain an approximation to round i.

Assume now that $f(\alpha)$ is a non-linear function. Again, it is obvious that the relation:

$$(C_h^{i-1} \oplus C_l^i)[f(\alpha)] = C_h^{i-1}[f(\alpha)] \oplus C_l^i[f(\alpha)] \tag{18}$$

will not generally hold.

In summary, one-round approximations that are not-linear in the output bits from $f(C_l^{i-1}, k_i)$ cannot be generally joined together. However, as noted by Knudsen and Robshaw in [15], they can be useful in a number of ways. For the first and last rounds of a cipher, the input to an approximation need not to be combined with any other approximation. Therefore, non-linear approximations can be used in these rounds without concern. Moreover, both the 1R- and 2R-methods of linear cryptanalysis proposed by Matsui (see [17]) require that some bits of the input to the second (or penultimate) round will be available to the cryptanalyst. By using this fact, non-linear approximations can be used in these rounds too.

The previous applications of non-linear approximations may allow the cryptanalyst to recover, in some circumstances, more key bits less plaintexts than required by linear techniques. As an example, in [15] it is shown how a non-linear cryptanalysis of reduced-round LOKI91 [2] allows to recover 7 additional key bits with less than 1/4 of the plaintexts required by linear techniques.

3   Heuristic Search for Non-linear Approximations

We wish to explore the space of possible approximations by searching for a solution that maximises the bias as defined by expression (7). The search space has a size of $\mathcal{O}(2^{2^n+2^m})$, n and m being the input and output sizes of the S-box, respectively. Obviously, this makes exhaustive search impossible (by standard computation means, at least) even for relatively low values of n and m.

In the experiments reported in this paper, we have used the well-established technique of Simulated Annealing [14]. A description of its operation is provided in Appendix B. In order to apply it to our problem, we have to provide three basic elements: a descriptive characterisation of the search space (i.e. a representation for solutions), a move function defining a neighbourhood for each element in the search space, and a fitness (or cost) function measuring how good a potential solution is. Next we describe these three components.

3.1   Solutions

The representation and operations for evolving Boolean functions used in this work are based on those provided by Fuller, Millan and Dawson in [10]. Each candidate approximation A is a structure comprising functions Γ_x and Γ_y, as given by Definition 1. Function Γ_x is represented by three elements:

- The number $\hat{n} < n$ of variables of the approximation.
- The TT of the non-linear approximation (of $\hat{n}$ variables).
- An injective function $J : \mathbb{Z}_{\hat{n}} \rightarrow \mathbb{Z}_n$ assigning each variable in the approximation to one input variable of the S-box. We shall represent J by a vector with its values: $(J(0), J(1), \ldots, J(\hat{n}-1))$. The terms $\hat{n}$ and $\dim(J)$ shall be used interchangeably.

The purpose of J is to *project* the non-linear function into some of the input bits of the S-box. This is necessary due to the reasons previously discussed –only approximations making use of some input and output bits are cryptanalytically useful.

The next example illustrates how this structure should be interpreted. Suppose S is a 4×8 S-box, $\hat{n} = 3$, and an approximation for the input variables is given by:

$$\Gamma(z_0, z_1, z_2) = z_0 \oplus z_0 z_1 \oplus z_0 z_1 z_2$$

Assuming $J_1 = (1, 3, 0)$, the projection of Γ into the input bits of S is given by:

$$\Gamma_x^{J_1}(x_0, x_1, x_2, x_3) = x_1 \oplus x_1 x_3 \oplus x_1 x_3 x_0$$

while for $J_2 = (0, 2, 1)$ it would be:

$$\Gamma_x^{J_2}(x_0, x_1, x_2, x_3) = x_0 \oplus x_0 x_2 \oplus x_0 x_2 x_1$$

Note that, given an input approximation and fixed values for n and $\hat{n}$, the number of different projections is exactly the number of permutations:

$$P_{\hat{n}}^n = \frac{n!}{(n - \hat{n})!} \tag{19}$$

As it will be discussed later, this value should be taken into account when defining some of the search parameters.

A similar representation could be provided to Γ_y. In our experiments, however, we decided to let it simply be a linear mask rather than a general non-linear function. This presents some remarkable advantages. From a complexity standpoint, it reduces the amount of space required to represent a solution, reduces considerably the search space, and also makes the search much faster (recall that linearity in one function implies a much simpler way of computing the bias of the whole approximation). Moreover, S-boxes are non-linear in their input bits, so it seems reasonable to look for approximations with non-linearity just in them, even though non-linear combinations of output bits might have some cryptanalytic interest too.

3.2 Move Function

To obtain a candidate $A' \in N(A)$ in the neighbourhood of A, we have defined a move function governed by five parameters:

- The values P_x and P_y, with:

$$0 \leq P_x \leq P_y \leq 1$$
$$P_x + P_y \leq 1$$

which define if a neighbour of A will be obtained by changing Γ_x, Γ_y or J, and keeping the other two unaltered. Specifically:
  - P_x is the probability of moving Γ_x,
  - $P_y - P_x$ is the probability of moving Γ_y, and
  - $1 - P_y$ is the probability of moving J.
- The parameters C_x, C_y and C_J control how many elements of Γ_x, Γ_y or J, respectively, will be mutated. In case of Γ_x and Γ_y, this is the number of bits to be flipped, while for J it defines the number of elements in the projection to be substituted by a new one. This new component is randomly generated and, as J is an injective function, must be different to each element already present in J.

The basic operation of the move function is described in Fig. 1.

3.3 Evaluation and Fitness

The fitness function used to guide the search is simply the bias achieved by the approximation:

$$F(A) = \epsilon \tag{20}$$

as defined by expression (7). This is computed by generating the 2^n different input values to the S-box. Each value is then translated to the corresponding projected value according to J, and the output value produced by Γ_x is stored. For the same (not projected) input value, the actual output produced by the S-box is computed and the mask Γ_y is applied. The resulting value is stored together with the previous one.

1	Pick $u \in [0, 1]$ with uniform probability
2	**if** $0 \leq u \leq P_x$ **then**
3	**for** $i = 1$ **to** C_x
4	Pick $r \in [0, 2^{\hat{n}} - 1]$ with uniform probability
5	Flip bit r in the TT of Γ_x
6	**else if** $P_x < u \leq P_y$ **then**
7	**for** $i = 1$ **to** C_y
8	Pick $r \in [0, m - 1]$ with uniform probability
9	Flip bit r in the TT of Γ_y
10	**else if** $P_y < u \leq 1$ **then**
11	**for** $i = 1$ **to** C_J
12	Pick $r \in [0, \hat{n} - 1]$ with uniform probability
13	Pick $v \in [0, n - 1]$ with uniform probability, and such that $v \notin J$
14	$J(r) \leftarrow v$

Fig. 1. Move function for evolving functions Γ_x and Γ_y, and projection J

Upon reaching the end of the process, the bias can be calculated as:

$$\epsilon = \left| \frac{E}{2^n} - \frac{1}{2} \right| \tag{21}$$

E being the number of times both outputs coincided. For subsequent analysis, it is also useful to define this magnitude in absolute terms:

$$Hits = |E - 2^{n-1}| \tag{22}$$

so the bias can be represented as $Hits/2^n$ (recall the maximum value for $Hits$ is 2^{n-1}). Here it is important to note that, under the assumption of random input, the expected hit rate is 0 due to the linearity of Γ_y. In a general case where Γ_x and Γ_y are both non-linear, the expected hit rate has to be computed as described in Section 2.2.

4   Experimental Results and Analysis

We have applied the technique briefly described above to the S-box included in the block cipher MARS [4]. MARS was IBM's candidate submission to the AES contest and, as such, is nowadays a quite well-studied scheme.

One of the core components of MARS is a 9×32 S-box with very specific combinatorial, differential and linear correlation properties. This S-box was the outcome of a search process that took IBM about a week of intensive computation. Burnett et al. showed in [3] that 9×32 S-boxes with cryptographic properties clearly superior to those of MARS S-box can be found with heuristic search methods in less than 2 hours on a single PC. In 2000, Robshaw and Yin [29] made some comments about the resistance of this S-box to linear cryptanalysis, challenging the claims by IBM's designers. Soon after, Knudsen and Raddum [16] found a large number of linear approximations with biases higher than 2^{-3}, again contradicting a conjecture made by the designers. Among the 871 linear masks found in that work, the best exhibits a bias of 82/512. Soon after, Aoki was able to compute the complete distribution of MARS's S-box linear masks [1], finding that the best linear approximation had a bias of 84/512.

In response to these criticisms, the MARS team issued some comments (see e.g. [12]) pointing out that some of their claims were poorly worded and hence easily misinterpreted. Basically, they argued that it is the MARS *core* function which actually has no bias higher than a given magnitude (2^{-3}), but not specific components such as, for example, the S-box itself.

One of the reasons for choosing MARS S-box to illustrate our approach is (apart from the fact that it was one of the five AES finalists) precisely the existence of linear masks with such a large bias. From this point of view, finding non-linear approximations with biases much larger than those –and in considerably less time– is an interesting challenge.

4.1   Parameterisation

In our preliminary experimentation, we found that the search dynamics is quite sensitive to the values $(P_x, P_y, C_x, C_y, C_j)$ required by the move function. For the last three, the best results are systematically found when the number of changes is low. This seems reasonable, since a strong mutation of a candidate implies an abrupt move in the search space, often resulting in a loss of many of the properties obtained so far.

Concerning the probabilities of mutation assigned to each component, the probability of moving J should be somehow related to the number of different projections as defined by expression (19). If the number of possible projections is large and the associated probability low, the search will be mostly devoted to find functions that "fits" a projection which is rarely changed. Alternatively, when the number of projections is relatively low, a high probability will result in an inefficient search eventually repeating candidate solutions.

In our experiments, the best results were obtained by using the parameterisation shown in Table 1. The probability of moving J is decreased as the number of possible permutations gets lower. A probability higher than 0.5 did not demonstrate better performance.

Each inner loop tries 10000 moves, and the number of inner loops is bounded by 1000. The search stops whenever this number is reached or after 250 consecutive inner loops without improvement.

4.2   Results

We ran 25 experiments for each value of $\hat{n}$ from 2 to 8. Each experiment takes around 1 hour in a PC (Intel Pentium 4 CPU 2.80 GHz with 1 GB RAM.)

For values of $\hat{n}$ from 2 to 4, no approximation better than the best known linear mask was found. In these cases, the functions found exhibit biases between 60 and 75. This is also the case for $\hat{n} = 5$, even though eventually we found two approximations with bias 83/512, i.e. very similar to the best linear mask.

In the case of approximations using 6, 7 or 8 out of the 9 input bits, the results were considerably better, achieving approximations with an average bias of around 89, 110

Table 1. Simulated Annealing parameters

GENERAL		MOVE FUNCTION
Max. No. inner loops	1000	$\hat{n} = 8 \rightarrow (P_x, P_y, C_x, C_y, C_J) = (0.25, 0.25, 1, 1, 1)$
Max. No. moves in inner loop	10000	$\hat{n} = 7 \rightarrow (P_x, P_y, C_x, C_y, C_J) = (0.30, 0.30, 1, 1, 1)$
Max. No. failed inner loops	250	$\hat{n} = 6 \rightarrow (P_x, P_y, C_x, C_y, C_J) = (0.35, 0.35, 1, 1, 1)$
Initial temperature	200	$\hat{n} = 5 \rightarrow (P_x, P_y, C_x, C_y, C_J) = (0.40, 0.40, 1, 1, 1)$
Cooling rate	0.99	$\hat{n} = 4 \rightarrow (P_x, P_y, C_x, C_y, C_J) = (0.45, 0.45, 1, 1, 1)$
EVALUATION: MARS S-box		$\hat{n} = 3 \rightarrow (P_x, P_y, C_x, C_y, C_J) = (0.45, 0.45, 1, 1, 1)$
n	9 bits	$\hat{n} = 2 \rightarrow (P_x, P_y, C_x, C_y, C_J) = (0.45, 0.45, 1, 1, 1)$
m	32 bits	

and 140, respectively. Fig. 2 shows the bias distribution for the 25 approximations found in each case.

The best among them constitute certainly an interesting result. For $\hat{n} = 8$, we found an approximation with bias 151/512, which translates to a deviation of around 0.29. Something similar occurs for $\hat{n} = 7$ and 6, for which the best approximations show deviations of 0.23 and 0.18, respectively. A remarkable point is the effectiveness of the search: even though finding the "best" approximation for a given number of input bits may require several runs, the search consistently provides good candidates.

Finally, the two best approximations found for 6, 7 and 8 variables are provided in Appendix A.

4.3 Effectiveness of the Heuristic

Now we provide a brief analysis concerning the statistical significance of our results. The purpose of this is to show that the search is indeed effective, i.e. it behaves considerably better than what should be expected from a pure blind search.

Computing the bias of a given approximation to MARS S-box can be seen as performing 512 experiments, each of which can result in a 1 (real and predicted output match) or 0 (when not). If we perform 512 independent experiments of a Binomial $B(1, 1/2)$, the result, by the additive property of the Binomial probability distribution, should behave as a $B(512, 1/2)$, whose standard deviation is $\sigma = \sqrt{0.5 \cdot 0.5 \cdot 512} = \sqrt{128}$. It is well known that in certain conditions ($n \geq 30$, $np \geq 5$ and $n(1 - p) \geq 5$), a Binomial $B(n, p)$ can be accurately approximated by a $N(np, np(1 - p))$. As these conditions clearly hold in our case, we can safely approximate a $B(512, 1/2)$ by a $N(256, 128)$; or, equivalently:

$$\frac{E - 256}{\sqrt{128}} = \frac{\pm Hits}{\sqrt{128}} \sim N(0, 1) \tag{23}$$

E being, as in (21), the number of times both outputs match. Even if the S-box were generated completely at random, if the number of experiments is high enough, a random search would find candidates increasingly "better." Therefore, the number of total

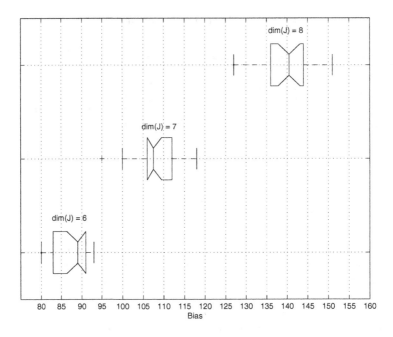

dim(J) = 6		dim(J) = 7		dim(J) = 8	
Bias	Number of Solutions	Bias	Number of Solutions	Bias	Number of Solutions
80	2	95	1	127	1
83	4	100	1	129	1
86	2	101	1	131	1
87	3	102	2	132	1
88	1	103	1	135	1
89	2	106	5	136	3
90	4	107	1	137	1
91	2	108	3	138	2
92	4	109	2	140	2
93	1	112	4	141	1
		115	2	143	3
		118	2	144	2
				145	2
				147	2
				148	1
				151	1

Fig. 2. Bias distribution for 25 experiments with $\dim(J) = 6$, 7 and 8. Each boxplot has lines at the lower quartile, median, and upper quartile values. The lines extending from each box show the extent of the rest of the data.

Table 2. Statistical significance for some bias values

Bias	Equivalent value observed in a $N(0,1)$	Probability
150/512	13.25	3.00×10^{-39}
140/512	12.37	2.36×10^{-34}
130/512	11.49	8.57×10^{-30}
120/512	10.61	1.43×10^{-25}
110/512	9.72	1.22×10^{-21}
100/512	8.84	4.28×10^{-18}
90/512	7.95	7.53×10^{-15}
80/512	7.07	5.58×10^{-12}

evaluations performed during the search process (say I) should be somehow incorporated into the analysis. For our purposes, if:

$$1 - \frac{1}{I} = \mathrm{erf}(\frac{n}{\sqrt{2}}) \tag{24}$$

where:

$$\mathrm{erf}(x) = \frac{2}{\sqrt{\pi}} \int_0^x e^{-t^2} dt \tag{25}$$

is the Gauss error function, then we should expect a number of hits at around n standard deviations from the mean.

In the case of the best approximations obtained for $\hat{n} = 8$, a bias of $151/512$ implies 151 hits, which translates to $151/\sigma \simeq 13.35$ standard deviations away from the mean. This is statistically equivalent to observe a value of around 13.35 coming from a $N(0,1)$, an extremely unusual event with an associated probability of occurrence of around 7.95×10^{-40}. This means that by following a pure blind search, the average number of evaluations required to yield this number is around 7.95×10^{40}. Recall that our search is bounded by 10^7 evaluations (actually, in most cases the stopping criterion is met at around 10^6; this shall be discussed below.)

Table 2 shows the equivalent value observed in a $N(0,1)$ and its associated probability for some values in the range of the best approximations found in our experimentation. From a simple inspection of these numbers, it should be clear that the heuristic is effectively achieving solutions that, otherwise, would not have been feasible to a random search.

4.4   Search Dynamics

Figure 3 shows the evolution of the fitness value associated with the best candidate found so far in a typical search. Around the first 500000 movements are completely random, during which the algorithm tries to locate a good candidate in the search space. In almost all the executions tried, the next behaviour has been identical: once an "appropriate" candidate is found, its fitness is considerably incremented in the next 500000 movements. This corresponds to the rapid growth observed in the curve, during which the resulting candidate is constantly refined until no more improvement can be done.

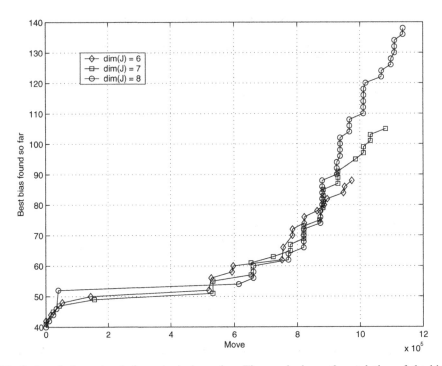

Fig. 3. Search dynamics during a typical running. The graph shows the evolution of the bias corresponding to the best approximation found so far.

In nearly all cases, the search stops improving solutions at around 1 or 1.2 millions of evaluations (i.e. 100 inner loops). After that, the next 250 cycles are completely useless and, therefore, the stop criterion is met and the search stops. This behaviour suggests that the stopping criterion can be greatly relaxed, stopping the search after 80 or 100 non-improving inner loops, instead of 250. This would result in searches of around 40 minutes of running time each, rather than the 1 hour on average pointed out previously.

4.5 Discussion

The heuristic is obviously working at improving the quality of the candidates, reaching approximations that might not have been found by a blind search. However, though effective, we are probably not using an optimal approach in the search. It is not clear to us whether the fitness function is appropriate for providing guidance, so alternative measures should be considered. Moreover, a restriction of the search space could be useful too. In a similar problem, Millan et al. [24] showed how searching over the space of ANFs of Boolean functions of degree less than or equal to $n/2$ can be used to provide Bent functions of highest possible degree. In [7], Clark et al. searched over the space of Walsh spectra for a permutation of an initial spectrum that corresponds to a Boolean function.

One of the most interesting features of both papers is that the form of representation used greatly facilitates the restriction of the possible solutions. In Millan et al.'s work, bent function theory shows that the maximal algebraic degree is $n/2$, and so restricting the space to functions with terms of or lower than this degree is clearly well motivated (and ANF is an obvious form to enforce this restriction).

In work of Clark et al., working over permutations of a supplied Walsh spectrum allows various Walsh values to be fixed at convenient values. This allows criteria such as balance, non-linearity and correlation immunity to be "assigned". Walsh spectra are transformed to polar vector form and it is the Boolean-ness that provides the guidance. Essentially, the function space is the set of all vectors induced under inversion by Walsh spectral permutations. Some are Boolean functions (with elements of $+1$ and -1) whilst others are not. In later work, Stanica et al. [30] applied further restrictions to the Walsh spectra by considering only rotation symmetric functions –again with hitherto unattained results. Both the above works demonstrate in different ways the usefulness of appropriate representations and restrictions.

5   Conclusions

In this work, we have shown how heuristic search methods can be useful to identify non-linear approximations to S-boxes. The technique seems to be quite effective and efficient, reaching approximations which can be very useful for mounting a more sophisticate linear attack than by simply using linear masks.

Here we have chosen a family of higher order approximations over the input bits of an S-box. Restricting Γ_y to be a linear function serves good practical purposes. Also, under the assumption of bit independence the expected bias for the overall approximation should be 0 whatever the characteristics of Γ_x. (This is not essential but is certainly convenient.) Allowing Γ_x –along with projection J– to roam free, as it were, over the space of higher order approximations is, we believe, almost forced by modern cryptographic design! The dangers of linearity are well documented and designers seek to design out linear attacks. In these cases, a good non-linear approximation can substantially improve a classical linear attack against the cipher.

An additional advantage of our technique is that the search complexity is bounded by the number of variables desired in the approximation, and not by the actual size of the S-box. According to our experience, the approximations with highest bias usually depends on a large number of variables, which certainly poses a problem for the case of very large S-boxes. However, this method allows a search to be attempted even in these cases, considering that the space explored depends on how many computational resources one can afford.

Acknowledgements

We would like to express our gratitude to the anonymous reviewers for their insights and comments, which have greatly contributed to enhance the quality of the original manuscript.

References

1. Aoki, K.: The Complete Distribution of Linear Probabilities of MARS' s-box. IACR Eprint Archive (2000), http://eprint.iacr.org/2000/033.pdf
2. Brown, L., Kwan, M., Pieprzyk, J., Seberry, J.: Improving Resistance to Differential Cryptanalysis and the Redesign of LOKI. In: Seberry, J., Pieprzyk, J.P. (eds.) AUSCRYPT 1990. LNCS, vol. 453, pp. 36–50. Springer, Heidelberg (1990)
3. Burnett, L., Carter, G., Dawson, E., Millan, W.: Efficient Methods for Generating MARS-like S-boxes. In: Schneier, B. (ed.) FSE 2000. LNCS, vol. 1978, pp. 50–57. Springer, Heidelberg (2001)
4. Burwick, C., Coppersmith, D., D'Avignon, E., Gennaro, R., Halevi, S., Jutla, C., Matyas Jr., S.M., O'Connor, L., Peyravian, M., Safford, D., Zunic, N.: MARS – A Candidate Cipher for AES. In: Proc. 1st AES Conference, NIST (1998)
5. Chaum, D., Evertse, J.-H.: Cryptanalysis of DES with a Reduced Number of Rounds; Sequences of Linear Factors in Block Ciphers. In: Williams, H.C. (ed.) CRYPTO 1985. LNCS, vol. 218, pp. 192–211. Springer, Heidelberg (1986)
6. Clark, J.A., Jacob, J.L.: Two Stage Optimisation in the Design of Boolean Functions. In: Clark, A., Boyd, C., Dawson, E.P. (eds.) ACISP 2000. LNCS, vol. 1841, pp. 242–254. Springer, Heidelberg (2000)
7. Clark, J.A., et al.: Almost Boolean Functions: The Design of Boolean Functions by Spectral Inversion. In: CEC 2003, IEEE Computer Society Press, Los Alamitos (2004)
8. Evertse, J.-H.: Linear Structures in Blockciphers. In: Price, W.L., Chaum, D. (eds.) EUROCRYPT 1987. LNCS, vol. 304, pp. 249–266. Springer, Heidelberg (1988)
9. Feistel, H.: Cryptography and Computer Privacy. Scientific American 228(5), 15–23 (1973)
10. Fuller, J., Millan, W., Dawson, E.: Efficient Algorithms for Analysis of Cryptographic Boolean Functions. In: AWOCA 2002, Frasier Island, Australia (2002)
11. Harpes, C., Kramer, G.G., Massey, J.L.: A Generalization of Linear Cryptanalysis and the Applicability of Matsui's Piling-Up Lemma. In: Guillou, L.C., Quisquater, J.J. (eds.) EUROCRYPT 1995. LNCS, vol. 921, pp. 24–38. Springer, Heidelberg (1995)
12. IBM MARS Team. Comments on MARS's Linear Analysis (2000), http://www.research.ibm.com/security/mars.html
13. Kaliski, B.S., Robshaw, M.J.B.: Linear Cryptnalaysis using Multiple Approximations. In: Desmedt, Y.G. (ed.) CRYPTO 1994. LNCS, vol. 839, pp. 26–39. Springer, Heidelberg (1994)
14. Kirkpatrick, S., Gelatt Jr., C.D., Vecchi, M.P.: Optimization by Simulated Annealing. Science 220(4598), 671–680 (1983)
15. Knudsen, L.R., Robshaw, M.J.B.: Non-Linear Approximations in Linear Cryptanalysis. In: Maurer, U.M. (ed.) EUROCRYPT 1996. LNCS, vol. 1070, pp. 224–236. Springer, Heidelberg (1996)
16. Knudsen, L.R., Raddum, H.: Linear Approximations to the MARS S-box. NESSIE Public Report NESSIE/DOC/UIB/001A/WP3 (2000)
17. Matsui, M.: Linear Cryptanalysis Method for DES cipher. In: Helleseth, T. (ed.) EUROCRYPT 1993. LNCS, vol. 765, pp. 386–397. Springer, Heidelberg (1994)
18. Matsui, M.: The First Experimental Cryptanalysis of the Data Encryption Standard. In: Desmedt, Y.G. (ed.) CRYPTO 1994. LNCS, vol. 839, pp. 1–11. Springer, Heidelberg (1994)
19. Matsui, M.: On Correlation between the Order of S-boxes and the Strength of DES. In: EUROCRYPT 1994. LNCS, vol. 850, pp. 366–375. Springer, Heidelberg (1995)
20. Millan, W., Clark, A., Dawson, E.: Smart Hill-Climbing Finds Better Boolean Functions. In: Han, Y., Quing, S. (eds.) ICICS 1997. LNCS, vol. 1334, pp. 149–158. Springer, Heidelberg (1997)

21. Millan, W., Clark, A., Dawson, E.: Heuristic Design of Cryptographically Strong Balanced Boolean Functions. In: Nyberg, K. (ed.) EUROCRYPT 1998. LNCS, vol. 1403, pp. 489–499. Springer, Heidelberg (1998)
22. Millan, W.: How to Improve the Non-Linearity of Bijective S-boxes. In: Boyd, C., Dawson, E. (eds.) ACISP 1998. LNCS, vol. 1438, pp. 181–192. Springer, Heidelberg (1998)
23. Millan, W., Burnett, L., Carter, G., Clark, A., Dawson, E.: Evolutionary Heuristics for Finding Cryptographically Strong S-boxes. In: Nyberg, K. (ed.) EUROCRYPT 1998. LNCS, vol. 1403, pp. 489–499. Springer, Heidelberg (1998)
24. Millan, W., et al.: Evolutionary Generation of Bent Functions for Cryptography. In: CEC 2003, IEEE Computer Society Press, Los Alamitos (2003)
25. Nakahara, J., Preneel, B., Vandewalle, J.: Experimental Non-Linear Cryptanalysis. COSIC Internal Report, 17 pages. Katholieke Universiteit Leuven (2003)
26. National Bureau of Standards (NBS). Data Encryption Standard. U.S. Department of Commerce, FIPS Publication 46 (January 1977)
27. National Institute of Standards and Technology (NIST). Data Encryption Standard. FIPS Publication 46-2. (December 30, 1993)
28. Reeds, J.A., Manferdelli, J.L.: DES Has no Per Round Linear Factors. In: Blakely, G.R., Chaum, D. (eds.) CRYPTO 1984. LNCS, vol. 196, pp. 377–389. Springer, Heidelberg (1985)
29. Robshaw, M., Yin, Y.L.: Potential Flaws in the Conjectured Resistance of MARS to Linear Cryptanalysis. In: Manuscript presented at the rump session in the 3rd AES Conference (2000)
30. Stanica, P., Maitra, S., Clark, J.A.: Results on Rotation Symmetric Bent and Correlation Immune Boolean Functions. In: Roy, B., Meier, W. (eds.) FSE 2004. LNCS, vol. 3017, pp. 161–177. Springer, Heidelberg (2004)

A   Best Approximations Found

dim(J) = 8
Γ_X = C3CB0E857D575014CE88552F31EC89AA02489173571719BB7EB48E96C366CD1F
J = (1, 5, 3, 8, 6, 7, 2, 0)
Γ_Y = 64393DD1
ϵ = 151/512
Γ_X = BDAE5075DB42A75D9279DD3358A4E8907A4C87D61B85D7D137BF7C6DBDF33B71
J = (2, 3, 4, 6, 7, 0, 8, 5)
Γ_Y = ECACB346
ϵ = 148/512

dim(J) = 7
Γ_X = AF65A106E589F470E043D55CFEAED634
J = (7, 0, 6, 4, 8, 1, 3)
Γ_Y = 1387BAF2
ϵ = 118/512
Γ_X = 1DE5BC329F1B44356E08BEEA44B48F86
J = (2, 4, 5, 6, 1, 3, 0)
Γ_Y = 3718E4F8
ϵ = 118/512

dim(J) = 6
Γ_X = 09778AA1F491AD47
J = (7, 8, 3, 4, 2, 0)
Γ_Y = 87AEA17C
ϵ = 93/512
Γ_X = F4E4ADD42AEAF2B9
J = (4, 7, 2, 3, 5, 1)
Γ_Y = CD6D89CC
ϵ = 92/512

B   Simulated Annealing

Simulated Annealing [14] is a search heuristic inspired by the cooling processes of molten metals. Basically, it can be seen as a basic hill-climbing coupled with the probabilistic acceptance of non-improving solutions. This mechanism allows a local search that eventually can escape from local optima.

The search starts at some initial state (solution) $S_0 \in \mathbb{S}$, where $\mathbb{S}$ denotes the solution space. The algorithm employs a control parameter $T \in \mathbb{R}^+$ known as the temperature. This starts at some positive value T_0 and is gradually lowered at each iteration, typically by geometric cooling: $T_{i+1} = \alpha T_i$, $\alpha \in (0, 1)$.

At each temperature, a number MIL (Moves in Inner Loop) of neighbour states are attempted. A candidate state C in the neighbourhood $N(S_i)$ of S_i is obtained by applying some move function to S_i. The new state is accepted if its better than S_i (as measured by a fitness function $F : \mathbb{S} \to \mathbb{R}$). To escape from local optima, the technique also accepts candidates which are slightly worse than S_i, meaning that its fitness is no more than $|T \ln U|$ lower, with U a uniform random variable in $(0, 1)$. As T becomes smaller, this term gets closer to 0, so as the temperature is gradually lowered it becomes harder to accept worse moves.

1 $S \leftarrow S_0$
2 $T \leftarrow T_0$
3 **repeat until** stopping criterion is met
4   **repeat MIL times**
5   Pick $C \in N(S)$ with uniform probability
6   Pick $U \in (0,1)$ with uniform probability
7   **if** $F(C) > F(S) + T \ln U$ **then**
8             $S \leftarrow C$
9     $T \leftarrow \alpha T$

Fig. 4. Basic Simulated Annealing for maximization problems

The algorithm terminates when some stopping criterion is met, usually after a fixed number MaxIL of inner loops have been executed, or when some maximum number MUL of consecutive inner loops without improvements have been reached. The basic algorithm is shown in Figure 4.

Cryptanalysis of the EPBC Authenticated Encryption Mode

Chris J. Mitchell

Information Security Group, Royal Holloway, University of London
Egham, Surrey TW20 0EX, UK
c.mitchell@rhul.ac.uk

Abstract. A large variety of methods for using block ciphers, so called
'modes of operation', have been proposed, including some designed to
provide both confidentiality and integrity protection. Such modes, usually known as 'authenticated encryption' modes, are increasingly important given the variety of issues now known with the use of unau-
thenticated encryption. In this paper we show that a mode known as
EPBC (Efficient error-Propagating Block Chaining), proposed in 1997
by Zúquete and Guedes, is insecure. Specifically we show that given a
modest amount of known plaintext for a single enciphered message, new
enciphered messages can be constructed which will pass tests for authenticity. That is, we demonstrate a message forgery attack.

1 Introduction

Traditionally, the recommended way to use a block cipher to provide both integrity and confidentiality protection for a message has been to encrypt the data
and then compute a CBC-MAC on the encrypted data, using two distinct secret
keys. This approach is rather unattractive for some applications because it requires each block of data to be processed twice. This observation has given rise
to a number of proposals for combining encryption and integrity protection (see,
for example, Sect. 9.6 of [1]).

At the same time, in recent years two major problems have been identified
which have highlighted the need for better-defined integrity and confidentiality
modes. Firstly, issues have been identified with certain combinations of encryption and use of a CBC-MAC — see, for example, Bellare, Kohno and Namprempre [2]. That is, it is vital to define precisely how the two operations are
combined, including the order of the computations; otherwise there is a danger
of possible compromise of the data. Secondly, even where integrity is not explicitly required by the application, if integrity is not provided then in some cases
padding oracle attacks may be used to compromise secret data (see, for example,
[3,4,5,6,7]).

This has given rise to a number of proposals for well-defined authenticated-
encryption modes, including OCB [8], EAX [9] and CCM [10,11]. These techniques are also the subject of ongoing international standardisation efforts —
the third committee draft of what is intended to become ISO/IEC 19772 on

S.D. Galbraith (Eds.): Cryptography and Coding 2007, LNCS 4887, pp. 118–128, 2007.
© Springer-Verlag Berlin Heidelberg 2007

authenticated encryption was published in June 2007 [12] (see also Dent and Mitchell, [13]).

In this paper we examine another authenticated-encryption mode, known as EPBC, which was introduced by Zúquete and Guedes in 1997 [14]. We show that this mode is subject to a message forgery attack using only a modest amount of known plaintext, and hence does not provide adequate integrity protection. When combined with other recent cryptanalyses of authenticated-encrypted modes [15], this emphasises the need to only use modes which have robust evidence for their security, e.g. OCB, EAX or CCM.

2 Integrity Protection

Before discussing any possible attacks, we need to explain how EPBC mode is intended to be used to provide both confidentiality and integrity protection. The idea is very simple. First divide the data to be encrypted into a sequence of n-bit blocks, padding as necessary, where n is the block length for the block cipher in use. Then append an additional n-bit block to the end of the message, where this block can be predicted by the decrypter (e.g. a fixed block); this is referred to as the *integrity control value* by Zúquete and Guedes [14]. When the message is decrypted, a check is made that the final block is the expected value and, if it is, the message is deemed authentic.

Before proceeding observe that this general approach possesses an intrinsic weakness. That is, suppose that a fixed final block (the *terminator block*) is used to detect message manipulations (as above). Then an attacker might be able to persuade the legitimate originator of protected messages to encrypt a message which contains the fixed terminator block somewhere in the middle of the message. The attacker will then be able to delete all ciphertext blocks following the encrypted terminator block, and such a change will not be detectable.

Despite this weakness, using an appropriate encryption mode combined with a method for adding verifiable redundancy to a message is still used for message integrity protection — e.g. in Kerberos (see, for example, [13]). As far as this paper is concerned we note that such an attack could be prevented by ensuring that the legitimate encrypter refuses to encrypt any plaintext message containing the terminator block. We further note that such an attack requires *chosen* plaintext, and the attack we demonstrate later in this paper requires only a limited amount of *known* plaintext.

3 The Zúquete-Guedes EPBC Mode

First suppose that the data is to be protected using an n-bit block cipher, i.e. a block cipher operating on plaintext and ciphertext blocks of n bits. We further suppose that n is even, and put $n = 2m$ (as is the case for all standardised block ciphers — see, for example, [16]). We write $e_K(P)$ for the result of block cipher encrypting the n-bit block P using the secret key K, and $d_K(C)$ for the result of block cipher decrypting the n-bit block C using the key K. Suppose the plaintext to be protected is divided into a sequence of n-bit blocks (if necessary, having first been padded): $P_1, P_2, \ldots, P_t$.

The scheme uses two secret n-bit Initialisation Vectors (IVs), denoted by F_0 and G_0. The EPBC encryption of the plaintext $P_1, P_2, \ldots, P_t$ is then defined as:

$$G_i = P_i \oplus F_{i-1}, \quad (1 \leq i \leq t), \tag{1}$$

$$F_i = e_K(G_i), \quad (1 \leq i \leq t), \tag{2}$$

$$C_i = F_i \oplus g(G_{i-1}), \quad (2 \leq i \leq t), \tag{3}$$

where $C_1 = F_1 \oplus G_0$, $\oplus$ denotes bit-wise exclusive-or, and g is a function that maps an n-bit block to an n-bit block, defined below. The operation of the mode (when used for encryption) is shown in Figure 1. Note that we refer to the values F_i and G_i as 'internal' values, as they are computed during encryption, but they do not constitute part of the ciphertext.

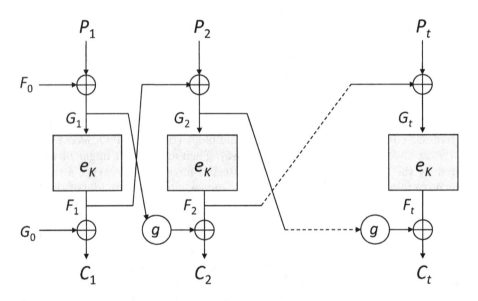

Fig. 1. EPBC encryption

The function g is defined as follows. Suppose X is an n-bit block, where $X = L||R$ and L and R are m-bit blocks (and, as throughout, $||$ denotes concatenation). Then

$$g(X) = (L \vee \overline{R})||(L \wedge \overline{R})$$

where $\vee$ denotes the bit-wise inclusive or operation, $\wedge$ denotes the bit-wise logical and operation, and $\overline{X}$ denotes the logical negation of X (i.e. changing every zero to a one and vice versa).

Finally, note that decryption operates similarly. We have:

$$F_i = C_i \oplus g(G_{i-1}), \quad (2 \leq i \leq t), \tag{4}$$

$$G_i = d_K(F_i), \quad (1 \leq i \leq t), \tag{5}$$
$$P_i = G_i \oplus F_{i-1}, \quad (1 \leq i \leq t). \tag{6}$$

and $F_1 = C_1 \oplus G_0$, where d denotes block cipher decryption.

4   Some Preliminary Observations

We first establish some simple results on the operation of the EPBC scheme. In particular, we consider the operation of the function g.

Lemma 1. *Suppose* $g(X) = L'||R'$, *where* X *is an n-bit block and we let* $L' = (\ell'_1, \ell'_2, \ldots, \ell'_m)$ *and* $R' = (r'_1, r'_2, \ldots, r'_m)$ *be m-bit blocks. Then, for every i* $(1 \leq i \leq m)$, *if* $\ell'_i = 0$ *then* $r'_i = 0$.

Proof. Let $X = L||R$, where $L = (\ell_1, \ell_2, \ldots, \ell_m)$ and $R = (r_1, r_2, \ldots, r_m)$. Suppose $\ell'_i = 0$ for some i. But, by definition, $\ell'_i = \ell_i \vee \overline{r_i}$, and hence $\ell_i = \overline{r_i} = 0$. Hence $r'_i = \ell_i \wedge \overline{r_i} = 0$. □

The above Lemma implies that output bit pairs (ℓ'_i, r'_i) can never be equal to $(0,1)$. In fact, we can obtain the following more general result which gives Lemma 1 as a special case.

Lemma 2. *Suppose that, as above,* $X = L||R$ *where* $L = (\ell_1, \ell_2, \ldots, \ell_m)$ *and* $R = (r_1, r_2, \ldots, r_m)$. *Suppose also that* $g(X) = L'||R'$ *where* $L' = (\ell'_1, \ell'_2, \ldots, \ell'_m)$ *and* $R' = (r'_1, r'_2, \ldots, r'_m)$. *Then if* $(\ell_i, r_i) \in A$ *then* $(\ell'_i, r'_i) \in B$, *where all possibilities for* A *and* B *are given in Table 1. Note that, for simplicity, in this table we write* xy *instad of* (x, y).

Table 1. Input/output possibilities for g

A (set of input pairs)	B (set of output pairs)
$\{00, 01, 10, 11\}$	$\{00, 10, 11\}$
$\{01, 10, 11\}$	$\{00, 10, 11\}$
$\{00, 10, 11\}$	$\{10, 11\}$
$\{00, 01, 11\}$	$\{00, 10\}$
$\{00, 01, 10\}$	$\{00, 10, 11\}$
$\{10, 11\}$	$\{10, 11\}$
$\{01, 11\}$	$\{00, 10\}$
$\{01, 10\}$	$\{00, 11\}$
$\{00, 11\}$	$\{10\}$
$\{00, 10\}$	$\{10, 11\}$
$\{00, 01\}$	$\{00, 10\}$
$\{11\}$	$\{10\}$
$\{10\}$	$\{11\}$
$\{01\}$	$\{00\}$
$\{00\}$	$\{10\}$

Proof. The result follows from a simple case by case analysis.   □

Before proceeding we make a general observation which underlies the attack procedure described below. That is, given a random set A of any particular size (not equal to 1), then the expected value of $|B|$ is always smaller than $|A|$.

5   Attack Stage 1 — Deducing Internal Pairs

The objective of this stage of the attack is to use knowledge of known plaintext/ciphertext pairs (P_i, C_i) to learn the values of corresponding 'internal pairs' (F_i, G_i). In the second stage of the attack, described in Sect. 6, we show how to use these internal values to complete a forgery attack on EPBC mode.

Suppose an attacker knows s consecutive pairs of plaintext/ciphertext blocks for some $s > 1$. That is, suppose

$$(P_j, C_j), (P_{j+1}, C_{j+1}), \ldots, (P_{j+s-1}, C_{j+s-1})$$

are known, where we also suppose that $j > 1$.

First observe that we know that $C_j = F_j \oplus g(G_{j-1})$ (since $j > 1$). From Lemma 1, we also know that, if $g(G_{j-1}) = L'||R'$ where $L' = (\ell'_1, \ell'_2, \ldots, \ell'_m)$ and $R' = (r'_1, r'_2, \ldots, r'_m)$, then (ℓ'_i, r'_i) can never equal $(0,1)$ for any i. Hence knowledge of C_j will immediately yield knowledge about F_j. Specifically, it will yield the information that certain bit pairs cannot occur in F_j, where each bit pair contains a bit from the left half and the corresponding bit from the right half. More precisely, given that there are m such bit pairs in F_j, and for each such pair one of the four possible bit pairs will be ruled out, the number of possibilities for F_j will be reduced from 2^{2m} to 3^m.

Using Lemma 2, we can extend this observation making use of subsequent plaintext/ciphertext block pairs. Since $G_{j+1} = P_{j+1} \oplus F_j$, information about forbidden bit pairs in F_j, combined with knowledge of P_{j+1}, gives information about forbidden bit pairs in G_{j+1}. This yields information about (potentially) even more forbidden bit pairs in $g(G_{j+1})$. Given that

$$C_{j+2} = F_{j+2} \oplus g(G_{j+1}),$$

and given knowledge of C_{j+2}, this gives even more information about forbidden bit pairs in F_{j+2}, and so on.

That is, it is possible to deduce increasing amounts of information about the sequence of n-bit blocks:

$$F_j, g(G_{j+1}), F_{j+2}, g(G_{j+3}), \ldots.$$

Hence, assuming that we know sufficiently many pairs to perform the calculations, for sufficiently large w there will only be one possibility for F_{j+2w}. Using knowledge of P_{j+2w+1}, this immediately gives certain knowledge of G_{j+2w+1}. I.e., for all sufficiently large values of w, complete knowledge can be obtained of F_{j+2w} and G_{j+2w+1}.

In the above discussion we did not use all the available information to make the deductions. In fact we only used knowledge of $C_j, C_{j+2}, C_{j+4}, \ldots$ and $P_{j+1}, P_{j+3}, P_{j+5}, \ldots$. We have also only shown how to derive information about $F_j, F_{j+2}, F_{j+4}, \ldots$ and $G_{j+1}, G_{j+3}, G_{j+5}, \ldots$.

However, we can simply repeat the above argument starting with F_{j+1}, using the remainder of the information available. That is, repeating the process starting one block later will enable the deduction of information about $F_{j+1}, F_{j+3}, F_{j+5}, \ldots$ and $G_{j+2}, G_{j+4}, G_{j+6}, \ldots$. Finally note that the above analysis does not make use of knowledge of P_j, only of C_j.

The above discussion has been rather informal, in that 'sufficiently large' has not been quantified. However, Lemma 2 enables us to make this more precise.

Consider any pair of bit positions in an n-bit block: $(i, i + m)$, say, where $1 \leq i \leq m$. Then, from Lemma 1, we know that $g(G_{j-1})$ cannot have $(0,1)$ in these two bit positions. Hence, given knowledge of C_j, we know that the bits in positions $(i, i + m)$ in $F_j = C_j \oplus g(G_{j-1})$ can only take three of the four possible values. Precisely which three possibilities will depend on the pair of bits in positions $(i, i + m)$ in C_j, which we assume are randomly distributed.

As a result we know that the bits in positions $(i, i+m)$ in G_{j+1} can only take three of the four possible values. From an examination of Table 1, the number of possibilities for the bits in positions $(i, i+m)$ in $g(G_{j+1})$ will either be two or three, depending on the three possibilities for the bit pair in G_{j+1}. Specifically, there is a 50% chance that there will only be two possibilities for the bit pair in G_{j+1}, and a 50% that there will be three possibilities for the bit pair in G_{j+1}.

Extending this analysis using standard probabilistic arguments for stochastic processes, it follows that the probability p that there will only be a single possibility for the bit pair after v iterations of the above process is equal to the bottom left entry in the vth power of the four by four matrix given in Figure 2. The entry in the i row and the jth column of this matrix represents the probability that a set A of size i will map to a set B of size j (as derived from Table 1). Some values of this matrix entry (i.e. of p) for various values of v are given in Table 2.

$$\begin{pmatrix} 1 & 0 & 0 & 0 \\ 1/6 & 5/6 & 0 & 0 \\ 0 & 1/2 & 1/2 & 0 \\ 0 & 0 & 1 & 0 \end{pmatrix}$$

Fig. 2. Transition probability matrix

That is, after 30 iterations, i.e. given knowledge of 60 consecutive plaintext/ciphertext pairs, the probability p that a bit pair will be known with certainty is 0.99241. We are actually interested in the probability q that an entire n-bit block will be known with certainty. It is straightforward to verify that

$$q = p^{n/2}.$$

Table 2. Probability of a unique possibility for a bit pair

v	prob. p	v	prob. p	v	prob. p
10	0.71027	20	0.95305	30	0.99241
40	0.99878	50	0.99980	60	0.99997

Table 3. Probability of a unique possibility for a 128-bit block

v	prob. q	v	prob. q	v	prob. q
10	very small	20	0.04607	30	0.61409
40	0.92485	50	0.98728	60	0.99808

In Table 3, this probability is tabulated for the same values of v as given in Table 2, assuming the use of a 128-bit block cipher.

Hence, for such a cipher, if 60 consecutive pairs of plaintext/ciphertext blocks are known (bearing in mind that each iteration involves alternate blocks), then there is a 60% chance that the final internal variables F_i and G_i will be completely known. This probability increases to nearly 99% if 100 consecutive block pairs are available.

6   Attack Stage 2 — Completing the Forgery

We now suppose that the attack procedure described in the previous section has been completed, i.e. matching pairs of consecutive plaintext/ciphertext blocks have been used to learn internal values G_i, for some i. We suppose also that the EPBC mode is being used to provide both confidentiality and integrity using a fixed n-bit integrity control value V. That is, when a message is decrypted, the final plaintext block must equal V if the message is to be accepted as genuine. To demonstrate a forgery attack, we therefore need to show how to construct a message for which this will be true.

We suppose that the following resources are available to an attacker.

- An encrypted message $C_1, C_2, \ldots, C_t$ for which the attacker knows the internal value G_s, for some $s \le t$.
- The final two blocks (C'_{u-1}, C'_u) of an encrypted message for which the attacker also knows the internal value G'_{u-2}. Note that we are assuming that $P'_u = V$, where P'_u is the final plaintext block corresponding to this enciphered message.

Note that we assume also that the same secret key K has been used to compute both the ciphertexts involved (in fact, the same message could be used to yield both of the sets of values required).

We now define the 'forged' ciphertext message $C_1^*, C_2^*, \ldots, C_{s+2}^*$ as follows:

$$C_i^* = C_i, \quad (1 \le i \le s),$$

$$C^*_{s+1} = C'_{u-1} \oplus g(G'_{u-2}) \oplus g(G_s), \quad \text{and}$$
$$C^*_{s+2} = C'_u.$$

It remains to show that, when the above forged message is decrypted, the final recovered plaintext block will equal V. Suppose that, when decrypting $C^*_1, C^*_2, \ldots, C^*_{s+2}$, the internal values are F^*_i and G^*_i $(1 \le i \le s+2)$.

We first note that it follows immediately from the definitions that $F^*_i = F_i$ and $G^*_i = G_i$ $(1 \le i \le s)$, where F_i and G_i are the internal values generated during the encryption process that yielded the ciphertext message $C_1, C_2, \ldots, C_t$. We now consider the decryption of C^*_{s+1}.

We have

$$
\begin{aligned}
F^*_{s+1} &= C^*_{s+1} \oplus g(G^*_s) \quad \text{(from Sect. 3)} \\
&= C'_{u-1} \oplus g(G'_{u-2}) \oplus g(G_s) \oplus g(G^*_s) \quad \text{(by defn. of } C^*_{s+1}) \\
&= C'_{u-1} \oplus g(G'_{u-2}) \quad \text{(since } G^*_s = G_s) \\
&= F'_{u-1}.
\end{aligned}
$$

Hence $G^*_{s+1} = G'_{u-1}$.

We now consider the decryption of C^*_{s+2}. We have

$$
\begin{aligned}
F^*_{s+2} &= C^*_{s+2} \oplus g(G^*_{s+1}) \quad \text{(from Sect. 3)} \\
&= C'_u \oplus g(G'_{u-1}) \quad \text{(by defn. of } C^*_{s+2}) \\
&= F'_u.
\end{aligned}
$$

Hence $G^*_{s+2} = G'_u$. Finally, we have

$$
\begin{aligned}
P^*_{s+2} &= G^*_{s+2} \oplus F^*_{s+1} \quad \text{(from Sect. 3)} \\
&= G'_u \oplus F'_{u-1} \quad \text{(from above)} \\
&= P'_u \\
&= V,
\end{aligned}
$$

as required.

7   Further Observations

7.1   Attack Performance

It is of interest to try to understand the overall complexity of the attack described above, i.e. to understand what information is required to complete an attack, and what computations need to be performed. For a 128-bit block cipher, we have shown in Sect. 5 that knowledge of 80 consecutive pairs of known plaintext/ciphertext blocks will be very likely to yield certain knowledge of the 'final' internal variable G_i.

Thus, if the attacker has access to two messages encrypted using the same key, and the attacker knows the final 80 plaintext blocks for both messages (as well as the ciphertext), then the attacker will almost certainly have sufficient information to perform the procedure described in Sect. 6. The computations involved in both stages of the attack are trivial — indeed, they involve only a small number of computations of the (very simple) function g, and some exclusive-or operations on pairs of blocks.

Thus the attack will have a relatively high success probability given only a modest amount of known plaintext, and the computations involved are completely trivial.

7.2   The Choice for the Function g

As should be clear from the discussions above, completing the attack as described in Sect. 6 requires knowledge of internal variables G_i. It is possible to learn these values as a result of the procedure described in Sect. 5; this procedure only works because the function g is non-bijective, and has a very simple structure. This raises two questions.

Firstly, why is the function g chosen to be non-bijective? Secondly, if there is a good reason to use a non-bijective function, then why not use one with less obvious structure?

The answer to the first question can be found in the original paper of Zúquete and Guedes [14]. As stated in [14], EPBC is very similar to a previously devised mode called IOBC, proposed by Recacha in 1996[1]. Indeed, the only difference is that the function g used in IOBC is bijective, involving some fixed bit permutations. However, as very briefly outlined in [14], IOBC is subject to known-plaintext attacks if the message being encrypted contains more than $n^2/4$ blocks, where n is the block cipher block length, i.e. around 4000 blocks for AES. The exact attack approach is not clear from [14], which states that a detailed description can be found in the 1996 paper of Recacha. Because of the existence of this attack, EPBC was designed to use a non-bijective function g, which (apparently) rules out the known-plaintext attacks applying to IOBC.

The answer to the second question is less clear. Obviously, g could be implemented using one or more block cipher encryptions, which, if done correctly, would certainly remove the simple structures exploited in Sect. 5. However, such an approach would significantly increase the complexity of the mode of operation, and one of the main design goals was to devise a scheme with minimal complexity. Designing a function g which is both sufficiently complex to prevent attack, but is nevertheless very fast to perform would, perhaps, be an interesting research question; however, given that highly efficient provably secure authenticated encryption modes are now known to exist (as discussed in Sect. 1), this is probably not likely to be a particularly fruitful line of enquiry.

---

[1] Unfortunately, the 1996 Recacha paper cited in [14] does not appear to be readily available.

8   Summary and Conclusions

In this paper we have demonstrated a forgery attack against EPBC mode when used to provide message integrity. This attack required only known plaintext (and no chosen plaintext). We can therefore conclude that this mode is unacceptably weak, and should therefore not be used. (Whilst it is probably an effective mode for encryption only, much simpler modes are known for this purpose).

If both confidentiality and integrity protection are required, then encryption and a MAC should be combined in an appropriate way, or a dedicated 'authenticated encryption' mode should be used — see, for example, ISO/IEC 19772 [12].

Acknowledgements

The author would like to thank Po Yau for his generous help in performing numerical computations, and anonymous referees for valuable comments which have improved the paper.

References

1. Menezes, A.J., van Oorschot, P.C., Vanstone, S.A.: Handbook of Applied Cryptography. CRC Press, Boca Raton (1997)
2. Bellare, M., Kohno, T., Namprempre, C.: Breaking and provably repairing the SSH authenticated encryption scheme: A case study of the encode-then-encrypt-and-MAC paradigm. ACM Transactions on Information and System Security 7, 206–241 (2004)
3. Black, J., Urtubia, H.: Side-channel attacks on symmetric encryption schemes: The case for authenticated encryption. In: USENIX 2002. Proceedings of the 11th USENIX Security Symposium, San Francisco, CA, USA August 5-9, 2002 pp. 327–338 (2002)
4. Canvel, B., Hiltgen, A., Vaudenay, S., Vuagnoux, M.: Password interception in a SSL/TLS channel. In: Boneh, D. (ed.) CRYPTO 2003. LNCS, vol. 2729, pp. 583–599. Springer, Heidelberg (2003)
5. Paterson, K.G., Yau, A.: Padding oracle attacks on the ISO CBC mode padding standard. In: Okamoto, T. (ed.) CT-RSA 2004. LNCS, vol. 2964, pp. 305–323. Springer, Heidelberg (2004)
6. Vaudenay, S.: Security flaws induced by CBC padding — Applications to SSL, IPSEC, WTLS. In: Knudsen, L.R. (ed.) EUROCRYPT 2002. LNCS, vol. 2332, pp. 534–545. Springer, Heidelberg (2002)
7. Yau, A.K.L., Paterson, K.G., Mitchell, C.J.: Padding oracle attacks on CBC-mode encryption with secret and random IVs. In: Gilbert, H., Handschuh, H. (eds.) FSE 2005. LNCS, vol. 3557, pp. 299–319. Springer, Heidelberg (2005)
8. Rogaway, P., Bellare, M., Black, J.: OCB: A block-cipher mode of operation for efficient authenticated encryption. ACM Transactions on Information and System Security 6, 365–403 (2003)
9. Bellare, M., Rogaway, P., Wagner, D.: The EAX mode of operation. In: Roy, B., Meier, W. (eds.) FSE 2004. LNCS, vol. 3017, pp. 389–407. Springer, Heidelberg (2004)

10. National Institute of Standards and Technology (NIST): NIST Special Publication 800-38C, Recommendation for Block Cipher Modes of Operation: The CCM Mode For Authentication and Confidentiality (2004)
11. Whiting, D., Housley, R., Ferguson, N.: RFC 3610, Counter with CBC-MAC (CCM). Internet Engineering Task Force (2003)
12. International Organization for Standardization Genève, Switzerland: ISO/IEC 3rd CD 19772, Information technology — Security techniques — Authenticated encryption mechanisms (2007)
13. Dent, A.W., Mitchell, C.J.: User's Guide to Cryptography and Standards. Artech House (2005)
14. Zuquete, A., Guedes, P.: Efficient error-propagating block chaining. In: Darnell, M. (ed.) Cryptography and Coding. LNCS, vol. 1355, pp. 323–334. Springer, Heidelberg (1997)
15. Mitchell, C.J.: Cryptanalysis of two variants of PCBC mode when used for message integrity. In: Boyd, C., González Nieto, J.M. (eds.) ACISP 2005. LNCS, vol. 3574, pp. 560–571. Springer, Heidelberg (2005)
16. International Organization for Standardization Genève, Switzerland: ISO/IEC 18033-3, Information technology — Security techniques — Encryption algorithms — Part 3: Block ciphers (2005)

Blockwise-Adaptive Chosen-Plaintext Attack and Online Modes of Encryption

Gregory V. Bard

Department of Mathematics, Fordham University, Bronx, NY, USA

Abstract. Here, we present a generalized notion of online modes of encryption that make one call to a pseudorandom permutation per block of plaintext. This generalization, called "Canonical Form," not only allows for modes of encryption to be written in a common format, but provides for easy proofs of blockwise-adaptive chosen-plaintext (BACPA) security/insecurity.

We also develop necessary and sufficient conditions for security of a mode of encryption in Canonical Form. As an application, we write ten modes of encryption in Canonical Form, and we prove the security status (under BACPA) of nine of them. While most of these modes already had proven BACPA security status in previously published papers, it is hoped the more general method specified here will be of use in writing simpler proofs for other modes, including modes of encryption yet to be developed.

BACPA is a model for adversaries slightly more powerful than those in traditional chosen-plaintext attack. In particular, instead of forcing the target to encrypt messages of his/her own choosing, the attacker can insert blocks of his/her own choosing into the target's messages [JMV02]. Some modes of encryption which are secure against traditional CPA, for example the ubiquitous Cipher Block Chaining (CBC), are insecure against BACPA. Several papers have been written to explore BACPA and modes of encryption under it.

Keywords: Modes of Encryption, Blockwise-Adaptive Chosen-Plaintext Attack, BACPA, MACPA, Online-ness, ECB, CBC, CTR, OFB, CFB, IGE, ABC, HCBC, HPCBC, XCBC.

1 Introduction

In 2002, [BKN02] and [JMV02], simultaneously discovered an attack on CBC. This attack was similar to but not within the scope of chosen plaintext attack (CPA). In [JMV02] the latter was renamed messagewise-adaptive CPA (MACPA), and a new notion, called blockwise-adaptive chosen-plaintext attack or BACPA was created.

The difference between MACPA and BACPA is simple: Instead of inserting messages of his/her own choosing into the target's message stream, the attacker can insert blocks of his/her own choosing into the target's messages. Some modes that are insecure against BACPA are secure against MACPA, for example, CBC

S.D. Galbraith (Eds.): Cryptography and Coding 2007, LNCS 4887, pp. 129–151, 2007.

and IGE. Others are secure against both, like CTR and OFB. But any method secure in the BACPA model is also secure in the MACPA model, as well as the known-ciphertext and known-plaintext models. Thus, BACPA serves as a convenient notion for expressing security against all adversaries less powerful than chosen-ciphertext attacks (CCA).

A natural question that may arise in discussion of any attack model of any kind is that of feasibility. So far, two papers have been published (that we are aware of) which demonstrate cryptographically feasible attacks against systems currently in use. One is the original paper which gave birth to the subject [BKN02] by showing the attack on the Secure Shell or SSH. This attack was placed in the environment of the Secure Sockets Layer (SSL) in [Bar06a]. See those papers for more details.

Many different types of modes of encryption exist but a crucial and large category are those which are online. Unfortunately, there are conflicting ideas of what constitutes an online cipher. A necessary requirement is that when encrypting a message $P_1, P_2, \ldots, P_n$, the corresponding ciphertexts $C_0, C_1, \ldots, C_n$ must be computable such that C_i does not depend on P_j for $j > i$. Note the value C_0 is the initialization vector. We discuss why this notion of online-ness is important in Section 2.2. In [BT04], it is suggested that ciphers which allow P_i to be decrypted upon receipt of $C_0, C_1, \ldots, C_i$, but before receipt of the whole message, be designated "online with decryption on the fly". We suggest denoting these notions as "almost online" and "fully online." But in this paper, we always mean the stricter condition of "fully online."

To examine the relationships between these two notions of "BACPA" and online-ness, we prove four theorems. But first we introduce a Canonical Form. Any mode of encryption which can be written with at most one call to the pseudorandom permutation per plaintext block can be written in Canonical Form. This format yields very simple proofs of security in the BACPA model. The first is that a BACPA attack in the left-or-right indistinguishability model need only have one block in which the left and right plaintexts are unequal. We call BACPA attacks which are restricted in this way "primitive", and show that primitive BACPA security is necessary and sufficient for BACPA security in general. We define three notions, the first two of which, entropy-preserving and collision-verifiable, we believe nearly all modes of encryption will have. The third is collision-resistance. In the second theorem, we show that if a scheme is entropy-preserving, then collision resistance is sufficient for BACPA security. In the third theorem, we show that if a scheme is collision-verifiable, then collision-resistance is necessary for BACPA security. It follows immediately that if a scheme is both collision-verifiable and entropy preserving, it is BACPA secure if and only if it is collision resistant, which is the fourth theorem.

We then examine actual modes of encryption. Ten are presented here. For example, two of them, Electronic Code Book (ECB) and Infinite Garble Extension (IGE), are insecure against MACPA and so they are obviously insecure against BACPA. On the other hand, CBC and certain forms of P-ABC (public accumulated block ciphers), are proven insecure against BACPA even though they are

MACPA secure. Counter Mode (CTR), Output Feedback Mode (OFB), Cipher Feedback Mode (CFB), Hash Cipher Block Chaining (HCBC), Hash Permuted Cipher Block Chaining (HPCBC), and certain forms of (Secret Accumulated Block Ciphers) S-ABC are proven BACPA secure. Finally, we were unable to determine the status of Extended Cipher Block Chaining, or XCBC, although we can write it in Canonical Form.

The rest of the paper is organized as follows. In Section 1.1, we discuss previous work on BACPA. We give some background and definitions in Sections 2 and 3 respectively. The theorems are stated, and Theorems 2–4 are proven, in Section 4. In Section 5, we show which modes are insecure against BACPA, including explicit attacks. The appendices contain proofs of security for several modes that are BACPA secure, the proof of Theorem 1, and a presentation of XCBC in Canonical Form, even though we cannot provide a security proof of it, nor an attack against it, at this time.

1.1   Previous Work on BACPA

As stated earlier, the first published paper in this topic was Bellare, Kohno, and Namprempre [BKN02], wherein the vulnerability in SSH was observed (and fixes proposed). Simultaneously to this, Joux, Martinet, and Valette further identified three additional operational attacks of similar form to Bellare's attack on SSH, naming this new class of attacks "blockwise-adaptive CPA" [JMV02]. Also introduced in that paper is "blockwise-adaptive CCA". It was in this paper that an ingeniously straight-forward response to BACPA was defined, namely Delayed-CBC mode. However, as explained in [BT04], it does not allow for decryption-on-the-fly, and so is not considered here.

Fouque, Joux, Martinet, and Valette next provided the first security definitions for blockwise-adaptive CPA and CCA [FJMV03]. The paper was geared to smart cards in authentication devices, and so is a somewhat different context than on-line schemes. They propose a generic model, Decrypt-then-Mask, proved to be blockwise-adaptive CCA secure under certain conditions, and an example of it, CBC-MAC. However, none of these schemes are on-line and so are not included here.

Boldyreva and Taesombut, focusing on on-line schemes in general, showed that neither HCBC nor any on-line scheme in general is CCA secure under the general definitions of Fouque, et al. [BT04]. Instead, they define a slightly different security definition, IND-BLK-CCA. They prove that the mode HPCBC, which is a variation on HCBC, is IND-BLK-CCA secure. We analyze both HPCBC and HCBC, and prove them blockwise-adaptive CPA secure as well, which confirms their results.

At almost the same time, Fouque, Martinet, and Poupard [FMP03] expanded the idea of Delayed-CBC presented in their previous paper to include a general notion of a delayed cipher. Moreover, these authors introduce an adversary with *concurrent* access to multiple blockwise oracles. All previous papers had limited the adversary to sequential access to the oracle, as does this paper. The authors proved that both delaying modes and CFB are secure under their model.

Fouque, Joux, and Poupard [FJP04] expanded the definitions of security in the messagewise case into the blockwise case. That paper examines the relationships between indistinguishability in the Left-Or-Right, Real-Or-Random and Find-Then-Guess senses, in the attack models of messagewise chosen-plaintext attack, and blockwise-adaptive chosen-plaintext attack with either sequential or concurrent adversaries. An analysis of these security goals for the messagewise case and their relationships is found in Katz and Yung [KY00]. The notion used for security in our paper is LORS-BCPA in their notation, meaning that the adversary is given blockwise capability in every stage of the attack, but for sequential messages (i.e. not for concurrent messages).

1.2   Previous Work on Modes of Encryption

The case of ECB is somewhat obvious. The insecurity of IGE (in the MCPA and thus BACPA context also) was demonstrated by [GD00] and [BBKN01]. Of the three modes that fail unconditionally, the case of CBC was the only one to be MCPA secure, and was the very first mode of encryption to be proven insecure in BACPA but not MACPA, simultaneously in [BKN02] and [JMV02].

Of the five modes that pass, the security of the first, or Counter Mode (CTR), was proven independently by [FJP04] and this author [Bar06b]. Next, Cipher Feedback Mode (CFB) was first proven independently by [BT04], [FMP03] and this author [Bar06b]. We believe we are the first to show the security of the next two modes, Output Feedback Mode (OFB) and Hash Cipher Block Chaining (HCBC) against BACPA in [Bar06b]. The proof of security of Hash Permuted Cipher Block Chaining (HPCBC) was proven by [BT04] as implied by proving what they denote IND-BLK-CCA, a modified form of BACPA that adds limited chosen-ciphertext capabilities to the adversary.

Finally, we believe that we are the first to analyze ABC in the BACPA context.

1.3   Notation

We denote by $x \xleftarrow{R} S$ that x is an element of the set S chosen uniformly at random. The algorithms in this paper are in terms of a secret key sk of length k, which was presumably generated by $sk \xleftarrow{R} \{0,1\}^k$.

The symbol $x||y$ means x concatenated to y.

We define modes of encryption in terms of a block cipher $F : \{0,1\}^k \times \{0,1\}^\ell \to \{0,1\}^\ell$. We will write $F(sk, P) = C$ as $F_{sk}(P) = C$. We assume (as is standard) that F is a pseudorandom permutation family.

2   Background

2.1   Blockwise Encryption Schemes

An encryption scheme is a quadruple of a positive integer and three algorithms $\{\ell, K, \mathfrak{E}, \mathfrak{D}\}$. The positive integer ℓ is the block size. The first algorithm is a key generation algorithm K, which does not concern us, and which outputs sk. Next an algorithm $\mathfrak{E}_{sk}(\boldsymbol{P}) \rightarrow \boldsymbol{C}$ takes a plaintext and produces a ciphertext. The plaintext is a bit string, with length a positive integer multiple of ℓ. The ciphertext is of the same length, but with a "prefix" of zero or one blocks in length. A decryption function also exists, $\mathfrak{D}_{sk}(\boldsymbol{C}) = \boldsymbol{P}$, to map ciphertexts back to their corresponding plaintexts. However, in some systems, $\mathfrak{D}_{sk}$ will output a special symbol $\perp$, which means that the ciphertext was invalid, forged, or improper in some way, and that it should be disregarded. Both of these algorithms are indexed by sk, the secret key, which can alternatively be thought of as an extra parameter.

All encryption schemes discussed in this paper are assumed to be correct. An encryption scheme is correct if and only if $\mathfrak{D}_{sk}(\mathfrak{E}_{sk}(\boldsymbol{P})) = \boldsymbol{P}$ for all messages $\boldsymbol{P}$, with probability one. It is a tautological consequence of this definition that $A \neq B$ implies $\mathfrak{E}_{sk}(A) \neq \mathfrak{E}_{sk}(B)$. The correctness property forces $\mathfrak{D}_{sk}$ to be deterministic, but $\mathfrak{E}_{sk}$ is quite often a randomized algorithm.

2.2 Online Modes of Encryption

Unfortunately, two different notions of online-ness exist in the literature.

The first requirement is that as $\mathfrak{E}_{sk}(P_1, P_2, \ldots, P_n) = C_0, C_1, \ldots, C_n$ is computed, it must be the case that C_i not depend on P_j for all $j > i$. Formally, it must be possible in polynomial time to compute C_i knowing only $P_1, \ldots, P_i$ and sk. At first, this seems like a strange distinction. But it is equivalent to saying that $\mathfrak{E}$ can be computed in one pass (even if it is not actually computed that way). This paper only deals with online schemes but most encryption schemes relating to block ciphers in the literature are online with this definition.

The primary reason for concern of this type is encryption in small devices like smart cards. In particular, a smart card might not have enough memory for storing all the plaintext of a message. Such storage would be needed in a two-pass algorithm, but in a one pass algorithm, C_i could be transmitted, and P_i discarded before C_{i+1} was calculated. Dependencies between blocks could be maintained by state variables. Another example is a tiny sensor in a large network. It may be expected to broadcast data over a month-long period. Using a two-pass scheme would require enormous storage. Finally consider a service that sells movies over the internet, which are encrypted. If the entire movie needed to be read from storage and encrypted, and only then transmitted, the user might lose patience. If transmission could occur a few seconds into the encryption process, and proceed as a stream, then this might be more tolerable for users.

The second requirement was identified by [BT04], but may have been mentioned earlier. It should be possible to decrypt C_i into P_i without C_j for $j > i$. Formally, it must be possible in polynomial time to compute P_i given only sk and $C_0, C_1, \ldots, C_i$. At first glance, this seems very similar to the previous definition. In [BT04], this is denoted "online with decryption on-the-fly," because online already implies encryption on-the-fly.

But consider the very common Encrypt-then-MAC strategy. The ciphertext is computed with a CPA-secure encryption scheme. Then that ciphertext is input to a message-authentication code (MAC). The tag outputted by the MAC is included as part of the ciphertext. Upon receipt, the decoder first verifies that the cipher is authentic by first computing his/her own tag and comparing it to the tag provided. If the tags match, only then is the decryption algorithm called. Otherwise the ciphertext is discarded as a forgery. This results in a CCA-secure scheme. Unfortunately, the entire ciphertext must be available in order for the computation of the tag to be completed.

Consider the internet movie provider mentioned above. The entire movie would have to be resident on the machine, and the tag checked, before the first frame could be shown. The normal buffering strategy even for services like U-tube [utu] is that a many minute movie might well take several minutes to download, but after several seconds (as a buffer) are downloaded, the movie starts and the download continues in the background. This would be impossible if a MAC was being used. Of course, the data can be packeted at a high-level and the packets can be MAC'ed, but with non-trivial overhead. Likewise, for the sensor network, the entire month's data would have to be present for the tag to be computed, or a packetization strategy used. A counter-argument to the packetization option comes to mind when considering how the movie viewing software or a sensor network's data aggregator should behave if the first packet passes and the second fails the MAC. Video footage shown to a user cannot be unshown. A sufficiently strong MAC undoes this objection.

We suggest that schemes which meet the first requirement only be called "online-1" or "almost-online." Schemes that meet both should be called "online-2" or "fully online." In this paper, we only consider schemes which satisfy both requirements.

2.3   Modes of Encryption

A mode of encryption is a set of algorithms, which when given some block cipher F, yield an encryption scheme. The model called "stateful encryption," mentioned in for example in [BRS02] but probably defined earlier, and used here in a slightly modified form, is descriptive of many modes in practice, and is described below. In Section 3, however, we will propose an alternative notation called "Canonical Form," that will enable our proofs.

It should be noted that modes which require two passes over the data, such as certain implementations of the "encrypt-then-MAC" paradigm, cannot be described in the "stateful encryption" model, which allows only one pass. However, a system that must be written with two or more passes is not online, because C_i must have some dependency on P_j for $j > i$ (otherwise it could be re-written to be executed in one pass).

Prior to encryption or decryption, the plaintext or ciphertext is divided into blocks of length ℓ, the block-length of the block cipher. We write $\boldsymbol{P} = P_1, P_2, \ldots, P_n$. In this paper, we are not concerned with padding, so we assume that the plaintext can always be evenly divided into blocks.

First, an algorithm $Begin_{sk}()$ will output an initial state s_1, and an initialization vector C_0. Second, an encryption algorithm $E_{sk}(P_i, s_i)$ will take as inputs the secret key, the plaintext, and the current state. It will output the next ciphertext block C_i, and the next state, s_{i+1}. The state is some binary string, the format and significance of which depends on the chosen mode of encryption. However, typical choices for the state are the previous ciphertext block, the previous plaintext block, or a counter.

For decryption, an algorithm $Setup_{sk}(C_0)$ will output an initial state s_1, based on C_0. From there, a decryption algorithm $D_{sk}(C_i, s_i)$ will return the next plaintext block P_i, and the state s_{i+1}. This function is called repeatedly until all the ciphertext blocks are decrypted. If any ciphertext block causes D_{sk} to yield $\perp$, then the entire ciphertext should be considered to decrypt to $\perp$. In this paper, we are not concerned with decryption and so we will not mention $Setup$ or D_{sk} again.

$\mathfrak{E}_{sk}(\boldsymbol{P})$
1. Divide $\boldsymbol{P}$ into $P_1, \ldots, P_n$, blocks of length ℓ bits.
2. $(C_0, s_1) \leftarrow Begin_{sk}$
3. Output C_0
4. For $i = 1, 2, \ldots, n$ do
   (a) $(C_i, s_{i+1}) \leftarrow E_{sk}(P_i, s_i)$
   (b) Output C_i

$\mathfrak{D}_{sk}(\boldsymbol{C})$
1. Divide $\boldsymbol{C}$ into $C_0, C_1, \ldots, C_n$, blocks of length ℓ bits.
2. $s_1 \leftarrow Setup_{sk}(C_0)$
3. For $i = 1, 2, \ldots, n$ do
   (a) $(P_i, s_{i+1}) \leftarrow D_{sk}(C_i, s_i)$
   (b) Output P_i
4. If any of $P_1, \ldots, P_n = \perp$, abort.

All of these algorithms are in terms of an abstract block cipher F. Only when a specific block cipher F is chosen, can they be used. Thus the choice of block cipher is a hidden input to the various functions, but to write that would only clutter the notation.

Given the above, and the assumption that no padding is ever required, it is very easy to see how to generate $\mathfrak{E}_{sk}$ and $\mathfrak{D}_{sk}$ from $Init_{sk}, E_{sk}, Setup_{sk}$ and D_{sk}. For key generation, $K : sk \xleftarrow{R} \{0,1\}^k$, where k is the key-length of the block cipher, is used. Thus every mode of encryption, coupled with a block cipher, produces an encryption scheme. Lastly, note that in practice, improper padding can lead to attacks, such as [Vau02].

This model is similar to that used in Bellare and Boldyreva's several papers [BKN02], [BBKN01], [BDJR97], [BT04] as well as [FJP04]. In Section 3.1 we will propose an alternative model, called "Canonical Form", which allows for very simple proofs of BACPA security.

2.4 The BACPA Game

Before we formally define the BACPA left-or-right indistinguishability game, we will generally describe the MCPA and MACPA left-or-right indistinguishability games to build intuition. Also, we will avoid the cumbersome phrase "left-or-right indistinguishability game" and simply say "game" for the remainder of the paper.

Loosely speaking, the messagewise-CPA game, for an encryption scheme $\mathfrak{E}$, proceeds as follows. A secret key is generated, $sk \xleftarrow{R} \{0,1\}^k$, and a secret coin is tossed: $b \xleftarrow{R} \{0,1\}$. The adversary is presented with an oracle that takes as input two messages $\boldsymbol{P}_0, \boldsymbol{P}_1$, and outputs $\mathfrak{E}_{sk}(\boldsymbol{P}_b)$. The adversary composes a set of challenges $(\boldsymbol{A}_0, \boldsymbol{A}_1); (\boldsymbol{B}_0, \boldsymbol{B}_1); \ldots$ and transmits them to the oracle. The oracle returns $\mathfrak{E}_{sk}(\boldsymbol{A}_b), \mathfrak{E}_{sk}(\boldsymbol{B}_b), \ldots$ Now the adversary makes a guess $\hat{b}$ of the value of b. If it turns out that $b = \hat{b}$ then adversary wins, and if not, loses. Of course, the adversary is limited to polynomial time.

Suppose that $\boldsymbol{A}_1$ was very long and that $\boldsymbol{A}_0$ was very short. Then $\mathfrak{E}(\boldsymbol{A}_b)$ would be easily determined to be either $\mathfrak{E}(\boldsymbol{A}_1)$ or $\mathfrak{E}(\boldsymbol{A}_0)$ by its length. Therefore, it is required that $|\boldsymbol{A}_1| = |\boldsymbol{A}_0|$, and so forth for all pairs, but it can be that $|\boldsymbol{A}_1| \neq |\boldsymbol{B}_1|$. Finally note that since the adversary runs in polynomial time (compared to k), it will always be the case that only polynomially many messages are transmitted and their total length is upper-bounded by a polynomial in terms of k.

In the messagewise-adaptive-CPA game, or MACPA, the procedure is slightly different. Here the adversary submits $\boldsymbol{A}_0, \boldsymbol{A}_1$ and receives back $\mathfrak{E}_{sk}(\boldsymbol{A}_b)$. Only then does he/she submit $\boldsymbol{B}_0, \boldsymbol{B}_1$. Information obtained from the first oracle output can be used to determine the second oracle inputs, and so forth. Thus the name adaptive. This is the traditional model of CPA in cryptography.

In the BACPA game, blocks rather than messages will be submitted. Since this game requires "blocks", it must be described using a more restrictive model than just a blockwise encryption scheme. We will use "stateful encryption" as the model, as is standard.

First, the secret key and fair coin will be generated. Then the adversary will submit (**start**, **start**) (a special symbol created for the game). For each later submission, the adversary can either submit (**start**, **start**) or some pair of blocks (A_0, A_1). Upon receipt of (**start**, **start**), the oracle begins a new message by using $Init_{sk}$ and outputs C_0, but not s_1. Upon receipt of (A_0, A_1), two plaintext blocks, the block A_b is selected, and $C = E_{sk}(A_b, s_i)$ is outputted, but not s_{i+1}. In effect, this means that A_b is appended as the next plaintext block in the currently-encrypted message. In either case, the adversary is shown the resulting ciphertext block, not the state variable, and is asked for the next query. This repeats until the adversary submits (**stop**, **stop**). The adversary must then provide a guess $\hat{b}$, and he/she wins if and only if $b = \hat{b}$.

Note that it is not permitted for the adversary to submit $(A_1, \textbf{start})$ or $(\textbf{start}, A_1)$ with $A_1 \neq \textbf{start}$ because this would allow messages of different length to be present, which cannot be tolerated for reasons explained above.

Definition 1. *The* **advantage of the adversary in the BACPA game** *is the absolute value of the difference between the probability that he/she wins and one-half.*

Definition 2. *For a mode of encryption S, if the maximum advantage over all probabilistic polynomial-time (PPT) adversaries, and all choices of F_{sk}, is negligible, then the mode of encryption is said to be* **BACPA-secure.**

Table 1. The BACPA Game

1. $b \xleftarrow{R} \{0,1\}$
2. $sk \xleftarrow{R} \{0,1\}^k$
3. Adversary submits (**start**, **start**)
4. $(C_0, s_1) \leftarrow Begin_{sk}$
5. Transmit C_0 to the adversary
6. $Done \leftarrow 0; i \leftarrow 1$
7. Do
    (a) Obtain a query (A_0, A_1) from the adversary
    (b) If $(A_0, A_1) = ($**start**, **start**$)$ then
        i. $(C_0, s_1) \leftarrow Begin_{sk}$
        ii. Transmit C_0 to the adversary
        iii. $i \leftarrow 1$
    (c) else if $A_0 = $ **start** or $A_1 = $ **start** then HALT
    (d) else if $A_0 = $ **stop** or $A_1 = $ **stop** then $Done \leftarrow 1$
    (e) else
        i. $(C_i, s_{i+1}) \leftarrow E_{sk}(A_b, s_i)$
        ii. Transmit C_i to the adversary
        iii. $i \leftarrow i + 1$
    Until $(Done = 1)$
8. Request a guess $\hat{b}$ from the adversary
9. If $b = \hat{b}$ then tell adversary he/she has won
10. else tell adversary he/she has lost

2.5   Online Encryption and CCA-Security Are Incompatible

This section can be found in the full version of the paper, [Bar06b].

3   Definitions

We introduce in this section several new concepts which enable security proofs for various modes of encryption.

3.1   Canonical Form

The model described in Section 2.3, called "stateful encryption", is descriptive of modes in practice. However, all modes of encryption we are aware of are based on a single block cipher, and make one call per single block of encryption. This additional fact allows one to be more specific. To model these modes of encryption, we define Canonical Form, a model which will enable very simple proofs of BACPA-security.

    A **mode of encryption in Canonical Form** is a quadruple of algorithms, {Init, Pre, Post, Update}. The general concept is that the function $E_{sk}(P_i, s_i)$ uses some pseudorandom permutation family member $F_{sk}()$ to calculate the ciphertext and next state. Therefore consider the following: Some function of the inputs of E_{sk} is the input to F_{sk}. Next, some function of the output of F_{sk}

and the inputs of E_{sk} becomes the ciphertext block. Finally, some function of the output of F_{sk} and the inputs of E_{sk} becomes the next state. This gives rise to the following quadruple.

- $\text{Init}_{sk}(k) \rightarrow (C_0, s_1)$
- $\text{Pre}(P_i, s_i) \rightarrow x_i$
- $\text{Post}(P_i, s_i, y_i) \rightarrow C_i$ where $y_i = F_{sk}(x) = F_{sk}(\text{Pre}(P_i, s_i))$.
- $\text{Update}(P_i, s_i, y_i) \rightarrow s_{i+1}$.

It should be clear that any mode of encryption, that makes exactly one call to the pseudorandom permutation (F_{sk}) per plaintext block, can be modeled in this way. Modes that make more than one call cannot be so modeled, but we are not aware of any such modes in practice.

$Begin_{sk}$ $E_{sk}(P_i, s_i)$
1 return $Init_{sk}$ 1 $x_i \leftarrow \text{Pre}(P_i, s_i)$
                               2 $y_i \leftarrow F_{sk}(x_i)$
                               3 $C_i \leftarrow \text{Post}(P_i, s_i, y_i)$
                               4 $s_{i+1} \leftarrow \text{Update}(y_i, P_i, s_i)$
                               5 return (C_i, s_{i+1})

Technical Note: These four functions are making use of F_{sk}, but are intended to define an abstract mode regardless of the choice of F, so long as it is a pseudorandom permutation family. This means that F_{sk} itself is actually a hidden parameter of each of these functions, but writing it that way would only clutter the notation.

Examples: Here we describe CBC, CTR, and OFB as three examples. Later, we will add ECB, CFB, ABC, IGE, HCBC, HPCBC and XCBC. All ten of these modes of encryption can be written in Canonical Form.

For CBC, one normally writes:
$$r_1 \xleftarrow{R} \{0,1\}^k ; \quad C_0 = r_1$$
$$C_i = F_{sk}(C_{i-1} \oplus P_i)$$

However, in Canonical Form:
$$\text{Init}_{sk}(k) : r \xleftarrow{R} \{0,1\}^k ; \quad C_0 = s_1 = r$$
$$\text{Pre}(P_i, s_i) = P_i \oplus s_i; \quad \text{Post}(P_i, s_i, y_i) = y_i; \quad \text{Update}(P_i, s_i, y) = y$$

Note: above the state s_i represents the previous ciphertext. For CTR, the standard notation is
$$\text{Init}_{sk}(k) : i_0 \xleftarrow{R} \{0,1\}^k ; \quad C_0 = F_{sk}(i_0)$$
$$C_i = F_{sk}(i + i_0) \oplus P_i$$

However, in Canonical Form:
$$\text{Init}_{sk}(k) : r \xleftarrow{R} \{0,1\}^k ; \quad C_0 = F_{sk}(r); \quad s_1 = r + 1$$
$$\text{Pre}(P_i, s_i) = s_i; \quad \text{Post}(P_i, s_i, y_i) = y \oplus P_i; \quad \text{Update}(P_i, s_i, y) = s_i + 1$$

Note above the state s_i represents a counter. For OFB, the classic form would be

$$x_0 \overset{R}{\leftarrow} \{0,1\}^k; \quad C_0 = x_0$$
$$x_{i+1} = F_{sk}(x_i); \quad C_i = x_i \oplus P_i$$

However, in Canonical Form:

$$\text{Init}_{sk}(k): r \overset{R}{\leftarrow} \{0,1\}^k; \quad C_0 = s_1 = r$$
$$\text{Pre}(P_i, s_i) = s_i; \quad \text{Post}(P_i, s_i, y_i) = y_i \oplus P_i; \quad \text{Update}(P_i, s_i, y_i) = y_i$$

Note above the state s_i represents the "recycled" output x_i of the pseudorandom permutation F_{sk}, which is fed back into F_{sk} during the next block.

From these examples, it should be clear that if $\{\text{Init}, \text{Pre}, \text{Post}, \text{Update}\}$ are known, and some F is chosen, that $\mathfrak{E}_{sk}$ and $\mathfrak{D}_{sk}$ are completely defined.

3.2 The Collision Game

Informally, the objective of this game is to create a collision in the outputs of Pre. Since the output of Pre serves as the sole input to F_{sk}, a collision on the output of Pre is also a collision on the output of F_{sk}, and thus provides a potential opportunity for attack. By collision, it is meant that that the output of Pre is equal for two distinct blocks, and the block-numbers of that pair are known. With this objective in mind, the definition is rather straight-forward.

The adversary is given access to a blockwise-encryption oracle $E_{sk}(\cdot)$. The adversary is not given the state s_i. This oracle is meant to represent a realistic blockwise attacker who can submit blocks for encryption and view their ciphertext, but who does not have access to the internal state of the cipher (though perhaps he/she may calculate it by other means). The adversary can submit **start** queries to begin new messages.

Formally, the game proceeds as follows. A call is made to $E_{sk}(\textbf{start})$ and the adversary is given C_0. Next, the adversary submits P_1, and receives C_1. This continues, submitting P_i and receiving C_i for polynomially many blocks. Some of the P_i's might equal **start** resulting in several, but still polynomially many, messages. Then the adversary must output a block-number t, and a potential plaintext $P*$. The adversary is successful if the collision would actually occur, namely $\text{Pre}(P*, s_i) = \text{Pre}(P_t, s_t)$.

To accommodate several messages, the plaintext blocks should be numbered only once, not resetting to 1 after each **start**. Otherwise it is unclear what P_1 refers to. To keep the indexes clean, we write s_i and C_i similarly.

Definition 3. *A mode of encryption is said to be* **collision-resistant** *if all* PPT*adversaries have negligible probability of success in the collision game.*

3.3 Two More Conditions

We show in Theorem 4 that if the following two security properties hold, then blockwise-adaptive chosen-plaintext security follows if and only if the mode of

Table 2. The Collision Game

1. $sk \xleftarrow{R} \{0,1\}^k$
2. Adversary submits (**start**, **start**)
3. $(C_0, s_1) \leftarrow Begin_{sk}$
4. Transmit C_0 to the adversary
5. $Done \leftarrow 0; i \leftarrow 1$
6. Do
   (a) Obtain a query P from the adversary
   (b) If $P = $ **start** then
      i. $(C_i, s_{i+1}) \leftarrow Begin_{sk}$
      ii. Transmit C_i to the adversary
      iii. $i \leftarrow i + 1$
   (c) else if $P = $ **stop** then $Done \leftarrow 1$
   (d) else
      i. $(C_i, s_{i+1}) \leftarrow E_{sk}(P)$
      ii. Transmit C_i to the adversary
      iii. $i \leftarrow i + 1$
   Until $(Done = 1)$
7. Request $(P*, t)$ from the adversary.
8. If it is true that $\text{Pre}(P*, s_i) = \text{Pre}(P_n, s_n)$ then tell the adversary he/she has won.
9. else tell the adversary he/she has lost.

encryption is collision-resistant. This is intuitive, because the pseudorandom permutation is the "work horse" of the encryption scheme, and if its input can be duplicated, then its output will be duplicated. If this condition can be detected, then an attack against the scheme can be built. (This is essentially how the original blockwise-adaptive attack operates against CBC mode, as discovered by Bellare, et al [BKN02]). We go further, and show conditions merely necessary or merely sufficient for security, see Theorems 3 and 2.

Definition 4. *Let* $S = \{Init, Pre, Post, Update\}$ *be a mode of encryption in Canonical Form. Then* S *is* **entropy-preserving** *if, when the plaintext and state are held constant, the function from* y *to the ciphertext is injective.*

All modes of encryption (that we know of) are entropy-preserving. One should note that XOR with a constant is an injective map, and this is the most common function choice for Post along with the identity map (which is also injective). Of the ten modes of encryption analyzed in this paper, all are entropy-preserving. Loosely speaking, the reason that this condition would essentially always be present is as follows:

The "data processing inequality" states that the entropy of $f(X)$, where f is a function and X is a random variable, is less than or equal to the entropy of X [CT06, Ch. 2]. The necessary and sufficient condition for equality is an injective function f. Assume for the moment that y is a random variable with entropy h, and the plaintext and state are held constant. If Post is injective, then the ciphertext has entropy h also. Moreover, if Post is not injective, then the ciphertext has entropy strictly less than h.

In the proof of Theorem 2, we exploit this property by noting that a uniform random variable has maximal entropy over its domain, and so if the input y is uniformly random, and the plaintext and state are fixed, then the ciphertext is also uniformly random (though possibly over a different set).

A related concept is collision-verifiability.

Definition 5. *A mode of encryption* $\{Init, Pre, Post, Update\}$ *is* **collision-verifiable** *if there exists an algorithm, such that when given a plaintext message* $\boldsymbol{P} = P_1, \ldots, P_n$, *a ciphertext message* $\boldsymbol{C} = C_0, \ldots, C_n$, *and a block number* i, *which will with success probability non-negligibly greater than one-half, output one when the outputs of Pre at blocks* i *and* n *are equal, (i.e.* $Pre(P_i, s_i) = Pre(P_n, s_n)$ *and thus* $y_i = y_n$) *and zero at all other times.*

All the modes of encryption analyzed in this paper but XCBC are collision-verifiable.

3.4   Primitive BACPA Security

In the MACPA case, it turns out only one query of the form $(\boldsymbol{P}, \boldsymbol{Q})$ with the messages $\boldsymbol{P} \neq \boldsymbol{Q}$ is required. It turns out in BACPA the same is true, that only one query need have (P, Q) with the blocks $P \neq Q$. The proof proceeds along similar lines (the hybrid argument). To describe this, we define the following notion.

Definition 6. *For a mode of encryption* S, *if the maximum advantage over all probabilistic polynomial-time (*PPT*) adversaries which happen to restrict themselves to only one query of the form* (P, Q) *with* $P \neq Q$, *and all choices of* F_{sk}, *is negligible, then the mode of encryption is said to be* **primitively BACPA-secure.**

The term **"primitive BACPA-game"** refers to the BACPA-game, but when the adversary is restricted to at most one query of the form (P, Q) with $P \neq Q$. If two such queries are submitted the game will react as if $(\mathbf{start}, Q)$ was submitted, with $Q \neq \mathbf{start}$.

4   Main Results

The main result of this paper is the following statement: If $\{Init, Pre, Post, Update\}$ is an entropy-preserving and collision-verifiable encryption scheme, then it is generally blockwise-adaptive chosen-plaintext secure if and only if it is collision-resistant. This will be proven in three steps, each of which is a theorem below. We also show that each of the last two conditions results in collision resistance becoming a sufficient/necessary condition, respectively, for BACPA-security.

Theorem 1. *Let* $S = \{Init, Pre, Post, Update\}$ *be a mode of encryption in Canonical Form. Then* S *is primitively blockwise-adaptive chosen-plaintext secure, if and only if it is generally blockwise-adaptive chosen-plaintext secure.*

Proof. See AppendixA. The proof is essentially the classical hybrid argument, (e.g. see [BDJR97], or many other papers).

Theorem 2. *Let S={Init, Pre, Post, Update} be a mode of encryption in Canonical Form. If S is collision-resistant and entropy-preserving, then it is primitively blockwise-adaptive chosen-plaintext secure.*

Proof. Assume S is an entropy-preserving encryption scheme that is collision resistant. Assume there exists a PPT Algorithm A, which can win the primitive blockwise-adaptive chosen-plaintext game, with non-negligible advantage δ_A. Then we will construct a PPT algorithm $Dist$ which will win the pseudorandom game against F_{sk} with non-negligible advantage. Since F_{sk} is assumed to be pseudorandom, this is a contradiction. Thus Algorithm A does not exist. Therefore {Init, Pre, Post, Update} is primitive BW-CPA secure (and by the first theorem, is also generally BW-CPA secure).

Algorithm $Dist$, attempting the pseudorandom game against F_{sk}, is given an oracle F', which is either a random function (case 0), or F_{sk} for a randomly chosen value of sk (case 1). Algorithm $Dist$ will use F' with {Init, Pre, Post, Update} acting as an encryption scheme. It will challenge Algorithm A, and perfectly simulate the primitive blockwise-adaptive chosen-plaintext game.

If Algorithm A wins the game, Algorithm $Dist$ will guess that $F' = F_{sk}$ for some sk in the key-space, and output 1. If Algorithm A loses the game, Algorithm $Dist$ will guess that F' is a random function, and output 0. Let us now analyze both of these cases individually.

The Random Case. Consider the outputs of Pre over the course of the game. We claim that the probability of any pair of them being equal is negligible. The reason for this is not obvious. Let the probability of at least one pair of outputs of Pre being equal be P_e. (Thus with probability $1 - P_e$ all the outputs of Pre are distinct).

Let n be the median[1] value of the block number of the latter block of the first matching pair, when at least one pair of outputs is equal. At most half the time, one block of the first pair will be before block n and its mate will be after block n. But at least half the time, both blocks of the first pair will be before or at block n. Thus with probability at least $\frac{P_e}{2}$ there is at least one pair of blocks before block n+1, which have equal outputs from Pre.

Algorithm $Coll$ will generate a random number r, taken uniformly from the set $\{2, \ldots, n\}$. It will attempt the Collision Game by running a perfect simulation of the primitive BW-CPA game on behalf of Algorithm A. However, it will halt Algorithm A immediately after it submits the query request for block r. With probability $\frac{1}{(n-1)} \frac{P_e}{2}$, the output of Pre during the encryption of block r will collide the output of Pre for a previous block.

---

[1] Since the random variable (the block number) is a positive integer, call it z, there exist many m such that $\Pr\{z \in [1, m]\} \geq 1/2$. Let the median be the least of all such m. Since any collection of positive integers has a lower bound, such a median exists.

Algorithm *Coll* doesn't know which block is the mate for block r, and so will guess a random value s uniformly from the set $\{1, \ldots, r-1\}$. However, this will be correct with probability at worse $\frac{1}{r-1}$. In the case when block r is indeed going to cause a collision, the probability that s will be the mate for r is given by:

$$\frac{1}{n-1} \sum_{r=2}^{r=n} \frac{1}{r-1} = \left(\frac{1}{n-1}\right)\left(1 + \frac{1}{2} + \frac{1}{3} + \frac{1}{4} + \frac{1}{5} + \cdots + \frac{1}{n}\right) \approx \frac{\ln(n) + \gamma}{n-1}$$

Where γ signifies the Euler-Mascheroni Constant[2]. The Algorithm *Coll* thus produces a correct collision at r while guessing the mate s, (and therefore winning the collision game) with probability approximately

$$\frac{P_e(\ln(n) + \gamma)}{2(n-1)^2}$$

However, we can simplify this by noting that $\frac{\ln(n) + \gamma}{2(n-1)^2} > \frac{1}{2n^2}$ and so Algorithm *Coll* will succeed with probability greater than $\frac{P_e}{2n^2}$. Since n must be at most polynomial to k, we know that this success probability will be non-negligible compared to k if P_e is non-negligible. However, by assumption, the mode is collision-resistant, and so therefore P_e is negligible.

Thus the outputs of Pre over all blocks of the game are distinct with all but negligible probability. Furthermore, a series of distinct inputs to a random function creates a sequence of independent, identically and uniformly distributed random outputs.

Recall that the inputs to Post are the plaintext to be encrypted, the state s, and y (which is the output of the pseudorandom function). Further, we required that Post be one-to-one (an injection) when the plaintext and state are fixed. Due to this condition, when the y's are a sequence of uniform independent and identically distributed variables, the outputs of Post (the ciphertexts) are also uniform independent and identically distributed variables, or white noise. These cannot convey any information about the plaintext, and the best that the Algorithm A can output is a fair coin.

Thus Algorithm A will output a 1 or 0 with probability $\frac{1}{2} + \epsilon$, where $|\epsilon|$ is negligible. This extra negligible advantage comes from the fact that the accidental collisions described in detail above occur with negligible but non-zero probability.

The Pseudorandom Case. Here Algorithm *Dist* is providing Algorithm A with a perfect simulation of the blockwise-adaptive chosen-plaintext game. Therefore Algorithm A will be correct with non-negligible advantage δ_A by assumption. Thus Algorithm *Dist* will be correct (output 1) with probability $\frac{1}{2} + \delta_A$.

The Advantage of Algorithm *Dist*: In the Pseudorandom Case an output of 1 occurs with probability $\frac{1}{2} + \delta_A$, and in the Random Case with probability $\frac{1}{2} + \epsilon$.

---

[2] Recall that $\Sigma_{i=1}^{i=n} \frac{1}{i} \approx ln(n) + \gamma$ for very large n.

Thus the advantage of the Algorithm $Dist$ is $\delta_A - \epsilon$, which is non-negligible. This is a contradiction, and so we know that Algorithm A must not exist. Therefore the encryption scheme is primitive blockwise-adaptive chosen-plaintext secure.

Theorem 3. *Let $S = \{Init, Pre, Post, Update\}$ be a mode of encryption in Canonical Form. If S is collision-verifiable, and is primitively blockwise-adaptive chosen-plaintext secure, then it is collision-resistant.*

Proof. Suppose S is collision-verifiable and primitively blockwise-adaptive chosen-plaintext secure. Suppose further that it is not collision-resistant. This means there exists a PPT Algorithm $Coll$ which can win the collision game with non-negligible probability δ_C. We will use that algorithm to construct an Algorithm A which wins the primitive blockwise-adaptive chosen-plaintext game with non-negligible advantage. This contradicts the security of $\{Init, Pre, Post, Update\}$ and therefore Algorithm $Coll$ does not exist, and the mode is collision-resistant.

Algorithm A begins by calling Algorithm $Coll$. Algorithm A will pass Algorithm $Coll$'s oracle queries $E_{sk}(P_i)$ to the left-or-right encryption oracle as $LRBW_{(b,sk)}(P_i, P_i)$, and return the ciphertexts C_i.

When Algorithm $Coll$ terminates, it outputs $P*$ and a block number t. Furthermore, if Algorithm $Coll$ wins the collision game, we know $\text{Pre}(P*, s_n) = \text{Pre}(P_t, s_t)$. Let the probability that it wins the collision game be δ_C.

The split query $LRBW_{(b,sk)}(Q, P*)$ is now submitted to the blockwise oracle, where $Q \neq P*$ is a random string of appropriate length. The output C_n is received. The collision verification algorithm, which we will denote Algorithm Ver, will now be called. Its inputs are the plaintext message $P_1 \ldots P_{n-1}P*$ as well as $C_0 \ldots C_n$ and the block number t. Let its success probability[3] be $1/2 + \delta_V$. Algorithm A will output the bit that Ver outputs.

If Algorithm $Coll$ has succeeded, and the bit is one, $P*$ was encrypted and so then $\text{Pre}(s_t, P_t) = \text{Pre}(s_n, P*)$, and therefore also $y_t = y_n$, because $y_i = F_{sk}(\text{Pre}(P_i, s_i))$. With probability $1/2 + \delta_V$, Algorithm Ver will output one, and Algorithm A will be correct. If the bit is zero, then the plaintext block Q and not $P*$ will be encrypted, and so $C_0 \ldots C_n$ will not decrypt to $P_1 \ldots P_{n-1}P*$ (since $C_0 \ldots C_n$ decrypts to $P_1 \ldots P_{n-1}Q$, and $A \neq B \Rightarrow \mathfrak{E}_{sk}(A) \neq \mathfrak{E}_{sk}(B)$). Thus, Algorithm Ver will output zero with probability $1/2 + \delta_V$, and Algorithm A will be correct.

If Algorithm $Coll$ has failed, then with probability $1/2 + \delta_V$, the verification algorithm will output zero. If the hidden bit is actually zero, then this is correct. If the bit is one, then Algorithm A will be wrong. So in this case, Algorithm A is correct with probability one-half. This is all summarized in Table 3.

$$Adv_A = |Pr[A = 1|b = 0] - Pr[A = 1|b = 1]|$$
$$= |[(1/2)(\delta_C)(1/2 - \delta_V) + (1/2)(1 - \delta_C)(1/2 - \delta_V)]$$
$$- [(1/2)(\delta_C)(1/2 + \delta_V) + (1/2)(1 - \delta_C)(1/2 - \delta_V)]|$$
$$= |\delta_C\delta_V|$$

---

[3] Note, the success probabilities of Algorithm $Coll$ and Algorithm Ver are independent.

Table 3. Summary of Probabilities in Theorem 4

Bit b	Alg Coll	Alg Ver	Is $y_n = y_t$?	Output	Probability
0	Succ	Succ	No	0	$(1/2)(\delta_C)(1/2 + \delta_V)$
0	Succ	Fail	No	1	$(1/2)(\delta_C)(1/2 - \delta_V)$
0	Fail	Succ	No	0	$(1/2)(1 - \delta_C)(1/2 + \delta_V)$
0	Fail	Fail	No	1	$(1/2)(1 - \delta_C)(1/2 - \delta_V)$
1	Succ	Succ	Yes	1	$(1/2)(\delta_C)(1/2 + \delta_V)$
1	Succ	Fail	Yes	0	$(1/2)(\delta_C)(1/2 - \delta_V)$
1	Fail	Succ	No	0	$(1/2)(1 - \delta_C)(1/2 + \delta_V)$
1	Fail	Fail	No	1	$(1/2)(1 - \delta_C)(1/2 - \delta_V)$

Since the system is collision-verifiable δ_V is non-negligible, and δ_C is non-negligible by assumption. Thus the advantage of Algorithm A is then obviously non-negligible, which contradicts our assumption that Algorithm A is primitive blockwise-adaptive chosen-plaintext secure. Thus our assumption that the scheme is not collision-resistant must be false.

Theorem 4. *Let $S=\{Init, Pre, Post, Update\}$ be a mode of encryption in Canonical Form. If S is entropy-preserving, and collision-verifiable, then it is generally blockwise-adaptive chosen-plaintext secure if and only if it is collision-resistant.*

Proof. Suppose the scheme is collision-resistant and entropy-preserving. By the second theorem it is primitively BACPA secure, and thus by the first theorem, generally blockwise-adaptive chosen-plaintext secure. Alternatively, suppose the scheme is generally blockwise-adaptive chosen-plaintext secure. Then it is primitively secure, by the first theorem, and since it is collision-verifiable, then by the third theorem it is collision-resistant.

5 Modes That Fail

Here we will show explicit BACPA attacks for ECB, CBC, IGE, and certain forms of ABC. In each case, the equality of two ciphertext blocks demonstrates an internal collision, namely on the input to the pseudorandom permutation F_{sk}. Thus, the role of collision-resistance and collision-verifiability is clear. In particular, each of these modes of encryption are not collision-resistant (because the attacks exist), and collision-verifiable by means of the equality check which reveals the collision.

5.1 Electronic Codebook, or ECB

Electronic Codebook Mode, or ECB, is extremely simple but insecure even against MACPA or MCPA. The definition is $C_i = F_{sk}(P_i)$, and there is no internal state, initialization vector, nor a C_0. We include it only for completeness. Consider the following BACPA-game.

1. Submit (**start**, **start**) and receive nothing.
2. Submit (P, Q), with $P \neq Q$ and receive C_1.
3. Submit (P, P) and receive C_2.
4. If $C_2 = C_1$ guess $b = 0$, else guess $b = 1$.

Clearly $C_2 = F_{sk}(P)$. If $b = 0$ then $C_1 = F_{sk}(P)$ otherwise $C_1 = F_{sk}(Q)$. Thus we see that $C_1 = C_2$ if and only if $b = 0$. Therefore, the adversary wins with probability one, and ECB is not a BACPA-secure mode of encryption.

For completeness, to write ECB in Canonical Form:

$\text{Init}_{sk} = \{\} \,; \text{Pre}(P_i, s_i) = P_i; \text{Post}(P_i, s_i, y_i) = y_i; \text{Update}(P_i, s_i, y) = \{\}.$

5.2   Cipher Block-Chaining, or CBC

The definition and Canonical Form of CBC were given in Section 3.1. Consider the following BACPA-game.

1. Submit (**start**, **start**) and receive C_0.
2. Submit (P, Q), with $Q \neq P$ and receive C_1.
3. Submit $(P \oplus C_0 \oplus C_1, R)$, with $R \neq P \oplus C_0 \oplus C_1$, and receive C_2.
4. If $C_2 = C_1$ guess $b = 0$, else guess $b = 1$.

Suppose $b = 0$. Then $C_1 = F_{sk}(C_0 \oplus P)$ by the definition of CBC. Then

$$
\begin{aligned}
C_2 &= F_{sk}(C_1 \oplus (P \oplus C_0 \oplus C_1)) \\
&= F_{sk}(P \oplus C_0) \\
&= C_1
\end{aligned}
$$

and so it is easy to see that the adversary wins with probability one, and so CBC is not a BACPA-secure mode of encryption.

5.3   Infinite Garble Extension, or IGE

Infinite Garble Extension was proposed by Campell in 1978 [Cam78]. This mode is not commonly used but it is a simple example of the broad class of modes called ABC, which follows as the next example. For IGE one normally writes

$\text{Init}_{sk}(k) : C_0 \overset{R}{\leftarrow} \{0, 1\}^k$

$C_i = F_{sk}(P_i \oplus C_{i-1}) \oplus P_{i-1}$

Here, the state is the previous plaintext and the previous ciphertext. The initial conditions are C_0 and P_0. There are several possibilities for P_0. It can be some string known to both parties and kept secret from the adversary (thus becoming part of the key, essentially), or it can be generated by Init_{sk} and transmitted along with C_0.

Assume that P_0 is secret, and observe the following attack works anyway.

1. Submit (**start**, **start**) and receive C_0.
2. Submit (Q, Q) and receive C_1.
3. Submit (Q, Q) and receive C_2.

4. Submit $(Q \oplus C_2 \oplus C_1, R)$ with $R \neq Q \oplus C_2 \oplus C_1$ and receive C_3.
5. If $C_2 = C_3$ guess $b = 0$ else guess $b = 1$.

To see why this works, assume $b = 0$ and observe,

$$
\begin{aligned}
C_1 &= F_{sk}(Q \oplus C_0) \oplus P_0 \\
C_2 &= F_{sk}(Q \oplus C_1) \oplus Q \\
C_3 &= F_{sk}((Q \oplus C_2 \oplus C_1) \oplus C_2) \oplus Q \\
&= F_{sk}(Q \oplus C_1) \oplus Q \\
&= C_2
\end{aligned}
$$

Therefore equality will be observed if $b = 0$. Now suppose the bit $b = 1$. Surely by the correctness property, (for the same choices of randomness or initialization vector) we know $\mathfrak{E}_{sk}(Q||Q||Q \oplus C_2 \oplus C_1) \neq \mathfrak{E}_{sk}(Q||Q||R)$, since $Q \oplus C_2 \oplus C_1 \neq R$. But by the online-ness property, this inequality cannot occur in the first or second block. This is because C_2 can only depend on P_1 or P_2 and not upon P_3, and those first two blocks are equal. Therefore, the inequality occurs in the third block, and so the equality check in the last step of the attack will always be false because $C_2 \oplus C_3$ will have the wrong value. Therefore, the adversary always wins with probability one, even though P_0 was unknown.

In Canonical Form, like ABC, there are actually two state variables, the previous ciphertext and plaintext. But one can think of these as two binary strings concatenated into a larger string, since they are always of a fixed and known size. Thus we write the state $s_i = s_i'||s_i''$, where s_i' is the previous ciphertext and s_i'' the previous plaintext.

$Init_{sk}(k) : s_1 \stackrel{R}{=\!\!=} \{0,1\}^k; \quad C_0 = s_1$
$Pre(P_i, s_i) = P_i \oplus s_i'; \quad Post(P_i, s_i, y_i) = y_i \oplus s_i'';$
$Update(P_i, s_i, y) = y \oplus s_i''||P_i$

5.4 Some Forms of Accumulated Block Ciphers, or ABC

Accumulated Block Ciphers are a class of modes of encryption, first proposed by Knudsen [Knu00], based on a function denoted h.

In the usual notation:

$P_i' \leftarrow P_i \oplus h(P_{i-1}')$
$C_i \leftarrow F_{sk}(P_i' \oplus C_{i-1}) \oplus P_{i-1}'$

Either h is publicly computable or not. If not, then for reasons of efficiency, it is most likely a keyed function with a secret key, distinct from or equal to the secret key of the block cipher. Otherwise an entire secret function would have to be arranged for, along with the secret key of the block cipher, which would be cumbersome.

The values $P_0 = P_0'$ and C_0 act as initialization vectors for the scheme. In particular, if P_0 or some key for h are secret, then one writes S-ABC. If instead P_0 is public, and h is publicly computable, then one writes P-ABC. The value of C_0 is always public.

Canonical Form of ABC. This Canonical Form is correct independent of the secrecy status of P_0 and h. Here, there are actually two state variables, the previous ciphertext and (primed) plaintext. But, we can think of these as two binary strings concatenated into a larger string, since they are always of a fixed and known size. Thus we write the state $s_i = s_i'||s_i''$, where s_i' is the previous ciphertext and s_i'' the previous (primed) plaintext.

$$Init_{sk}(k) : r \xleftarrow{R} \{0,1\}^k; \quad C_0 = r; \quad s_1 = C_0||P_0$$
$$Pre(P_i, s_i) = P_i \oplus s_i' \oplus h(s_i''); \quad Post(P_i, s_i, y_i) = y_i \oplus s_i'';$$
$$Update(P_i, s_i, y) = y \oplus s_i''||P_i \oplus h(s_i')$$

P-ABC. Here, h is publicly computable, and both C_0 and P_0 are known to the adversary.

1. Submit (**start**, **start**) and receive C_0.
2. Submit (P_1, P_1), and receive C_1.
   (Comment) At this point, $P_1' = P_1 \oplus h(P_0') = P_1 \oplus h(P_0)$.
   (Comment) That further implies $C_1 = F_{sk}(P_1' \oplus C_0) \oplus P_0'$ or $C_1 = F_{sk}(P_1 \oplus h(P_0) \oplus C_0) \oplus P_0$.
3. Calculate $P_2 = P_1 \oplus C_0 \oplus h(P_0) \oplus h(P_1 \oplus h(P_0)) \oplus C_1$.
4. Submit (P_2, Q), with $Q \neq P_2$, and receive C_2.
5. If $C_2 \oplus C_1 = P_1 \oplus h(P_0) \oplus P_0$ then guess $b = 0$, else guess $b = 1$.

Suppose the secret bit is $b = 0$. Then,

$$\begin{aligned}
P_2' &= P_2 \oplus h(P_1') \\
&= P_2 \oplus h(P_1 \oplus h(P_0)) \\
&= (P_1 \oplus C_0 \oplus h(P_0) \oplus h(P_1 \oplus h(P_0)) \oplus C_1) \oplus h(P_1 \oplus h(P_0)) \\
&= P_1 \oplus C_0 \oplus h(P_0) \oplus C_1 \\
C_2 &= F_{sk}(P_2' \oplus C_1) \oplus P_1' \\
&= F_{sk}(P_1 \oplus C_0 \oplus h(P_0) \oplus C_1 \oplus C_1) \oplus P_1' \\
&= F_{sk}(P_1 \oplus C_0 \oplus h(P_0)) \oplus P_1' \\
C_2 \oplus C_1 &= P_1' \oplus P_0 \\
&= P_1 \oplus h(P_0) \oplus P_0
\end{aligned}$$

Thus upon receipt of C_2, the adversary should merely compute $C_2 \oplus C_1$ and compare it to $P_1 \oplus h(P_0) \oplus P_0$. If they are equal, he/she knows the bit $b = 0$. They will be unequal in the case $b = 1$ for the reasons given in the discussion of IGE. Note, here we are making explicit use of the fact that the adversary knows P_0 and that h is publicly computable.

Public h but secret P_0. In the case of h being the function that always returns zero, we have IGE. We saw in Section 5.3 that the scheme was breakable even if P_0 was not known. Thus, the secrecy of P_0 is insufficient alone to guarantee security. It is also known that if h is a linear function, i.e. $h(x \oplus y) = h(x) \oplus h(y)$, that attacks exist even if h is not publicly computable (see [BBKN01]).

References

[Bar06a]   Bard, G.: A challenging but feasible blockwise-adaptive chosen-plaintext attack on ssl. In: ICETE/SECRYPT 2006. Proc. IEEE International Conference on Security and Cryptography (2006),
http://eprint.iacr.org/2006/136

[Bar06b]   Bard, G.: Modes of encryption secure against blockwise-adaptive chosen-plaintext attack. Cryptology ePrint Archive, Report 2006/271 (2006),
http://eprint.iacr.org/2006/271 and
http://www.math.umd.edu/~bardg (March 26, 2004)

[BBKN01] Bellare, M., Boldyreva, A., Knudsen, L., Namprempre, C.: On-line ciphers, and the hash-cbc constructions. In: Kilian, J. (ed.) CRYPTO 2001. LNCS, vol. 2139, Springer, Heidelberg (2001)

[BDJR97] Bellare, M., Desai, A., Jokipii, E., Rogaway, P.: A concrete security treatment of symmetric encryption: Analysis of the des modes of operation. In: FOCS 1997. Proc. of the 38th Annual IEEE Symposium on Foundations of Computer Science (1997)

[BKN02]   Bellare, M., Kohno, T., Namprempe, C.: Authenticated encryption in ssh: Provably fixing the ssh binary packet protocol. In: CCS 2002. Proc. Ninth ACM Conference on Computer And Communications Security (2002)

[BRS02]   Black, J., Rogaway, P., Shrimpton, T.: Encryption-scheme security in the presence of key-dependent messages. In: Nyberg, K., Heys, H.M. (eds.) SAC 2002. LNCS, vol. 2595, Springer, Heidelberg (2003)

[BT04]   Boldyreva, A., Taesombut, N.: on-line encryption schemes: New security notions and constructions. In: RSA 2004. Proc. RSA Conference, Cryptographer's Track (2004)

[Cam78]   Campbell, C.: Design and specification of cryptographic capabilities. Technical report, US Department of Commerce (1978)

[CT06]   Cover, T., Thomas, J.: Information Theory, 2nd edn. Wiley-Interscience, Chichester (2006)

[FJMV03] Fouque, P., Joux, A., Martinet, G., Valette, F.: Authenticated on-line encryption. In: Matsui, M., Zuccherato, R.J. (eds.) SAC 2003. LNCS, vol. 3006, Springer, Heidelberg (2004)

[FJP04]   Fouque, P., Joux, A., Poupard, G.: Blockwise adversarial model for on-line ciphers and symmetric encryption schemes. In: Handschuh, H., Hasan, M.A. (eds.) SAC 2004. LNCS, vol. 3357, Springer, Heidelberg (2004)

[FMP03]   Fouque, P., Martinet, G., Poupard, G.: Practical symmetric on-line encryption. In: Johansson, T. (ed.) FSE 2003. LNCS, vol. 2887, Springer, Heidelberg (2003)

[GD00]   Gligor, V., Donescu, P.: On message integrity in symmetric encryption. In: Proc. 1st NIST Workshop on AES Modes of Operation (2000)

[JMV02]   Joux, A., Martinet, G., Valette, F.: Blockwise-adaptive attacks. revisiting the (in)security of some provably secure encryption modes: Cbc, gem, iacbc. In: Yung, M. (ed.) CRYPTO 2002. LNCS, vol. 2442, Springer, Heidelberg (2002)

[Knu00]   Knudsen, L.: Block chaining modes of operation. In: Proc. Symmetric Key Block Cipher Modes of Operation Workshop (2000)

[KY00]   Katz, J., Yung, M.: Complete characterization of security notions for probabilistic private-key encryption. In: Proc. 32nd ACM Annual Symposium on Theory of Computing (2000)

[utu]   The u-tube website, http://www.utube.com
[Vau02]   Vaudeney, S.: Security flaws induced by cbc padding - applications to ssl,
            ipsec, wtls.. In: Knudsen, L.R. (ed.) EUROCRYPT 2002. LNCS, vol. 2332,
            Springer, Heidelberg (2002)

Here, in the full version of the paper, would appear a discussion of those modes of encryption which pass the standard for being blockwise-adaptive chosen-plaintext secure, as well as XCBC. This can be downloaded from e-print [Bar06b].

A   Equivalence of Primitive and General Games

Theorem 5. *An encryption scheme is general blockwise-adaptive chosen-plaintext secure, if and only if it is secure in the primitive blockwise-adaptive chosen-plaintext game.*

Proof. Note, in this proof, we use the term "general BACPA game" to be the BACPA game without the restriction given in Definition 6, and "primitive BACPA game" to be the game with that restriction.

Since the primitive blockwise-adaptive game is merely a special case of the general game, an adversary who can win the primitive game can win the general game, and so (by contrapositive) general security implies primitive security as well.

To show that primitive security implies general security, we will demonstrate that if there exists an adversary who can win the general blockwise-adaptive chosen-plaintext game, then there exists an adversary who can win the primitive blockwise-adaptive chosen-plaintext game (both in polynomial time with non-negligible advantage).

Suppose there exists a cryptosystem secure against the *primitive* blockwise-adaptive game, which means there is no adversary which can achieve non-negligible advantage in polynomial time. Suppose further, there is an adversary *Gen* who can achieve non-negligible advantage in the *general* blockwise-adaptive game, again in polynomial time. We will demonstrate that this leads to a contradiction.

Let N be the random variable that is the number of queries that *Gen* will make. Let n be such that $Pr[N \leq n] \geq 1/2$. To see that such an n exists, see the footnote in the proof of Theorem 2. It is a requirement of the primitive and general blockwise-adaptive games that the secret bit b remain constant throughout the entire game. However, we will violate this rule, and create a series of games $G_1, G_2, \ldots, G_{n+1}$. The ith oracle query in Game G_j will work as follows:

$$LRBW_{(b,sk)}(P,Q) = \begin{cases} i < j & LRBW_{(b,sk)}(P,P) \\ j \leq i \leq n & LRBW_{(b,sk)}(Q,Q) \\ i > n & abort \end{cases}$$

Thus the game G_j will pretend as if the secret bit were 0 until and not including the jth query. All queries after and including the jth, until and including the nth, will be as if the secret bit were 1. Note that during the game G_1, it is as if the secret bit were always 1, and during the game G_{n+1}, it is as if the secret bit were always 0.

One cannot expect the output of adversary Gen to have any of its original properties since the behavior of the oracle has been changed. But nonetheless, its output is a random variable on the domain $\{0, 1\}$. Suppose that the outputs (as random variables) during games G_x and G_{x+1} are computationally distinguishable, for some x in the range $1, 2, \ldots n$.

This means there is an Algorithm Diff which can, with probability $1/2 + \delta$, correctly guess which game has been played, and that δ is non-negligible. Now Algorithm $Prim$ will play the primitive blockwise-adaptive game as follows. First, it will execute Algorithm Gen, and receive its queries in the form (P, Q). It will submit a query to its own oracle for each of these. For queries $1, 2, \ldots, x-1$, it will be (P, P). For query x it will be honest, or (P, Q), and for queries $x, x + 1, \ldots, n$ it will be (Q, Q). Finally if an $n + 1$th query is given, it will give up and guess a bit equal to the value of a fair coin. It is easy to see that $Prim$ only makes at most one split query, and so does not violate the rules of the primitive game. If the algorithm does not abort, Algorithm Gen will report a guess, and this will be passed to the distinguisher Algorithm Diff.

Observe that if the secret bit of the primitive game is actually one, then G_x has been played. And if the secret bit of the primitive game is actually zero, then G_{x+1} has been played. Since Algorithm Diff can distinguish between these correctly with probability $1/2 + \delta$, then Algorithm $Prim$ should guess 0 if Diff returns G_{x+1} and 1 if Diff returns G_x. Obviously $Prim$ will be correct if Diff is correct and there is no abortion, or correct half the time if there is an abortion.

Since the probability of an abortion is at most $1/2$, then the advantage of $Prim$ is at least $\delta/2$. This is non-negligible, and so $Prim$ can win the primitive blockwise-adaptive game, which is a contradiction. Therefore G_x and G_{x+1} are computationally indistinguishable, for all x in the range $1, 2, \ldots, n$.

Note further that computational indistinguishability is transitive so long as the sequence of objects compared is polynomial in length. Since Gen runs in polynomial time, there are polynomially many queries, and so n is upper-bounded by a polynomial. Therefore we can conclude, by transitivity, that G_1 is computationally indistinguishable from G_{n+1}.

But note, that G_1 is the general blockwise-adaptive game with the secret bit set to 1, and G_{n+1} is the same with the secret bit set to zero. Since Gen wins the general blockwise-adaptive game with non-negligible advantage, it does in fact distinguish between G_1 and G_{n+1} in polynomial time and with probability non-negligibly different from one-half. This is the required contradiction.

Therefore no such algorithm Gen can exist, and any system secure against the primitive blockwise-adaptive game is secure against the general blockwise-adaptive game as well.

Algebraic Cryptanalysis of the Data Encryption Standard

Nicolas T. Courtois[1] and Gregory V. Bard[2]

[1] University College of London, Gower Street, London, UK,
n.courtois@ucl.ac.uk
[2] Department of Mathematics, Fordham University, Bronx, NY, 10458, USA
Gregory.Bard@ieee.org

Abstract. In spite of growing importance of the Advanced Encryption Standard (AES), the Data Encryption Standard (DES) is by no means obsolete. DES has never been broken from the practical point of view. The variant "triple DES" is believed very secure, is widely used, especially in the financial sector, and should remain so for many many years to come. In addition, some doubts have been risen whether its replacement AES is secure, given the extreme level of "algebraic vulnerability" of the AES S-boxes (their low I/O degree and exceptionally large number of quadratic I/O equations).

Is DES secure from the point of view of algebraic cryptanalysis? We do not really hope to break it, but just to advance the field of cryptanalysis. At a first glance, DES seems to be a very poor target — as there is (apparently) no strong algebraic structure of any kind in DES. However in [15] it was shown that "small" S-boxes always have a low I/O degree (cubic for DES as we show below). In addition, due to their low gate count requirements, by introducing additional variables, we can always get an extremely sparse system of quadratic equations.

To assess the algebraic vulnerabilities of DES is the easy part, that may appear unproductive. In this paper we demonstrate that in this way, several interesting attacks on a real-life "industrial" block cipher can be found. One of our attacks is the fastest known algebraic attack on 6 rounds of DES. It requires only **one single known plaintext** (instead of a very large quantity) which is quite interesting in itself.

Our attacks will recover the key using an ordinary PC, for only six rounds. Furthermore, in a much weaker sense, we can also attack 12 rounds of DES. These results are very interesting because DES is known to be a very robust cipher, and our methods are very generic. We discuss how they can be applied to DES with modified S-boxes, and potentially other reduced-round block ciphers.

Keywords: block ciphers, algebraic cryptanalysis, DES, s^5DES, AES, solving overdefined and sparse systems of multivariate equations, Elim-Lin algorithm, Gröbner bases, logical cryptanalysis, SAT solvers.

1 Introduction

According to Claude Shannon, breaking a good cipher should require "as much work as solving a system of simultaneous equations in a large number of

S.D. Galbraith (Eds.): Cryptography and Coding 2007, LNCS 4887, pp. 152–169, 2007.

unknowns of a complex type" (see [43]). For example, the problem of key recovery in AES given one known plaintext can be written as solving a system of 4000 multivariate quadratic equations, see [14,15]. In general, this problem (called the MQ problem) is NP-hard, and solving this particular system remains a very ambitious goal. Nevertheless, there is a growing body of positive results: systems of equations that arise in the cryptanalysis of block, stream and public-key encryption schemes, turn out to be — for some specific reason — efficiently solvable, see [38,13,12,24,27,17,21,19,20], to quote only some major results. Yet the potential of efficiently solving certain multivariate systems of equations with special properties is still underestimated in scientific community. For example, in 2002, Courtois and Pieprzyk, have conjectured that sparse systems of equations are in general much easier to solve than dense systems of the same size. In 2006, Courtois Bard and Jefferson have discovered that SAT solvers, but also known Gröbner bases algorithms such as F4, can in fact solve efficiently very sparse systems of multivariate quadratic equations (dense MQ is a known NP-hard problem) [1,2]. To the best of our knowledge no researcher have so far demonstrated such working cryptanalytic attacks on systems of multivariate equations of comparable size. In this paper use very similar methods, but instead of randomly generated sparse systems, we use systems of equations derived from a real-life block cipher. With our methods, several interesting systems can be solved in practice, despite their very large size.

The rest of the paper is organized as follows: In the next section we study several methods of writing equations for DES. In Section 3 we summarise our attacks, explain in detail important previous and related work, and give a complete description of a couple of (best to date) attacks we did perform. In Section 4 we compare algebraic cryptanalysis of DES to AES, and algebraic cryptanalysis to differential and linear cryptanalysis. In Section 5 we show one example showing attacks one of our attacks becomes easier, and can solve a system of equations derived from as many as 12 full rounds of DES, when it has a large number of solutions. We conclude in Section 6.

2   Algebraic Vulnerabilities of DES S-Boxes

Unlike AES, there is no special algebraic structure in DES S-boxes that makes them particularly vulnerable. In most of this work, we treat them exactly as any other S-box of the same size. These attacks should therefore also work on DES with any modified set of S-boxes. For example, in Section 3.3 we give an example of algebraic attack on s⁵DES, a clone of DES [29]. Though s⁵DES has been designed to be more resistant than DES against all previously known attacks [29]), it appears to be visibly weaker against one of our attacks.

The S-boxes in DES have $n = 6$ inputs and $m = 4$ outputs. There are many ways in which one can write I/O equations for these S-boxes. The speed and the success of the algebraic attack will greatly depend on how this is done. In our work we consider the following three classes of equations that, heuristically, seem to be relevant to algebraic cryptanalysis:

- Class 1. Low-degree multivariate I/O relations (cf. definition below),
- Class 2. I/O equations with a small number monomials (can be of high or of low degree),
- Class 3. Equations of very low degree (between 1 and 2), low non-linearity and extreme sparsity that one can obtain by adding additional variables.

We have tried several types of equations falling in one of the above categories, as well as a number of their combinations (merging equations from several classes). We have computed and tested all the equations we consider in this paper, (and some others), and most of them can be found and downloaded from [8].

Very little is known about what approach would make an algebraic attack efficient and why. In our simulations, though Class number 3 seems to be the best choice, all the three do in fact give solvable systems of equations for several rounds of DES. This in spite of the fact that some resulting systems of equations are substantially larger in size. We anticipate that better methods for writing DES as a system of equations should be proposed in the future. We consider that the question which representation of a symmetric cipher is the most suitable for algebraic attacks and why, is an important research topic in itself.

2.1   Low-Degree Multivariate I/O Relations

The following notion plays an essential role in algebraic attacks on LFSR-based stream ciphers, see [17,9] as well as for (at least) certain insecure block ciphers [28,21].

Definition 1 (The I/O degree, [9,3]). *Consider a function $f : GF(2)^n \to GF(2)^m$, $f(x) = y$, with $x = (x_0, \ldots, x_{n-1})$, $y = (y_0, \ldots, y_{m-1})$.*
The I/O degree of f is the smallest degree of the algebraic relation

$$g(x_0, \ldots, x_{n-1}; y_0, \ldots, y_{m-1}) = 0$$

that holds with certainty, i.e. for every pair (x, y) such that $y = f(x)$.

The minimum number (and frequently the exact number) of equations of some type that do exist for one S-box can be obtained by applying the following theorem:

Theorem 1 (Courtois [15,17,20]). *For any $n \times m$ S-box, $F : (x_1, \ldots, x_n) \mapsto (y_1, \ldots, y_m)$, and for any subset T of t out of 2^{m+n} possible monomials in the x_i and y_j, if $t > 2^n$, there are at least $t - 2^n$ linearly independent I/O equations (algebraic relations) involving (only) monomials in T, and that hold with probability 1, i.e. for every (x, y) such that $y = F(x)$.*

Proof (sketch). All the monomials can be rewritten as a function of n variables and their Algebraic Normal Form (ANF) belong to a $GF(2)$-linear space of dimension 2^n. If the number of monomials in T is bigger than this dimension, there will be at least $t - 2^n$ linear dependencies among these ANF in n variables, and the same linear dependencies will also hold for the original monomials.   □

First Example of Application of Theorem 1 to DES

For example, we can consider the equations of the following type:

$$\sum \alpha_{ijk} x_i y_j y_k + \sum \beta_{ijk} x_i x_j y_k + \sum \gamma_{ij} x_i y_j + \sum \delta_i x_i + \sum \epsilon_i y_i + \eta = 0$$

These equations are of degree 3. The total number of monomials that arise in these equations is $t = n \cdot m(m-1)/2 + m \cdot n(n-1)/2 + nm + n + m + 1 = 131$. By straightforward application of Theorem 1 we get:

Corollary 1. *For any $F(x_1, \ldots, x_6) = y_1, \ldots, y_4$ (i.e. a 6×4 S-box), the number of linearly independent equations of this type (belonging to Class 1) is at least:*

$$r \geq t - 2^n = 67.$$

Thus, for any 6×4 S-box (not only a DES S-box) there are at least $r \geq t - 2^n = 67$ such equations. In practice, for DES S-boxes, we get sometimes 67, sometimes 68:

Table 1. The Actual Number of Equations Observed for Different S-boxes

DES S-box	1	2	3	4	5	6	7	8
$r =$	67	67	67	67	68	68	67	67

Fully Cubic Equations

We also consider fully cubic equations in the 10 variables x_i and y_i. We have

$$t = 1 + (n+m) + (n+m)(n+m-1)/2 + (n+m)(n+m-1)(n+m-2)/6 = 176,$$

and thus $r \geq t - 2^n = 112$. Computer simulations give exactly 112 for all the 8 S-boxes of DES. The same numbers are obtained for s^5DES [29].

Remark: Three fully functional examples of equations for 6, 8 and 16-round DES based on these (cubic) polynomials can be downloaded from [8].

I/O Equations of Degree 4

We have $t = 386$, $r \geq t - 2^n = 322$. We obtain exactly this many for each S-box.

Dense Class 1 Equations, Discussion

The equations we considered so far, are dense (as opposed to sparse) equations in Class 1 of degree 2-4. Apparently the S-boxes of DES behave more or less as random S-boxes of the same size. With the first type equations it seems that we can still "distinguish" them from random. For fully cubic and higher degree equations, no difference was observed, and we expect that there is no differen also for other types of I/O low degree equations where t is approximately above 131.

This is an important remark because it means that we do not expect that algebraic attacks that use dense Class 1 equations, will be much less or much more efficient on DES itself compared to versions with modified or random S-boxes. However, for specific S-boxes, the equations may be very special, for example more sparse.

Sparse Cubic Equations

We can observe that not all cubic equations we found are dense. In this section we give the number of very sparse cubic equations that we found for different DES S-boxes. These equations have up to 8 monomials. We expect that even more sparse cubic equations can be found. It is possible that replacing our 112 equations by a smaller but particularly sparse subsystem of equations (if it uniquely defines the S-box) gives better results in some attacks. Currently we do not know of a convincing example.

Table 2. The Number of Sparse Cubic Equations Found for Different S-boxes

DES S-box	1	2	3	4	5	6	7	8
$HW \leq 6, r =$	2	4	6	1	15	1	0	2
$HW \leq 8, r =$	3	8	11	13	17	8	1	7

2.2 Quadratic Equations

Though no theorem guarantees their existence, for certain S-boxes, there also exist quadratic I/O equations. Their number is not very large and they cannot alone be used to mount an algebraic attack. In comparison, for s^5DES [29], there are more quadratic equations, but the number remains quite small and they these equations taken alone do not uniquely define the S-boxes.

Table 3. Quadratic Equations Observed for Different S-boxes

S-box	1	2	3	4	5	6	7	8
DES	1	0	0	5	1	0	0	0
s^5DES	3	3	3	4	3	3	3	3

Remark. For all above mentioned types of low-degree equations, it is possible to delete some equations, for example taking every second equation. This leads to systems that are smaller and less over-determined. This is expected to give worse results in Gröbner basis attacks. However in some SAT attacks, such smaller systems seems to give slightly faster attacks, but we cannot say for certain.

2.3 Relations with a Very Small Number Monomials

These equations were first proposed and studied in [18]. First, we study equations that can be of arbitrary degree but that contain only one monomial. These are called monomial equations in [18]. For example $x_1x_2x_5y_3y_4 = 0$. One should note that we count 1 as a monomial and the equation $x_1x_2x_5y_3y_4 = 1$ would be counted as a binomial equation. We have studied and computed equations with 1, 2, 3 and 4 monomials (cf. Table 4). Before we present the results, several things should be noted. Since linear combinations may ruin the sparsity of equations (maintaining sparsity is our focus), all these equations do not have to be linearly independent. Still, from our count of binomial equations we exclude those that are trivial because they are linear combinations of simpler, monomial equations. Similarly, from our count of trinomial equations we exclude equations that would be a XOR of one monomial and one binomial equation, etc.. The number of equations with 4 monomials is getting already quite large. However it is possible to select among these equations, a smaller subset of equations and preferably of lower degree (their degree is 10 at maximum). We have decided to limit the sum of the degrees of the 4 monomials to 15 which also forces the overall degree to be ≤ 4 and to have at least one monomial of degree 3. For example, for DES S-box S1, we have the following equation $0 = x[1]x[5]x[32] + x[1]x[2]x[5] + x[1]x[3]x[4]x[5] + x[1]x[5]y[31]$, Here, the bits are numbered not according to their position in the S-box, but from 1 to 32, according to their position in the whole round function of DES. The sum of degrees in this equation is $3 + 3 + 4 + 3 = 13$.

In Table 4 we give the number of equations of each type found for DES, and compare it with results obtained for several randomly generated S-boxes of the same size.

Remark 1. We observe that for a random S-box, the number of equations of different types is rather strongly variable, On the contrary, all the DES S-boxes give quite similar results and clearly these equations are a good method to distinguish the DES S-boxes from a random function. We note also that monomial equations have a curious property that, for a random S-box, it is not totally unusual to have 0 such equations.

Table 4. Equations that Contain a Small Number of Monomials in DES

	random S-box	DES S-box 1	2	3	4	5	6	7	8
1 monomial	$0 - 463$	170	140	179	145	207	154	153	173
2 monomials	$233 - 524$	360	385	322	362	303	345	379	329
3 monomials	$1 - 112$	123	125	56	66	74	115	81	99
4 monomials	$1880 - 6106$	716	608	771	567	484	543	750	448
4 m; $\sum \deg \leq 15$	$250 - 1053$	87	73	104	57	86	104	94	75

Remark 2. When equations of this type are used alone to describe DES (especially with a single plaintext/ciphertext pair), and the key is computed by an algebraic attack, they typically will *not* uniquely define the solution to the system. This is because typically, when all $y_i = 0$ and regardless the value of x, these equations will all be satisfied (!). Though in some cases (by accident) we still were able to recover the right key in our attacks, we advocate the usage of these equations in conjunction with some other equations that permit the removal of spurious solutions to systems of equations. We observed that in some cases, mixing these equations with our (more traditional and dense) Class 1 I/O equations of degree 3, gave faster attacks than with our cubic equations alone, but the improvement was contained within the natural variability margin, therefore the interest of this method remains unclear.

2.4   Equations with Additional Variables

Equations Related to Efficient Hardware Implementation. By adding up to 52 additional variables per S-box and per round, it is possible to dramatically reduce the size of equations, increase their sparsity and decrease their non-linearity. All equations will have either 0 or 1 nonlinear monomial. There are many different methods to achieve this, and ours is directly derived from the low-gate count non-standard representation of DES that has been developed by Matthew Kwan, see [30]. These are our "favorite equations" so far, and one example of system of equations that contains all these (exact) equations we use can be downloaded from [8]). In practice, we have observed a speedup factor between 2 and 20 compared to the same attack done with the sets with 112 cubic equations per S-box.

Quadratic Representations with a Minimum Number of Added Variables. In the previous version, we add as many as up to 52 additional variables. One can do much better and it is possible to see that, due to the size of the DES S-boxes, by adding just one variable the degree of the equations collapses from 3 to 2. More generally we have:

Theorem 2. *For every S-box with 6 input bits and 4 output bits, if we add* **any** *additional variable that is defined by an arbitrary Boolean function of 6 input bits, the number of linearly independent quadratic equations of degree 2 with these* $4 + 6 + 1 = 11$ *variables is at least 3.*

Proof (sketch). Following the same argument as in Theorem 1, with $10+1$ variables, $\binom{11}{2} + 11 + 1 = 67$ quadratic or lower degree monomials, while the number of cases is 64. Therefore there are at least 3 quadratic equations.

We do not know a satisfactory choice for the additional variable to be added. More research about quadratic representation of DES S-boxes is needed.

3   Our Attacks on DES

3.1   Summary

From our equations on the S-boxes, it is easy to write a system of multivariate equations that describe the whole cipher. This system will be of degree 2, 3, 4 or more, depending on which equations we use for the S-boxes. This system should have a unique solution (if it is not the case one should either fix some variables or use some extra equations). Examples of such systems of equations can be downloaded from [8].

Interestingly, though almost all researchers in cryptography we know, as well as some computer algebraists, believe that there is no method whatsoever capable of solving (in practice) such systems of equations, we have discovered two totally different families of methods (that are of very different nature) that both work quite well.

1. The first is a particularly simple elimination algorithm called ElimLin, which can be see as a very simplified version of known Gröbner bases algorithms. It is fully described in Section 3.3.
2. The second is a simple and straightforward ANF to CNF conversion method following [1,2]. For each monomial in the equations we add a dummy variable, and CNF equations that relate it logically to variables and sub-monomials. To encode long XORs we use additional dummy variables and obtain shorter XORs. When the conversion is done, we obtain a large SAT problem, on which we run MiniSat 2.0, a very efficient and one of the latest SAT solvers, that is freely available on the internet with source code [35].

These methods can also be combined as follows: first we derive additional equations (not always sparse) by ElimLin (or by using other methods such as F5 [24]), then we add these new equations with the initial (very sparse) equations, then we run the ANF to CNF conversion and then MiniSat.

3.2 Related Work and What's New

In the past, many researchers quite naturally wondered if DES could be broken by solving a system of Boolean equations, see for example [44,26] and Section 4.3.2. of [23]. The idea was known as a method of "formal" coding. Unhappily, most people worked with a "functional" approach to describing S-boxes and whole rounds of the cipher. This is a very strong limitation that overlooks a wide range of attacks, see [9,15,38,28]. Nevertheless, at Crypto'85 Chaum and Evertse looked at bits (and their linear combinations) inside the DES encryption, that do not depend on some key bits (or their linear combinations), see [5]. If a bit can be found that computed in the forward direction from the plaintext, and computed from the ciphertext in the backwards direction, this bit gives an equation that does not depend on some key bits. Such equations can be used to speed-up the exhaustive search and for 6 rounds of DES, an attack 2^2 times faster than brute force is reported. This can be seen as the first algebraic attack on a reduced version of DES (our best attack will be faster).

The modern concept of algebraic cryptanalysis using arbitrary algebraic relations, see [9,15,38,28] is much richer in possibilities and working attacks. Our results should be compared with previous work on solving very large systems of multivariate equations and to previous successful attacks on general block ciphers with no special/algebraic properties such as in [5]. None of our solving methods is completely new. The use of Gröbner bases for solving systems of equations derived from a cipher has become very popular since [15], yet no convincing attacks on block ciphers were reported so far. The use of SAT solvers to break 3 rounds of DES have previously been shown to be feasible by Massacci and Marraro [33]. The authors of [33] call it "logical cryptanalysis" to emphasise the "automated reasoning" view. We consider this to be a part of "algebraic cryptanalysis" especially that we do not write SAT systems directly, but first write multivariate low-degree equations, then work on general-purpose conversion. We also consider that the methods of abstract algebra include and go beyond classical logic and reasoning. Unlike as in [33], our method — write equations, convert and solve — is very general and applicable to any block or stream cipher. It has an interesting property that the equations can be combined with any set of "additional" equations that are typically derived in Gröbner bases-like and related algorithms. SAT solvers may then be also used as a tool to complete any algebraic attack that does not work (sufficiently well) by itself, and this could be interesting because SAT solvers make heuristic guesses based on "non-algebraic" (but rather statistical) criteria.

Admittedly, the attack on 3 rounds of DES described in [33] is a very weak attack (even knowing that it requires one single plaintext), and the authors report "an abrupt jump in complexity" at 4 rounds. Maybe for this reason the result remained almost unnoticed in the cryptographic community. Some time after a preprint of this paper was written and circulated, Raddum and Semaev proposed yet another, new and very different approach to algebraic cryptanalysis, see [39,40]. So far (as of September 2007, see [39]), their attack works only for up to 4 rounds of DES, and it runs out of memory for 5 rounds. Our attacks on DES are the first to be be faster than the (older) algebraic attack on 6 round of DES [5]. Our methods are also clearly of much broader applicability.

The immediate contribution of this paper is to show that some very sparse systems of multivariate low-degree equations over small finite fields derived from industrial block ciphers can be solved in a matter of seconds on a PC. This by both our conversion to SAT, as well as techniques in the line of Gröbner bases (in fact we only worked with extremely simple monomial elimination tools that were however highly optimised in terms of memory management, and the order of operations was rearranged to conserve sparsity). One can wonder to what extent the systems we are solving here are special (i.e. weak)? It is very hard to know what exactly makes systems efficiently solvable, but it appears that sparsity alone will make systems efficiently solvable, both by SAT solvers and classical Gröbner bases methods, see [1,2]. One may notice that in the past, a reduction from the MQ problem to SAT, has been used to show that MQ was NP-hard. And now, we see that the very same reduction method can be used to solve very large instances of MQ that were believed intractable to handle.

3.3   Examples of Working Attacks on DES — Elimination Attacks

We start with a very simple yet remarkable algebraic attack that we call Elim-Lin. The ElimLin function works as follows: we take the initial system (that is of degree 2 or 3) and look if there are linear equations in the linear span of the equations. If so we can eliminate several variables, by simple substitution (by a linear expression). Then, quite surprisingly, new linear equations can be obtained, and this can go on for many, many iterations. This process is repeated until no more linear equations can be found. The order of variables is such that the variables that appear in the smallest number of equations are eliminated first, which helps to preserve sparsity. In addition, key variables are eliminated only when no other variable can be eliminated.

ElimLin alone gives very good results, given its extreme simplicity. We write a system of 112 fully cubic equations per S-box following Section 2.1, for 4 full rounds of DES, and for one known plaintext. We fix first 19 key bits to their actual values. And 37 remain to be determined. The time to compute 2^{36} times 4 rounds of DES on our 1.6 GHz Centrino CPU can be estimated to be about 8000 seconds. Instead, ElimLin takes only 8 seconds to find the correct solution. Attacks on 5 rounds can still be (marginally) faster than brute force. For example, with 3 known plaintexts and 23 variables fixed, we compute the key in 173 seconds, compared to about 540 s that would be needed by a brute force attack.

With eliminate ElimLin we did not go very far, but still we do two more rounds than in [32]. We observed that strictly better results (in terms of feasibility) can be obtained with XL algorithm and the so called T' method [14,15,22], or algorithms such as F4 or F5, however we do not report any results with these, as they do not really go much further, and we feel that our implementation of these still needs improvement, and the T' method is not optimal (in particular it computes the same equations many times). We have also tried ready packages such as MAGMA [31] and Singular [41], and found that these systematically run out of memory on our examples due to (apparently) lack of adequate support for sparse elimination on large systems. In fact, this occurred even on some simple examples we could solve completely with ElimLin in less than 1 hour.

Comparison to s⁵DES. The same attacks work on s⁵DES and the attack on 5 rounds with 3 chosen plaintexts is about 8 times faster. This might be due to the fact that for s⁵DES, a large subset of equations we use here are in fact of degree 2, see Section 2.2.

3.4   Examples of Working Attacks — Attacks with Conversion to SAT

With a very simple early version of our ANF to CNF converter, we write a system of quadratic equations with additional variables as described in section 2.4. We do it for full 6 rounds of DES, fix 20 key variables (it does not really matter which) and do the conversion that takes few seconds. Then with the latest version of MiniSat 2.0. with pre-conditioning we compute the key in 68 seconds while the exhaustive search would take about 4000 s. The complexity to recover full 56-bit key by this attack is about 2^{48} applications of reduced DES (feasible in practice).

Remark: We have tried if either MAGMA [31] or Singular [41] could solve this system of equations that we solve in 68 s. Both crash with out of memory message after allocating nearly 2 Gbytes. The memory usage reported by MiniSat is 9 Mbytes.

Comparison to s⁵DES. Unhappily, we cannot apply this attack to s⁵DES, because it is based on a special low gate count representation developed for DES [30], and no such representation of s⁵DES is known. It would be interesting to know if s⁵DES that has been specifically designed to be more resistant than DES against all previously known attacks [29]), is weaker than DES against algebraic attacks as our results in Section 3.3 seem to indicate, however it is certainly too early to draw such a conclusion and more research on this topic is needed.

4   Algebraic Cryptanalysis: The Great Challenge

4.1   Are Large Systems of Very Sparse Low-Degree Equations Solvable?

In our (best) system of equations in section 3.4 above, we have 2900 variables, 3056 equations and 4331 monomials. [1] The system is very sparse and compact, it has on average less than 1 non-linear monomial per equation. It is solved in 68 seconds.

We believe to be the first to show that such large systems of equations generated from a real-life cipher structure can be efficiently solvable. Obviously, not every system with similar parameters is efficiently solvable, and clearly the security of DES (as probably for any other cipher) against our attacks does quickly increase with the number of rounds.

Comparison to AES. Nevertheless, the following question can be asked, can we hope to break, say 6 rounds of AES by using SAT solvers? In comparison to ours, the binary system of equations proposed by Courtois and Pieprzyk in [15] has 4000 equations and 1600 variables: it is in fact overdefined and may seem easier to solve. Very unhappily, this system has substantially more monomials, about $137 \cdot 200 = 27400$, much more than a few thousands. [2]

4.2   Algebraic vs. Linear and Differential Cryptanalysis

Our vision of cryptanalysis changes each time a new cipher is considered, and each time we discover a new powerful attack. In the past DES has been thoroughly cryptanalysed by linear and differential cryptanalyses for up to 16 rounds. In this context our results may appear quite insignificant. On the contrary, we believe that, our results are interesting and this for several reasons.

First, we can recover the key given one single known plaintext. A tiny amount of data needed by the attacker is maybe the most striking feature of algebraic cryptanalysis. This is a rare and strong property of algebraic attacks, very few attacks that nave this property were ever proposed. It is precisely the reason why algebraic attacks are potentially very devastating, and this however immature and inefficient they are today. For example, from one single MAC computed by

---

[1] Some equations are linear and if we eliminated them, we would have 1298 variables, 1326 equations and 10369 monomials. It would become less sparse (15 monomials per equation on average) but still very sparse. We don't do this, it makes the attack run slower.

[2] Another system of equations that describes the whole AES have been proposed by Murphy and Robshaw [37], and it contains on average less than one non-linear monomial per equation. This is very similar to ours, however their system is over $GF(256)$, not over $GF(2)$.

an EMV bank card with a chip that is printed on a customer receipt, one would recover the key of the card, and from this single key, the master key of the issuing bank that could be used to make false bank cards. Luckily, there is no reason to believe that this could happen in a foreseeable future.

Nevertheless, we contend that it is inappropriate to compare algebraic cryptanalysis with linear and differential cryptanalysis and claim it is slower. In a certain way, linear and differential cryptanalysis became the reference as a by-product of our incapacity to ever find any attack on DES, that would be better than exhaustive search in a realistic setting. Algebraic cryptanalysis, while still not very powerful and unable to break full DES, does slowly emerge as more or less the only branch of cryptanalysis that may work in real life (i.e. when very few known plaintexts are available, not 2^{40} plaintexts). We suggest that attacks that require only a very small number of known plaintexts should be considered as a research topic of its own right. They should mainly be compared only to other attacks of this type. Moreover, if we can agree that for DES algebraic cryptanalysis is currently no match compared to classical attacks, we may as well argue that actually none of these attacks are of practical importance. Both types of attacks (algebraic vs. classical ones) represent the current state of research in cryptology, and yet it is the algebraic cryptanalysis that is new and can still be improved a lot. (It will already improve just by using better SAT solvers and more powerful computers. For some systems we have observed a speed-up of a factor 8 between MiniSat version 1.4 and 2.0.)

One should also note that, the situation that we have for DES could be very different for AES. Since AES is, by design, very strong against differential and linear cryptanalysis, the number of rounds is accordingly quite small in AES, and the threat is indeed that some form of algebraic cryptanalysis could give better results for this cipher (comparatively to linear and differential attacks). However, since the initial attack proposal [14,15], it seems that no visible progress is being made in this direction. Our feeling that, before attacking AES, we need to learn much more about algebraic cryptanalysis, and try it on many other ciphers. This was the main motivation of the present paper.

5   Algebraic Cryptanalysis as a Tool for Studying Ciphers

In this paper we demonstrated that algebraic (and logical) cryptanalysis is a tool for key recovery capable of finding the best known attack on 6 rounds of DES given 1 known plaintext. There are other interesting applications. One should be able to use it to solve many other problems that arise in cipher design such as detecting weaknesses, special properties, weak keys, finding collisions, second pre-images, long-range impossible differentials etc.. In the past, these tasks were done manually by a cryptanalyst. In the very near future, these should be automated.

We provide one simple example.

5.1 A Special-Property Finder on Full 12 Rounds of DES

Let '0123456789ABCDEF' be a fixed DES key (one that we did not choose to be weak or have some special property). We want to find an "educational" example of differential cryptanalysis for the first 12 rounds of DES with difference ('00196000', '00000000'), that comes from the best existing differential characteristic for DES, see [16]. It is known that this difference is reproduced after two rounds with probability exactly 2^{-8}, regardless the value of the key. The naive method to find a sample plaintext for which this difference holds throughout the whole computation is exhaustive search. For 10 consecutive rounds it requires about 2^{41} reduced DES computations and we estimate that it would take about 4 days on our laptop. For 12 consecutive rounds it requires about 2^{49} reduced DES computations which would last for about 3 years.

An algebraic approach to this problem is obvious: we can write this problem as a system of equations that has many solutions (we expect approximately 2^{24} and 2^{16}, respectively). We have tried this approach. By using our (last) quadratic and very sparse representation of the S-box, and by converting it to SAT, we have managed to find a desired solution. For 10 rounds this is particularly easy, we do it in 50 seconds while in addition fixing 6 additional variables to values chosen by us (many different solutions can thus be obtained). For 12 rounds it is harder to do, and the solution was found in 6 hours (instead of 3 years). For example, one can verify that the plaintext '4385AF6C49362B58' is a solution to this problem for 12 rounds and the key '0123456789ABCDEF'.

Thus we are able to find a special property of 12 rounds of DES within a time much much smaller than the inverse of the probability of this property. This is a nice and unexpected result with unclear ramifications. The system of equations is very similar that in key recovery attacks, yet due to the multiplicity of solutions, it is much easier to solve and we could do it on a laptop PC for as many as 12 rounds. It is not a key recovery attack, but could be treated as a weak "certificational" algebraic attack on 12 rounds of DES.

5.2 Discussion

This attack is open to interpretation and discussion. How do we perceive and interpret a cryptographic attack greatly depends on how it compares with other attacks. This perception may change when we discover new attacks (for example, it would change if somebody have found another attack that would achieve a similar speed-up). Here is our current interpretation of this result. We encourage other researchers to challenge this interpretation.

DES with 12-rounds can be treated and used as a pseudo-random permutation generator. We have found a new weakness of this generator and this w.r.t. attackers disposing of a very low computing power (e.g. only 50 seconds on a PC for 10 rounds).

Thus far it was known that DES had a particular property w.r.t. differential cryptanalysis that happens with a small probability and that can be detected when treating it as a "black box". In a "glass box" scenario, when the key and the algorithm is known to the attacker, plaintexts that have these properties can be detected and generated much faster. DES with 12 rounds cannot be treated as a "black box" or "random oracle" or "random cipher". We expect that our attack works for every possible DES key.

From differential cryptanalysis (a basic and naive application of it, could be improved) we already knew that DES with 12 rounds cannot be treated as a "black box" by an adversary that can do 2^{49} queries to the oracle. We also knew that it cannot be treated as a "black box" when the adversary can carefully choose the key — this is because DES has some weak keys (and there is also the complementation property). Here we learn that it cannot be treated as a "black box" when the key is random, and known to the adversary: the adversary can do more things than just implement 12 rounds of DES and experiment with it. The adversary does not have to be very powerful. He doesn't need to make 2^{49} queries to the oracle, and he needs a rather small computing power that is no match for computing answers for these 2^{49} queries. To summarize, DES with up to 12 rounds is not a very good permutation generator even against adversaries with very limited computing power.

5.3   Future Research

We believe that many other results of this kind can be obtained by our (and similar) methods. In particular, it appears that SAT solvers are particularly efficient in solving problems that have many solutions as demonstrated in recent work on hash function cryptanalysis with SAT solvers [36]. In general, we expect that it should be also possible to break certain hash functions and MACs by algebraic and/or logical attacks similar to ours, or in combination with other methods. It should be also possible to answer many questions such as, for example: given a block cipher, is there a choice of a subset of key bits to be fixed such that there will be a differential true with probability 1 for 4 rounds. In some cases the attack would just consist of running ready computer packages designed to efficiently solve SAT instances or/and systems of multivariate equations and may require very little human intervention.

6   Conclusion

In this paper we show that in many interesting cases, it is possible to solve in practice very large systems of multivariate equations with more than 1000 unknowns derived from a contemporary block cipher such as DES. Several methods were considered, and our best key-recovery attack allows one to break in practice, up to 6 complete rounds of DES and given **only 1** known plaintext. Very few attacks on realistic ciphers are known that work given such a low quantity of plaintext material. At the same time, our approach is extremely general. It is clearly possible to use it to find algebraic attacks of this type in an automated way starting from the very description of a symmetric cipher, and without the necessity to find any strong property or particular weakness. This opens new avenues of research which is rich in possibilities (there are many different representations of the S-boxes) and in which experimentation is an essential ingredient.

Until now, direct attempts to attack block ciphers with Gröbner bases have given very poor results. In 2006 Courtois proposed a general strategy for "fast" algebraic attacks on block ciphers [10]. We need to avoid methods such as Gröbner bases that expand systems of equations to a larger degree (e.g. 4 or 5) and then solve them. Instead, we need to find methods to produce systems of equations that, though may be much larger in size, can be nevertheless much easier to solve and by much simpler techniques, without time and memory-consuming expansion. Here, linear algebra and known elimination techniques need to be complemented with heuristics that take advantage of and (to some degree) preserve sparsity. Then, for attacks such as [10] and in the present paper, it appears that current Gröbner bases techniques are no match compared much simpler techniques such as ElimLin.

For DES (and also for KeeLoq, see [11]) it appears that the fastest algebraic attacks currently known are those obtained with modern SAT solvers. Our specific approach is to write problems algebraically and work on conversion. This allows methods from both families to be combined in many ways. By just the few simple working examples we give in this paper, we have considerably enlarged the family of algebraic cryptanalytic methods that are available to researchers.

Another interesting contribution of this paper is to point out that while the performance of algebraic elimination methods is usually greatly degraded when the system of equations has many solutions, SAT solvers in fact can benefit from it. This potential remains largely unexplored, and may lead to interesting results in cryptanalysis of hash functions and MACs. As an illustration we computed a special property of 12 full rounds of DES.

It should be noted that we ignore why some systems of equations are efficiently solvable. We just demonstrate that they are. It is certainly an important topic for further research to understand why these attacks actually work, but it would be wrong to believe that only attacks that are well understood should be studied in cryptology. This is because the number of possible algebraic attacks that can be envisaged is very large: one finite DES S-box can be described by a system of algebraic equations in an infinite number of ways, and the attacks that should be studied in priority are the fastest ones, not the ones for which a nice mathematical theory already exists such as Gröbner bases. Moreover, if one does not experiment, or if one only studies attacks that are faster than linear and differential cryptanalysis, then some important attacks on block ciphers will never be discovered.

References

1. Bard, G.: Algorithms for Solving Linear and Polynomial Systems of Equations over Finite Fields with Applications to Cryptanalysis. PhD Thesis, University of Maryland at College Park (April 30, 2007)

2. Bard, G.V., Courtois, N.T., Jefferson, C.: Efficient Methods for Conversion and Solution of Sparse Systems of Low-Degree Multivariate Polynomials over GF(2) via SAT-Solvers, http://eprint.iacr.org/2007/024/

3. Augot, D., Biryukov, A., Canteaut, A., Cid, C., Courtois, N., Cannière, C.D., Gilbert, H., Lauradoux, C., Parker, M., Preneel, B., Robshaw, M., Seurin, Y.: AES Security Report, D.STVL.2 report, IST-2002-507932 ECRYPT European Network of Excellence in Cryptology, www.ecrypt.eu.org/documents/D.STVL.2-1.0.pdf

4. Biham, E., Shamir, A.: Differential Cryptanalysis of DES-like Cryptosystems. Journal of Cryptology (IACR) 4, 3–72 (1991)

5. Chaum, D., Evertse, J.-H.: Cryptanalysis of DES with a Reduced Number of Rounds. In: Williams, H.C. (ed.) CRYPTO 1985. LNCS, vol. 218, pp. 192–211. Springer, Heidelberg (1986)

6. Tardy-Corfdir, A., Gilbert, H.: A Known Plaintext Attack of FEAL-4 and FEAL-6. In: Feigenbaum, J. (ed.) CRYPTO 1991. LNCS, vol. 576, pp. 172–181. Springer, Heidelberg (1992)

7. Coppersmith, D.: The development of DES, Invited Talk. In: Bellare, M. (ed.) CRYPTO 2000. LNCS, vol. 1880, Springer, Heidelberg (2000)

8. Courtois, N.: Examples of equations generated for experiments with algebraic cryptanalysis of DES, http://www.cryptosystem.net/aes/toyciphers.html

9. Courtois, N.: General Principles of Algebraic Attacks and New Design Criteria for Components of Symmetric Ciphers. In: Dobbertin, H., Rijmen, V., Sowa, A. (eds.) AES 2005. LNCS, vol. 3373, pp. 67–83. Springer, Heidelberg (2005)

10. Courtois, N.T.: How Fast can be Algebraic Attacks on Block Ciphers? In: Biham, E., Handschuh, H., Lucks, S., Rijmen, V. (eds.) Symmetric Cryptography (January 07-12, 2007), http://drops.dagstuhl.de/portals/index.php?semnr=07021

11. Courtois, N., Bard, G.V., Wagner, D.: Algebraic and Slide Attacks on KeeLoq, (preprint) http://eprint.iacr.org/2007/062/

12. Courtois, N., Shamir, A., Patarin, J., Klimov, A.: Efficient Algorithms for solving Overdefined Systems of Multivariate Polynomial Equations. In: Preneel, B. (ed.) EUROCRYPT 2000. LNCS, vol. 1807, pp. 392–407. Springer, Heidelberg (2000)

13. Courtois, N.: The security of Hidden Field Equations (HFE). In: Naccache, D. (ed.) CT-RSA 2001. LNCS, vol. 2020, pp. 266–281. Springer, Heidelberg (2001)

14. Courtois, N., Pieprzyk, J.: Cryptanalysis of Block Ciphers with Overdefined Systems of Equations. In: Zheng, Y. (ed.) ASIACRYPT 2002. LNCS, vol. 2501, pp. 267–287. Springer, Heidelberg (2002)

15. Courtois, N., Pieprzyk, J.: Cryptanalysis of Block Ciphers with Overdefined Systems of Equations, http://eprint.iacr.org/2002/044/

16. Courtois, N.: The Best Differential Characteristics and Subtleties of the Biham-Shamir Attacks on DES, http://eprint.iacr.org/2005/202

17. Courtois, N., Meier, W.: Algebraic Attacks on Stream Ciphers with Linear Feedback. In: Biham, E. (ed.) Eurocrypt 2003. LNCS, vol. 2656, pp. 345–359. Springer, Heidelberg (2003)

18. Courtois, N., Castagnos, G., Goubin, L.: What do DES S-boxes Say to Each Other? http://eprint.iacr.org/2003/184/

19. Courtois, N.: Fast Algebraic Attacks on Stream Ciphers with Linear Feedback. In: Boneh, D. (ed.) CRYPTO 2003. LNCS, vol. 2729, pp. 177–194. Springer, Heidelberg (2003)

20. Courtois, N.: Algebraic Attacks on Combiners with Memory and Several Outputs. In: Park, C.-s., Chee, S. (eds.) ICISC 2004. LNCS, vol. 3506, Springer, Heidelberg (2005), http://eprint.iacr.org/2003/125/

21. Courtois, N.: The Inverse S-box, Non-linear Polynomial Relations and Cryptanalysis of Block Ciphers. In: Dobbertin, H., Rijmen, V., Sowa, A. (eds.) AES 4 Conference, Bonn. LNCS, vol. 3373, pp. 170–188. Springer, Heidelberg (2005)

22. Courtois, N., Patarin, J.: About the XL Algorithm over $GF(2)$, Cryptographers. In: Joye, M. (ed.) CT-RSA 2003. LNCS, vol. 2612, pp. 141–157. Springer, Heidelberg (2003)

23. Davio, M., Desmedt, Y., Fosseprez, M., Govaerts, R., Hulsbosch, J., Neutjens, P., Piret, P., Quisquater, J.-J., Vandewalle, J., Wouters, P.: Analytical Characteristics of the DES. In: Crypto 1983, pp. 171–202. Plenum Press, New York (1984)

24. Faugère, J.C.: A new efficient algorithm for computing Gröbner bases without reduction to zero (F5). In: Workshop on Applications of Commutative Algebra, Catania, Italy, 3-6 April 2002, ACM Press, New York (2002)

25. Data Encryption Standard (DES), Federal Information Processing Standards Publication (FIPS PUB) 46-3, National Bureau of Standards, Gaithersburg, MD (1999),
http://csrc.nist.gov/publications/fips/fips46-3/fips46-3.pdf

26. Hulsbosch, J.: Analyse van de zwakheden van het DES-algoritme door middel van formele codering, Master thesis, K. U. Leuven, Belgium (1982)

27. Joux, A., Faugère, J.-C.: Algebraic Cryptanalysis of Hidden Field Equation (HFE) Cryptosystems Using Gröbner Bases. In: Boneh, D. (ed.) CRYPTO 2003. LNCS, vol. 2729, pp. 44–60. Springer, Heidelberg (2003)

28. Jakobsen, T.: Cryptanalysis of Block Ciphers with Probabilistic Non-Linear Relations of Low Degree. In: Krawczyk, H. (ed.) CRYPTO 1998. LNCS, vol. 1462, pp. 212–222. Springer, Heidelberg (1998)

29. Kim, K., Lee, S., Park, S., Lee, D.: Securing DES S-boxes against Three Robust Cryptanalysis. In: Nyberg, K., Heys, H.M. (eds.) SAC 2002. LNCS, vol. 2595, pp. 145–157. Springer, Heidelberg (2003)

30. Kwan, M.: Reducing the Gate Count of Bitslice DES,
http://eprint.iacr.org/2000/051,
equations: http://www.darkside.com.au/bitslice/nonstd.c

31. MAGMA, High performance software for Algebra, Number Theory, and Geometry, — a large commercial software package: http://magma.maths.usyd.edu.au/

32. Massacci, F.: Using Walk-SAT and Rel-SAT for Cryptographic Key Search. In: IJCAI 1999. International Joint Conference on Artifical Intelligence, pp. 290–295 (1999)

33. Massacci, F., Marraro, L.: Logical cryptanalysis as a SAT-problem: Encoding and analysis of the U.SS. Data Encryption Standard. Journal of Automated Reasoning 24, 165–203 (2000). And In: Gent, J., van Maaren, H., Walsh, T. (eds.) The proceedings of SAT-2000 conference, Highlights of Satisfiability Research at the Year 2000, pp. 343–376. IOS Press, Amsterdam (2000)

34. Matsui, M.: Linear Cryptanalysis Method for DES Cipher. In: Helleseth, T. (ed.) EUROCRYPT 1993. LNCS, vol. 765, pp. 386–397. Springer, Heidelberg (1994)

35. Eén, N., Sörensson, N.: MiniSat 2.0. An open-source SAT solver package,
http://www.cs.chalmers.se/Cs/Research/FormalMethods/MiniSat/

36. Mironov, I., Zhang, L.: Applications of SAT Solvers to Cryptanalysis of Hash Functions. In: Biere, A., Gomes, C.P. (eds.) SAT 2006. LNCS, vol. 4121, pp. 102–115. Springer, Heidelberg (2006), http://eprint.iacr.org/2006/254

37. Murphy, S., Robshaw, M.: Essential Algebraic Structure within the AES. In: Yung, M. (ed.) CRYPTO 2002. LNCS, vol. 2442, Springer, Heidelberg (2002)

38. Patarin, J.: Cryptanalysis of the Matsumoto and Imai Public Key Scheme of Eurocrypt 1988. In: Coppersmith, D. (ed.) CRYPTO 1995. LNCS, vol. 963, pp. 248–261. Springer, Heidelberg (1995)
39. Raddum, H., Semaev, I.: New Technique for Solving Sparse Equation Systems, ECRYPT STVL, http://eprint.iacr.org/2006/475/
40. Raddum, H., Semaev, I.: Solving MRHS linear equations. In: ECRYPT Tools for Cryptanalysis workshop, Kraków, Poland (September 24-25, 2007)(accepted)
41. Singular: A Free Computer Algebra System for polynomial computations. http://www.singular.uni-kl.de/
42. Shamir, A.: On the security of DES. In: Williams, H.C. (ed.) CRYPTO 1985. LNCS, vol. 218, pp. 280–281. Springer, Heidelberg (1986)
43. Shannon, C.E.: Communication theory of secrecy systems. Bell System Technical Journal 28, 704 (1949)
44. Schaumuller-Bichl, I.: Cryptanalysis of the Data Encryption Standard by the Method of Formal Coding. In: Beth, T. (ed.) Cryptography. LNCS, vol. 149, Springer, Heidelberg (1983)

Cryptographic Side-Channels from Low-Power Cache Memory*

Philipp Grabher, Johann Großschädl, and Dan Page

University of Bristol, Department of Computer Science,
Merchant Venturers Building, Woodland Road, Bristol, BS8 1UB, U.K.
{grabher,johann,page}@cs.bris.ac.uk

Abstract. To deliver real world cryptographic applications, we are increasingly reliant on security guarantees from both the underlying mathematics and physical implementation. The micro-processors that execute such applications are often designed with a focus on performance, area or power consumption. This strategy neglects physical security, a fact that has recently been exploited by a new breed of micro-architectural side-channel attacks. We introduce a new attack within this class which targets the use of low power cache memories. Although such caches offer an attractive compromise between performance and power consumption within mobile computing devices, we show that they permit attack where a more considered design strategy would not.

1 Introduction

Side-channel Analysis. Advances in cryptanalysis are often produced by mathematicians who seek techniques to unravel the hard problems on which modern cryptosystems are based. Attacks based on the concept of physical security move the art of cryptanalysis from the mathematical domain into the practical domain of implementation. By considering the implementation of cryptosystems rather than purely their specification, researchers have found they can mount physical attacks which are of low cost, in terms of time and equipment, and are highly successful in extracting useful results.

Side-channel attacks are based on the assumption that one can observe an algorithm being executed on a micro-processor, for example, and infer details about the internal state of computation from the features that occur. Ignoring the field of active fault injection attacks, a typical side-channel attack consists of a passive collection phase, which provides the attacker with profiles of execution, and an analysis phase which recovers otherwise secret information from the

* The work described in this paper has been supported by the EPSRC under grant EP/E001556/1 and, in part, by the European Commission through the IST Programme under contract IST-2002-507932 ECRYPT. The information in this paper reflects only the authors' views, is provided as is and no guarantee or warranty is given that the information is fit for any particular purpose. The user thereof uses the information at its sole risk and liability.

S.D. Galbraith (Eds.): Cryptography and Coding 2007, LNCS 4887, pp. 170–184, 2007.

profiles. The execution profile might be collected via mediums such as timing variation [20], power consumption [21] or electromagnetic emission [8]. Considering power consumption as the collection medium from here on, attack methods can be split into two main classes. Simple Power Analysis (SPA) is where the attacker can only collect one profile and is required to recover secret information by focusing mainly on the operation being executed. In contrast, Differential Power Analysis (DPA) uses statistical methods to form a correlation between a number of profiles and secret information by focusing mainly on the data items being processed. Approaches to defending against both these attack paradigms are increasingly well understood on a case-per-case basis, although the potential for delivery of a panacea via provably side-channel secure implementation is still unclear. It is obvious however that a trade-off exists between efficiency and security. That is, the methods that are viewed as most secure in this setting often introduce the highest performance overhead. Therefore, as in the study of theoretical cryptography, reasoning about and achieving required security levels for the lowest computational cost is an ongoing research challenge.

Micro-architectural Side-channels. The design of micro-processors can be very coarsely split into two separate tasks: the design of an Instruction Set Architecture (ISA) which defines an interface between a program and the processor, and the design of a micro-architecture by which a particular processor implementation realises the ISA. This separation implies that there are many valid micro-architectural choices for a single ISA; example ISA designs such as MIPS32 exploit this to allow compatible processors with vastly different characteristics for different markets.

The freedom to make micro-architectural decisions that provide an advantage in terms of a particular design goal is a powerful tool. However, an advantage for one goal often implies a disadvantage for another; for example it is hard to improve performance without an impact on area or power consumption. In the resulting struggle for compromise, physical security is often relegated to a second-class goal without direct consideration. As understanding of attack and defence techniques has evolved, the field of side-channel analysis has recently been extended to include use of micro-architectural features in the processor under attack to directly provide execution profiles. For example, one might make the micro-architectural decision to use cache memory within a processor design. This feature is transparent to a program running on the processor but can produce data dependent timing delays which are both observable and usable by an attacker.

This is both an interesting and a worrying development which has sparked a slew of recent research. It impacts directly on systems, such as those providing trusted computing or Digital Rights Management (DRM), which might be otherwise strictly analysed and protected against security vulnerabilities via sand-boxing techniques.

Main Contributions. The increasing ubiquity of mobile computing devices has presented practical cryptography with a problem. On one hand mobile devices are required to be as compact and resource conservative as possible; on the other hand they are increasingly required to perform complex computational tasks in a secure way. This challenge has resulted in a number of micro-architectural proposals to reduce power consumption while maximising performance. Focusing on low power cache memories, the main contribution of this paper is a micro-architectural side-channel attack that targets this compromise. In simple terms the result shows that deploying certain low power cache memories within a micro-architectural design allows the attack, using a standard cache memory does not. We posit that in order to nullify this genre of attack, it is vital to accept and adopt an altered processor design strategy that may result in an impact on performance but will consider security as a first class requirement.

The paper is organised as follows. In Section 2 we give a thorough overview of current research on micro-architectural side-channel attack and defence. In Section 3 we recap on the implementation and physical security of RSA. We then describe two styles of low power cache memory design in Section 4, each of which has been feted for use in the same sort of mobile computing devices most open to side-channel attack. Finally we present our attack methodology, a set of simulated results and proposals for countermeasures in Section 5, followed by concluding remarks in Section 6.

2   A Survey of Micro-architectural Side-Channels

2.1   Active and Passive Monitoring

Traditionally, side-channel attacks have been distinguished from fault attacks by the role of the attacker. In a side-channel attack the attacker passively monitors a target processor to collect execution profiles that occur normally, in a fault attack the attacker actively tampers with the target device to disrupt execution. The passive monitoring of side-channel attacks is usually external to the processor. That is, no access (beyond perhaps depackaging) is required to the internals of the device via either software or hardware; for example, monitoring power consumption can be achieved by tapping the power source outside the package.

Some micro-architectural side-channel attacks have focused on a new monitoring paradigm that has proved to be both interesting and powerful. Roughly speaking a so-called spy process S is executed on the target device concurrently with the process P that is under attack. A multitasking Operating System (OS) rapidly switches between S and P allowing S to inspect the micro-architectural state of the processor after execution of P; this is primarily because said state is shared between all running processes. If any changes in the shared state are correlated to secret information held in P, it is possible that S can recover it. This attack paradigm might be viewed separately from traditional side-channel attacks in the sense that it is no longer entirely passive. That is, the attacker must have some necessarily active means by which to execute S on the target device. This alone might limit the appeal of the paradigm in that such a system might already be viewed as insecure at this stage. Furthermore, execution of S places an artificial computational load on the target device. This load is of a fairly special form: S typically runs in a tight loop performing few or no system or library calls. This is somewhat atypical behaviour on relevant platforms, for example web-servers, and as a result one might expect to successfully deploy some form of intrusion detection.

To distinguish the two, we term the method of collecting execution profiles used by traditional side-channel attacks and those using a concurrent spy process passive-monitoring and active-monitoring respectively.

2.2   Cache Memory Based Attacks

A cache is a small area of fast storage and associated control logic which is placed between the processor and main memory; for an in depth description see [26, Chapter 5]. The area of storage is typically organised as a number of cache lines, each of which comprise a number of sub-words that are used to store contiguous addresses from main memory. Since the cache is smaller than main memory, it stores a sub-set of the memory content. As a result of locality in the incoming address stream, the cache reduces the load on the rest of the memory hierarchy by holding the current working set of data and instructions. Accesses that are serviced by the cache are termed cache-hits and are completed very quickly; accesses that are not held by the cache are termed cache-misses and take much longer to complete since main memory must be accessed. Since locality should guarantee more cache-hits than cache-misses, performance of the average case application is improved. However, given that many addresses in memory can map to the same location in the cache, data items can compete for space and evict each other; this is termed cache interference or contention.

Historic notes on the implications of data dependent timing variation as a result of cache behaviour can be found in the literature [15,28]. However, the study of cache behaviour in the context of security gained pace with the birth of side-channel analysis as a research topic within cryptography. Specifically, one finds that Kocher [20] looked at the effect of memory access on execution time, and Kelsey et al. [18] predicted that the timing dependent behaviour of S-Box access in Blowfish, CAST and Khufu could leak information to an attacker. This was followed with more concrete attacks on DES by Page [25], who assumed cache behaviour would be visible in a profile of power consumption, and Tsunoo et al. [29,30] who simply required that an attacker timed the cipher over many executions. Further break-throughs were made by Bertoni et al. [10] and Bernstein [9] who applied power and timing analysis attacks to AES. The former work shows cache behaviour is observable in a power trace (although using some simplifying assumptions), the latter shows that attacks can be mounted remotely; both further magnify the danger of cache attacks in an operational context. The issue of exploiting cache based vulnerabilities remotely was further studied by Acıiçmez et al. [6]. Finally, Percival [27] demonstrated an attack against CRT based RSA utilising the Hyper-Threading capability of Intel Pentium 4 processors but essentially relying on cache behaviour as a means of leaking information. Osvik et al. [23] extended this by applying the active-monitor attack paradigm (rather than concurrent hyper-threads), which allows one process to spy on the cache behaviour of another and thus make deductions about internal, secret state. Although various defence methods have been proposed [24,13], attack methods using both data caches [12,11,3] and instruction caches [1] are an active research area.

2.3   Branch Prediction Based Attacks

As a result of pipelined processor design, branches in control flow (such as loops or conditionals) represent a significant performance penalty. During the period between when a branch instruction is fetched into the pipeline and when it is executed, the fetch unit potentially stalls because it cannot be sure where to fetch instructions from: either execution progresses to the next instruction or the (unknown) branch target. The use of speculative execution helps to mask the resulting performance penalty. Essentially the processor makes a prediction of whether the branch will be taken and if so, where the branch target is. If the guess is correct then execution can continue with no penalty, if the branch is incorrect the pipeline content must be emptied of invalid instructions at some cost. A central requirement of efficient processor design is effective prediction of whether if branch will be taken or untaken, and prediction of the branch target address where execution continues if the branch is taken; for an in depth description see [26, Chapter 3].

Prediction of the branch target is often achieved using a Branch Target Buffer (BTB) which is a specialised form of cache memory: for a given address (i.e. where a branch is located) it stores the likely branch target. Like cache memory, use of the BTB can produce data dependent timing variation during program execution. Namely, if a BTB entry for a given branch is not present the processor is forced to make an uninformed prediction and potentially updating the BTB. Recent work by Acıiçmez et al. [5,4] takes advantage of this feature by forcibly evicting BTB entries and using the resulting behaviour to reason about the direction of the branch and hence the program control-flow. Their work is similar in concept to Osvik et al. [23] in that they use a spy process to inspect the state of the branch prediction mechanism. Unlike attacks using the data and instruction caches, attacks against the branch prediction mechanism are easier to defend against using program transformations; see for example proposals by Agosta and Pelosi [7] and Acıiçmez et al. [2].

3   Power Analysis Attacks on RSA

Many public key cryptosystems make use of arithmetic modulo some number N. In particular, encryption and decryption operations in RSA is achieved using a modular exponentiation composed from a series of modular multiplications. The so-called Montgomery representation [22] offers an efficient way to perform such arithmetic. To define the Montgomery representation of x, denoted x_M, one selects an $R = b^t > N$ for some integer t; the representation then specifies

---

Algorithm 1. The CIOS method for Montgomery multiplication

Input: An l-bit modulus $N = (n_{s-1}, n_{s-2}, ..., n_0)$, operands $A = (a_{s-1}, a_{s-2}, ..., a_0)$
  and $B = (b_{s-1}, b_{s-2}, ..., b_0)$ in the range $[0, N-1]$, precomputed factor $n'_0 = -n_0^{-1} \bmod 2^w$.

Output: Montgomery product $P = A \cdot B \cdot 2^{-l} \bmod N$.

  1: $P \leftarrow 0$
  2: **for** $i = 0$ to $s - 1$ **do**
  3:     $u \leftarrow 0$
  4:     **for** $j = 0$ to $s - 1$ **do**
  5:         $(u, v) \leftarrow p_j + a_j \cdot b_i + u$
  6:         $p_j \leftarrow v$
  7:     **end for**
  8:     $(u, v) \leftarrow p_s + u$
  9:     $p_s \leftarrow v$
 10:     $p_{s+1} \leftarrow u$
 11:     $q \leftarrow p_0 \cdot n'_0 \bmod 2^w$
 12:     $(u, v) \leftarrow p_0 + n_0 \cdot q$
 13:     **for** $j = 1$ to $s - 1$ **do**
 14:         $(u, v) \leftarrow p_j + n_j \cdot q + u$
 15:         $p_{j-1} \leftarrow v$
 16:     **end for**
 17:     $(u, v) \leftarrow p_s + u$
 18:     $p_{s-1} \leftarrow v$
 19:     $p_s \leftarrow p_{s+1} + u$
 20: **end for**
 21: **if** $P \geq N$ **then** $P \leftarrow P - N$ **end if**
 22: **return** P

---

that $x_M \equiv xR \pmod{N}$. To compute the product of x_M and y_M held in Montgomery representation, one interleaves a standard integer multiplication with an efficient reduction technique tied to the choice of R. We term the conglomerate operation Montgomery multiplication. Algorithm 1 outlines the so-called Coarsely Integrated Operand Scanning (CIOS) method which is regarded as the most efficient way to realise Montgomery multiplication. In the following, long integer numbers are represented as an s-word array of word size w.

To perform RSA encryption, one uses a public key e and an l-bit modulus N to transform a plaintext message P into a ciphertext C by computing $C = P^e \bmod N$. To decrypt, the corresponding private key d is used to recover the plaintext by computing $P = C^d \bmod N$. A basic technique to realise the modular exponentiation operation, in combination with Montgomery arithmetic, is the square-and-multiply method in Algorithm 2. Note that only when $k_i = 1$ is the multiplication step performed; since k is secret during RSA decryption this can present a problem in that the control flow depends on the exponent. In [34], the authors describe differential power analysis attacks to reveal k when using the square-and-multiply technique within RSA.

Algorithm 2. The square-and-multiply algorithm for exponentiation

---

Input: An integer M and an l-bit exponent $k = (k_{l-1}, k_{l-2}, \ldots, k_0)$.
Output: $X = M^k$.
  1: $X \leftarrow M$
  2: **for** $i = l - 2$ to 0 **do**
  3:     $X \leftarrow X^2$
  4:     **if** $k_i = 1$ **then**
  5:       $X \leftarrow X \cdot M$
  6: **end for**
  7: **return** X

3.1 A Single-Exponent, Multiple-Data (SEMD) Attack on RSA

The SEMD attack [34] assumes that the cryptographic device performs modular exponentiation with both the secret exponent under attack and some publicly known exponent. Comparing the resulting power profiles allows the attacker to draw conclusions about the secret parameter. In order to average out the noise components on a real cryptographic device, the attacker collects m profiles, denoted by S_i, and computes the mean power profile

$$\bar{S} = \frac{1}{m} \sum_{i=0}^{m} S_i.$$

where $\bar{S}$ and $\bar{P}$ denote the mean power profiles for exponentiation with the secret exponent and the known exponent respectively. A zero value in the difference signal $\bar{D} = \bar{S} - \bar{P}$ indicates that the same operation is performed within the exponentiation, whereas a nonzero value means that different exponentiation operations are executed.

3.2 Multiple-Exponent, Single-Data (MESD) Attack on RSA

In a MESD attack [34], the secret exponent is revealed using an iterative, bit-by-bit method. We assume that the attacker is in a position to perform modular exponentiation of a constant value M with arbitrary exponents of his choice. The basic idea for the MESD attack is sketched in Algorithm 3.

In the first step, we collect the power profile $\bar{S}_M$ of a modular exponentiation using the secret exponent. In each iteration we try to recover the i-th bit of the secret exponent under the assumption that we have successfully recovered the first $i - 1$ bits. Variable g keeps track of the correct guesses until the $(i - 1)$-th bit. Next, the attacker sets the i-th bit of g to 0 and 1 respectively, and collects the corresponding power profiles $\bar{S}_0$ and $\bar{S}_1$. Depending on which of these two power profiles correlates best with $\bar{S}_M$ reveals another bit of the secret exponent. For instance, the attacker could compute the difference signals $\bar{D}_0$ and $\bar{D}_1$ to correlate the power profiles:

$$\bar{D}_0 = \bar{S}_M - \bar{S}_0 \text{ and } \bar{D}_1 = \bar{S}_M - \bar{S}_1$$

Algorithm 3. An algorithm for a Multiple-Exponent, Single-Data (MESD) attack against the square-and-multiply algorithm

---

1: $M \leftarrow$ random integer value
2: Collect $\bar{S}_M$
3: $g \leftarrow 0$
4: **for** $i = l - 2$ to 0 **do**
5:     Set $g_i = 0$ and collect $\bar{S}_0$
6:     Set $g_i = 1$ and collect $\bar{S}_1$
7:     Compare $\bar{S}_0$ and $\bar{S}_1$ to $\bar{S}_M$
8:     Decide which guess correlates best with $\bar{S}_M$
9:     Update g with the correct guess
10: **end for**
11: Return g

In so doing, the difference signal obtained with the correct guess remains zero for a longer time. At the end of the attack, the variable g holds the secret exponent.

3.3 Zero-Exponent, Multiple-Data (ZEMD) Attack on RSA

It is assumed that the attacker can exponentiate random messages M with the secret exponent e and collect the resulting power traces [34]. The attack itself proceeds in a similar fashion as the MESD attack. In each step we guess the i-th bit of the secret exponent, assuming that the first $i - 1$ bits have been recovered correctly. Unlike in a MESD attack, where the reference power profiles for both guesses are generated with the real cryptographic device, the attacker has to be able to simulate the intermediate results of the square-and-multiply algorithm.

4 Low Power Cache Memory Design

Beyond the threat of power analysis attacks, power consumption plays an important role in mobile and ubiquitous computing devices. For a processor within a mobile telephone, low power consumption is a good selling point; within a device such as a sensor node it is a vital operational characteristic. This has lead to significant research into low power design and manufacturing techniques. The design of low power memory hierarchies is a specific area of interest given that memory, cache memories in particular, form a significant component in overall power consumption.

The overall power consumption in SRAM caches can be divided into dynamic and static components. Dynamic power dissipation occurs when a logic gate switches from one state to another only; static power is dissipated by every logic gate independent of its switching activity. Originally the dynamic component dominated the overall power consumption. However, as transistors have become smaller, and hence their density has increased, static power dissipation has started to have a major impact on the overall power consumption. In fact,

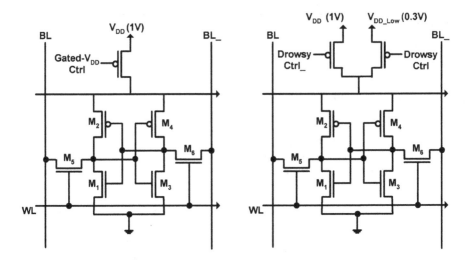

Fig. 1. Gated-V_{dd} memory cell   **Fig. 2.** Drowsy memory cell

static leakage is expected to dominate the total power consumption of future processor generations.

Although there are many approaches to reduce power dissipation in cache memories, the gated-V_{dd} technique suggested by Powell et al. [31] and the drowsy cache design of Flautner et al. [14,19] are attractive. Considering data caches only from here on, memory cells in a gated-V_{dd} structure can be forced into a low power mode using an extra high threshold transistor that disconnects the power supply to the cell. Figure 1 shows the implementation of a conventional 6-T memory cell using a PMOS gated V_{dd} transistor in the power supply path. In case that the memory cell is not used, the gated V_{dd} transistor simply turns off the power supply to the memory cell. While this technique is effective in minimising power dissipation, the information stored in a cell is not preserved; the design is referred to as a non-state-preserving cache. In [14,19], Flautner et al. introduce an alternative, state-preserving technique. Lines within a drowsy cache are constructed using memory cells capable of operating in a normal mode, where they can be freely read or written to, and a drowsy, low power mode where they retain their state but need to be woken before any access can occur. Such a drowsy memory cell is realised by introducing two extra PMOS transistors in the power supply path of a standard 6-T memory cell as illustrated in Figure 2. Depending whether the memory cell is in use or not, either the normal supply voltage or the low supply voltage is passed to the memory cell. Both gated-V_{dd} and drowsy caches switch cache lines into the low power according to some policy [14, Section 2]. Two basic approaches can be identified: the "simple" policy powers down all cache lines in a periodic manner, the "no access" policy only switches caches lines into low power mode that have not been accessed in a specified number of cycles.

Unlike in a gated-V_{dd} structure, where a deactivated cell line is fully disconnected from the power supply, a cell in a drowsy cache switches from the power supply V_{dd} to a lower power supply when put into low power mode. Therefore, a cache line in low power mode dissipates less leakage with the gated-V_{dd} technique. However, due to its non-state-preserving nature accessing a powered down cache line in a gated-V_{dd} structure causes a cache miss. Such a cache miss involves access to the second level cache, which in turn is costly in terms of power dissipation and execution time. As a consequence, one might assume that drowsy caches outperform gated-V_{dd} caches thanks to its state-preserving property. However, Li et al. [32] disprove this assumption by comparing the two techniques under discussion. The results indicate, that under certain operating conditions its preferable to use the non-state-preserving technique.

5 Low Power Cache Memory as a Side-Channel

Assume that some mobile computing device with a low power cache uses the square-and-multiply method to realise RSA encryption and decryption operations. We claim that such a device is vulnerable to a power analysis based side-channel attack as a direct result of the low power cache behaviour. We use simulated devices equipped with gated-V_{dd} and drowsy caches as targets to investigate two attacks.

5.1 Using the Gated-V_{dd} Cache for Attack

To provide a simulation platform for the non-state-preserving, gated-V_{dd} cache we used Sim-Panalyzer [33], an architectural level cycle accurate power estimator built using the SimpleScalar [36] framework. We modified Sim-Panalyzer to simulate a L1 gated-V_{dd} data cache using the "simple" policy. That is, all cache lines are periodically switched into a low power mode independent of the data access pattern. To best model the type of processor within a mobile computing device, we forced Sim-Panalyzer to issue instructions in-order; the L1 data cache consisted of 512 lines, each of 16 bytes, and was 4-way set-associative. Based on the CIOS method for Montgomery multiplication, we implemented RSA exponentiation for $l = 1024$ with a word size $w = 32$.

Figure 3 shows the output of our simulation platform while executing such an exponentiation. In this fragment, two sequential squarings are executed at the start of the exponentiation; we have that the exponent is of the form $e = (1, 0, \ldots)$. One can identify a periodic group of peaks in power consumption that are caused by memory accesses after all cache lines have been put into a low power mode. In Figure 4 the same fragment of the power profile is pictured except that the exponent is altered so that a multiplication operation is executed where before we had a squaring; we have that the exponent is of the form $e = (1, 1, \ldots)$. Comparing these two power profiles to each other, one can clearly detect eight additional peaks, labelled $\delta_1 \ldots \delta_8$, in Figure 4. Algorithm 1 processes one operand on a word-by-word basis, each word is then multiplied with

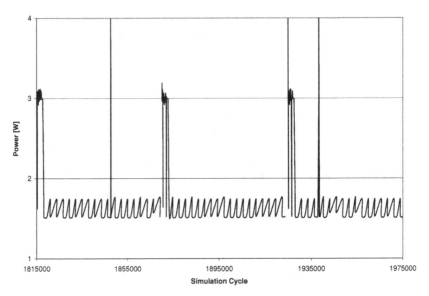

Fig. 3. A power profile fragment for two squarings in sequence

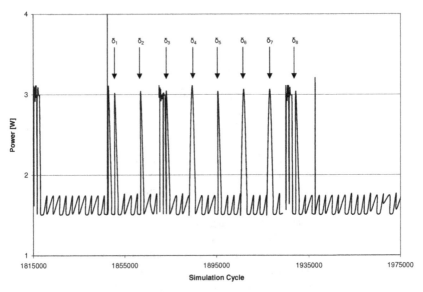

Fig. 4. A power profile fragment for a squaring followed by a multiplication

all the words of the other operand to create the partial products. Assume all cache lines are switched into low power mode before we compute the CIOS algorithm. Initially only one operand is loaded into the data cache. In contrast to a squaring operation, the multiplication requires a second operand to be loaded consecutively into the cache as the algorithm proceeds. Hence, the peaks $\delta_1 \ldots \delta_8$ are caused by eight accesses to the L2 cache so that the second operand can be loaded into the L1 cache.

Based on this observation that the cache offers a way to distinguish between squarings and multiplications, one can easily mount a side-channel attack to recover the secret exponent. We wrote a script that performs the steps in Algorithm 3 using the output of our simulation platform to provide power profiles. The noiseless environment allows us to collect useful power profiles $\bar{S}_i$ for an arbitrary exponent with just one simulation. At each iteration, we assume that the first $i - 1$ bits of the exponent are correct and generate two power profiles $\bar{S}_0$ and $\bar{S}_1$ with Sim-Panalyzer where the i-th bit is set to 1 and 0 respectively. We compare both power profiles to the power profile $\bar{S}_M$ of the secret exponent using the correlation function in Matlab, repeating the process until we have successfully recovered the whole secret exponent.

5.2 Using the Drowsy Cache for Attack

Recall that with a state-preserving cache no cache misses occur when accessing cache lines in low power mode. In order to attack such a cache structure we study the static power dissipation as opposed to non-state-preserving caches where we focus on the dynamic component of the power consumption. In this context we used the Hotleakage [37] simulator tool, a cycle accurate architectural model for leakage power, to output the static power consumption of the cache. To match the simulation platform from above, we equipped Hotleakage with a 4-way set-associative drowsy L1 data cache with 512 lines, of 16 bytes each, and force the simulator to issue instructions in-order. The status of the cache lines is updated according to the "simple" policy.

Because a squaring needs one operand and a multiplication needs two, more cache lines are in a drowsy state when performing a squaring operation than when executing a multiplication operation. Therefore, one would expect that static power dissipation would be higher during a multiplication due to the larger number of active cache lines. However, Kim et al. [19] state that the leakage for a cell in drowsy mode is 6.24 nW, whereas a cell not in low power mode dissipates 78 nW. If we map these values onto our cache architecture, a cache line in low power mode consumes 0.8 μW whereas an active cache line contributes approximately 2 μW to the total leakage power. Hence a cache line in drowsy mode makes a difference of 1.2 μW in the power profile, a difference that cannot be accurately measured in practise. Although this means the attack is not realistic at the present, as the contribution of static power dissipation increases within modern processor designs and RSA operands sizes increase, this attack scenario could be applicable in the future.

5.3 Countermeasures Against Attack

There is no silver-bullet solution for defence against side-channel attack. Thus, in common with other micro-architectural side-channel attacks, here one can deploy defence techniques at the algorithmic and architectural level. Effective

---

Algorithm 4. The square-and-multiply-always algorithm for exponentiation

---

Input: An integer M and an l-bit exponent $k = (k_{l-1}, k_{l-2}, \ldots, k_0)$.
Output: $X[2] = M^k$.
1: $X[2] \leftarrow M$
2: **for** $i = l - 2$ to 0 **do**
3:     $X[0] \leftarrow X[2]^2$
4:     $X[1] \leftarrow X[2] \cdot M$
5:     $X[2] \leftarrow X[k_i]$
6: **end for**
7: Return $X[2]$

---

examples at the algorithmic level include the square-and-multiply-always method in Algorithm 4. Essentially the multiplication operation is always performed so there is no advantage in distinguishing it from a squaring. However, although this countermeasure is simple to implement, it has a significant performance penalty.

We suggest that a far better approach would be to act at the architectural level by adopting a processor design strategy that considers security as a first-class goal. An example of such a strategy would be to make the cache operation visible to the programmer as part of the ISA rather than invisible as part of the micro-architecture. For example, the Intel i960 [16] and XScale [17] architectures both employ schemes whereby the contents of the instruction cache can be protected against replacement (or locked) under the control of the program being executed. Through a simple instruction set extension, a similar scheme in the context of low power caches would see sections of the cache (or potentially the whole cache) locked into an active state during execution of security critical code. This would prevent cache lines being periodically forced into a low power mode and hence prevent the data dependant behaviour which permits our attack.

A typical criticism of this sort of approach is that it can impact adversely on other design goals (such as performance or area); our suggestion is that without some action in this and other areas, micro-architectural vulnerabilities will remain unaddressed and start to cause more significant problems. That is, computer architects need to change their focus: if security is a critical in hosted applications, it is always worthwhile making a compromise to honour that fact rather ignoring the issue and concentrating on more traditional design goals.

6   Conclusion

In this paper, we have investigated a side-channel vulnerability introduced by the use of low power cache memory. Our results indicate that the power profile resulting from use of a non-state-preserving L1 data cache allows one to draw conclusions about control flow within the square-and-multiply algorithm. Exploiting this fact, we launched a successful MESD side-channel attack to recover the entire secret exponent of a typical RSA implementation. Running the RSA exponentiation with a state-preserving L1 data cache, no assumptions about the control flow can be made; this is due to the low signal-to-noise ratio. However, we argue that such an attack could pose a threat in future architectures as both the static power dissipation and the operand length of RSA increase.

Acknowledgements

The authors would like to thank various anonymous referees for their comments.

References

1. Acıiçmez, O.: Yet Another MicroArchitectural Attack: Exploiting I-cache. In: Cryptology ePrint Archive, Report 2007/164 (2007)
2. Acıiçmez, O., Gueron, S., Seifert, J-P.: New Branch Prediction Vulnerabilities in OpenSSL and Necessary Software Countermeasures. In: Cryptology ePrint Archive, Report 2007/039 (2007)
3. Acıiçmez, O., Koç, Ç.K.: Trace-Driven Cache Attacks on AES. In: Cryptology ePrint Archive, Report 2006/138 (2006)
4. Acıiçmez, O., Koç, Ç.K., Seifert, J-P.: On the Power of Simple Branch Prediction Analysis. Cryptology ePrint Archive Report 2006/351 (2006)
5. Acıiçmez, O., Seifert, J-P., Koç, Ç.K.: Predicting Secret Keys via Branch Prediction. In: Abe, M. (ed.) CT-RSA 2007. LNCS, vol. 4377, pp. 225–242. Springer, Heidelberg (2006)
6. Acıiçmez, O., Schindler, W., Koç, Ç.K.: Cache Based Remote Timing Attacks on the AES. In: Abe, M. (ed.) CT-RSA 2007. LNCS, vol. 4377, pp. 271–286. Springer, Heidelberg (2006)
7. Agosta, G., Pelosi, G.: Countermeasures for the Simple Branch Prediction Analysis. In: Cryptology ePrint Archive, Report 2006/482 (2006)
8. Agrawal, D., Archambeault, B., Rao, J.R., Rohatgi, P.: The EM Side-Channel(s). In: Kaliski Jr., B.S., Koç, Ç.K., Paar, C. (eds.) CHES 2002. LNCS, vol. 2523, pp. 29–45. Springer, Heidelberg (2003)
9. Bernstein, D.J.: Cache-timing Attacks on AES.
   http://cr.yp.to/antiforgery/cachetiming-20050414.pdf
10. Bertoni, G., Zaccaria, V., Breveglieri, L., Monchiero, M., Palermo, G.: AES Power Attack Based on Induced Cache Miss and Countermeasure. In: ITCC. IEEE Conference on Information Technology: Coding and Computing (2005)
11. Bonneau, J.: Robust Final-Round Cache-Trace Attacks Against AES. In: Cryptology ePrint Archive, Report 2006/374 (2006)
12. Bonneau, J., Mironov, I.: Cache-Collision Timing Attacks Against AES. In: Goubin, L., Matsui, M. (eds.) CHES 2006. LNCS, vol. 4249, pp. 201–215. Springer, Heidelberg (2006)
13. Brickell, E., Graunke, G., Neve, M., Seifert, J-P.: Software Mitigations to Hedge AES Against Cache-based Software Side Channel Vulnerabilities. In: Cryptology ePrint Archive, Report 2006/052 (2006)
14. Flautner, K., Kim, N.S., Martin, S., Blaauw, D., Mudge, T.N.: Drowsy Caches: Simple Techniques for Reducing Leakage Power. In: ISCA. International Symposium on Computer Architecture, pp. 148–157 (2002)
15. Hu, W.M.: Lattice Scheduling and Covert Channels. In: IEEE Symposium on Security and Privicy, pp. 52–61. IEEE Computer Society Press, Los Alamitos (1992)
16. Intel Corporation. Intel i960 Jx Processor Documentation.
    http://www.intel.com/design/i960/documentation/

17. Intel Corporation. Intel XScale Processor Documentation.
    http://www.intel.com/design/intelxscale/
18. Kelsey, J., Schneier, B., Wagner, D., Hall, C.: Side Channel Cryptanalysis of Product Ciphers. Journal of Computer Security 8(2-3), 141–158 (2000)
19. Kim, N.S., Flautner, K., Blaauw, D., Mudge, T.N.: Drowsy Instruction Caches: Leakage Power Reduction using Dynamic Voltage Scaling and Cache Sub-bank Prediction. In: MICRO. International Symposium on Microarchitecture, pp. 219–230 (2002)
20. Kocher, P.C.: Timing Attacks on Implementations of Diffie-Hellman, RSA, DSS, and Other Systems. In: Koblitz, N. (ed.) CRYPTO 1996. LNCS, vol. 1109, pp. 104–113. Springer, Heidelberg (1996)
21. Kocher, P.C., Jaffe, J., Jun, B.: Differential Power Analysis. In: Wiener, M.J. (ed.) CRYPTO 1999. LNCS, vol. 1666, pp. 388–397. Springer, Heidelberg (1999)
22. Montgomery, P.L.: Modular Multiplication Without Trial Division. Mathematics of Computation 44, 519–521 (1985)
23. Osvik, D.A., Shamir, A., Tromer, E.: Cache attacks and Countermeasures: the Case of AES. Cryptology ePrint Archive, Report 2005/271 (2005)
24. Page, D.: Defending Against Cache Based Side-Channel Attacks. Information Security Technical Report, 8 (1), 30–44 (2003)
25. Page, D.: Theoretical Use of Cache Memory as a Cryptanalytic Side-Channel. Cryptology ePrint Archive, Report 2002/169 (2002)
26. Patterson, D.A., Hennessy, J.L.: Computer Architecture: A Quantitative Approach. Morgan Kaufmann, San Francisco (2006)
27. Percival, C.: Cache Missing For Fun And Profit.
    http://www.daemonology.net/papers/htt.pdf
28. Trostle, J.T.: Timing Attacks Against Trusted Path. In: IEEE Symposium on Security and Privicy, pp. 125–134. IEEE Computer Society Press, Los Alamitos (1998)
29. Tsunoo, Y., Saito, T., Suzaki, T., Shigeri, M., Miyauchi, H.: Cryptanalysis of DES Implemented on Computers with Cache. In: D.Walter, C., Koç, Ç.K., Paar, C. (eds.) CHES 2003. LNCS, vol. 2779, pp. 62–76. Springer, Heidelberg (2003)
30. Tsunoo, Y., Tsujihara, E., Minematsu, K., Miyauchi, H.: Cryptanalysis of Block Ciphers Implemented on Computers with Cache. In: ISITA. International Symposium on Information Theory and Its Applications (2002)
31. Powell, M., Yang, S.-H., Falsafi, B., Roy, K., Vijaykumar, T.N.: Gated-Vdd: A circuit technique to reduce leakage in deep-submicron cache memories. In: Proc. of Int. Symp. Low Power Electronics and Design (2000)
32. Li, Y., Parikh, D., Zhang, Y., Sankaranarayanan, K., Stan, M., Skadron, K.: State-Preserving vs. Non-State-Preserving Leakage Control in Caches. Design, Automation and Test in Europe (DATE), 22–29 (2004)
33. University of Michigan Sim-Panalyzer 2.0.3,
    http://www.eecs.umich.edu/~panalyzer/
34. Messerges, T.S., Dabbish, E.A., Sloan, R.H.: Power Analysis Attacks of Modular Exponentiation in Smartcards. In: Koç, Ç.K., Paar, C. (eds.) CHES 1999. LNCS, vol. 1717, pp. 144–157. Springer, Heidelberg (1999)
35. Koç, Ç.K., Acar, T., Kaliski, B.S.: Analyzing and Comparing Montgomery Multiplication Algorithms. IEEE Micro 16(3), 26–33 (1996)
36. Burger, D., Austin, T.M.: The SimpleScalar Tool Set Version 2.0. Computer Architecture News (1997)
37. Zhang, Y., Parikh, D., Sankaranarayanan, K., Skadron, K., Stan, M.: Hotleakage: A temperature-aware model of subthreshold and gate leakage for architects,
    http://lava.cs.virginia.edu/HotLeakage/

New Branch Prediction Vulnerabilities in OpenSSL and Necessary Software Countermeasures

Onur Acıiçmez[1], Shay Gueron[2,3], and Jean-Pierre Seifert[1,4]

[1] Samsung Information Systems America, San Jose, 95134, USA
[2] Department of Mathematics, University of Haifa, Haifa, 31905, Israel
[3] Intel Corporation, IDC, Israel
[4] Institute for Computer Science, University of Innsbruck, 6020 Innsbruck, Austria
onur.aciicmez@gmail.com, shay@math.haifa.ac.il,
jeanpierreseifert@yahoo.com

Abstract. Software based side-channel attacks allow an unprivileged spy process to extract secret information from a victim (cryptosystem) process by exploiting some indirect leakage of "side-channel" information. It has been realized that some components of modern computer microarchitectures leak certain side-channel information and can create unforeseen security risks. An example of such MicroArchitectural Side-Channel Analysis is the Cache Attack — a group of attacks that exploit information leaks from cache latencies [4,7,13,15,18]. Public awareness of Cache Attack vulnerabilities lead software writers of OpenSSL (version 0.9.8a and subsequent versions) to incorporate countermeasures for preventing these attacks. In this paper, we present a new and yet unforeseen side channel attack that is enabled by the recently published Simple Branch Prediction Analysis (SBPA) which is another type of MicroArchitectural Analysis, cf. [2,3]. We show that modular inversion — a critical primitive in public key cryptography — is a natural target of SBPA attacks because it typically uses the Binary Extended Euclidean algorithm whose nature is an input-centric sequence of conditional branches. Our results show that SBPA can be used to extract secret parameters during the execution of the Binary Extended Euclidean algorithm. This poses a new potential risk to crypto-applications such as OpenSSL, which already employs Cache Attack countermeasures. Thus, it is necessary to develop new software mitigation techniques for BPA and incorporate them with cache analysis countermeasures in security applications. To mitigate this new risk in full generality, we apply a security-aware algorithm design methodology and propose some changes to the CRT-RSA algorithm flow. These changes either avoid some of the steps that require modular inversion, or remove the critical information leak from this procedure. In addition, we also show by example that, independently of the required changes in the algorithms, careful software analysis is also required in order to assure that the software implementation does not inadvertently introduce branches that may expose the application to SBPA attacks. These offer several simple ways for modifying OpenSSL in order to mitigate Branch Prediction Attacks.

Keywords: Side channel attacks, branch prediction attacks, cache eviction attacks, Binary Extended Euclidean Algorithm, modular inversion, software mitigation methods, OpenSSL, RSA, CRT.

S.D. Galbraith (Eds.): Cryptography and Coding 2007, LNCS 4887, pp. 185–203, 2007.
© Springer-Verlag Berlin Heidelberg 2007

1   Introduction

Side channel attacks are methods by which an attacker can extract secret information from an implementation of a cryptographic algorithm. They come in various flavors, exploiting different security weaknesses of both the cryptographic implementations and the environments on which the cryptographic applications run. MicroArchitectural Side-Channel attacks are a special new class of attacks that exploit the microarchitectural throughput-oriented behaviour of modern processor components. These attacks capitalize on situations where several applications can share the same processor resources, which allows a spy process running in parallel to a victim process to extract critical information.

Cache Attacks are one example of MicroArchitectural Side-Channel attacks. In one flavor of Cache Attacks, the adversary takes advantage of a specifically crafted spy process. This spy first fills the processor's cache before the victim process (i.e., cipher process) takes over at context switch. In the subsequent context switch the spy process experiences cache evictions that were caused by the operation of the victim process, and the eviction patterns can be used for extracting secret information. This concept was used ([18]) for demonstrating an attack on OpenSSL-0.9.7. The attack focused on the Sliding Windows Exponentiation (SWE) algorithm for computing modular exponentiation - part of the RSA decryption phase that uses the server's private key.

As a result, the subsequent version OpenSSL-0.9.8 included mitigations against this attack. The SWE algorithm was replaced with a Fixed Window Exponentiation (FWE) algorithm, using window length 5. In this algorithm, a sequence of 5 repeated modular squares is followed by a modular multiplication by the appropriate power of the base, depending on the value of the 5 scanned bits of the exponent. The powers of the base are computed once, before the actual exponentiation, and stored in a table. Each entry of this table is stored in memory in a way to span multiple cache lines. This way, the eviction of any table entry from the cache does not expose information on the actual the table index of this entry. Another incorporated mitigation is "base blinding": the exponentiation base is multiplied by some factor (unknown to the adversary), and the final exponentiation result is then multiplied by a corresponding factor that cancels out the undesired effect. This technique eliminates the "chosen plaintext" scenario that enables remote timing attacks such as [5,9].

Branch Prediction Analysis (BPA) and Simple Branch Prediction Analysis (SBPA) attacks are a new type of MicroArchitectural attacks that have been recently published by Acıiçmez et al. [1,2,3,6]. These attacks exploit the branch prediction mechanism, which is nowadays a part of all general purpose processors. Microprocessors speed up their performance by using prediction algorithms to guess the most probable code path to execute, and fill the pipeline with the corresponding instructions. When the speculatively executed instruction flow turns out to be wrong, the execution has to start over from the correct one. Good prediction mechanisms/algorithms are those that have high correct prediction rates, and the development of such mechanisms is part of the technological advances in microprocessors. Furthermore, deeper pipelines enhance the average performance. Measurable timing differences between a correct and incorrect prediction are the inevitable outcome of such performance optimization, which is exactly what the BPA/SPBA attacks capitalize on. In theory, a spy process running on the target machine, together with the victim process, can use these timing differences as a side-channel information and deduce the precise execution flow performed by the victim process. This can potentially lead to a complete break of the system if the software implementation of the victim process is written in a way that knowledge of the execution flow provides the attacker with useful information.

Acıiçmez et al. showed that BPA and SBPA can be used to allow a spy process to extract the execution flow of an RSA implementation using the Square and Multiply (S&M) exponentiation algorithm. They demonstrate the results on the S&M algorithm implemented in OpenSSL-0.9.7. The initially published BPA attack required statistical analysis from many runs and could be viewed as difficult to implement in practice. However, the subsequent demonstration of the SBPA attack showed that measurements taken from a *single* run of the S&M exponentiation is sufficient to extract almost all of the RSA secret exponent bits.

The S&M algorithm is not frequently used in practice because there are more efficient exponentiation algorithms such as SWE, which is used by OpenSSL. The attack was mounted on OpenSSL-0.9.7 where the window size was changed from the default value $w = 5$ to $w = 1$ which degenerates SWE to S&M. Attacking the S&M algorithm was only a case study that showed the potential of SBPA. However, as we show in this paper, the actual scope of SBPA attacks is much broader. We identify a novel side-channel attack which is especially enabled by the SBPA methodology.

Obviously, it is unreasonable to handle the new (and also the old) threats by deactivating branch prediction or by disabling multi-process capabilities in general purpose processors and in operating systems. Thus, the conclusion is that cryptographic software applications that run on general platforms need to be (re-)written in an SBPA-aware style. To this end, another focus of this paper is on a software mitigation methodology for protecting applications against SBPA attacks.

The paper is organized as follows. Section 2 briefly recalls some basic facts and details of RSA and of its implementation in OpenSSL. The next section presents then the central contribution of the present paper. This is our novel SBPA side-channel against the BEEA, which enables full reconstruction of the input parameters to the BEEA — even if we assume that they are all completely unknown. Section 4 illustrates new vulnerabilities which we found in OpenSSL in the presence of potential SBPA attacks. In this section we also explain why these attacks are potentially feasible and can be mounted on almost all platforms. In Section 6 we explain the necessary software countermeasures to protect the openSSL CRT-RSA library against branch prediction attacks. The paper finishes with our conclusions on the BPA story as we currently see it.

2   Preliminaries and Notations

To facilitate smooth reading of the details in the paper, we provide a very brief description of the RSA cryptosystem and its corresponding implementation in OpenSSL. For a more detailed exposition of both topics we refer the reader to [12] and to [16].

2.1 RSA Parameters

RSA key generation starts by generating two large (secret) random primes p and q, where their n-bit long product $N = p\,q$ is used as a public modulus. Then, a public exponent e is chosen ($e = 2^{16}+1$ by default in OpenSSL), for which $d = e^{-1} \bmod (p-1)(q-1)$ is the private exponent. Figure 1 shows the RSA key generation flow.

RSA is used for both encryption and digital signature, where signing and decrypting messages require the use of the private exponent and modulus, while signature verification and encrypting messages require the used on only the public exponent and modulus. Factoring N (i.e., obtaining p and q) immediately reveals d. If an adversary obtains the secret value d, he can read all of the encrypted messages and impersonate the owner of the key for signatures. Therefore, the main purpose of cryptanalytic attacks on RSA is to reveal either p, q, or d.

---

input: null
output: $(N, p, q, e, d, d_p, d_q, C_2)$

procedure RSA_KeyGen:

(1) generate random primes p and q
(2) compute $N = p * q$
(3) choose $e \in \mathbb{Z}_N^*$
(4) d = Mod_Inverse(e, $((p-1)(q-1))$)   // compute ($d = e^{-1} \bmod (p-1)(q-1)$)
(5) (null, d_p) = Division(d, $p-1$)   // compute $d_p = d \bmod (p-1)$
(6) (null, d_q) = Division(d, $q-1$)   // compute $d_q = d \bmod (q-1)$
(7) C_2 = Mod_Inverse(q, p)   // compute ($C_2 = q^{-1} \bmod p$)

Fig. 1. RSA key generation procedure

2.2 RSA Implementation Using Chinese Remainder Theorem

OpenSSL, as many other implementations, uses the Chinese Remainder Theorem (CRT). This allows to replace modular exponentiation with full-length modulus N with two modular exponentiations with half-length modulus p and q. This speedups the modular exponentiation by a factor of approximately 4.

In order to employ CRT, the following additional values are computed during the key generation procedure: $d_p = d \bmod (p-1)$, $d_q = d \bmod (q-1)$, and $C_2 = q^{-1} \bmod p$. To decrypt a message M, the CRT procedure starts with reducing M modulo p and q, calculating $M_p = M \bmod p$, $M_q = M \bmod q$. Then, the two modular exponents $S_p = M_p^{d_p} \bmod p$ and $S_q = M_q^{d_q} \bmod q$ are calculated, and the desired result S is obtained bye the so-called "Garner re-combination" $S = S_q + (((S_p - S_q) * C_2) \bmod p) * q$. Figure 2 shows the CRT exponentiation flow as used in the current OpenSSL version (OpenSSL-0.9.8).

---

input: ciphertext M, RSA key N, p, q, d, d_p, d_q, C_2
output: decrypted plaintext $A = M^d \bmod N$

procedure RSA_Mod_Exp():

// exponentiation modulo q
(1) (null, M_q) = Division(M, q)   // $M_q = M \bmod q$
(2) S_q = Mod_Exp(M_q, d_q, q)   // $S_q = M_q^{d_q} \bmod q$

// exponentiation modulo p
(3) (null, M_p) = Division(M, p)   // $M_p = M \bmod p$
(4) S_p = Mod_Exp(M_p, d_p, p)   // $S_p = M_p^{d_p} \bmod p$

// combine S_p and S_q using Garner's method:
// compute $S = S_q + (((S_p - S_q) * C_2) \bmod p) * q$
(5) $S = S_p - S_q$
(6) if $S < 0$ then $S = S + p$
(7) S = Multiplication(S, C_2)
(8) (null, S) = Division(S, p)
(9) S = Multiplication(S, q)
(10) $S = S + M_q$
(11) return $A = S$

Fig. 2. Modular RSA exponentiation procedure with CRT

Modular exponentiation is based on a sequence of modular multiplications. These are typically, and in particular in OpenSSL, performed by using the Montgomery Multiplication (MMUL) algorithm.

2.3 Fixed Window Exponentiation

OpenSSL-0.9.8 uses the so called Fixed Window Exponentiation (FWE) algorithm with a window size of $w = 5$. To compute $M^{d_p} \bmod p$ (analogously hereafter also for q), the algorithm first computes a table of the $2^w - 1$ vectors, $V_i = M^i \bmod p$, for $1 \leq i \leq 2^w - 1$. The exponent d is then scanned in groups of w bits to read the value of the corresponding window i, $1 \leq i \leq 2^w - 1$. For each window, it performs w consecutive modular squarings followed by one modular multiplication with V_i, c.f. Figure 3.

2.4 Base Blinding

To thwart statistical side-channel attacks (e.g., [9]), OpenSSL-0.9.8 also applies a base blinding technique, as illustrated in Figure 4. A pair (X, Y) is generated, where X is a random number and $Y = X^{-e} \bmod N$. Then, instead of directly computing $M^d \bmod$

---

input: M, d, N ($M < N$, N is odd, and d is $k * w$ bits long)
output: $M^d \bmod N$

notation: $d = d[k * w - 1] \ldots d[0]$, where $d[0]$ is the least significant bit of d

Procedure:
// computation of the table values
$V_1 = M$
for i from 2 to $2^w - 1$
    $V_i = V_{i-1} * M \ \bmod N$

// actual exponentiation phase
$S = 1$
for i from $k - 1$ to 0 do
    $S = S^{2^w} \ (\bmod N)$
    // scanning the window value
    $wvalue = 0$
    for j from $w - 1$ to 0 do
        $wvalue = wvalue * 2$
        if $d[i * w + j] = 1$ then $wvalue = wvalue + 1$
    // multiplication with the table entry
    $S = S * V_{wvalue} \ \bmod N$
return S

Fig. 3. Fixed Window Exponentiation Algorithm

Input: c (ciphertext), rsa (RSA parameters: $N, p, q, d, d_p, d_q, C_2$)
Output: m (decrypted plaintext)

Procedure:
(1) $(X, Y) = $ Get_Blinding(rsa)   // X is a random; $Y = X^{-e} \bmod N$
(2) $c = $ Blinding_Convert(X, c)   // $c = Y * c \bmod N$
(3) $m = $ RSA_Mod_Exp(c, rsa)   // $m = c^d \bmod N$
(4) $m = $ Blinding_Invert(Y, m)   // $m = X * m \bmod N$
(5) padding m and output m

Fig. 4. RSA decryption procedure with base blinding, as implemented in OpenSSL

N, the base M is first modular-multiplied by Y and the exponentiation computes $(M * Y)^d \bmod N$ (e.g., by using the CRT). After the exponentiation, the result is modular-multiplied by X to obtain the desired result, because $X * (M * Y)^d \bmod N = M^d \bmod N$. Note that if (X, Y) is a proper pair, then $(X^2 \bmod N, Y^2 \bmod N)$ is also a proper pair. Therefore, once a proper blinding pair is generated, subsequent new pairs can be obtained for subsequent message decryptions, by performing only two modular squaring operations. By default, OpenSSL refreshes the blinding pair every 32 exponentiations.

The multiprecision operations of OpenSSL are handled by a multiprecision arithmetic library called BIGNUM. The implementations in BIGNUM library also introduces SBPA vulnerabilities to OpenSSL. The complete details of BIGNUM library are out of the scope in this paper. Yet, we outline the SBPA vulnerabilities introduced by this library after presenting a new attack on the BEEA in the next section.

3   The Main Result: Modular Inversion Via Binary Extended Euclidean Algorithm Succumbs to SBPA

Modular inversion operation is at the heart of public key cryptography. The most frequently used algorithm for modular inversion is the well known Extended Euclidean Algorithm (EEA). Due to the "unpleasant" division operations which are heavily used in the EEA, it is often substituted by another variant called the Binary Extended Euclidean Algorithm (BEEA), cf. [12]. BEEA replaces the complicated divisions of the EEA by simple right shift operations. It achieves performance advantages over the classical EEA and is especially efficient for large key-lengths (around 1024-2048 bits). The BEEA is indeed used by OpenSSL for these bitlengths.

The performance advantage of BEEA over the classical EEA is obtained via a "bitwise-scanning" of its input parameters. As we show here, in the presence of SBPA, this property opens a new side-channel that allows an easy reconstruction of unknown input parameters to the BEEA.

To derive our new SBPA-based side-channel attack against the BEEA, we start from the classical BEEA as described in [12] and illustrated in Figure 5.

The correctness of the BEEA (and also of the EEA) relies on the fact that the following equations

$$x \cdot A + y \cdot B = u \tag{1}$$
$$x \cdot C + y \cdot D = v \tag{2}$$

hold at any iteration step of the algorithm. In the case where $\gcd(x, y) = 1$, we also have at the termination

$$x \cdot A \equiv 0 \bmod y \tag{3}$$
$$x \cdot a \equiv 1 \bmod y, \tag{4}$$

which means that a is the inverse to x modulo y.

We note here that the algorithm may terminate with a $C < 0$, which would need in practice one additional step, namely $a := x + C$, to assure that $a \in \{0, \ldots, y - 1\}$. However, since this conditional final end addition is of no further interest because as it does not affect our unknown input reconstruction problem, we ignore it for the rest

input: integers x, y
output: integers a, b, v such that $ax + by = v$, where $v = \gcd(x, y)$

1. $g := 1$;
2. while x and y are even **do**
  $x := x/2, y := y/2, g := 2g$;
3. $u := x, v := y, A := 1, B := 0, C := 0, D := 1$;
4. while u is even **do**
  $u := u/2$;
  if $A \equiv B \equiv 0 \bmod 2$ **then**
    $A := A/2, B := B/2$
  else
    $A := (A + y)/2, B := (B - x)/2$;
5. while v is even **do**
  $v := v/2$;
  if $C \equiv D \equiv 0 \bmod 2$ **then**
    $C := C/2, D := D/2$
  else
    $C := (C + y)/2, D := (D - x)/2$;
6. if $u \geq v$ **then**
  $u := u - v, A := A - C, B := B - D$
else
  $v := v - u, C := C - A, D := D - B$;
7. if $u = 0$ **then**
  $a := C, b := D$, and **return**$(a, b, g \cdot v)$
else
  goto 4;

Fig. 5. Binary Extended Euclidean Algorithm

of our discussion. Another important detail to point out is that for computing modular inversion, i.e., $\gcd(x, y) = 1$, the **while**-loop in Step 2 is never executed.

To derive our SBPA-based side-channel attack, we extract from the BEEA flow only the serial internal information flow regarding the branches that depend on the input values u, v, and their difference $u - v =: z$. This flow is visualized in Figure 6, where we simply "serialized" all the steps that are relevant to our case.

From Figure 6, we identify that the 4 critical input-dependent branches that produce 4 important information leaks. These can be put into two groups:

1. Number of right-shift operations performed on u.
2. Number of right-shift operations performed on v.
3. Number of subtractions $u := u - v$.
4. Number of subtractions $v := v - u$.

Thus, assuming that the branches in the BEEA flow are known due to a spy employing SBPA, the backward reconstruction of u and v from the termination values $u = 0$ and $v = 1$ is possible, as in Figure 3.

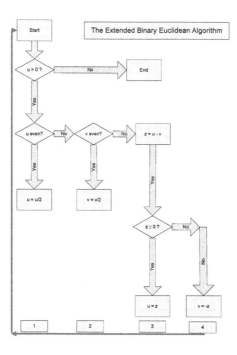

Fig. 6. The information flow of the BEEA regarding u and v

Branch	Reconstruction operation
1	$u := 2 \cdot u$
2	$v := 2 \cdot v$
3	$u := u + v$
4	$v := v + u$

To come up with a concrete reconstruction algorithm, we confine the 4 branches 1, 2, 3, and 4 into successive information leakage groups. A single iteration group comprises of the information whether branch 3 or branch 4 was executed, and additionally the information on how often branch 1 and branch 2 was executed in this group. This constitutes one group — the next group starts again with entering branch 3 or branch 4. That is, for $i = 1, \ldots, \ell$ and $j = 1, 2$, we define

$$\mathcal{SHIFTS}_j[i] := \#\{\text{group } i \text{ iterations spend in branch } j\},$$

and

$$\mathcal{SUBS}[i] := \begin{cases} \text{``}u\text{''} & : \text{if branch 3 is taken} \\ \text{``}v\text{''} & : \text{if branch 4 is taken} \end{cases}$$

Armed we the leaked information $\mathcal{SHIFTS}_j[\cdot]$ and $\mathcal{SUBS}[\cdot]$ which are obtained by a (perfect) SBPA, our task is now to reconstruct x, y and W, X, Y, Z from the known final values $u_{\mathrm{f}} = 0$ and $v_{\mathrm{f}} = 1$ such that

$$u_{\mathrm{f}} \cdot W + v_{\mathrm{f}} \cdot X = x$$
$$u_{\mathrm{f}} \cdot Y + v_{\mathrm{f}} \cdot Z = y,$$

Here, for the sake of clarity, we explicitly keep the value u_{f} and W and Y, although they actually vanish due to $u_{\mathrm{f}} = 0$. The corresponding algorithm, illustrated in Figure 7, accomplishes the task.

---

input: arrays $\mathcal{SHIFTS}[\cdot]$ and $\mathcal{SUBS}[\cdot]$ and ℓ
output: integers W, X, Y, Z as above

1. $W := 1, X := 0, Y := 1, Z := 0$;
2. $i := \ell$;
3. for $i = \ell$ **down to** 2 **do**
    if $\mathcal{SUBS}[i] =$ "u" **then**
        $W := W << \mathcal{SHIFTS}_1[i], X := X << \mathcal{SHIFTS}_2[i]$,
        $W := W + Y, X := X + Z$
    else
        $Y := Y << \mathcal{SHIFTS}_1[i], Z := Z << \mathcal{SHIFTS}_2[i]$,
        $Y := W + Y, Z := X + Z$;
4. $Y := Y << \mathcal{SHIFTS}_1[1], Z := Z << \mathcal{SHIFTS}_2[1]$;
5. return(W, X, Y, Z);

Fig. 7. BEEA reconstruction from SBPA information

Thus, by the above exposition we have given a proof of the main contribution of the present paper.

Theorem 1. *For unknown numbers x and y with $\gcd(x, y) = 1$ as inputs to the BEEA, the side-channel information $\mathcal{SHIFTS}_j[\cdot]$ and $\mathcal{SUBS}[\cdot]$ (which can be obtained by a perfect SBPA) can be used to completely reconstruct, in polynomial time, both x and y.*

In the following sections, we show how this SBPA-enabled reconstruction theorem can be used to compromise the security of the OpenSSL RSA-CRT version in several points of the algorithm, and moreover — even in the presence of CA side-channel mitigations.

4   New SBPA Vulnerabilities in OpenSSL

4.1   On the Granularity of SBPA And the Actual Threat of SBPA For Almost All Platforms

To illustrate the feasibility of attacking the BEEA algorithm with an SBPA spy or other code constructions inside openSSL, we provide here some details on how, and at what granularity, a spy process (or processes) can actually exploit the environment and extract information.

Acıiçmez et al. showed that a carefully written spy process running simultaneously with an RSA process, is able to collect almost all of the secret key bits during a *single*

RSA decryption execution. They demonstrated the attack on a simplified S&M RSA implementation, which means that the branches in that demonstration were separated by thousands of cycles from each other. Moreover, this attack ran a spy that relied on the simultaneous-multithreading (SMT) capability of the tested platform.

However, we point out here that the actual power of SBPA is most probably not limited to this basic application where branches are widely separated, and not limited to SMT capable platforms. In general, SBPA has the potential to reveal the entire execution flow of a target process in *almost any* execution environment — with or without SMT. We now explain the grounds for this claim.

MicroArchitectural attacks exploit shared microprocessor components to compromise security systems by using spy processes. They require a quasi-parallel execution of the spy and the "crypto" processes on the same processor. Although a hardware-assisted SMT feature seems to be mandatory to achieve this, recent studies indicate the opposite [13,14]. Neve et al. showed that it is possible to exploit some of the scheduling functionality of operating system (OS) to accomplish the same effect on a non-SMT platform — that is, as if the spy and crypto processes ran simultaneously on an SMT platform. In fact, the idea of exploiting OS scheduling for side-channel attacks by using a spy process and a trojan process, was published already in 1992 [10].

Their method of transferring SMT-based MicroArchitectural attacks to non-SMT systems rely on the preemptive OS scheduling property, which is the way that on OS handles multitasking capability. Multitasking operating systems allow for the execution of multiple processes on the same computer, concurrently. In other words, each process is given permission to use the processor's resources, including the cache, branch prediction unit, and other MicroArchitectural resources. When several different processes run actively on the system, the OS assigns a certain maximum execution time (called quantum) to the process that gets his turn to execute.

The preemptive OS scheduling property gives a process the ability to trigger a context switch, i.e., yielding the CPU to another process, at any desired time before consuming the entire quantum. The interesting and security-critical point is what the OS does afterwards. It indeed schedules the next process in the queue, but lets it run only for the duration of the remaining part of the quantum that was mostly consumed by the prior process. Neve et. al. used this OS functionality to stretch the very short execution time of the AES over several quantums and could therefore obtain spy measurements precisely centered around a very small number of instructions.

Using this result, we can safely speculate that it would be possible to apply SBPA attacks at a fine-granularity of even a few hundred cycles and — in an idealized scenario — to detect *all* the branches of an attacked application. This dramatically broadens the scope of SBPA attacks, for example to attack the BEEA procedure.

We illustrate a possible attack path. A spy process can run until the end of a quantum, but just before the end of this quantum — e.g., a couple of hundred cycles — trigger a context switch. Then, the OS would let the crypto process run for a very short time — the remaining time till the end of that quantum — and give the execution back to the spy. In fact, an attacker can even use another "trojan process" to handle switching spying independently.

This method still does not allow us to analyze branches at the granularity of a few cycles. However, note that the arithmetical and logical operations in RSA are multi-precision, meaning that the operands are several machine-words long. Any such computation in RSA requires execution of tens to hundreds of instructions. Furthermore, the actual implementation of the BBEA algorithm in OpenSSL involves function calls, which inherently consume overhead cycles. At the minimum, these functions declare and initiate some variables, and need to execute some steps (in a loop). All these steps together most probably will leave a sufficiently large window for a successful SBPA attack.

Therefore — although we have not developed code to demonstrate this attack — we strongly conjecture that an SBPA attack, coupled with this "OS trick", has the potential of revealing the entire execution flow of the BEEA on almost any platform. As we show in the following section, if such an "ideal SBPA attack" is possible, and if a spy process successfully enters the attacked system, this can compromise the security of the RSA implementation of OpenSSL at several points.

4.2   New SBPA Vulnerabilities in OpenSSL Due to Theorem 1

In this section we describe several vulnerabilities that we have identified in OpenSSL, in the context of SBPA attacks and Theorem 1 that shows how the inputs to the modular inversion procedure can be extracted by an SBPA spy. Referring back to Section 2, we can see that modular inverses are computed during the RSA computations when:

1. the private exponent $d = e^{-1} \bmod (p-1)(q-1)$ is computed,
2. the CRT parameter $q^{-1} \bmod p$ is computed,
3. the Montgomery constant $-p^{-1} \bmod m$ (e.g. with $m = 2^{32}$) is computed, and
4. the masking pair $(X, Y = X^{-e} \bmod N)$ is computed.

Now, the implications of Theorem 1 become clear. In cases 1-3, an SBPA attack can directly compromise the secret RSA keys. In case 4, an SBPA attack can compromise the blinding mitigation. If the blinding factor is revealed, the blinding mitigation technique to thwart statistical attacks collapses. In such a scenario, other and simpler attacks could be launched. If the SBPA spy indeed compromises the blinding mitigation, the situation is effectively a chosen-plaintext scenario, where the timing depends on the chosen-plaintext and the secret primes. Clearly, a remote attack can now be mounted on the application. We mention here two further attack scenarios: First, remote timing attacks [9] that exploit timing differences due to the End Reduction (ER) step in the Montgomery multiplications, and second remote timing attacks that exploit the early exit short cut taken in the division algorithm. We add some details to explain the second attack in the following.

Remote timing attacks that exploit the early exit short cut taken in the division algorithm. The first step in the CRT algorithm, is reducing the message M modulo p and q. This reduction is carried out by the BIGNUM Division function. There are two branches in the Division procedure of OpenSSL-0.9.8, that leak information that could potentially compromise the private key. Given A and B, Division (A, B) performs the following:

a. If $A < B$, Division (A, B) immediately returns the quotient 0, remainder B, and exits.
b. If the divisor B has t words, and the first t words of A form a number which exceeds B, the following is done, where b is the bitlength of B: $2^{tw} < A < 2^{b+tw}$, and if it is satisfies, Division (A, B) resorts to some shortcut path.

It follows that the execution time of Division is data dependent and could be used very simple in a binary-search oriented algorithm to find p or q.

4.3   Vulnerability Due to Exponent Scanning in The FWE Procedure

The FWE algorithm is in theory immune to SBPA. Unlike S&M or SWE algorithms, it does not involve exponent-dependent execution because the modular multiplications/squares are independent of the exponent value. However, the actual code implementation of FWE in OpenSSL-0.9.8 makes it potentially susceptible to SPBA attack. The routine that constructs the "window value", while scanning the exponent string does it via a bit-by-bit, i.e., an if-then-else method. The code invokes the function BN_is_bit_set(a, n) whose task is to read the value of the nth bit of a. Actually, BN_is_bit_set is currently implemented as follows:

```
int BN_is_bit_set(const BIGNUM *a, int n)
{
    int i,j;

    bn_check_top(a);
    if (n < 0) return 0;
    i=n/BN_BITS2;
    j=n%BN_BITS2;
    if (a->top <= i) return 0;
    return((a->d[i]&(((BN_ULONG)1)<<j))?1:0);
}
```

If the conditional statement in the last line of this function is actually performed (i.e., not removed by the compiler), it can be exploited by an SBPA spy to completely discover the secret exponent that is being scanned.

5   Software Mitigations to Protect OpenSSL Against the SBPA Vulnerabilities

Our general methodology for creating an SBPA-aware RSA implementation is to first assume that SBPA can reveal the entire flow of the execution. Under this assumption, we scan through all steps of the algorithm to detect the conditional branches. For each branch we consider the following question: *if an attacker obtains the complete history of this branch, would this provide him with useful information?* Whenever the result to this question is positive, we consider fixing the problem by either changing the algorithm flow to eliminate the conditional execution, or by assuming that this branch is executed in an off-line environment (i.e., before any spy can enter the system).

5.1   Fixing the Exponent Scanning Vulnerability

Mitigating this vulnerability is simple: the conditional statement in the last line BN_is_bit_set(a, n) needs to be removed. Since BN_is_bit_set(a, n) checks whether the input bit is 0 or 1, the function can simply return the value of that bit. We therefore propose the following change in the code:

- Current Version:

```
return((a->d[i]&(((BN_ULONG)1)<<j))?1:0);
```

- Our Proposal:

```
return((a->d[i] >> j) & 0x01);
```

Remarks:

1. A more efficient way to handle the exponent scanning is to avoid the call to BN_is_bit_set function. The window values can be constructed by directly copying the corresponding section of the exponent string at once (using shifts and logical operations) instead of doing it in a bit-by-bit fashion.
2. The secret exponents d_p and d_q are fixed for a given RSA key. Thus, the window values can be computed once, during the key generation phase, stored, and then used per each exponentiation without re-scanning the exponent string.
3. A good compiler, particularly in an optimized mode, should be able to identify and to automatically eliminate the discussed branch. In fact, when OpenSSL-0.9.8 is compiled with a gcc compiler (using Linux operating system), the compiler indeed removes the branches. However, this compiler optimizing property cannot be guaranteed for any compiler in any mode, and should not be considered as an automatic mitigation.

5.2   A Minimal Set of Mitigation for CRT-RSA and Key Generation

In this section, we outline a few pin-pointed changes to OpenSSL-0.9.8, that can be used for achieving (what seems to be) an SBPA-protected implementation.

The proposed changes are in the Big Numbers library of OpenSSL, particularly in the functions
BN_mod_inverse(in, a, n) and BN_div (dv, rm, num, $divisor$).

1. **Avoiding the use of BEEA for modular inversion in BN_mod_inverse.** For odd n, whose bit-length is less than 450 (for a 32-bit machine) or less than 2048 bit (for a 64-bit machine), the function BN_mod_inverse(in, a, n) uses the BEEA. In other cases, it uses a general inversion algorithm that uses the division as implemented in BN_div(dv, rm, num, $divisor$). An easy mitigation technique is to eliminate the branch that allows for selecting BEEA, thus to always use the general inversion via Division.

2. **Avoiding other branches during modular inversion that uses BN_div.** To optimize performance the general inversion algorithm first checks if the divisor and dividend are close and in that case does not invoke an BN_Division (performing the division in another method). Further, in intermediate steps that require multiplication, the function first checks if the multiplier is between 1 and 4 and in that case performs additions instead of invoking BN_mul. These shortcuts can be eliminated, in order to avoid leaking any information associated with the relative sizes of the operands (which could potentially facilitate an attack). The early exit step in the division algorithm should also be eliminated. We point out that these branch elimination steps can be spared if the base blinding mitigations are assumed to successfully mask the side-channel information.

We emphasize that in general, a minimal quick-fix approach for handling the security of a cryptographic implementation is a less recommended approach than a fundamental revision of the CRT-RSA procedures which should be seriously considered. We propose such an approach in the paper's next part.

5.3 Intrinsically BPA-Protected CRT-RSA Implementation — Smooth CRT-RSA

Our goal here is to outline a CRT-RSA implementation that does not contain conditional branches. To this end, we need to eliminate modular reductions, divisions, and the conditional ER step from the computation of RSA at all. Our proposed approach, which we call "Smooth CRT-RSA" achieves this objective, and is therefore an intrinsically SBPA-protected CRT-RSA implementation.

There are several methods one can use to eliminate the conditional ER step (e.g., [11], [20]). The reason why we need to perform the ER step at the end of a Montgomery multiplication is that the result of the algorithm returns a result that is smaller than twice the modulus, but not necessarily smaller than the modulus itself. Thus, the result sometimes needs to be reduced by means of one subtraction of the modulus (hence called an ER step). However, if the data structures used in an CRT-RSA implementation can accommodate the possibility of storing values between p and $2p$ (q and $2q$, resp.), then the ER step can be simply avoided. There are two ways to achieve this:

1. Increasing the size of the data structures by one word.
2. Decreasing the size of p and q by two bits.

Note that if the ER step is eliminated, and if the procedure that implements MMUL is written in a way that does not introduce superfluous branches, then the base blinding mitigation is not necessary.

To eliminate the need for modular reductions and divisions in CRT-RSA computation, we propose to change the CRT-RSA flow, and introduce three new variables H_{3p}, H_{3q}, and MC_2:

$$H_{3p} = 2^{3k} \bmod p, \quad H_{3q} = 2^{3k} \bmod q, \quad MC_2 = q^{-1} * 2^k \bmod p \qquad (5)$$

where k is defined by $2k = n$.

Input:	M (ciphertext)	
Input:	N (modulus; n=2k bits long)	
Input:	p (secret prime; k bits long)	
Input:	q (secret prime; k bits long)	
Input:	d_p (secret CRT exponent for the modulus p)	
Input:	d_q (secret CRT exponent for the modulus q)	
Input:	$MC_2 (= q^{-1} * 2^k \ mod \ p)$	
Output:	$A = M^d \ mod \ N$	// d is the RSA private exponent

Pre-computed constants:
    $H_{3p} = 2^{3k} \ mod \ p$
    $H_{3q} = 2^{3k} \ mod \ q$
    // Procedure: $MontExp(A, X, N) = A^X 2^{-n(X-1)} \ (mod \ N)$
    // (modular exp steps with modular multiplications replaced by $MMUL$ operations)
    // Procedure: $MontReduce$ — Montgomery reduction.

Setup:
    Conversion into Montgomery Domain:

(1)	$R_1 =$MontReduce (M, p)	// $R_1 = M * 2^{-k} \ mod \ p$
(2)	$R_1 =$MMUL (R_1, H_{3p}, p)	// $R_1 = M * 2^k \ mod \ p$
(4)	$R_2 =$MontReduce (M, q)	// $R_2 = M * 2^{-k} \ mod \ q$
(5)	$R_2 =$MMUL (R_2, H_{3q}, q)	// $R_2 = M * 2^k \ mod \ q$

Exponentiations:

(3)	$M_1 =$MontExp (R_1, d_p, p)	// $M_1 = R_1^{d_p} 2^{-k(d_p-1)} \ mod \ p$
(6)	$M_2 =$MontExp (R_2, d_q, q)	// $M_2 = R_2^{d_q} 2^{-k(d_q-1)} \ mod \ q$
	Conversion from Montgomery Domain:	
(5)	$M_1 =$MMUL $(M_1, 1, p)$	// $M_1 = M^d \ mod \ p$
(5)	$M_2 =$MMUL $(M_2, 1, q)$	// $M_2 = M^d \ mod \ q$

Recombination with MMUL:

(7)	$M_2 = M_2 - M_1$	//
(8)	if$(M_2 < 0) \ M_2 = M_2 + p$	//
(9)	$R_1 =$MMUL(M_2, MC_2, p)	//
(10)	$R_1 = M_2 * p$	//
(11)	$A = R_1 + M_1$	//
(12)	return A	//

Fig. 8. Smooth CRT-RSA

These parameters can be easily derived from the traditional Montgomery constants $H_p = 2^{2k} \ mod \ p$, $H_q = 2^{2k} \ mod \ q$, and from the CRT parameter $C_2 = q^{-1} \ mod \ p$ which are currently used in OpenSSL for converting the exponentiation base from the integer to the Montgomery domain. Only one Montgomery multiplication is required for this derivation:

$$H_{3p} = 2^{3k} \bmod p \qquad = MMUL(H_p, H_p, p)$$
$$H_{3q} = 2^{3k} \bmod q \qquad = MMUL(H_q, H_q, q)$$
$$MC_2 = q^{-1} * 2^k \bmod p = MMUL(C_2, H_p, p)$$

These computations can be included in the Montgomery initialization phase, and the new parameters can replace the old ones. Figure 8 illustrates the new proposed smooth CRT-RSA flow. Interestingly, the Smooth CRT-RSA has a better performance than the standard CRT-RSA that is implemented in OpenSSL-0.9.8.

6 Conclusions

Our main result in this paper was the unexpected novel SBPA side-channel attack against the public-key cryptography primitive BEEA that is used for modular inversion. This started a research path for fully exploring the potential risk of SBPA attacks.

Since deactivating branch prediction units on all general purpose platforms, or disabling multiprocessing capabilities in the OS, is clearly an unattractive mitigation approach, one conclusion is that cryptographic software that run on general platforms needs to be (re-)written in an SBPA-aware style. Thus, another focus of our paper was on a software mitigation methodology for protecting applications against SBPA attacks. In order to do so, we also highlighted — on top of the central SBPA-enabled BEEA side-channel attack — some other very obvious OpenSSL code constructs which are susceptible to SBPA attacks. We then proposed several simple techniques to mitigate these vulnerabilities, which could be applied to future versions of OpenSSL towards an SBPA-aware version.

The minimal set of mitigations that were detailed in Section 5.2 have already been implemented by Intel Corporation's security experts, and have been provided to the open source community, in order to facilitate their quick deployment into a new version of OpenSSL. These modifications carry only a small performance penalty, and are by now already implemented in the current OpenSSL version, [17].

In addition, we have also im implemented the Smooth RSA proposal from Section 5.3 on top of the minimal set. As expected, Smooth RSA and the minimal set of mitigations not only protects OpenSSL 0.9.8a from the SBPA attacks, but also *improves* the performance due to the elimination of some operations such as the ER step in the Montgomery multiplications. The details, including the comparative performance results, will soon be available in a subsequent paper.

Still, we also point out that the present paper does not claim an exhaustive SBPA side-channel security analysis with a full proof on corresponding software countermeasures for OpenSSL. It rather indicates that this task is required, especially since SBPA attacks potentially threatens every step of the OpenSSL code. However, the present paper made a major step towards this goal.

Another conclusion of the paper is that the PC-oriented side-channel attack field is very subtle due to the high MicroArchitectural complexity of today's processors. Thus, we expect that new MicroArchitectural attacks will inevitably be discovered in the future.

We also point out here that Smooth RSA has the important security advantage of eliminating the ER step at all. Since SBPA has the potential to reveal the entire execution flow of the target process, it could be also used for identifying whether or not a given Montgomery multiplication requires the ER step. Taking [19] into account, one can realize that this attack, which is based on tracking the ER steps, can be actually launched against the OpenSSL implementation if the ER steps are not eliminated (even while OpenSSL is using the FWE algorithm). We also mention that the ER steps can be identified by other methods, such as cache eviction based spying, and not only using SBPA. For this reason, we recommend that the ER step is eliminated at all, and suggest using Smooth RSA to this end. We feel that moving towards the Smooth RSA would be the best solution to get rid off many ER related attacks and also to avoid new twists of known and future ones.

Last but not least, we summarize the current situation as follows. Probably the best defense against current and future MicroArchitectural attacks is to let the cryptographic software community become aware of the security implications of writing cryptographic software that is going to be executed on throughput-optimized general purpose processors. This will help software be written using a proper side-channel attack aware methodology. A first step into this direction are the recent changes in OpenSSL fixing the vulnerabilities pointed out in this paper, [17].

Acknowledgements

The authors are members of the Applied Security Research Group at The Center for Computational Mathematics and Scientific Computation within the Faculty of Science and Science Education at the University of Haifa (Israel).

References

1. Acıiçmez, O., Koç, Ç.K., Seifert, J.-P.: Predicting Secret Keys via Branch Prediction. In: Abe, M. (ed.) CT-RSA 2007. LNCS, vol. 4377, pp. 225–242. Springer, Heidelberg (2006)
2. Acıiçmez, O., Koç, Ç.K., Seifert, J.-P.: On The Power of Simple Branch Prediction Analysis. In: ASIACCS 2007. ACM Symposium on Information, Computer and Communications Security (to appear, 2007)
3. Acıiçmez, O., Koç, Ç.K., Seifert, J.-P.: On The Power of Simple Branch Prediction Analysis. Cryptology ePrint Archive, Report 2006/351 (October 2006)
4. Acıiçmez, O., Schindler, W., Koç, Ç.K.: Cache Based Remote Timing Attack on the AES. In: Abe, M. (ed.) CT-RSA 2007. LNCS, vol. 4377, pp. 271–286. Springer, Heidelberg (2006)
5. Acıiçmez, O., Schindler, W., Koç, Ç.K.: Improving Brumley and Boneh Timing Attack on Unprotected SSL Implementations. In: Meadows, C., Syverson, P. (eds.) Proceedings of the 12th ACM Conference on Computer and Communications Security, pp. 139–146. ACM Press, New York (2005)
6. Acıiçmez, O., Seifert, J.-P., Koç, Ç.K.: Predicting Secret Keys via Branch Prediction. Cryptology ePrint Archive, Report 2006/288 (August 2006)
7. Bernstein, D.J.: Cache-timing attacks on AES. Technical Report, 37 pages (April 2005), http://cr.yp.to/antiforgery/cachetiming-20050414.pdf
8. Bleichenbacher, D.: Chosen Ciphertext Attacks Against Protocols Based on the RSA Encryption Standard PKCS #1. In: Krawczyk, H. (ed.) CRYPTO 1998. LNCS, vol. 1462, pp. 1–12. Springer, Heidelberg (1998)

9. Brumley, D., Boneh, D.: Remote Timing Attacks are Practical. In: Proceedings of the 12th Usenix Security Symposium, pp. 1–14 (2003)

10. Hu, W.M.: Lattice scheduling and covert channels. In: Proceedings of IEEE Symposium on Security and Privacy, pp. 52–61. IEEE Press, Los Alamitos (1992)

11. Gueron, S.: Enhanced Montgomery Multiplication. In: Kaliski Jr., B.S., Koç, Ç.K., Paar, C. (eds.) CHES 2002. LNCS, vol. 2523, pp. 46–56. Springer, Heidelberg (2003)

12. Menezes, A.J., van Oorschot, P., Vanstone, S.: Handbook of Applied Cryptography. CRC Press, New York (1997)

13. Neve, M., Seifert, J.-P.: Advances on Access-driven Cache Attacks on AES. In: SAC 2006. Selected Areas of Cryptography (2006)

14. Neve, M.: Cache-based Vulnerabilities and SPAM Analysis. Ph.D. Thesis, Applied Science, UCL (July 2006)

15. Osvik, D.A., Shamir, A., Tromer, E.: Cache Attacks and Countermeasures: The Case of AES. In: Pointcheval, D. (ed.) CT-RSA 2006. LNCS, vol. 3860, pp. 1–20. Springer, Heidelberg (2006)

16. OpenSSL: the open-source toolkit for ssl / tls

17. OpenSSL (Changelog) Changes between 0.9.8f and 0.9.9,
   `http://www.openssl.org/news/changelog.html`

18. Percival, C.: Cache missing for fun and profit. In: BSDCan 2005, Ottawa (2005),
   `http://www.daemonology.net/hyperthreading-considered-harmful/`

19. Schindler, W., Walter, C.D.: More Detail for a Combined Timing and Power Attack against Implementations of RSA. In: Paterson, K.G. (ed.) Cryptography and Coding. LNCS, vol. 2898, pp. 245–263. Springer, Heidelberg (2003)

20. Walter, C.D.: Montgomery exponentiation needs no final subtractions. Electronics Letters 35, 1831–1832 (2001)

Remarks on the New Attack on the Filter Generator and the Role of High Order Complexity

Panagiotis Rizomiliotis[*]

K.U.Leuven, ESAT/COSIC,
Kasteelpark Arenberg 10,
B-3001 Leuven-Heverlee, Belgium
Panagiotis.Rizomiliotis@esat.kuleuven.be

Abstract. Filter generators are important building blocks of stream ciphers and have been studied extensively. Recently, a new attack has been proposed. In this paper, we analyze this attack using the trace representation of the output sequence y and we prove that the attack does not work always as expected. We propose a new algorithm that covers the cases that the attack cannot be applied. The new attack is as efficient as the original attack. Finally, trying to motivate the research on the nonlinear complexity of binary sequences, we present a scenario where the knowledge of the quadratic complexity of a sequence can decrease significantly the necessary for the attack amount of known keystream bits.

1 Introduction

Stream ciphers are symmetric ciphers where the plaintext is combined with a pseudorandom bit stream. Most of the designs consist of a generator that produces the keystream which is added modulo 2 with the plaintext. Stream ciphers have been the center of interest of the cryptographic society the last few years, due to the eStream contest organized by the European Network of Excellence ECRYPT ([3]).

One of the most important building blocks of many stream ciphers is the *filter generator* ([6]). The filter generator consists of a linear feedback shift register (LFSR) with primitive feedback polynomial and a Boolean function that filters the state and produces the output sequence y. Filter generators have been extensively studied and many attacks have been proposed, leading to a variety of restrictions on the choice of the feedback polynomial and the filter function. In [1] and [2] you can find a very good overview on the most powerful attacks, namely the correlation and the algebraic attacks.

Recently, a new efficient attack has been proposed ([9]). This attack discloses the initial state of the filter generator, when $D(d) = \sum_{i=0}^{d} \binom{n}{i}$ bits of data

---

[*] The author's work is funded by the FWO-Flanders project nr. G.0317.06.

S.D. Galbraith (Eds.): Cryptography and Coding 2007, LNCS 4887, pp. 204–219, 2007.

are available, where n is the size of the LFSR and d is the degree of the filter function. The new attack requires $O(D(d)log_2(D(d))^3)$ pre-computation and $O(D(d))$ computation complexity for the calculation of the initial state. The attack is based on the fact that when the output bits are expressed as a function of the initial state, the coefficients of the monomials have a specific form.

The attack does not work when the feedback polynomial of the LFSR is not factor of the feedback polynomial of the shortest LFSR that produces the same output sequence y. In that case, the authors propose a modified, less efficient, version of their attack, where a system of nonlinear equations is solved using the linearization method.

In this paper, we interpret the results of [9] in terms of the trace representation of sequence y. Using this approach we show that the modified version of the attack does not work as expected in [9]. The nonlinear system is not overdefined and thus the linearization method can not be applied. We propose a new modified version of the attack to cover the case that the original attack cannot be applied. The attack has the same complexity as the original attack disproving the conjecture from [9], that filter generators producing sequences with smaller linear complexity than the maximum possible can resist their attack.

Finally, we describe a scenario where, given the quadratic complexity of y, the amount of data needed for the attack decreases. Though very efficient, the original attack requires amount of known keystream bits equal to the upper bound of the linear complexity. This amount of data usually is not available, as it was demonstrated in [10]. We want to stress out that our main goal is to warm over the research interest on high order nonlinear complexity and motivate researchers to work in this area, and not to present a specific attack.

The paper is organized as follows. In Section 2, we present the necessary background and we briefly describe the attack proposed in [9]. In Section 3, we introduce a different interpretation of the attack and we show that the modified version of the attack does not work. In Section 4, we present the new modified version of the attack for the cases that the original attack does not cover. Finally, in Section 5, we exhibit the role of the high order nonlinear complexity in such an attack.

2   Preliminaries

Let α be a primitive element of $\mathbb{F}_{2^n}$ and let C_s be the cyclotomic coset modulo $N = 2^n - 1$ defined as

$$C_s = \{s, 2s, \cdots, 2^{n_s-1}s\}$$

where n_s is the smallest integer such that $s \equiv s2^{n_s} \mod N$. The integer s is the smallest in the coset C_s and is called the coset leader. The polynomial

$$h_s(z) = \prod_{j \in C_s} (z + \alpha^j)$$

is irreducible over $\mathbb{F}_2$ of degree n_s and n_s divides n. Let I be the set of all coset leaders and $I(d)$ the set of coset leaders whose binary representation has Hamming weight at most d.

Consider a periodic binary sequence $\boldsymbol{y} = \{y_t\}_{t \geq 0}$ with period N. The length of the shortest linear feedback shift register (LFSR) that generates $\boldsymbol{y}$ defines the linear complexity of the sequence $\boldsymbol{y}$, denoted by $LC(\boldsymbol{y})$. The feedback polynomial $h(z)$ of the LFSR can be uniquely written as the product of irreducible polynomials ([5]),

$$h(z) = \prod_{j \in \hat{I}} h_j(z),$$

where $\hat{I} \subseteq I$. The linear complexity is given by

$$LC(\boldsymbol{y}) = \sum_{j \in \hat{I}} n_j.$$

Sequence $\boldsymbol{y}$ admits the trace representation ([11])

$$y_t = \sum_{\kappa \in I} tr_1^{n_\kappa}(\gamma_\kappa \alpha^{\kappa t})$$

where $\gamma_\kappa \in \mathbb{F}_{2^{n_\kappa}}^*$, if the coset leader of $C_{-\kappa}$ belongs to $\hat{I}$ and $\gamma_\kappa = 0$, otherwise. $\mathbb{F}_{2^{n_\kappa}}^*$ is the multiplicative group of $\mathbb{F}_{2^{n_\kappa}}$. The function

$$tr_1^n(z) = z + z^2 + z^4 + \cdots + z^{2^{n-1}}$$

is the trace function and it maps elements of $\mathbb{F}_{2^n}$ onto the prime subfield $\mathbb{F}_2$ ([5]).

In other words the sequence $\boldsymbol{y}$ can be written as the sum of sequences $y_i^\kappa = tr_1^{n_\kappa}(\gamma_\kappa \alpha^{\kappa i})$. Each sequence $\boldsymbol{y}^\kappa$ has linear complexity n_κ and period $\frac{2^n - 1}{\gcd(\kappa, 2^n - 1)}$. Since $h_\kappa(\alpha^\kappa) = 0$, we have

$$\sum_{j=0}^{n_\kappa} h_j^\kappa y_{j+t}^\kappa = tr_1^{n_\kappa}(h_\kappa(\alpha^\kappa)\gamma_\kappa \alpha^{\kappa t}) = 0 \qquad (1)$$

where $h_\kappa(z) = \sum_{i=0}^{n_\kappa} h_i^\kappa z^i$. When $\gcd(\kappa, 2^n - 1) = 1$, $\boldsymbol{y}^\kappa$ is an m-sequence ([4]).

The LFSR with feedback polynomial the primitive polynomial $h_{-\kappa}(z)$ generates the m-sequence $y_t^\kappa = tr_1^n(\alpha^{\kappa t_0} \alpha^{\kappa t})$. The initial state $\boldsymbol{X}_{t_0}$ of the LFSR determines a different phase t_0 of the m-sequence. The phase $t_0 = 0$ is called the *characteristic phase* of the m-sequence. We define as

$$\boldsymbol{X}_c = \left(tr_1^n(1), tr_1^n(\alpha^\kappa), \cdots, tr_1^n(\alpha^{\kappa(n-1)})\right)$$

the *characteristic state* of the LFSR, that is the initial state corresponding to the characteristic phase $t_0 = 0$ of the m-sequence. Let L_κ be the $n \times n$ state transition matrix of the LFSR defined as

$$L_\kappa = \begin{pmatrix} 0 & 1 & 0 & \cdots & 0 \\ 0 & 0 & 1 & \cdots & 0 \\ \vdots & \vdots & \vdots & & \vdots \\ 0 & 0 & 0 & \cdots & 1 \\ h_0^\kappa & h_1^\kappa & h_2^\kappa & \cdots & h_{n-1}^\kappa \end{pmatrix}.$$

Each state can be written as

$$(\boldsymbol{X}_{t_0})^T = L_\kappa^{t_0} \cdot (\boldsymbol{X}_c)^T \tag{2}$$

where $L_\kappa^{t_0}$ is t_0-th power of L_κ and T denotes the transpose of the matrix. That means that, using $\boldsymbol{X}_{t_0}$ as initial state the t_0 shift from the characteristic phase of the m-sequence is produced, i.e.,

$$tr_1^n(\alpha^{\kappa t_0}\alpha^{\kappa t}) = \boldsymbol{g} \cdot L_\kappa^{t+t_0} \cdot \boldsymbol{X}_c^T = \boldsymbol{g} \cdot L_\kappa^t \cdot \boldsymbol{X}_{t_0}^T \tag{3}$$

where $\boldsymbol{g} = (1, 0, \cdots, 0)$. It is straightforward to check that

$$\begin{pmatrix} \boldsymbol{g} \\ \boldsymbol{g} \cdot L_\kappa \\ \vdots \\ \boldsymbol{g} \cdot L_\kappa^{n-1} \end{pmatrix} = \begin{pmatrix} 1 & 0 & \cdots & 0 \\ 0 & 1 & \cdots & 0 \\ \vdots & \vdots & & \vdots \\ 0 & 0 & \cdots & 1 \end{pmatrix}.$$

From (2), between two states $\boldsymbol{X}_{t_0}$ and $\boldsymbol{X}_{t_1}$ we have the relationship

$$(\boldsymbol{X}_{t_1})^T = L_\kappa^{t_1 - t_0} \cdot (\boldsymbol{X}_{t_0})^T = \begin{pmatrix} \boldsymbol{g} \\ \boldsymbol{g} \cdot L_\kappa \\ \vdots \\ \boldsymbol{g} \cdot L_\kappa^{n-1} \end{pmatrix} \cdot L_\kappa^{t_1 - t_0} \cdot (\boldsymbol{X}_{t_0})^T$$

$$= \begin{pmatrix} \hat{\boldsymbol{g}} \\ \hat{\boldsymbol{g}} \cdot L_\kappa \\ \vdots \\ \hat{\boldsymbol{g}} \cdot L_\kappa^{n-1} \end{pmatrix} \cdot (\boldsymbol{X}_{t_0})^T \tag{4}$$

where $\hat{\boldsymbol{g}} = \boldsymbol{g} \cdot L_\kappa^{t_1 - t_0}$. That is,

$$L_\kappa^{t_1 - t_0} = \begin{pmatrix} \hat{\boldsymbol{g}} \\ \hat{\boldsymbol{g}} \cdot L_\kappa \\ \vdots \\ \hat{\boldsymbol{g}} \cdot L_\kappa^{n-1} \end{pmatrix}. \tag{5}$$

Solving the system of equations (4) we have

$$\hat{\boldsymbol{g}} = \boldsymbol{X}_{t_1} \cdot \left((\boldsymbol{X}_{t_0})^T, L_\kappa \cdot (\boldsymbol{X}_{t_0})^T, \cdots, L_\kappa^{n-1} \cdot (\boldsymbol{X}_{t_0})^T \right)^{-1}. \tag{6}$$

A Boolean function $f : \mathbb{F}_2^n \to \mathbb{F}_2$ can be written in the so–called *algebraic normal form* (ANF) as follows

$$f(z_1, \ldots, z_n) = a_0 + a_1 z_1 + \cdots + a_n z_n + a_{1,2} z_1 z_2 + \cdots + a_{n-1,n} z_{n-1} z_n$$
$$+ \cdots + a_{1,\ldots,n} z_1 \cdots z_n.$$

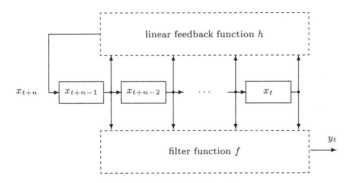

Fig. 1. The block diagram of a filter generator

All coefficients are in $\mathbb{F}_2$. The degree of f is defined to be the maximum of the orders of the product terms appearing in the algebraic normal form with nonzero coefficients.

A filter generator consists of two parts, namely a LFSR with primitive feedback polynomial h and a Boolean function f that filters the state of the LFSR (Fig. 1). We assume that both of them are publicly known. Let $\boldsymbol{x} = \{x_t\}_{t\geq 0}$ be the sequence produced by the LFSR. The sequence $\boldsymbol{x}$ is an m-sequence. The keystream sequence $\boldsymbol{y} = \{y_t\}_{t\geq 0}$ generated by the filter generator is given by

$$y_t = f(x_t, \cdots, x_{t+n-1}), \ t \geq 0.$$

Let d be the degree of the filter function f. Then, the linear complexity is upper bounded by

$$LC(\boldsymbol{y}) \leq D(d),$$

where $D(d) = \sum_{i=1}^{d} \binom{n}{i}$.

In [9] a new attack against the filter generator is proposed. The goal of the attack is to retrieve the initial state $\boldsymbol{X} = (x_0, \cdots, x_{n-1})$ of the LFSR. The output of the filter generator at time t is expressed as a function of the initial state, i.e.

$$y_t = f(x_t, \cdots, x_{t+n-1}) = f_t(x_0, \cdots, x_{n-1}) = \sum_A m_A \cdot K_{A,t} \qquad (7)$$

where $K_{A,t}$ is the sequence of coefficients of the monomial $m_A = x_{a_0} x_{a_1} \cdots x_{a_{r-1}}$, and $A = \{a_0, a_1, \ldots, a_{r-1}\} \subseteq \{0, 1, \ldots, n-1\}$. The degree of the monomial m_A is $|A| \leq d$.

The attack is based on the observation that the characteristic polynomial p that generates all the coefficient sequences $K_{A,t}$, for $|A| \geq 2$ can be efficiently computed with complexity $O(D(d) \log_2(D(d))^3)$. This is the first step of the pre-computation stage. The degree of the polynomial p is $D(d) - n$. In the second step of the pre-computation stage, n linear equations are determined, applying the recursion defined by p, i.e.

$$y_t^* = \sum_{i=0}^{D(d)-n} p_i y_{t+i}, \tag{8}$$

and from (7) we have

$$y_t^* = \sum_{i=0}^{D(d)-n} p_i f_{t+i}(x_0, \cdots, x_{n-1}) = r_t^{(1)}(x_0, \cdots, x_{n-1}), \tag{9}$$

where $\deg(r_t^{(1)}) \leq 1$, since all the high order terms vanish. This step has complexity roughly $n \cdot D(d)$.

For each new keystream the first n bits of the sequence $\boldsymbol{y}^*$ are computed using the recursion (8). For $0 \leq t \leq n-1$, the linear system (9) is solved with complexity less than n^3 and the initial state is recovered. The necessary amount of data for the attack is $D(d)$ bits.

The attack exhibits two problems. First, as it is mentioned in [9], the attack does not work when $r_0^{(1)}$ is the zero function. In that case, a different polynomial p is chosen, leading to a system of quadratic equations, making the attack less efficient. The authors in [9] believe that by decreasing the linear complexity of the sequence $\boldsymbol{y}$ in a certain way the filter generator appears to be more resistant against their attack. In Section 3, we demonstrate that the modified version of the attack does not work as expected in [9]. In Section 4, we propose an efficient attack for the case when $r_0^{(1)}$ is the zero function and we disprove the claim on the reduced linear complexity.

The second problem concerns the amount of data needed for the attack. Though, the complexity of the attack is low, the amount of data needed is large (almost equal the linear complexity) and usually not available. This is demonstrated in [10] where an attack against the WG stream cipher is presented. The attack is a direct generalization of the above attack against filter generators over $\mathbb{F}_{2^n}$. The attack is more efficient than exhaustive key search, but the required data exceeds the keystream length allowed by the cipher's specifications for the same key and IV pair. In Section 5, we try to motivate the research on high order nonlinear complexity of sequences. We show that when the quadratic complexity is small the attack can be improved in terms of the data complexity.

3   The Trace Representation Interpretation

In this section we describe the attack in terms of the trace representation of the sequence $\boldsymbol{y}$, i.e.

$$y_t = \sum_{\kappa \in I(d)} tr_1^{n_\kappa}(\gamma_\kappa \alpha^{\kappa(t+k)}) \tag{10}$$

where d is the degree of the filter function f and α^k depends on the initial state of the filter generator.

Let $\hat{I}(d) \subseteq I(d)$ and let $p(z) = \prod_{\kappa \in \hat{I}(d)} h_\kappa(z)$. The degree of p is $\deg(p) = \sum_{\kappa \in \hat{I}(d)} n_\kappa$. From (1) and (10) we have

$$
\begin{aligned}
y_t^* &= \sum_{j=0}^{\deg(p)} p_j y_{j+t} = \sum_{\kappa \in I(d)} tr_1^{n_\kappa}(p(\alpha^\kappa)\gamma_\kappa \alpha^{\kappa(t+k)}) \\
&= \sum_{\kappa \in I(d) \setminus \hat{I}(d)} tr_1^{n_\kappa}(p(\alpha^\kappa)\gamma_\kappa \alpha^{\kappa(t+k)}) + \sum_{\kappa \in \hat{I}(d)} tr_1^{n_\kappa}(p(\alpha^\kappa)\gamma_\kappa \alpha^{\kappa(t+k)}) = \\
&= \sum_{\kappa \in I(d) \setminus \hat{I}(d)} tr_1^{n_\kappa}(p(\alpha^\kappa)\gamma_\kappa \alpha^{\kappa(t+k)}) + 0.
\end{aligned}
\tag{11}
$$

In [9] polynomial p is chosen as

$$
p(z) = \prod_{2 \leq j \leq d} g_j(z)
$$

where $g_j(z) = \prod_{\kappa \in I, \mathrm{wt}(\kappa)=j} h_\kappa(z)$. That is $\hat{I}(d) = I(d) \setminus \{1\}$. Thus, from (11) we get

$$
y_t^* = tr_1^n(p(\alpha)\gamma_1 \alpha^{(t+k)}).
$$

That is, $\boldsymbol{y}^*$ is a cyclic shift of $\boldsymbol{x}$, the m-sequence produced by the same LFSR and initial state as the one used in the under attack filter generator. Solving system (4) we retrieve the initial state $\boldsymbol{X}_{t_0}$, using $\boldsymbol{X}_{t_1} = (y_0^*, \cdots, y_{n-1}^*)$ and $\hat{\boldsymbol{g}} = \boldsymbol{r}_0^{(1)}$, where the vector $\boldsymbol{r}_0^{(1)}$ is defined in (9). In other words, the problem of retrieving the initial state has been transformed to the retrieval of the initial state of the former described LFSR. For details on the attack please refer to [9].

The problem is that the attack does not work when $\gamma_1 = 0$. The authors in [9] suggest to replace p by $p(z) = \prod_{3 \leq j \leq d} g_j(z)$. Using the new polynomial p all the terms of order greater than 2 vanish. From (7), we have

$$
y_t^* = \sum_{i=0}^{\deg(p)} p_i f_{t+i}(x_0, \cdots, x_{n-1}) = r_t^{(2)}(x_0, \cdots, x_{n-1})
\tag{12}
$$

where $\deg(r_t^{(2)}) = 2$. The attacker has now to solve a quadratic system. If $r_t^{(2)}$ is zero, a new polynomial $p(z) = \prod_{4 \leq j \leq d} g_j(z)$ is chosen, and so on. Generally, let λ_0 be the smaller integer such that $p(z) = \prod_{\lambda_0+1 \leq j \leq d} g_j(z)$ and $r_t^{(\lambda_0)}$ is a Boolean function different from the zero function given by

$$
y_t^* = \sum_{i=0}^{\deg(p)} p_i f_{t+i}(x_0, \cdots, x_{n-1}) = r_t^{(\lambda_0)}(x_0, \cdots, x_{n-1}).
\tag{13}
$$

The attacker has to solve the system (13) which has degree λ_0. Sequence $\boldsymbol{y}$ has linear complexity at most $D(d) - D(\lambda_0)$. The authors in [9] believe that this decrease of the linear complexity protects the filter generator from their attack, since the degree λ_0, of the system the attacker has to solve, increases.

In order to solve the quadratic system (12) using the linearization method, as proposed in [9], the system has to be overdefined. That is, each monomial is treated as a new variable and the system is solved with Gauss elimination given that we have $D(2)$ linearly independent equations. In what follows we will show that the system is not overdefined for $\lambda_0 \geq 2$.

Sequence $\boldsymbol{y}^*$ is given by

$$y_t^* = \sum_{\kappa \in I(2)\setminus\{1\}} tr_1^{n_\kappa}(p(\alpha^\kappa)\gamma_\kappa \alpha^{\kappa(t+k)})$$

since $\gamma_1 = 0$. The linear complexity of $\boldsymbol{y}^*$ equals

$$LC(\boldsymbol{y}^*) \leq \sum_{\kappa \in I(2)\setminus\{1\}} n_k = D(2) - D(1).$$

That means that, only $LC(\boldsymbol{y}^*) < D(2)$ linearly independent equations are available. It is easy to check that the situation gets worse as λ_0 increases, as $D(\lambda_0)$ linearly independent equations will be needed, while $D(\lambda_0) - D(\lambda_0 - 1)$ are available at most. In Section 4 we present a solution for this problem.

4   Computing the Initial State When $\gamma_1 = 0$

In this section we present a method for attacking the filter generator when $\gamma_1 = 0$. In the previous section we showed that when $\gamma_1 \neq 0$ the computation of the initial state of the cipher can become equivalent to the evaluation of the initial state of an LFSR that produces an m-sequence. In what follows, we will exploit the same idea.

We assume that $\gamma_1 = 0$. Let $s \in I(d)$ be an integer such that $\gcd(s, N) = 1$ and $\gamma_s \neq 0$. Let $\hat{s}$ be the inverse of s, that is $\hat{s}s \equiv 1 \mod N$, and let $p(z) = \prod_{\kappa \in I(d)\setminus\{1,s\}} h_\kappa(z)$, where $\deg(p) = D(d) - 2n$. Thus, from (10) we have

$$y_t^* = \sum_{i=0}^{D(d)-2n} p_i y_{i+t} = tr_1^n(p(\alpha^s)\gamma_s \alpha^{sk}\alpha^{st}). \tag{14}$$

Sequence $\boldsymbol{y}^*$ is an m-sequence. Observing n output bits y_t^*, $0 \leq t \leq n - 1$, a unique element α^δ is defined such that $y_t^* = tr_1^n(\alpha^\delta \alpha^{st})$. This leads to

$$\alpha^\delta = \alpha^\beta \cdot \alpha^{sk},$$

where $\alpha^\beta = p(\alpha^s)\gamma_s$. Thus, the initial state α^k is given by

$$k = (\delta - \beta)s^{\phi(N)-1} \mod N,$$

where $\phi(\cdot)$ is the Euler's totient function.

In other words, it would be easy to retrieve the initial state of the filter generator if we knew the trace representation of y and we could compute efficiently elements addition in the extension field $\mathbb{F}_{2^n}$. Next we present an algorithm where only the knowledge of the feedback polynomial and the Boolean function are known, as in the original attack, i.e. we do not need the knowledge of γ_s and $p(\alpha^s)$. This proposed attack is as efficient as the original attack for $\gamma_1 \neq 0$. We will use the trace representation form to explain our reasoning.

Let $LFSR_1$ be the LFSR used by the filter generator and let L_1 be its state transition matrix, and $LFSR_s$ be a second LFSR with feedback polynomial the primitive polynomial $h_{-s}(z)$ and state transition matrix L_s. The $LFSR_s$ produces a phase shift of the m-sequence $tr_1^n(\alpha^{sk}\alpha^{st})$, where the phase k is determined by the initial phase of the $LFSR_s$.

Pick an initial state $\boldsymbol{X}^{(0)}$ for $LFSR_1$. Then, there is a corresponding phase k_0 such that

$$x_t = tr_1^n(\alpha^{k_0}\alpha^t)$$

and $(\boldsymbol{X}^{(0)})^T = L_1^{k_0}(\boldsymbol{X}_{c,1})^T$, where $\boldsymbol{X}_{c,1}$ is the characteristic state of $LFSR_1$. Decimating $\boldsymbol{x}$ by s we get

$$x_t^* = tr_1^n(\alpha^{k_0}\alpha^{st}) = tr_1^n(\alpha^{s(\hat{s}k_0)}\alpha^{st}).$$

Let $\boldsymbol{Z}^* = (x_0^*, x_1^*, \cdots, x_{n-1}^*)$. $LFSR_s$ produces the sequence $\boldsymbol{x}^*$ and

$$(\boldsymbol{Z}^*)^T = L_s^{\hat{s}k_0}(\boldsymbol{X}_{c,s})^T, \tag{15}$$

where $\boldsymbol{X}_{c,s}$ is the characteristic state of $LFSR_s$.

Initializing the filter generator with $\boldsymbol{X}^{(0)}$ we compute from (14) the first n bits of $\boldsymbol{y}_0^*$

$$y_{0,t}^* = tr_1^n(\alpha^{s\beta'}\alpha^{sk_0}\alpha^{st}), \ 0 \leq t \leq n-1,$$

where $\alpha^{s\beta'} = p(\alpha^s)\gamma_s$. Let $\boldsymbol{Z}^{(0)} = (y_{0,0}^*, y_{0,1}^*, \cdots, y_{0,n-1}^*)$. $LFSR_s$ produces the sequence $\boldsymbol{y}_0^*$ and

$$(\boldsymbol{Z}^{(0)})^T = L_s^{\beta'+k_0}(\boldsymbol{X}_{c,s})^T. \tag{16}$$

Let $\boldsymbol{X}^{(1)}$ be the initial state of the filter cipher we want to retrieve. Then, there is a phase k_1 such that $(\boldsymbol{X}^{(1)})^T = L_1^{k_1}(\boldsymbol{X}_{c,1})^T$. We compute

$$y_{1,t}^* = tr_1^n(\alpha^{s\beta'}\alpha^{sk_1}\alpha^{st}), \ 0 \leq t \leq n-1.$$

Let $\boldsymbol{Z}^{(1)} = (y_{1,0}^*, y_{1,1}^*, \cdots, y_{1,n-1}^*)$. Then,

$$(\boldsymbol{Z}^{(1)})^T = L_s^{\beta'+k_1}(\boldsymbol{X}_{c,s})^T. \tag{17}$$

From (16) and (17), it holds,

$$(\boldsymbol{Z}^{(1)})^T = L_s^{k_1-k_0}(\boldsymbol{Z}^{(0)})^T.$$

From (6) we compute the vector $\hat{g}$ and from (5)

$$G_0 = \begin{pmatrix} \hat{g} \\ \hat{g} \cdot L_\kappa \\ \vdots \\ \hat{g} \cdot L_\kappa^{n-1} \end{pmatrix} = L_s^{k_1 - k_0}.$$

Let $G = G_0^{\hat{s}} = L_s^{\hat{s}(k_1 - k_0)}$. Then, from (15), the state $(\boldsymbol{Z})^T = G \cdot (\boldsymbol{Z}^*)^T = L_s^{\hat{s}k_1} \cdot \boldsymbol{X}_{c,s}$ used as initial state of the $LFSR_s$ produces the sequence

$$y_t^* = tr_1^n(\alpha^{s(\hat{s}k_1)}\alpha^{st}) = tr_1^n(\alpha^{k_1}\alpha^{st}).$$

Decimating $\boldsymbol{y}^*$ by $\hat{s}$ we get the k_1 phase shift from the characteristic phase of the m-sequence $\boldsymbol{x}$ produced by the $LFSR_1$, i.e. $x_t = tr_1^n(\alpha^{k_1}\alpha^t)$. Thus, the initial state we want to retrieve is given by

$$\boldsymbol{X}^{(1)} = (x_0, \cdots, x_{n-1}) = \left(y_0^*, y_{\hat{s}}^*, \cdots, y_{\hat{s}(n-1)}^* \right).$$

The algorithm is described as follows.

1. *Pre-computation. We chose $\boldsymbol{X}^{(0)} = (0, 0, \cdots, 0, 1)$.*
   (a) *Compute the polynomial p.*
   (b) *Using $\boldsymbol{X}^{(0)}$ for initial state compute the first n bits of the decimated by s output of $LFSR_1$ and formulate the vector $\boldsymbol{Z}^* = \left(x_0^*, x_1^*, \cdots, x_{n-1}^* \right).$*
   (c) *Clock the filter generator $(D(d) - n)$ times using $\boldsymbol{X}^{(0)}$ for initial state and compute from (14) the vector $\boldsymbol{Z}^{(0)} = \left(y_{0,0}^*, y_{0,1}^*, \cdots, y_{0,n-1}^* \right).$*
2. *Computation.*
   (a) *Observe $(D(d) - n)$ output bits of the filter cipher and compute from (14) the vector $\boldsymbol{Z}^{(1)} = \left(y_{1,0}^*, y_{1,1}^*, \cdots, y_{1,n-1}^* \right).$*
   (b) *Solve the system (6) and compute G_0.*
   (c) *Compute $G = G_0^{\hat{s}}$.*
   (d) *Compute $(\boldsymbol{Z})^T = G \cdot (\boldsymbol{Z}^*)^T$.*
   (e) *Using $\boldsymbol{Z}$ for initial state, compute the first n bits of the decimated by $\hat{s}$ output $\boldsymbol{y}^*$ of $LFSR_s$. The initial vector we are looking for is given by $\boldsymbol{X}^{(1)} = \left(y_0^*, y_{\hat{s}}^*, \cdots, y_{\hat{s}(n-1)}^* \right).$*

The complexity of each step is summarized as follows.

1. *Pre-computation.*
   (a) *From [9] this step requires $O(D(d)log_2(D(d))^3)$ complexity.*
   (b) *If s is small then we can just run the $LFSR_1$ and decimate it. Otherwise, we compute L_1^s, using the square-and-multiply method for exponentiation. Then, the n bits of the decimated sequence are given by*

$$x_t^* = (1, 0, 0, 0, 0) \cdot (L_1^s)^t \cdot (0, 0, \cdots, 0, 1)^T.$$

The complexity is at most $O(n^3)$.

(c) *The computation of* $y_{0,t}^*$, $0 \leq t \leq n-1$, *if it is done in parallel, has complexity* $O(D(d) - n)$.

2. *Computation.*

   (a) *The computation of* $y_{1,t}^*$, $0 \leq t \leq n-1$, *if it is done in parallel, has complexity* $O(D(d) - n)$.

   (b) *For solving the system and computing* G_0 *matrix, the complexity is approximately* $O(n^3)$.

   (c) *For computing the* $\hat{s}$ *power of* G_0 *we need approximately* $O(log_2(\hat{s})n^2)$.

   (d) *For computing* Z *the complexity is approximately* $O(n^3)$.

   (e) *Using similar argumentation with the step (b) of the Pre-computation stage, the complexity is* $O(n^3)$ *at most.*

In conclusion, both in the precomputation stage and the computation stage the calculation of the n bits of $\hat{y}^{(0)}$ and $\hat{y}^{(1)}$ respectively, is dominant, and the complexity of the attack is $O(D(d) - n)$. The data complexity is $O(D(d) - n)$ bits. All the other steps include inversions and multiplications of $n \times n$ binary matrices which is relatively costless. The attack has the same complexity as the one in [9] for $\gamma_1 \neq 0$. In Appendix, we propose an alternative attack which is less efficient.

Next we present a simple example to illustrate the proposed algorithm.

Example 1. Let $h_1(z) = 1 + z^2 + z^5$ be a primitive polynomial and let $\alpha \in \mathbb{F}_{2^5}$ such that $h_1(\alpha) = 0$. The $LFSR_1$ has feedback the polynomial $h_{-1}(z) = 1 + z^3 + z^5$ and produces an m-sequence. The state of $LFSR_1$ is filtered by the balanced Boolean function $f(x_0, x_1, x_2, x_3, x_4) = x_1 + x_3 + x_0x_2 + x_0x_3 + x_1x_2 + x_2x_4$. The output sequence y of the filter generator is balanced with period $N = 2^5 - 1 = 31$ and linear complexity $LC(y) \leq D(2) = 15$.

Following the attack proposed in [9] we have $p(z) = g_2(z) = h_3(z)h_5(z) = 1 + z + z^6 + z^7 + z^{10}$, where $h_3(z) = 1 + z^2 + z^3 + z^4 + z^5$ and $h_5(z) = 1 + z + z^2 + z^4 + z^5$. From (9), we get $r_t^{(1)}(z) = 0$, for all $t \geq 0$. Thus, the attack cannot be applied.

Next, we will follow our attack. Let $s = 3$. Then, $p(z) = h_5(z) = 1 + z + z^2 + z^4 + z^5$. We use $X^{(0)} = (0,0,0,0,1)$ as initial state and we decimate the output of $LFSR_1$ by $s = 3$. Then,

$$Z^* = (0,0,0,1,0).$$

Using $X^{(0)}$ also as the initial state of the filter generator we produce $D(2) - n = 15 - 5 = 10$ output bits and from (14) we compute

$$Z^{(0)} = (0,1,0,1,0).$$

The attack requires $D(2) - n = 10$ bits of y to be observed. Let

$$(y_0, \cdots, y_9) = (0,0,1,0,1,0,1,0,1,0).$$

From (14) and the observed bits of y we compute

$$Z^{(1)} = (0,0,1,0,1).$$

From (6) we compute $\hat{g}$ we get

$$\hat{g} = \boldsymbol{Z}^{(1)} \cdot \left((\boldsymbol{Z}^{(0)})^T, L_3 \cdot (\boldsymbol{Z}^{(0)})^T, \cdots, L_3^4 \cdot (\boldsymbol{Z}^{(0)})^T \right)^{-1} = (0,1,1,1,1),$$

where $L_3 = \begin{pmatrix} 0 & 1 & 0 & 0 & 0 \\ 0 & 0 & 1 & 0 & 0 \\ 0 & 0 & 0 & 1 & 0 \\ 0 & 0 & 0 & 0 & 1 \\ 1 & 0 & 1 & 1 & 1 \end{pmatrix}$. Thus, $G_0 = \begin{pmatrix} 0 & 1 & 1 & 1 & 1 \\ 1 & 0 & 0 & 0 & 0 \\ 0 & 1 & 0 & 0 & 0 \\ 0 & 0 & 1 & 0 & 0 \\ 0 & 0 & 0 & 1 & 0 \end{pmatrix}$.

The inverse of s is $\hat{s} = 21$ and $G = G_0^{21} = \begin{pmatrix} 0 & 1 & 0 & 1 & 0 \\ 0 & 0 & 1 & 0 & 1 \\ 1 & 0 & 1 & 0 & 1 \\ 1 & 1 & 1 & 0 & 1 \\ 1 & 1 & 0 & 0 & 1 \end{pmatrix}$.

The state $\boldsymbol{Z} = G \cdot (0,0,0,1,0) = (1,0,0,0,0)$. Since, $\hat{s} = 21$ is rather big, we compute

$$(L_3)^{21} = L_3 \cdot (L_3)^4 \cdot (L_3)^{16} = \begin{pmatrix} 0 & 1 & 1 & 0 & 0 \\ 0 & 0 & 1 & 1 & 0 \\ 0 & 0 & 0 & 1 & 1 \\ 1 & 0 & 1 & 1 & 0 \\ 0 & 1 & 0 & 1 & 1 \end{pmatrix}.$$

Thus, the state we are looking is given by $\boldsymbol{X}^{(1)} = (x_0, x_1, x_2, x_3, x_4)$, where

$$x_t = (1,0,0,0,0) \cdot (L_3^{21})^t (1,0,0,0,0)^T.$$

We compute $\boldsymbol{X}^{(1)} = (1,0,0,0,0)$.

5   The Role of High Order Complexity

In this section we present an attack scenario that requires fewer keystream bits. The price we must pay is some additional pre-computation complexity. The main objective of this section is to motivate the research on high order nonlinear complexity of binary sequences.

Definition 1. ([7], [8]) The length of the shortest feedback shift register (FSR) having a feedback function of degree at most k that generates y defines the k-th order nonlinearity of sequence y.

For $k = 1$ we have the linear complexity of the sequence, and for $k = 2$ the quadratic span.

In Section 3, we showed that the attack is based on the transformation of the state recovery problem to a much easier problem as the one of the recovery of the initial state of the LFSR that produces an m-sequence. Next we describe a generalization of the attack. We follow the notation of Section 3.

Assume that $\hat{I}(d) = I(d) \setminus I(d')$, where $d' < d$ and that $\gamma_\kappa \neq 0$, for all $\kappa \in I(d')$. For $d' = 1$ we have the attack described in [9]. From (11), it holds

$$y_t^* = \sum_{\kappa \in I(d')} tr_1^{n_\kappa}(p(\alpha^\kappa)\gamma_\kappa \alpha^{\kappa(t+k)})$$

where $p(z) = \prod_{\kappa \in \hat{I}(d)} h_\kappa(z)$ and $\deg(p) = D(d) - D(d')$.

Following the same reasoning as before, we pre-compute the functions $r_t^{(d')}$

$$y_t^* = \sum_{i=0}^{D(d)-D(d')} p_i f_{t+i}(x_0, \cdots, x_{n-1}) = r_t^{(d')}(x_0, \cdots, x_{n-1}). \qquad (18)$$

Since $\gamma_\kappa \neq 0$, for all $\kappa \in I(d')$, the linear complexity of $\boldsymbol{y}^*$ is $D(d')$ and the system (18) consists of $D(d')$ linearly independent equations. Using the linearization method we can solve (18) with complexity $D(d')^{2.7}$. We need $D(d')$ bits of $\boldsymbol{y}^*$. The computation of each one of these bits requires the knowledge of some $(D(d) - D(d') + 1)$ bits of the output sequence $\boldsymbol{y}$, resulting to an overall data complexity $O(D(d))$.

Suppose that the quadratic span $QS(\boldsymbol{y}^*)$ of $\boldsymbol{y}^*$ is known, as well as the quadratic feedback function q of the shortest FSR that generates the sequence $\boldsymbol{y}^*$. That means that, if we know $QS(\boldsymbol{y}^*)$ bits of the sequence $\boldsymbol{y}^*$, we can use it as the initial state of the FSR and produce the whole sequence. So, when the polynomial q is known, let's say as part of the pre-computation, we can use $QS(\boldsymbol{y}^*)$ bits of $\boldsymbol{y}^*$ to compute the $D(d')$ bits needed for the attack. Thus, the data complexity reduces to $(D(d) - D(d') + QS(\boldsymbol{y}^*))$.

In the above scenario, a trade-off between the computational and the data complexity appears. Increasing the value d' we increase the attack complexity $O((D(d'))^{2.7})$, while we decrease the data complexity to

$$O\left(D(d) - D(d') + QS(\boldsymbol{y}^*)\right).$$

In the above analysis we assume that the quadratic span for most of the sequences with the same period is almost the same, as it happens with the linear complexity. It is straightforward to see that we can replace quadratic span with any of the k-order nonlinear complexities we can compute. Unfortunately, there is not an efficient algorithm for computing the high order nonlinearity of a binary sequence or a analysis on the expected value of the k-order nonlinear complexity of a random periodic sequence ([7]). We hope that the attack we just described will motivate researchers to work in the field.

6   Conclusion

In this paper, we analyzed the attack against the filter generator proposed in [9]. Using trace representation of the output sequence $\boldsymbol{y}$ we proved that the attack does not work always as expected. We proposed a new algorithm for the cases that the original attack does not cover which is as efficient as the original attack. The main problem of all the former mentioned attacks, original and modified version, is the high data complexity. We showed the significance of nonlinear complexity of binary sequences in such an attack. We presented a scenario where the knowledge of the quadratic complexity can decrease significantly the data complexity. It remains an open problem the computation of k-order nonlinear complexity of a given binary sequence, which we believe is a very good direction for further research.

References

1. Canteaut, A.: Fast Correlation Attacks Against Stream Ciphers and Related Open Problems. In: IEEE Information Theory Workshop on Theory and Practice in Information-Theoretic Security, pp. 49–54 (2005)
2. Canteaut, A.: Open problems related to algebraic attacks on stream ciphers. In: Ytrehus, Ø. (ed.) WCC 2005. LNCS, vol. 3969, pp. 1–11. Springer, Heidelberg (2006)
3. eSTREAM. http://www.ecrypt.eu.org/stream/
4. Golomb, S.W.: Shift Register Sequences. Holden–Day Inc, San Francisco (1967)
5. Lidl, R., Niederreiter, H.: Finite Fields. Encyclopedia of Mathematics and its Applications, 2nd edn., vol. 20. Cambridge University Press, Cambridge (1996)
6. Menezes, A.J., Van Oorschot, P.C., Vanstone, S.A.: Handbook of applied cryptography. CRC Press, Boca Raton (1996)
7. Niederreiter, H.: Linear Complexity and Related Complexity Measures for Sequences. In: Johansson, T., Maitra, S. (eds.) INDOCRYPT 2003. LNCS, vol. 2904, pp. 1–17. Springer, Heidelberg (2003)
8. Rizomiliotis, P., Kalouptsidis, N.: Results on the nonlinear span of binary sequences. IEEE Trans. Inform. Theory 51, 1555–1563 (2005)
9. Ronjom, S., Helleseth, T.: A new attack on the filter generator. IEEE Trans. Inform. Theory 53, 1752–1758 (2007)
10. Ronjom, S., Helleseth, T.: S. Ronjom and T. Helleseth: Attacking the Filter Generator over $GF(2^m)$. In: SASC 2007. The State of the Art of Stream Ciphers (2007)
11. Rueppel, R.A.: Analysis and Design of Stream Ciphers. In: Communications and Control Engineering Series, Springer, Heidelberg (1986)

A   Appendix

In this Appendix we present a less efficient attack scenario for the case when $\gamma_1 = 0$. We follow the notation of Section 4.

Let $\hat{y}$ be the decimation of y^* by $\hat{s}$. Then, from (14) we have

$$\hat{y}_t = y^*_{\hat{s}t} = tr^n_1(p(\alpha^s)\gamma_s\alpha^{sk}\alpha^t) = tr^n_1(\alpha^\beta\alpha^{sk}\alpha^t) \tag{19}$$

where $\alpha^\beta = p(\alpha^s)\gamma_s$. The sequence $\hat{y}$ is produced by the linear feedback shift register $LFSR_1$ used by the filter generator.

Pick an initial state $\boldsymbol{X}^{(0)}$ and using the output of the filter generator compute from (19) the sequence $\hat{\boldsymbol{y}}^{(0)}$. Then, there is a shift k_0 such that

$$\hat{y}_t^{(0)} = tr_1^n(\alpha^{\beta+sk_0}\alpha^t),$$

and $\boldsymbol{X}^{(0)} = L_1^{k_0}(\boldsymbol{X}_{c,1})^T$, where $\boldsymbol{X}_{c,1}$ is the characteristic state of $LFSR_1$ and L_1 is the state transition matrix.

Let $\boldsymbol{Z}^{(0)} = \left(\hat{y}_0^{(0)}, \hat{y}_1^{(0)}, \cdots, \hat{y}_{n-1}^{(0)}\right)$. Since $LFSR_1$ produces the sequence $\hat{\boldsymbol{y}}^{(0)}$, we have

$$(\boldsymbol{Z}^{(0)})^T = L_1^{\beta+sk_0}(\boldsymbol{X}_{c,1})^T. \tag{20}$$

Let $\boldsymbol{X}^{(1)}$ be the initial state of the filter generator that we want to disclose. Then, again from (19) we compute sequence $\hat{\boldsymbol{y}}^{(1)}$, i.e.

$$\hat{y}_t^{(1)} = tr_1^n(\alpha^{\beta+sk_1}\alpha^t),$$

where $\boldsymbol{X}^{(1)} = L_1^{k_1}(\boldsymbol{X}_{c,1})^T$. Let $\boldsymbol{Z}^{(1)} = \left(\hat{y}_0^{(1)}, \hat{y}_1^{(1)}, \cdots, \hat{y}_{n-1}^{(1)}\right)$, then

$$(\boldsymbol{Z}^{(1)})^T = L_1^{\beta+sk_1}(\boldsymbol{X}_{c,1})^T. \tag{21}$$

From (20) and (21), it holds

$$(\boldsymbol{Z}^{(1)})^T = L_1^{s(k_1-k_0)}(\boldsymbol{Z}^{(0)})^T.$$

From (6) we compute the vector $\hat{\boldsymbol{g}}$ and from (5)

$$G_0 = \begin{pmatrix} \hat{\boldsymbol{g}} \\ \hat{\boldsymbol{g}} \cdot L_1 \\ \vdots \\ \hat{\boldsymbol{g}} \cdot L_1^{n-1} \end{pmatrix} = L_1^{s(k_1-k_0)}.$$

Let $G = G_0^{\hat{s}} = L_1^{k_1-k_0}$. Then, the initial state $\boldsymbol{X}^{(1)}$ is given by

$$G \cdot (\boldsymbol{X}^{(0)})^T = L_1^{k_1-k_0} \cdot L_1^{k_0} \cdot (\boldsymbol{X}_{c,1})^T = (\boldsymbol{X}^{(1)})^T.$$

The algorithm is described as follows.

1. *Pre-computation. We chose* $\boldsymbol{X}^{(0)} = (0,0,\cdots,0,1)$. *Generally it is convenient to chose an initial state with Hamming weight one.*
   (a) *Compute the polynomial p.*
   (b) *Clock the filter generator* $D(d) - 2n + \hat{s}(n-1)$ *times using* $\boldsymbol{X}^{(0)}$ *as input. From (19) compute n bits of* $\hat{\boldsymbol{y}}^{(0)}$ *and formulate the vector* $\boldsymbol{Z}^{(0)} = \left(\hat{y}_0^{(0)}, \cdots, \hat{y}_{n-1}^{(0)}\right)$.
2. *Computation.*
   (a) *Observe* $D(d) - 2n + \hat{s}(n-1)$ *output bits of the filter cipher. From (19) compute n bits of* $\hat{\boldsymbol{y}}^{(1)}$ *and formulate the vector* $\boldsymbol{Z}^{(1)} = \left(\hat{y}_0^{(1)}, \cdots, \hat{y}_{n-1}^{(1)}\right)$.
   (b) *Solve the system (6) and compute* G_0.
   (c) *Compute* $G = G_0^{\hat{s}}$.
   (d) *Output the last column of G.*

Next we give an estimation of the complexity.

1. *Pre-computation.*
   (a) *From [9] this step requires $O(D(d)log_2(D(d))^3)$ computational complexity.*
   (b) *The computation of $\hat{y}_t^{(0)}$, $0 \le t \le n-1$, if it is done in parallel has complexity $O(D(d) - 2n + \hat{s}(n-1))$.*
2. *Computation.*
   (a) *The computation of $\hat{y}_t^{(1)}$, $0 \le t \le n-1$, if it is done in parallel has complexity $O(D(d) - 2n + \hat{s}(n-1))$.*
   (b) *For solving the system (6) and computing the $n \times n$ matrix G_0, the complexity is approximately $O(n^3)$.*
   (c) *For computing the $\hat{s}$ power of G_0, we use the square-and-multiply method, i.e. we first compute the powers of two $G_0^{2^i}$, $1 \le i \le n-1$, and then, if $\hat{s} = \sum_{i=0}^{n-1} \hat{s}_i 2^i$, $\hat{s}_i \in \mathbb{F}_2$, $G_0^{\hat{s}} = \prod_{i=0}^{n-1} G_0^{\hat{s}_i 2^i}$. The complexity of this step is approximately $O(log_2(\hat{s})n^2)$.*
   (d) *The last step is for free for the specific choice of $\boldsymbol{X}^{(0)}$.*

Both in the pre-computation stage and the computation stage the calculation of the n bits of $\hat{\boldsymbol{y}}^{(0)}$ and $\hat{\boldsymbol{y}}^{(1)}$ respectively, is dominant, rising the complexity of the attack to $O(D(d) - 2n + \hat{s}(n-1))$. All the other steps are relatively costless. The data complexity is $O(D(d) - 2n + \hat{s}(n-1))$.

Note 1. The same attack can be applied by first decimating $\boldsymbol{y}$ by $\hat{s}$. Then,

$$y_t^* = y_{\hat{s}t} = \sum_{\kappa \in I(d)} tr_1^{n_\kappa}(\gamma_\kappa \alpha^{\kappa t_0} \alpha^{\kappa \hat{s}t}). \tag{22}$$

We define the subset of coset leaders

$$I_{\hat{s}} = \{\hat{\kappa} | \hat{\kappa} = 2^j \cdot \kappa \cdot \hat{s} \mod N, \ \kappa \in I(d), \ \hat{\kappa} \in I\}.$$

It is easy to see that $1 \in I_{\hat{s}}$. Let $p(z) = \prod_{\hat{\kappa} \in I_{\hat{s}} \setminus \{1\}} h_{\hat{\kappa}}(z)$. Applying the recursion p, from (22) we have,

$$\hat{y}_t = tr_1^n(p(\gamma_s) \cdot \alpha^{st_0} \alpha^t).$$

This approach requires $(D(n) - 2n)\hat{s} + n - 1$ bits of data.

Modified Berlekamp-Massey Algorithm for Approximating the k-Error Linear Complexity of Binary Sequences

Alexandra Alecu and Ana Sălăgean

Department of Computer Science,
Loughborough University,
Loughborough, UK
{A.Alecu, A.M.Salagean}@lboro.ac.uk
http://www.lboro.ac.uk

Abstract. Some cryptographical applications use pseudorandom sequences and require that the sequences are secure in the sense that they cannot be recovered by only knowing a small amount of consecutive terms. Such sequences should therefore have a large linear complexity and also a large k-error linear complexity. Efficient algorithms for computing the k-error linear complexity of a sequence only exist for sequences of period equal to a power of the characteristic of the field. It is therefore useful to find a general and efficient algorithm to compute a good approximation of the k-error linear complexity. We show that the Berlekamp-Massey Algorithm, which computes the linear complexity of a sequence, can be adapted to approximate the k-error linear complexity profile for a general sequence over a finite field. While the complexity of this algorithm is still exponential, it is considerably more efficient than the exhaustive search.

Keywords: pseudorandom sequences, stream ciphers, linear complexity, k-error linear complexity.

1   Introduction

The k-error linear complexity of a sequence is a generalisation of the notion of linear complexity. While the linear complexity of a sequence is defined as the length of the smallest linear recurrence relation which generates that sequence, the k-error linear complexity is the length of the smallest linear recurrence relation which generates a sequence which differs from the original sequence in at most k positions.

When designing a stream cipher, the keystream sequence has to have a large linear complexity. Using the Berlekamp-Massey Algorithm, a sequence can be efficiently recovered by knowing a number of consecutive terms equal to twice its linear complexity. Sequences with low linear complexity would therefore be vulnerable to known plaintext attacks. Similarly, sequences with low k-error linear

S.D. Galbraith (Eds.): Cryptography and Coding 2007, LNCS 4887, pp. 220–232, 2007.

complexity for small values of k could also be vulnerable if the corresponding linear recurrence relation was found.

An exact algorithm to compute the k-error linear complexity only exists for periodic sequences over a finite field $GF(p^m)$ and with period a power of p, p being prime and $m \geq 1$ (see Stamp and Martin [10], Lauder and Paterson [5] for $p = 2$ and Kaida, Uehara and Imamura [4] for an arbitrary p). These algorithms are based on the algorithms of Games and Chan [3] and Ding, Xiao, Shan [2] for computing the linear complexity of such sequences, and they work only when a full period of the sequence is known, i.e. the whole sequence is known, which is not the case in cryptanalysis applications.

We propose adapting the Berlekamp-Massey Algorithm (Berlekamp [1], Massey [7]) which computes the linear complexity, in order to approximate the k-error linear complexity profile for a general sequence over a finite field. The main idea is to devise a heuristic algorithm which explores only some of all the possible error patterns. The choice of the positions of the errors is guided by the steps of the Berlekamp-Massey Algorithm in which the complexity is increased.

While the proposed algorithm has an exponential complexity, the base of the exponential function is smaller than for an exhaustive search; moreover, the base for the proposed algorithm is independent of the size of the field, while for an exhaustive search it increases with the size of the field.

In this paper we consider mainly binary sequences but the proposed algorithm can be extended to arbitrary finite fields.

2   Background

Definition 1. *Given an infinite sequence* $s = s_0, s_1, \ldots$ *(or a finite sequence* $s = s_0, s_1, \ldots, s_{t-1}$ *) with elements in a field K, we say that s is a* linear recurrent sequence *if it satisfies a relation of the form* $s_j + c_{L-1}s_{j-1} + \ldots + c_1 s_{j-L+1} + c_0 s_{j-L} = 0$ *for all* $j = L, L+1, \ldots$ *(or for all* $j = L, L+1, \ldots t-1$, *respectively), where* $c_0, c_1, \ldots, c_{L-1} \in K$ *are constants. The associated* characteristic *polynomial is* $C(X) = X^L + c_{L-1}X^{L-1} + \ldots + c_1 X + c_0$. *If L is minimal for the given sequence, we call L the* linear complexity *of s, denoted $L(s)$.*

The notion of linear complexity has been generalised to k-error linear complexity by Stamp and Martin [10] (see also Ding, Xiao, Shan [2]). In the following $w_H(s)$ denotes the Hamming weight i.e. the number of non-zero terms of s.

Definition 2. *For a given infinite sequence* $s = s_0, s_1, \ldots$ *of period N, with elements in a field K and for a fixed integer k, $0 \leq k \leq w_H((s_0, \ldots, s_{N-1}))$, the* k-error linear complexity *of the sequence s is defined as*

$$L_k(s) = \min\{L(s+e) \mid e \text{ is a sequence of period } N \text{ over } K, w_H((e_0, e_1, \ldots, e_{N-1})) \leq k\}$$

For a given finite sequence $s = s_0, s_1, \ldots, s_{t-1}$ *with elements in a field K and for a fixed integer k, $0 \leq k \leq w_H(s)$, the k-error linear complexity of the sequence s is defined as* $L_k(s) = \min\{L(s+e) \mid e \in K^t, w_H(e) \leq k\}$. *The sequences e are* *called error sequences or error patterns. The k-error linear complexity profile of the sequence is defined as being the set of pairs $(k, L_k(s))$, for all k with $0 \leq k \leq w_H(s)$.*

Property 1. Given a (finite or infinite) sequence s with elements in a finite field $GF(q)$, we have $L_i(s) \geq L_j(s)$, for all $i < j$.

The Berlekamp-Massey Algorithm ([1],[7]) computes the characteristic polynomial and the linear complexity of a sequence over any field. It is iteratively processing each term of a finite sequence $s_0, s_1, \ldots, s_{t-1}$, adjusting the characteristic polynomial when necessary. At each step n the current minimal characteristic polynomial $C^{(n)}(X)$ generates the n sequence terms $s_0, s_1, \ldots, s_{n-1}$ processed so far. In addition, the last characteristic polynomial $C^{(m)}(X)$ of degree strictly smaller than the degree of $C^{(n)}(X)$ is also stored. We denote $L^{(i)} = \deg(C^{(i)})$. The discrepancy $d^{(n)}$

$$d^{(n)} = s_n + \sum_{i=0}^{L^{(n)}-1} c_i^{(n)} s_{i+n-L^{(n)}} \tag{1}$$

is the difference between the term which is expected using the current characteristic polynomial and the actual term s_n which is currently processed. Three possible cases are identified:

1. If $d^{(n)} \neq 0$ then s_n cannot be generated using $C^{(n)}(X)$:
   a) If $2L^{(n)} > n$ then the new characteristic polynomial is computed as
      $C^{(n+1)}(X) \leftarrow C^{(n)}(X) - \frac{d^{(n)}}{d^{(m)}} \cdot X^{(m-L^{(m)})-(n-L^{(n)})} \cdot C^{(m)}(X)$ and it has the same degree as the previous one;
   b) If $2L^{(n)} \leq n$ then the new characteristic polynomial is computed as
      $C^{(n+1)}(X) \leftarrow X^{(n-L^{(n)})-(m-L^{(m)})} \cdot C^{(n)}(X) - \frac{d^{(n)}}{d^{(m)}} \cdot C^{(m)}(X)$ and it has a higher degree than the previous one, namely $L^{(n+1)} = n+1-L^{(n)}$; m is updated to n.
2. If $d^{(n)} = 0$ then s_n can be generated using $C^{(n)}(X)$, so the characteristic polynomial stays unchanged $C^{(n+1)}(X) = C^{(n)}(X)$.

We initialise $C^{(i)}(X) \leftarrow 1$ for $i = 0, \ldots, j$, $C^{(j+1)}(X) \leftarrow X^{j+1}$ and $m \leftarrow j$, where s_j is the first non-zero term of the sequence. At the end of the algorithm, $L^{(t)}$ is the linear complexity of the sequence and $C^{(t)}(X)$ is a minimal characteristic polynomial (which is unique if $2L^{(t)} \leq t$, otherwise it may not be unique).

3   The Modified Berlekamp-Massey Algorithm

Determining the k-error linear complexity of a finite binary sequence of length t by an exhaustive search approach would mean investigating all the $\sum_{i=0}^{k} \binom{t}{i}$ patterns of up to k errors and computing the linear complexity of each of the sequences obtained by adding these error patterns to the original sequence. Some computational savings can be made by taking advantage of the incremental nature of the Berlekamp-Massey Algorithm; for error patterns which coincide on the first say i positions, reuse the computations made on the first i terms of the sequence. We implemented this more efficient version of an exhaustive search for the binary case and used it as a reference (we denote this algorithm the Efficient Exhaustive Search Algorithm).

A heuristic approach would only explore some of all the possible error patterns. Our proposed heuristic will use the Berlekamp-Massey Algorithm to choose these patterns. Namely, during the algorithm, only the case when the discrepancy $d^{(n)} \neq 0$ and $2L^{(n)} \leq n$ (case (1b) in Section 2) yields an increase in the current complexity of the sequence. It seems therefore natural to concentrate on what would happen if the current term of the sequence, which creates this increase in complexity, would be changed in such a way as to make the discrepancy zero, and therefore make an increase in complexity unnecessary. If we made these changes to the sequence early in the algorithm, we would soon run out of the k allowed errors, and we would not be able to explore the effect of errors on later terms of the sequence. Whenever case (1b) occurs in the algorithm we do therefore consider both possibilities: changing the current term of the sequence, or not changing it, and we continue exploring both branches. A tree of recursive calls is thus obtained.

Our approach is not guaranteed to give the exact result for the k-error linear complexity, as the error pattern that decreases the complexity the most may well not have the errors in those positions suggested by the Berlekamp-Massey Algorithm. Since we investigate only some of all the possible error patterns, our results will always be larger or equal to the optimum ones. We investigate experimentally in Section 4 how close the approximation is to the actual k-error complexity. Unfortunately we were unable to prove a theoretical bound on the approximation quality.

We firstly illustrate our algorithm with an example:

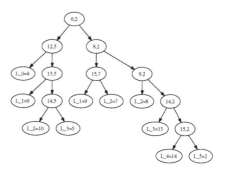

Fig. 1. Example of the Modified Berlekamp-Massey Algorithm tree of error and no-error recursive calls for the sequence $s = 0110111101110101$

Example 1. Applying the Berlekamp-Massey Algorithm to s=0110111101110101 (length 16), the degree of the intermediate characteristic polynomials changes at positions 6 and 12 (ignoring the initial change from degree 0). The linear complexity is 8 and the characteristic polynomial $C(X) = X^8 + X^6 + X^5 + X^4 + X + 1$.

Figure 1 shows the tree of recursive calls for the MBM Algorithm. The internal nodes show the index of the element of the sequence currently processed and the current linear complexity. The left child of each internal node corresponds to not forcing an error and the right child corresponds to introducing the error. The leaves in the tree show the final result on each path in the tree: the number of errors which were introduced and the corresponding k-error linear complexity. In our example the first change of complexity happens when s_6 is processed. The error sequence can be built using a bottom-up technique. For example, for the 3-error linear complexity, the errors will be at indices 14,13 and 12, so the error sequence is 0000000000001110.

By taking the minimum value of the linear complexity for each number of errors, the results in the tree in Figure 1 give an incomplete approximate k-error linear complexity profile as being $\{(0,8),(1,9),(2,7),(3,5),(5,2)\}$. Applying Property 1 in Section 2 and using the fact that $L_{w_H(s)}(s) = 0$ the full approximate k-error linear complexity profile is found:$\{(0,8),(1,8),(2,7),(3,5),(4,5),(5,2),(6,2),(7,2),(8,2),(9,2),(10,2),(11,0)\}$. The exact k-error linear complexity profile obtained by an exhaustive search algorithm is: $\{(0,8),(1,7),(2,6),(3,4),(4,2),(5,1),(6,1),(7,1),(8,1),(9,1),(10,1),(11,0)\}$.

We suppose that the input sequence has at least one non-zero element otherwise calculating the k-error linear complexity would be immediate since $L_k(\mathbf{0}) = 0$, for all $k \geq 0$, where $\mathbf{0}$ is an all-zero sequence of any length.

In practice we might be interested in the k-error complexities only for small values of k, say a certain percentage of the length of the sequence, so we introduce an extra parameter, k_{Max} (default value $w_H(s)$) in the algorithm and we compute the truncated k-error linear complexity profile $\{(i, L_i(s))|0 \leq i \leq k_{Max}\}$.

One might also be interested in the minimum number of errors needed to achieve a linear complexity below a certain set value L_{Max} (see Sălăgean [9]). This again would make some of the recursive calls unnecessary, when the current complexity is already equal to or below L_{Max}. The default value for L_{Max} is t. In this case the algorithm will return the profile $\{(i, \max\{L_i(s), L_{Max}\})|0 \leq i \leq k_{Max}\}$.

We implemented the algorithm in Algorithms 1. and 2. based on a recursive version of the Berlekamp-Massey Algorithm. Two new variables are needed to accommodate the current number of errors, k and the current error pattern, e. We denote the k-error linear complexity profile as sol and we define each element as a collection of three components $sol_k = \{sol\_L_k, sol\_C_k(X), sol\_err_k\}$, for $k = 0, 1, \ldots, \max\{k_{Max}, w_H(s) - 1\}$, where sol_L_k is the k-error linear complexity, $sol\_C_k(X)$ is the characteristic polynomial and sol_err_k is the error sequence corresponding to the k-error linear complexity.

There are some immediate improvements that can be performed on the Modified Berlekamp-Massey Algorithm. First of all the stop condition can be adjusted. Some of the paths taken by the recursion calls might get to a k-error linear complexity which is bigger or equal to the currently stored solution (sol_L_k) so they can be abandoned. Secondly, the currently stored k-error linear complexity profile can be maintained whenever a new solution is found, using Property 1 in Section 2. Finally, we can combine iteration and recursion in order to minimize the stack size.

Algorithm 1. Recursive Modified Berlekamp-Massey Algorithm

1: **Input:** A finite non-zero sequence $s = s_0, s_1, \ldots, s_{t-1}$; k_{Max}; L_{Max}
2: **Output:** The approximate k-error linear complexity profile, sol
3: $v \leftarrow max(k_{Max}, w_H(s) - 1)$
4: **for** $i = 0, 1, \ldots, v$ **do**
5:     $sol_i \leftarrow \{t, X^t, \underbrace{(0, 0, \ldots, 0)}_{t\ times}\}$
6: **end for**
7: $k \leftarrow 0$
8: $e = \underbrace{(0, 0, \ldots, 0)}_{t\ times}$
9: $n \leftarrow 0$
10: **while** $s_n = 0$ and $n < t - 1$ **do** $\triangleright$ go over the initial zeros
11:     $n \leftarrow n + 1$
12: **end while** $\triangleright$ Initialize the details corresponding to 'last degree change' position m
13: $m \leftarrow n$
14: $C_m(X) \leftarrow 1$
15: $d_m \leftarrow s_n$
16: $n \leftarrow n + 1$
17: $C_n(X) \leftarrow X^n$ $\triangleright$ Initialize the details corresponding to current position n
18: **call** $mbmR(sol, m, C_m(X), d_m, C_n(X), n, k, e)$
19: **return** sol

3.1 Algorithm Analysis

The correctness of the linear complexity and characteristic polynomial for each number of errors k and the corresponding error sequence stored in the solution array at the end of the algorithm, results from the way the algorithm was built and from the correctness of the Berlekamp-Massey Algorithm (Massey [7]).

For analysing the complexity of the algorithms we will use the trees described in Section 3 and estimate their number of nodes using the following Lemma:

Lemma 1. *A binary tree of depth n and with at most k right branches on any path from the root to a leaf has a maximum of $\sum_{i=0}^{k} \binom{n+1}{i+1}$ vertices.*

Proof. We associate to each node a description of the path from the root to that node, i.e. a sequence over the alphabet $\{L, R\}$, where L signifies a left branch and R a right branch. The number of nodes will therefore be equal to the number of sequences of lengths between 0 and n, each sequence containing at most k occurrences of R. For any fixed number i of R's we have

$$\sum_{m=i}^{n} \binom{m}{i} = \binom{n+1}{i+1}$$

such sequences, so the total for all i from 0 to k will be

$$\sum_{i=0}^{k} \binom{n+1}{i+1}.$$

Algorithm 2. The $mbmR$ procedure

1: **procedure** MBMR($sol, m, C_m(X), d_m, C_n(X), n, k, e$)
2:     **if** $(n = t)$ or $(k > k_{Max})$ or $(\deg(C_n(X)) \leq L_{Max})$ **then**
3:         **if** $((n = t)$ and $(sol\_L_k > \deg(C_n(X)))$ and $k \leq k_{Max}$ **then**
4:             $sol_k \leftarrow \{\deg(C_n(X)), C_n(X), e\}$
5:         **end if**
6:     **else**
7:         $d_n \leftarrow (s + e)_n + \sum_{i=0}^{\deg(C_n(X))-1} c_n \cdot (s + e)_{i+n-\deg(C_n(X))}$
8:         **if** $d_n \neq 0$ **then**
9:             **if** $2 \cdot \deg(C_n(X)) > n$ **then**   ▷ (1a) the complexity does not change
10:                 $C_n(X) \leftarrow C_n(X) - \frac{d_n}{d_m} \cdot X^{(m-\deg(C_m(X)))-(n-\deg(C_n(X)))} \cdot C_m(X)$
11:                 $mbmR(sol, m, C_m(X), d_m, C_n(X), n + 1, k, e)$
12:             **else**   ▷ (1b) the complexity does change
13:                 $T(X) \leftarrow C_n(X)$
14:                 $C_n(X) \leftarrow X^{(n-\deg(C_n(X)))-(m-\deg(C_m(X)))} \cdot C_n(X) - \frac{d_n}{d_m} \cdot C_m(X)$
15:                 $mbmR(sol, n, T(X), d_n, C_n(X), n + 1, k, e)$
16:                 **if** $e_n = 0$ **then**
17:                     $mbmR(sol, m, C_m(X), d_m, C_n(X), n+1, k + 1, (e - d_n I_n))$
18:                 **end if**
19:             **end if**
20:         **else**   ▷ (2) the current characteristic polynomial does not change
21:             $mbmR(sol, m, C_m(X), d_m, C_n(X), n + 1, k, e)$
22:         **end if**
23:     **end if**
24: **end procedure**

There is no closed form for sums of the form $\sum_{i=0}^{k} \binom{n}{i}$, so we will use bounds:

Lemma 2. *The following bound stands*

$$\sum_{i=0}^{k} \binom{n}{i} \leq \begin{cases} 2\binom{n}{k}, & if\ k \leq \left\lfloor \frac{n+1}{3} \right\rfloor, \\ \left(k - \left\lfloor \frac{n+1}{3} \right\rfloor + 2\right)\binom{n}{k}, & if\ \left\lfloor \frac{n+1}{3} \right\rfloor < k \leq \left\lfloor \frac{n-1}{2} \right\rfloor \end{cases}$$

Proof. The first case follows by induction on k, using the fact that $\binom{n}{k} = \binom{n}{k-1} \cdot \frac{n-k+1}{k}$ for all n and k. Also $\frac{n-k+1}{k} \geq 2$ if $k \leq \lfloor (n+1)/3 \rfloor$. The remaining inequalities follow from the first using elementary properties of the binomial coefficients.

We will approximate binomial coefficients using the following (see [6, Lemma 7, Chapter 10])

$$\binom{n}{k} \approx c \frac{1}{\sqrt{n\alpha(1-\alpha)}} \left(\frac{1}{\alpha^\alpha (1-\alpha)^{1-\alpha}} \right)^n \quad (2)$$

where $0 < k < n$, $\alpha = k/n$ and c is a constant, $1/\sqrt{8} \leq c \leq 1/\sqrt{2\pi}$.

When assessing exponential time complexities of algorithms we will also use the fact that for any $a > 1$ and $i > 0$ we have $n^i a^n \in \mathcal{O}((a + \varepsilon)^n)$ with $\varepsilon > 0$ an arbitrarily small constant.

We are now ready to estimate the complexity of the algorithms presented.

Theorem 1. *The worst case time complexity of the Efficient Exhaustive Search Algorithm for sequences of length t and number of errors at most $k_{Max} = vt$ with $0 < v < 1/3$ is $\mathcal{O}(\sqrt{t}\lambda^t)$ where $\lambda = \frac{1}{v^v(1-v)^{1-v}}$. This can also be expressed as $\mathcal{O}((\lambda + \varepsilon)^t)$ with $\varepsilon > 0$ an arbitrarily small constant. For a typical value of $v = 0.1$ (i.e. errors in at most 10% of the positions) the time complexity is $\mathcal{O}(\sqrt{t}1.384145^t)$.*

Proof. The Efficient Exhaustive Search Algorithm will construct a tree of depth t and at most k_{Max} right branches on any path from the root to a leaf. By Lemma 1, this tree will have at most $\sum_{i=0}^{k_{Max}} \binom{t+1}{i+1}$ nodes. So the number of nodes is bounded by $2\binom{t+1}{k_{Max}+1}$, by Lemma 2. In any node we compute a discrepancy and possibly adjust the characteristic polynomial, so there are $\mathcal{O}(t)$ computational steps. Therefore the complexity is $\mathcal{O}(t\binom{t+1}{k_{Max}+1})$.

Using (2) we obtain the following approximation:

$$2t\binom{t+1}{k_{Max}+1} = 2\frac{t(t+1)}{k_{Max}+1}\binom{t}{k_{Max}} =$$

$$= 2\frac{t(t+1)}{vt+1}\binom{t}{vt}$$

$$\approx 2c\frac{t(t+1)}{vt+1}\frac{1}{\sqrt{tv(1-v)}}\left(\frac{1}{v^v(1-v)^{1-v}}\right)^t$$

$$\approx \frac{2c}{v\sqrt{v(1-v)}}\sqrt{t}\left(\frac{1}{v^v(1-v)^{1-v}}\right)^t$$

which is $\mathcal{O}(\sqrt{t}\lambda^t)$ where $\lambda = \frac{1}{v^v(1-v)^{1-v}}$.

For the Modified Berlekamp Massey Algorithm it is harder to estimate the depth of the tree, as the number of terms processed in between two decision points will vary depending on the particular sequence. We will assume that an average of u terms are processed between two decision points, i.e. between two points where the Berlekamp-Massey algorithm would prescribe an increase in the current complexity of the sequence. In [8, Chapter 4] it is shown that for random binary sequences the average number of bits that have to be processed between two changes in complexity is 4 and the change in complexity has an average of 2. While the sequences used in the cryptographic applications are not truly random, using a value of $u = 4$ for the average number of terms between two changes of complexity seems reasonable.

Theorem 2. *The worst case time complexity of the Modified Berlekamp Massey Algorithm for sequences of length t, an average of u terms of the sequence processed between two changes in complexity, and a number of errors at most $k_{Max} = vt$ with $0 < v < \frac{1}{u}$ is*

$$\begin{cases} \mathcal{O}(\sqrt{t}\lambda_1^t) & \text{if } v < \frac{1}{3u} & \text{where } \lambda_1 = \frac{1}{uv^v(1-uv)^{\frac{1}{u}-v}}, \\ \mathcal{O}(t\sqrt{t}\lambda_1^t) & \text{if } \frac{1}{3u} \le v < \frac{1}{2u} & \text{where } \lambda_1 = \frac{1}{uv^v(1-uv)^{\frac{1}{u}-v}}, \\ \mathcal{O}(t\lambda_2^t) & \text{if } \frac{1}{2u} \le v \le \frac{1}{u} & \text{where } \lambda_2 = \sqrt[u]{2}. \end{cases}$$

In all cases the complexity can also be written as $\mathcal{O}((\lambda_i + \varepsilon)^t)$ where $\varepsilon > 0$ is an arbitrarily small constant. For a typical value of $v = 0.1$ (i.e. errors in at most 10% of positions) and $u = 4$ the complexity is $\mathcal{O}(t\sqrt{t}1.189208^t)$.

Proof. Since u is the number of terms between two decision points and t is the total number of terms, the depth of the tree will be t/u. We bound the number of vertices in the tree by $\sum_{i=0}^{k_{Max}} \binom{\frac{t}{u}+1}{i+1}$, using Lemma 1. When the number of right branches on any path, k_{Max}, is at most half the depth of the tree, by applying the first or the second bound in Lemma 2 (depending on whether k_{Max} is smaller or greater than a third of t/u), followed by the estimation (2), we obtain the first two computational complexities $\mathcal{O}$ of the Theorem in a similar way as in the proof of Theorem 1.

When the number of right branches allowed in the tree approaches the depth of the tree, i.e. k_{Max} approaches t/u, we will bound the number of nodes by $2^{\frac{t}{u}+1}-1$ (the number of nodes in a complete binary tree of depth t/u). Combining this with $\mathcal{O}(t)$ operations in each node gives the third $\mathcal{O}$ of the theorem.

The proposed algorithm has the advantage that even when the field has more than two elements, there are still only two choices that are investigated: introducing no error, or introducing an error of magnitude $-d^{(n)}$, where $d^{(n)}$ is the discrepancy; an exhaustive search approach would have to investigate all the possible error magnitudes for each error position, i.e. $\sum_{i=0}^{k} \binom{t}{i}(w-1)^i$ possibilities for a field of w elements. Both the complexities in Theorems 1 and 2 will increase by a factor of $(\log w)^2$ to account for the more costly operations in a field of w elements. However, the exponential part in the $\mathcal{O}$ estimate will remain unchanged in Theorem 2 (Modified Berlakamp-Massey Algorithm), whereas in Theorem 1 (Efficient Exhaustive Search), λ^t will be replaced by $(\lambda(w-1)^v)^t$.

For a typical value of $v = 0.1$ (i.e. errors in at most 10% of the positions) and an alphabet of $w = 16$ elements the worst case time complexity is $\mathcal{O}(\sqrt{t}1.826^t)$ for exhaustive search as compared to $\mathcal{O}(t\sqrt{t}1.189208^t)$ for the proposed modified Berlekamp-Massey algorithm.

4   Tests and Results

In order to estimate the efficiency and the accuracy of the algorithm, a comparison has been done between the optimised Modified Berlekamp-Massey (MBM) Algorithm and the Efficient Exhaustive Search (EES) Algorithm.

We define the accuracy, $ACC_k(s)$, as the ratio between $L_{MBM,k}(s)$, the approximate value of the k-error linear complexity obtained using the Modified Berlekamp-Massey Algorithm and $L_{EES,k}(s)$, the exact value obtained using the Efficient Exhaustive Search Algorithm, $L_{MBM,k}(s)/L_{EES,k}(s)$.

The running time improvement was computed as the ratio between the time taken by the Efficient Exhaustive Search Algorithm and the time taken by the Modified Berlekamp-Massey Algorithm on the same processor, $time_{EES}/time_{MBM}$.

The first test has involved running both algorithms on a number of 70 randomly chosen sequences of length 64 (each bit is generated with the C `rand()` linear congruential generator function).

Figure 2 presents the average, best and worst value of ACC_k over the 70 sequences tested. These results are detailed in Table 1 for $1 \leq k \leq 9$. For small values of k we notice that on average the k-error linear complexity obtained by the Modified Berlekamp-Massey Algorithm is pretty close to the actual value, being higher by only 3.37% for 1 error, increasing to 16.45% for 6 errors (i.e. errors in about 10% of the terms) and by 25.92% for 9 errors (i.e. about 15% of the terms). As k increases, the quality of the results obtained by the Modified Berlekamp-Massey Algorithm deteriorates. Note however that the small values of k are the ones of practical interest.

Table 1. The average accuracy of the results of the MBM Algorithm

Number of errors k	1	2	3	4	5	6	7	8	9
Average ACC_k	1.03	1.06	1.09	1.11	1.14	1.16	1.19	1.22	1.25
Best ACC_k	1	1	1	1	1	1	1	1	1
Worst ACC_k	1.14	1.2	1.21	1.31	1.5	1.35	1.37	1.5	1.66

The average running time improvement was 12691, i.e. the MBM Algorithm was nearly 13000 times faster than the EES Algorithm. Even better time improvements are obtained when imposing limits for the number of errors and/or the maximum linear complexity. For example for k_{Max} equal to 15% of the length of the sequence and L_{Max} approx. 1/3 of the length, the time improvement is 24017.

A second experiment involved running the Modified Berlekamp-Massey Algorithm for sequences of different lengths. We used 20 random sequences for each even length between 8 and 64. The time improvement shows an exponential

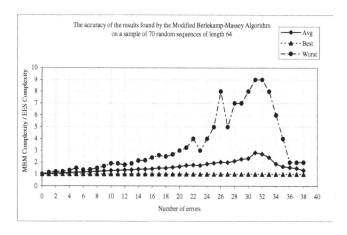

Fig. 2. The accuracy of the MBM Algorithm on a sample of 70 random sequences of length 64

increase with the length of the sequence (no limitations were imposed on the parameters k_{Max} and L_{Max}). See figure 3 for the results.

The quality of the approximation was measured for each sequence at different levels of error: 5%, 10% and 15% of the length of the sequence. The results are summarised in figure 4. We note that the approximate value of the k-error complexity found by the modified Berlekamp-Massey Algorithm is consistently good on all lengths tested and it deteriorates only slightly as k increases as a percentage of the length of the sequence. For 5% errors (i.e. k is 5% of the length), the k-error linear complexity found by the MBM algorithm is on average not more than 10% higher than the actual value, for 10% errors it is at most 20% higher and for 15% it is at most 30% higher.

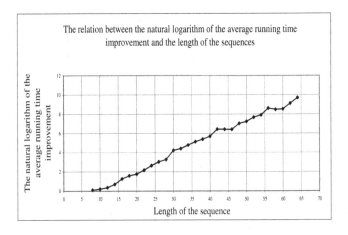

Fig. 3. The relation between the natural logarithm of the average running time improvement and the length of the sequences

For evaluating the accuracy of the MBM algorithm for sequences of higher length the actual k-error linear complexity could no longer be computed using exhaustive search due to hardware limitations. Instead, we carried out a controlled experiment where we took 50 sequences s of length 100, generated by a randomly chosen recurrence of size 33 (1/3 of the length). We computed the linear complexity $L(s)$ of each sequence s (this can be lower than 33). We artificially applied an error sequence e of weight k, such that the linear complexity of $s' = s + e$ is higher than $L(s)$. Obviously, $L_k(s') \le L(s)$, so even though we do not know the exact k-error complexity of s', we do have a good upper bound. We then applied the MBM Algorithm to s' and computed the ratio $L_{MBM,k}(s')/L(s)$. This time the ratio can be less than 1 because $L(s)$ is an upper bound rather than the exact value of $L_k(s')$. Figure 5 presents the distribution of the values of this ratio in each interval of length 0.1. Four cases were considered, depending on the choice of k: random values up to 15% of the length of the sequence, or fixed values of 5%, 10% and 15%, respectively. We notice that a high proportion of the ratios are below 1.1, i.e. the value found by

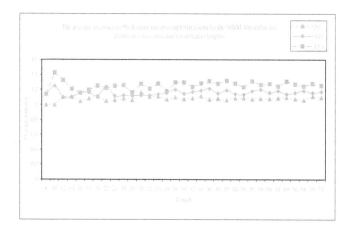

Fig. 4. The average accuracy of the k-error linear complexity found by the MBM Algorithm for different values of k and for different lengths

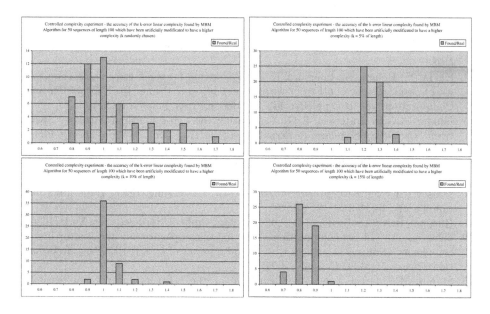

Fig. 5. The accuracy of the results found by MBM Algorithm on 50 sequences of length 100, when the sequences were artificially modified with errors sequences of weight : (a) random; (b) $k = 5\%$ of the length; (c) $k = 10\%$ of the length; (d) $k = 15\%$ of the length

the MBM algorithm is close, or even lower than the original complexity, $L(s)$. The results improve when k represents a higher proportion of the length of the sequence.

5   Conclusion

We propose a heuristic algorithm for approximating the k-error linear complexity, based on the Berlekamp-Massey Algorithm. We implemented and tested this algorithm and the results are encouraging. The k-error linear complexity is approximated pretty close: on average it is only 16% higher than the exact value, for up to 6 errors on our test set of 70 random sequences of length 64. While the time complexity of the proposed algorithm is still exponential, it is considerably faster than an exhaustive search (on average about 13000 times faster for the sequences above). Even higher efficiency gains are expected in the non-binary case. Future work will investigate the possibility of further reducing the search space with minimal accuracy loss.

References

1. Berlekamp, E.R.: Algebraic Coding Theory. McGraw-Hill, New York (1968)
2. Ding, C., Xiao, G., Shan, W.: The Stability Theory of Stream Ciphers. Springer, Heidelberg (1992)
3. Games, R.A., Chan, A.H.: A Fast Algorithm for Determining the Complexity of a Binary Sequence with Period 2^n. IEEE Trans. Information Theory 29(1), 144–146 (1983)
4. Kaida, T., Uehara, S., Imamura, K.: An Algorithm for the k-error linear complexity of Sequences over $GF(p^m)$ with Period p^n, p a Prime. In: Information and Computation, vol. 151, pp. 134–147. Academic Press, London (1999)
5. Lauder, A.G.B., Paterson, K.G.: Computing the Error Linear Complexity Spectrum of a Binary Sequence of Period 2^n. IEEE Trans. Information Theory 49(1), 273–283 (2003)
6. MacWilliams, F.J., Sloane, N.J.A.: The Theory of Error-correcting Codes. North Holland, Amsterdam (1977)
7. Massey, J.L.: Shift-Register Synthesis and BCH Decoding. IEEE Trans. Information Theory 15(1), 122–127 (1969)
8. Rueppel, R.A.: Analysis and Design of Stream Ciphers. Springer, New York (1986)
9. Salagean, A.: On the computation of the linear complexity and the k-error linear complexity of binary sequences with period a power of two. IEEE Trans. Information Theory 51(3), 1145–1150 (2005)
10. Stamp, M., Martin, C.F.: An Algorithm for the k-Error Linear Complexity of Binary Sequences with Period 2^n. IEEE Trans. Information Theory 39(4), 1398–1401 (1993)

Efficient KEMs with Partial Message Recovery

Tor E. Bjørstad[1], Alex W. Dent[2], and Nigel P. Smart[3]

[1] The Selmer Center, Department of Informatics,
University of Bergen, Pb. 7800, N-5020 Bergen, Norway
`tor.bjorstad@ii.uib.no`
[2] Information Security Group, Royal Holloway, University of London,
Egham, Surrey, TW20 0EX, United Kingdom
`a.dent@rhul.ac.uk`
[3] Department Computer Science, University of Bristol,
Merchant Venturers Building, Woodland Road,
Bristol, BS8 1UB, United Kingdom
`nigel@cs.bris.ac.uk`

Abstract. Constructing efficient and secure encryption schemes is an important motivation for modern cryptographic research. We propose simple and secure constructions of hybrid encryption schemes that aim to keep message expansion to a minimum, in particular for RSA-based protocols. We show that one can encrypt using RSA a message of length $|m|$ bits, at a security level equivalent to a block cipher of κ bits in security, in $|m| + 4\kappa + 2$ bits. This is therefore independent of how large the RSA key length grows as a function of κ. Our constructions are natural and highly practical, but do not appear to have been given any previous formal treatment.

1 Introduction

There are many important factors to consider when choosing a practical encryption scheme, including speed, provable security, code size, and bandwidth efficiency. Bandwidth efficiency is often only considered as an afterthought, but for many real world problems this can be as important as keeping the computational and implementational complexity low. In particular this is the case for wireless settings, where power consumption often is a limiting factor, and transmitting data is a major power drain compared to the cost of doing some extra (offline) computation [18].

Suppose that we wish to encrypt $|m|$-bit messages with a security level of κ bits, for example $\kappa = 128$. The aim of this paper is to suggest secure protocols which achieves the desired security level, while keeping the length of the ciphertexts as close to $|m|$ bits as possible. The existing ISO/IEC standard for public-key encryption [17] only considers bandwidth efficiency for short messages and does not give an RSA-based scheme that is efficient for arbitrary-length messages and arbitrary RSA key sizes.

To explain our starting point in more concrete terms, let R_κ denote the size of an RSA key equivalent to κ bits of security, and similarly let E_κ be the

S.D. Galbraith (Eds.): Cryptography and Coding 2007, LNCS 4887, pp. 233–256, 2007.
© Springer-Verlag Berlin Heidelberg 2007

corresponding key size for elliptic curve based systems. At the time of writing, it is widely believed that approximate values for these parameters are roughly $R_{80} = 1024$, $R_{128} = 3072$, and $E_\kappa = 2\kappa$ [14].

We first examine the problem by analogy with digital signature schemes. Using the ECDSA signature scheme one obtains a short signature scheme with appendix, of total length $|m| + 2E_\kappa$ bits, that is, in addition to the original message we get a signature appendix consisting of two group elements. With pairings, it is possible to shrink the size of signatures to $|m| + E_\kappa$ bits using the BLS scheme [9]. Standard RSA-based schemes such as RSA-FDH [5] or RSA-PSS [7], give signatures of length $|m| + R_\kappa$. For signature schemes based on RSA one should therefore consider schemes that provide some kind of message recovery, such as the message recovering variant of RSA-PSS, if bandwidth is a limiting factor. Indeed, there may even be situations where bandwidth is so precious that it is necessary to use elliptic-curve based schemes with message recovery, such as [15].

We consider the similar situation for public key encryption schemes. The standard elliptic curve encryption scheme for arbitrary-length messages is ECIES, which is a hybrid scheme modelled in the KEM+DEM framework [10]. This produces ciphertexts of length $E_\kappa + |m| + \kappa$. For RSA the hybrid scheme of choice is RSA-KEM [16], which leads to ciphertexts of length $R_\kappa + |m| + \kappa$. While the hybrid constructions are suitable for long messages, sufficiently short messages may be encrypted using only a single group element using a purely asymmetric encryption scheme, such as RSA-OAEP, [6,17]. With RSA-OAEP, a message of maximal length $|m|$ is encoded together with two strings, each of length that of a hash function output, as a single group element of size $R_\kappa = |m| + 4\kappa + 2$ bits.

Our initial question was whether the bandwidth requirements for hybrid encryption schemes in the KEM+DEM framework may be reduced, in particular that of the RSA-based schemes. In the usual KEM+DEM framework, the KEM is used to encrypt a symmetric key, while the DEM is used to encrypt the message itself (using a symmetric encryption scheme and the key from the KEM). However, existing KEMs typically encode the symmetric key as a group element, and hence require either R_κ or E_κ bits.

This disparity in expansion rate grows when the security level increases, as the size of the groups grows much faster than the size of symmetric components. This is not so much a problem for elliptic curve systems where the growth is linear, but for factoring based systems the growth is quite pronounced, as shown in Fig. 1.

For "long" messages, the overhead inherent in the key encapsulation is quite negligible. However, in constrained environments it may be of significance, particularly when the messages being sent are on (roughly) the same order of magnitude as the size of the group element representation. This motivates the design and analysis of protocols that focus on keeping the message expansion to a minimum, at the expense of some additional algorithmic complexity.

Since R_κ in particular grows much faster than κ, it is natural to consider whether we may embed part of the message as part of the key encapsulation, and recover it during the decapsulation process. Such an approach was suggested for

Symmetric key length (κ)	Size of RSA modulus (R_κ)	Size of elliptic curve (E_κ)
80	1024	160
112	2048	224
128	3072	256
192	7680	384
256	15360	512

Fig. 1. Suggested parameter sizes (in bits) to maintain comparable security levels between different primitives, as recommended by NIST [14]

the specific case of RSA-OAEP in [16], but to the best of the authors' knowledge no formal analysis of that composition has been published, nor is it mentioned in the final ISO/IEC standard for public-key encryption [17].

In this paper we present a general concept of KEMs with message recovery, so-called RKEMs. We define security models for RKEMs, and prove a composition theorem which shows that a secure RKEM may be combined with an IND-CCA and INT-CCA secure DEM to obtain an IND-CCA secure public key encryption scheme. We then present a concrete example of an RKEM based on the RSA-OAEP construction, which is secure in the random oracle model. In combination with a standard Encrypt-then-MAC DEM, this results in a public key encryption scheme with ciphertexts of length as low as $|m| + 5\kappa + 3$ bits, i.e. a scheme whose messages does not depend on the size of the RSA modulus while providing κ-bit security. We then extend the concept of RKEMs to the Tag-KEM framework proposed by Abe et.al. [1] and propose a tag-RKEM based on the RSA-OAEP construction, which is secure in the random oracle model and will encrypt a message of length $|m|$ as a ciphertext of length $|m| + 4\kappa + 2$. This is the most space-efficient RSA-based construction known for long messages.

2   Definitions

2.1   Public Key Encryption

Our goal is to create bandwidth-efficient public key encryption schemes for arbitrary-length messages.

Definition 1 (Public-Key Encryption Scheme). *We define a public-key encryption scheme $PKE = (PKE.Gen, PKE.Enc, PKE.Dec)$ as an ordered tuple of three algorithms:*

1. *A probabilistic key generation algorithm $PKE.Gen$. It takes as input a security parameter 1^κ, and outputs a private/public keypair (sk, pk). As part of the public key there is a parameter $PKE.msglen$ that specifies the maximum length of messages that can be encrypted in a single invocation; this value may be infinite.*

2. *A probabilistic encryption algorithm PKE.Enc. It takes as input a public key pk and a message m of length at most PKE.msglen, and outputs a ciphertext c.*

3. *A deterministic decryption algorithm PKE.Dec. It takes as input a private key sk and a ciphertext c, and outputs either a message m or the unique error symbol $\perp$.*

An encryption scheme must be sound, in the sense that the identity $m = PKE.Dec(sk, PKE.Enc(pk, m))$ holds for valid keypairs $(sk, pk)=PKE.Gen(1^\kappa)$. Although we assume throughout this paper that our cryptographic primitives are perfectly sound, extension of our results to the case of non-perfect soundness is straightforward, with standard techniques.

To discuss the security of an encryption scheme, we must define a formal notion of what it means to "break" the scheme. Encryption schemes are designed with the goal of *confidentiality* in mind: an adversary should not be able to learn any useful information about encrypted messages. An encryption scheme is said to be IND-secure if there does not exist an efficient adversary who, given an encryption of two equal length messages, can determine which messages was encrypted to form the ciphertext.

In practice, we want to construct public-key encryption schemes that maintain confidentiality even under adaptive chosen-ciphertext attack (-CCA), which means that the adversary is allowed adaptive access to a decryption oracle. To quantify the concept of security, we compare the success of an adversary with the trivial "attack" of flipping a coin and guessing at random, and require that the (asymptotic) gain must be a *negligible* function in the security parameter 1^κ.

Definition 2 (Negligible function). *A function $f : \mathbb{N} \to \mathbb{R}$ is said to be negligible if for every polynomial p in $\mathbb{N}[x]$, there exists an $x_0 \in \mathbb{N}$ such that $f(n) \leq \frac{1}{|p(x)|}$ for all $x > x_0$.*

We now define the IND-CCA attack game for public-key encryption.

Definition 3 (IND-CCA Game for PKE). *The IND-CCA game for a given public-key encryption scheme PKE is played between the challenger and a two-stage adversary $\mathcal{A} = (\mathcal{A}_1, \mathcal{A}_2)$, which is a pair of probabilistic Turing machines. For a specified security parameter 1^κ, the game proceeds as follows:*

1. *The challenger generates a private/public keypair $(sk, pk) = PKE.Gen(1^\kappa)$.*
2. *The adversary runs $\mathcal{A}_1$ on the input pk. During its execution, $\mathcal{A}_1$ may query a decryption oracle $\mathcal{O}_D$, that takes a ciphertext c as input, and outputs PKE.Dec(sk, c). The algorithm terminates by outputting state information s and two messages m_0 and m_1 of equal length.*
3. *The challenger picks a bit $b \xleftarrow{R} \{0,1\}$ uniformly at random, and computes $c^* = PKE.Enc(pk, m_b)$.*
4. *The adversary runs $\mathcal{A}_2$ on the input (c^*, s). During its execution, $\mathcal{A}_2$ has access to the decryption oracle as before, with the limitation that it may not ask for the decryption of the challenge ciphertext c^*. The algorithm terminates by outputting a guess b' for the value of b.*

We say that $\mathcal{A}$ wins the IND-CCA game whenever $b = b'$. The advantage of $\mathcal{A}$ is the probability

$$\mathrm{Adv}_{\mathrm{PKE}}^{\mathrm{IND-CCA}}(\mathcal{A}) = \left| Pr[\mathcal{A} \; wins] - 1/2 \right|.$$

A scheme is said to be IND-CCA secure if the advantage of every polynomial-time adversary is negligible (as a function of the security parameter). For specific schemes, we rarely have unconditional security, and speak instead of security with respect to some well-known reference problem that is thought to be difficult (such as the RSA or Diffie-Hellman problems).

2.2   Hybrid Encryption

Hybrid encryption is the practice of constructing public-key encryption schemes using a symmetric-key cipher as a building block. Although this adds a certain amount of complexity, it also enables PKE schemes to handle messages of more or less unrestricted length, at a low computational cost. The standard construction paradigm for hybrid encryption schemes is the KEM+DEM framework [10], in which a scheme is divided into two parts: an asymmetric *Key Encapsulation Mechanism* (KEM) and a symmetric *Data Encapsulation Mechanism* (DEM). Not only does this allow the encryption of arbitrary length messages but it also means that the PKE scheme obtained, by generically combining a secure KEM and a secure DEM, is itself secure [10]. This means that the components can be analysed separately and combined in a mix-and-match fashion.

Various modifications and variations of the basic KEM+DEM construction have also been proposed. In particular, we note the Tag-KEM schemes proposed in [1]. In Appendix A we present the standard definitions of a KEM and a DEM, how they are combined, and the appropriate security models for each.

3   KEMs with Partial Message Recovery

3.1   KEMs with Partial Message Recovery

In this section we introduce the notion of a KEM with partial message recovery (or RKEM). A KEM with partial message recovery is, quite simply, the use of public-key encryption to transmit a symmetric key and some partial message data in a secure manner. An obvious instantiation of such a scheme is to use a secure non-hybrid PKE to encrypt the "payload", although alternate implementations may also be possible. We will show that the obvious construction is indeed secure, and that the resulting RKEM+DEM hybrid encryption produces shorter messages than the traditional KEM+DEM construction.

Definition 4 (RKEM). *We define a KEM with partial message recovery to be an ordered tuple RKEM = (RKEM.Gen, RKEM.Encap, RKEM.Decap) consisting of the following algorithms:*

1. *A probabilistic key generation algorithm RKEM.Gen. It takes a security parameter 1^κ as input, and outputs a private/public keypair (sk, pk). As part of the public key there are two parameters, RKEM.msglen and RKEM.keylen. The value RKEM.msglen, which we assume is finite, denotes the maximum amount of message data that may be stored in an encapsulation, and RKEM.keylen denotes the fixed length of the symmetric key generated by the RKEM.*
2. *A probabilistic key encapsulation algorithm RKEM.Encap. It takes as input a public key pk, and a message m of length at most RKEM.msglen. The algorithm terminates by outputting a symmetric key k of length RKEM.keylen and an encapsulation ψ.*
3. *A deterministic key decapsulation algorithm RKEM.Decap. It takes as input a private key sk and an encapsulation ψ. The algorithm outputs either the unique error symbol $\perp$, or a pair (k, m) consisting of a symmetric key k of RKEM.keylen bits, and a message m of RKEM.msglen bits.*

If *RKEM.Encap* and *RKEM.Decap* are run using a valid keypair (sk, pk) and the ψ output by the encapsulation algorithm is used as input to the decapsulation algorithm, then the probability of failure is assumed to be zero. Furthermore, the decapsulation algorithm is required to output the same values of k and m as were associated with the encapsulation algorithm.

3.2 Security Definitions for RKEMs

Since the plaintext message is given as input to the encapsulation algorithm, it is necessary to adopt two separate security requirements for such RKEMs. First, we define the IND-CCA game for an RKEM similarly to that for a regular KEM, in which the adversary tries to distinguish whether a given key is the one embedded in a specified encapsulation.

Definition 5 (IND-CCA for RKEM). *The IND-CCA game for a given RKEM is played between a challenger and an adversary $\mathcal{A} = (\mathcal{A}_1, \mathcal{A}_2)$. For a particular security parameter 1^κ, the game runs as follows.*

1. *The challenger generates a keypair $(sk, pk) = RKEM.Gen(1^\kappa)$.*
2. *The adversary runs $\mathcal{A}_1$ on the input pk. During its execution, $\mathcal{A}_1$ may query a decapsulation oracle $\mathcal{O}_D$ that takes an encapsulation ψ as input, and outputs the result of computing $RKEM.Decap(sk, \psi)$. The algorithm terminates by outputting a message m of length at most RKEM.msglen bits, as well as some state information s.*
3. *The challenger computes $(k_0, \psi^*) = RKEM.Encap(pk, m)$, and draws another key $k_1 \xleftarrow{R} \{0, 1\}^{RKEM.keylen}$ as well as a random bit $b \xleftarrow{R} \{0, 1\}$ uniformly at random*
4. *The adversary runs $\mathcal{A}_2$ on the input (s, k_b, ψ^*). During its execution, $\mathcal{A}_2$ has access to the decapsulation oracle as before, with the restriction that it may not ask for the decapsulation of the challenge ψ^*. The algorithm terminates by outputting a guess b' for the value of b.*

We say that A wins the game whenever the guess was correct, i.e. $b = b'$. The advantage of A is given as

$$\text{Adv}_{\text{RKEM}}^{\text{IND-CCA}}(A) = \left| Pr[A \; wins] - 1/2 \right|.$$

The other criterion relates to the confidentiality of the message used as input, and is represented by adopting the notion of RoR-CCA security from [3,13]. The term *RoR* stands for real-or-random, which is because in this security definition an adversary is unable to tell a valid encryption of a message, from a random ciphertext. It can be shown that RoR-CCA security is equivalent to indistinguishability with respect to the message, but we shall not apply this equivalence directly. Instead, we only require the RoR-CCA property of the RKEM to imply that the full hybrid encryption scheme is IND-CCA secure.

Definition 6 (RoR-CCA for RKEM). *The RoR-CCA game for KEMs with partial message recovery is defined as follows:*

1. *The challenger generates a keypair* $(sk, pk) = RKEM.Gen(1^\kappa)$.
2. *The adversary runs A_1 on the input pk. During its execution, A_1 may query a decapsulation oracle $\mathcal{O}_D$ that takes an encapsulation ψ as input, and outputs the result of computing $RKEM.Decap(sk, \psi)$. The algorithm terminates by outputting a message m_0 of length at most $RKEM.msglen$ bits, as well as some state information s.*
3. *The challenger generates a random message m_1, which is of the same length as m_0, a random bit $b \xleftarrow{R} \{0, 1\}$, and computes $(k^*, \psi^*)=RKEM.Encap(pk, m_b)$.*
4. *The adversary runs A_2 on the input (s, k^*, ψ^*). During its execution, A_2 has access to the decapsulation oracle as before, with the restriction that it may not ask for the decapsulation of the challenge ψ^*. The algorithm terminates by outputting a guess b' for the value of b.*

We say that A wins the game whenever the guess was correct, i.e. $b = b'$. In each case the advantage of A is

$$\text{Adv}_{\text{RKEM}}^{\text{RoR-CCA}}(A) = \left| Pr[A \; wins] - 1/2 \right|.$$

Note that IND-CCA security definition really is about the ability of the adversary to determine whether a specified key is real or random, and RoR-CCA security is about the ability of the adversary to determine whether the embedded message is real or random. Hence, a more accurate nomenclature would be K-RoR-CCA and M-RoR-CCA, but we use the above nomenclature to stress the link with prior security definitions for standard KEMs.

3.3 Security of the Composition of an IND-CCA and RoR-CCA Secure RKEM and an IND-PA and INT-CTXT Secure DEM

Combining an RKEM with a DEM is done in the straightforward manner:

Definition 7 (RKEM+DEM Construction). *Given an RKEM and a DEM where the keys output by the RKEM are of the correct length for use with the DEM, i.e. RKEM.keylen = DEM.keylen, we construct a hybrid PKE scheme as follows.*

- *The key generation algorithm PKE.Gen executes RKEM.Gen to produce a private / public keypair, and appends any necessary information about the operation of the DEM.*
- *The encryption algorithm PKE.Enc is implemented as follows.*
  1. *The message m is padded to be of size at least RKEM.msglen − 1.*
  2. *The message m is then split into two component $m^{(0)}$ and $m^{(1)}$, i.e. $m = m^{(0)}||m^{(1)}$, where $m^{(0)}$ is of length RKEM.msglen − 1.*
  3. *Set $v = 1$, unless $m^{(1)} = \emptyset$, in which case we set $v = 0$.*
  4. *Compute a key/encapsulation pair $(k, \psi) = RKEM.Encap(pk, m^{(0)}||v)$.*
  5. *If $v = 1$ then encrypt the remaining part of the message to obtain a ciphertext $\chi = DEM.Enc_K(m^{(1)})$, otherwise set $\chi = \emptyset$.*
  6. *Output the ciphertext $c = (\psi, \chi)$.*
- *The decryption algorithm PKE.Dec is implemented as follows.*
  1. *Parse the ciphertext to obtain $(\psi, \chi) = c$.*
  2. *Recover the key and message fragment from ψ by computing $(k, m^{(0)}||v) = RKEM.Decap(sk, \psi)$.*
  3. *If $k = \bot$, return $\bot$ and halt.*
  4. *If $v = 1$ and $\chi \neq \emptyset$, return $\bot$ and halt.*
  5. *If $v = 0$, return $m^{(0)}$ and halt.*
  6. *Compute $m^{(1)} = DEM.Dec_k(\chi)$.*
  7. *If $m^{(1)} = \bot$, return $\bot$ and halt.*
  8. *Output $m^{(0)}||m^{(1)}$.*

The soundness of the RKEM+DEM construction follows from the soundness of the individual RKEM and DEM.

In the case where $|m| \leq RKEM.msglen$, there are few practical reasons to use the hybrid setup at all, and this is included mainly to avoid placing any artificial restrictions on our allowable message space. Our definition is no longer optimal in this case, since there is no reason to encapsulate a symmetric key k at all. We note that an alternate definition could specify that RKEM.Decap returns a binary string s instead of k and $m^{(1)}||v$, which may then be parsed and interpreted depending on the value of v. The distinction is not important for our analysis, and is omitted in the further discussion for the sake of clarity.

Theorem 1 (Security of RKEM+DEM). *If the underlying RKEM is both IND-CCA and RoR-CCA secure and the DEM is IND-PA and INT-CTXT secure[1], then the above composition is IND-CCA secure.*

More precisely we have, that if there is an adversary $\mathcal{A}$ against the above public key scheme, then there are polynomial-time adversaries $\mathcal{B}_1, \mathcal{B}_2, \mathcal{B}_3$ and $\mathcal{B}_4$ such that

---

[1] The security definitions for DEMs are given in the Appendix.

$$\text{Adv}_{\text{PKE}}^{\text{IND-CCA}}(\mathcal{A}) \leq 2 \cdot \text{Adv}_{\text{RKEM}}^{\text{IND-CCA}}(\mathcal{B}_1) + q_D \cdot \text{Adv}_{\text{DEM}}^{\text{INT-CTXT}}(\mathcal{B}_2)$$
$$+ 2 \cdot \text{Adv}_{\text{RKEM}}^{\text{RoR-CCA}}(\mathcal{B}_3) + \text{Adv}_{\text{DEM}}^{\text{IND-PA}}(\mathcal{B}_4),$$

where q_D is an upper bound on the number of decryption queries made by $\mathcal{A}$.

Proof. Let $\mathcal{A}$ denote our adversary against the hybrid PKE system and let Game 0 be the standard IND-CCA game for a PKE. We prove the security by successively modifying the game in which $\mathcal{A}$ operates. In Game i, we let T_i denote the event that $b = b'$. Hence

$$\text{Adv}_{\text{PKE}}^{\text{IND-CCA}}(\mathcal{A}) = |\Pr[T_0] - 1/2|.$$

Let Game 1 be the same as Game 0 except that if the challenger is asked by the adversary to decrypt a ciphertext (ψ^*, χ), where ψ^* is equal to the encapsulation-part of the challenge, then it uses the key k^* output by the encapsulation function when it decrypts χ, i.e. it only uses the valid decryption algorithm associated to the RKEM to obtain $m^{(0)}$. Since we assume that our algorithms are perfectly sound this is purely, at this stage, a conceptual difference, i.e.

$$\Pr[T_0] = \Pr[T_1].$$

Game 2 proceeds identically to Game 1, except for in the computation of the second component χ^* of the challenge ciphertext, where a *random* key k' is used instead of the key k^* that was returned by *RKEM.Encap*. It is clear that there exists a machine $\mathcal{B}_1$, whose running time is essentially that of $\mathcal{A}$, which can turn a distinguisher between the two games into an adversary against the IND-CCA property of the RKEM. We have

$$|\Pr[T_1] - \Pr[T_2]| \leq 2 \cdot \text{Adv}_{\text{RKEM}}^{\text{IND-CCA}}(\mathcal{B}_1).$$

Let Game 3 be the same as Game 2 except that when the challenger is asked by the adversary to decrypt a ciphertext (ψ^*, χ), where ψ^* is equal to the encapsulation-part of the challenge, then it simply rejects the ciphertext. It is clear that there exists a machine $\mathcal{B}_2$, whose running time is essentially that of $\mathcal{A}$, which can turn a distinguisher between the two games into an adversary against the INT-CTXT property of the DEM. We have

$$|\Pr[T_2] - \Pr[T_3]| \leq q_D \cdot \text{Adv}_{\text{DEM}}^{\text{INT-CTXT}}(\mathcal{B}_2).$$

In Game 4, we change the computation of the encapsulation so that it encapsulates a random string instead of the first part of the message $m^{(0)}$, but we encrypt the second part of the message as in Game 2. Again, it is clear that there exists a machine $\mathcal{B}_3$, whose running time is essentially that of $\mathcal{A}$, which can turn a distinguisher between the two games into an adversary against the RoR-CCA property of the RKEM. We have

$$|\Pr[T_3] - \Pr[T_4]| \leq 2 \cdot \text{Adv}_{\text{RKEM}}^{\text{RoR-CCA}}(\mathcal{B}_3).$$

Finally, in Game 4 we note that the first component of the ciphertext is completely random and independent of any message, and that the second part of the ciphertext is an encryption under a completely random key k^*. Hence, the adversary in Game 4 is essentially just an algorithm $\mathcal{B}_4$ which is attacking the IND-PA property of the DEM, i.e.

$$|\Pr[T_4] - 1/2| \leq \mathrm{Adv}_{\mathrm{DEM}}^{\mathrm{IND-PA}}(\mathcal{B}_4).$$

Putting the above equalities together we obtain the stated result

$$\begin{aligned}
\mathrm{Adv}_{\mathrm{PKE}}^{\mathrm{IND-CCA}}(\mathcal{A}) &= |\Pr[T_0] - 1/2| = |\Pr[T_1] - 1/2| \\
&= |(\Pr[T_1] - \Pr[T_2]) + (\Pr[T_2] - \Pr[T_3]) \\
&\quad + (\Pr[T_3] - \Pr[T_4]) + (\Pr[T_4] - 1/2)| \\
&\leq |\Pr[T_1] - \Pr[T_2]| + |\Pr[T_2] - \Pr[T_3]| \\
&\quad + |\Pr[T_3] - \Pr[T_4]| + |\Pr[T_4] - 1/2|| \\
&\leq 2\mathrm{Adv}_{\mathrm{RKEM}}^{\mathrm{IND-CCA}}(\mathcal{B}_1) + q_D \cdot \mathrm{Adv}_{\mathrm{DEM}}^{\mathrm{INT-CTXT}}(\mathcal{B}_2) \\
&\quad + 2\mathrm{Adv}_{\mathrm{RKEM}}^{\mathrm{RoR-CCA}}(\mathcal{B}_3) + \mathrm{Adv}_{\mathrm{DEM}}^{\mathrm{IND-PA}}(\mathcal{B}_4).
\end{aligned}$$

□

3.4   Constructions of RKEMs

A secure RKEM may be instantiated from an IND-CCA secure PKE in the obvious manner: use the PKE to encrypt the state bit, the κ-bit symmetric session key, and $PKE.msglen - \kappa - 1$ bits of message payload. It is easy to show that this construction is both IND-CCA and RoR-CCA secure, and that this is tightly related to the IND-CCA security of the underlying PKE [2]. If we implement this trivial scheme using RSA-OAEP, the total length of ciphertexts will be $|m| + 6\kappa + 3$; in the particular case of $\kappa = 128$ we save roughly 2400 bits in comparison with RSA-KEM.

However, the efficiency of our construction can be improved even further for a particular class of IND-CCA secure public-key encryption schemes. Let $c = \mathcal{E}(pk, m; r)$ denote a public key encryption algorithm taking a random string r as auxiliary input, with associated decryption function $m = \mathcal{D}(sk, c)$. We say that a public key algorithm is *randomness recovering*, if the decryption algorithm $\mathcal{D}(sk, c)$ can be modified so that it returns not only m but also the randomness used to construct c, i.e. we have that if $c = \mathcal{E}(pk, m; r)$ then $(m, r) = \mathcal{D}(sk, c)$. Such a scheme is said to be secure if it is IND-CCA secure with respect to the message m, and is OW-CPA secure with respect to the pair (m, r).

---

[2] In particular, the adversary is being given a challenge for the RKEM consisting of the encapsulation ψ^*, and a decapsulation oracle that computes $(k, m^{(0)}||v) = RKEM.Decap(sk, \psi)$, where $k|m^{(0)}|v = PKE.Dec(sk, c)$. In both the IND-CCA and ROR-CCA games the goal of the adversary is to determine some property of part of the "plaintext", either k or $m^{(0)}$. Whenever PKE is itself IND-CCA this is clearly not feasible.

There exist various practical public key encryption schemes that are securely randomness recovering, including RSA-OAEP and any scheme constructed using the Fujisaki-Okamoto transform [12]. In both of these constructions the IND-CCA security is standard, whilst the OW-CPA security with respect to the pair (m, r) follows from the OW-CPA security of the underlying primitive. We can use this to create a RKEM which incurs less overhead compared to the maximal message length of the underlying PKE.

Our IND-CCA and RoR-CCA secure RKEM is constructed as follows:

– *RKEM.Gen* is defined to be the key generation of the public key scheme, plus the specification of a hash function H. The parameter *RKEM.msglen* is a single bit less than the maximum message length of the public key scheme, and *RKEM.keylen* is the output length of H.
– *RKEM.Encap* takes a message m of length *RKEM.msglen*. It then computes $\psi = \mathcal{E}(pk, m; r)$ for some randomness r, $k = H(m||r)$ and returns (k, ψ).
– *RKEM.Decap* takes the encapsulation ψ and decrypts it using $\mathcal{D}(sk, \psi)$. If $\mathcal{D}(sk, \psi)$ returns $\perp$, then the decapsulation algorithm returns $\perp$ and halts. Otherwise the pair (m, r) is obtained and the algorithm proceeds to compute $k = H(m||r)$ and returns (m, k).

The RoR-CCA security of the above RKEM construction follows from the IND-CCA security of the underlying public key encryption scheme. The IND-CCA security follows, in the random oracle model, from the OW-CPA security of the underlying primitive with respect to the pair (m, r), using essentially the same proof of security as for standard RSA-KEM [16].

As mentioned in the introduction, by using RSA-OAEP in this construction one can obtain a public key encryption algorithm for messages of length m, which outputs ciphertexts of length $|m| + 5\kappa + 3$ bits for a given security parameter κ. This breaks down as follows: the RSA-OAEP encryption scheme (as defined in the ISO/IEC standard for public-key encryption [17]) has overhead of $2\,Hash.len + 2$ bits, where *Hash.len* is the length of the output from a cryptographic hash function, commonly taken to be 2κ bits[3]. Furthermore, a single state bit is used to keep track of the message length inside the RKEM. Finally, the usual method of constructing a DEM that is INT-CTXT requires a κ-bit message authentication code (MAC). In comparison with RSA-KEM using $\kappa = 128$, this scheme saves more than 2500 bits per message!

As we see, the above construction gives ciphertexts that are independent of the size of the RSA modulus used, being linear in the security parameter. Furthermore, we are able to extend the limited message space of the underlying RSA-based primitive "optimally", with only $\kappa + 1$ bits of overhead!

4   Tag-KEMs with Partial Ciphertext Recovery

After having discussed KEMs with partial message recovery it is natural to look at other formal models that exist for hybrid encryption schemes. In this section

---

[3] Although it may appear from the original OAEP paper that this should be only κ bits, it is necessary to use 2κ bits to deal with a more realistic attack model [2].

we consider Tag-KEMs [1], which have recently come into prominence as an attractive alternative to traditional KEMs. The main difference from regular KEMs is that the key encapsulation is used to preserve the integrity of the ciphertext, in addition to confidentiality of the symmetric key. The main result of [1] is that the use of Tag-KEMs makes it possible to create secure encryption schemes using a DEM that is only IND-PA. We define Tag-KEMs with partial ciphertext recovery (tag-RKEM) by direct extension of the previous definition in [1].

Definition 8 (Tag-KEM with Partial Ciphertext Recovery). *A Tag-KEM with partial ciphertext recovery (tag-RKEM) is defined as an ordered tuple of four algorithms.*

1. *A probabilistic algorithm TKEM.Gen used to generate public keys. It takes a security parameter 1^κ as input, and outputs a private/public keypair (sk, pk). The public key includes all information needed for users of the scheme, including parameters specifying the length of the symmetric keys used (TKEM.keylen) and the size of the internal message space (TKEM.msglen).*
2. *A probabilistic algorithm TKEM.Sym used to generate one-time symmetric keys. It takes a public key pk as input, and outputs a symmetric encryption key k and a string of internal state information s.*
3. *A probabilistic algorithm TKEM.Encap used to encapsulate the symmetric key and part of the ciphertext. It takes some state information s as well as a tag τ as input, and outputs a key encapsulation ψ together with a suffix string $\tau^{(1)}$ of $\tau = \tau^{(0)}||\tau^{(1)}$ (consisting of the part of τ that may not be recovered from ψ).*
4. *A deterministic algorithm TKEM.Decap used to recover the encapsulated key and ciphertext fragment from an encapsulation. It takes a private key sk, an encapsulation ψ and a partial tag $\tau^{(1)}$ as input, and outputs either a key k and the complete tag τ, or the unique error symbol $\perp$.*

A Tag-KEM is required to be sound in the obvious manner, i.e. for any τ and keypair (sk, pk) we have that $TKEM.Decap(sk, \psi, \tau^{(1)}) = (k, \tau)$ whenever $(\psi, \tau^{(1)}) = TKEM.Encap(\omega, \tau)$ and $(k, \omega) = TKEM.Sym(pk)$. We note that the definition collapses to that of [1] if we set $TKEM.msglen = 0$.

4.1 Security Definition for Tag-KEMs with Partial Ciphertext Recovery

A Tag-RKEM is said to be IND-CCA secure if there does not exist an adversary who can distinguish whether a given key k^* is the one embedded in an encapsulation ψ^*. The adversary has access to a decapsulation oracle, and is allowed to choose the tag τ^* used in the challenge encapsulation adaptively, but may not query the decapsulation oracle on the corresponding $(\psi^*, \tau^{(1)*})$. This corresponds to the notion of IND-CCA security for RKEMs used in the previous section, and is directly analogous to the security definitions for regular Tag-KEMs [1].

Definition 9 (IND-CCA Security of Tag-RKEM). *For a given security parameter* 1^κ, *the IND-CCA game played between the challenger and an adversary* $\mathcal{A} = (\mathcal{A}_1, \mathcal{A}_2, \mathcal{A}_3)$ *runs as follows.*

1. *The challenger generates a private / public keypair* $(sk, pk)=TKEM.Gen(1^\kappa)$.
2. *The adversary runs* $\mathcal{A}_1$ *on the input* pk. *During its execution,* $\mathcal{A}_1$ *may query a decapsulation oracle* $\mathcal{O}_{Decap}(\cdot, \cdot)$ *which takes an encapsulation* ψ *and a tag* $\tau^{(1)}$ *as input, and returns the result of computing* $TKEM.Decap(sk, \psi, \tau^{(1)})$. *The algorithm terminates by outputting some state information* s_1.
3. *The challenger computes* $(k_0, \omega) = TKEM.Sym(pk)$, *and samples another key* $k_1 \xleftarrow{R} \{0,1\}^{TKEM.keylen}$ *uniformly at random. He then selects a random bit* $b \xleftarrow{R} \{0,1\}$.
4. *The adversary runs* $\mathcal{A}_2$ *on the input* (s_1, k_b). *During its execution,* $\mathcal{A}_2$ *has access to the same oracle as before. The algorithm terminates by outputting some state information* s_2 *and a tag* τ^*.
5. *The challenger generates a challenge encapsulation*

$$(\psi^*, \tau^{(1)*}) = TKEM.Encap(s_1, \tau^*).$$

6. *The adversary runs* $\mathcal{A}_3$ *on the input* $(s_2, \psi^*, \tau^{(1)*})$. *During its execution,* $\mathcal{A}_3$ *has access to the same oracle as before, with the restriction that the challenge* $(\psi^*, \tau^{(1)*})$ *may not be queried. The algorithm terminates by outputting a guess* b' *of the value of* b.

We say that $\mathcal{A}$ *wins the game whenever the guess was correct, i.e.* $b = b'$. *The advantage of* $\mathcal{A}$ *is given as*

$$\mathrm{Adv}_{TKEM}^{IND-CCA}(\mathcal{A}) = \left| Pr[\mathcal{A} \ wins] - 1/2 \right|.$$

The security definition for Tag-RKEMs is versatile, in the sense that for a Tag-RKEM to be IND-CCA it must not only ensure that its symmetric keys are indistinguishable from random with respect to their encapsulations, but also enforce certain non-malleability conditions with respect to τ. In particular, since the adversary is able to submit decapsulation oracle queries adaptively on ψ and $\tau^{(1)}$, the decapsulation procedure must be non-malleable in the sense that oracle queries such as $(\psi^*, \tau^{(1)})$ or $(\psi, \tau^{(1)*})$ reveal no information about k^*.

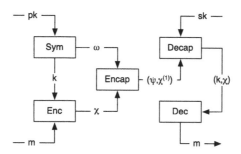

Fig. 2. Data flow in the Tag-KEM + DEM construction

4.2 Security of the Composition of an IND-CCA Secure Tag-RKEM and an IND-PA Secure DEM

Combining a Tag-RKEM with a DEM is done by using the key from $TKEM.Sym$ with $DEM.Enc$ to produce a ciphertext, and using the resulting ciphertext as the tag for $TKEM.Encap$. The overall data-flow is illustrated in Fig. 2.

Definition 10 (TKEM+DEM Construction). *Given a TKEM and a DEM where the keys output by the TKEM are of correct length for use with the DEM, we construct a hybrid PKE scheme as follows.*

- *The key generation algorithm PKE.Gen is implemented by using RKEM.Gen and appending any necessary information about the DEM to the public key.*
- *The encryption algorithm PKE.Enc is implemented as follows.*
  1. *Compute a symmetric key $(k, \omega) = TKEM.Sym(pk)$.*
  2. *Compute the symmetric ciphertext $\chi = DEM.Enc_k(m)$.*
  3. *Create a key encapsulation using the ciphertext χ as the tag, by computing $(\psi, \chi^{(1)}) = TKEM.Encap(\omega, \chi)$.*
  4. *Output the ciphertext $c = (\psi, \chi^{(1)})$.*
- *The decryption algorithm PKE.Dec is implemented as follows.*
  1. *Parse the ciphertext to obtain $(\psi, \chi^{(1)}) = c$.*
  2. *Recover k and χ from ψ and $\chi^{(1)}$ by computing $(k, \chi) = TKEM.Decap(sk, \psi, \chi^{(1)})$.*
  3. *If TKEM.Decap returned $\perp$, return $\perp$ and halt.*
  4. *Recover the original message by running $m = DEM.Dec_k(\chi)$.*
  5. *If DEM.Dec returned $\perp$, return $\perp$ and halt.*
  6. *Output m.*

The soundness of the above construction follows from the soundness of the individual tag-RKEM and DEM. We note that this construction embeds part of the symmetric *ciphertext* rather than plaintext in the encapsulation, which explains why we no longer require RoR-CCA security (with respect to the message). This fact simplifies security analysis a great deal.

Theorem 2 (Security of TKEM+DEM). *If the underlying Tag-KEM with partial ciphertext recovery is IND-CCA secure and the DEM is IND-PA secure, then the TKEM+DEM composition is secure.*

More precisely we have, that if there is an adversary $\mathcal{A}$ against the above hybrid public key scheme, then there are polynomial-time adversaries $\mathcal{B}_1, \mathcal{B}_2$ such that

$$\mathrm{Adv}_{\mathrm{PKE}}^{\mathrm{IND-CCA}}(\mathcal{A}) \leq 2 \cdot \mathrm{Adv}_{\mathrm{TKEM}}^{\mathrm{IND-CCA}}(\mathcal{B}_1) + \mathrm{Adv}_{\mathrm{DEM}}^{\mathrm{IND-PA}}(\mathcal{B}_2).$$

Proof. Let $\mathcal{A}$ denote our adversary against the hybrid PKE system, and let Game 0 be the standard IND-CCA game for a PKE. We prove the security by making a single modification to Game 0, which causes the maximal advantage of any $\mathcal{A}$ to be clearly bounded. We define T_0 and T_1 to be the event that $b = b'$ in Games 0 and 1.

Let Game 1 be the same as Game 0, except that the challenger creates the symmetric ciphertext using a key k^* picked uniformly at random, rather than the key output by $TKEM.Sym$. It is clear that there exists a machine $\mathcal{B}_1$ whose running time is essentially that of $\mathcal{A}$, which can turn a distinguisher between the two games into an adversary against the IND-CCA property of the TKEM. Hence we have that

$$|\Pr[T_0] - \Pr[T_1]| = 2 \cdot \mathrm{Adv}_{\mathrm{TKEM}}^{\mathrm{IND-CCA}}(\mathcal{B}_1).$$

However, in Game 1 the value of the challenge encapsulation ψ^* reveals *no* information about the key k^* used to create the ciphertext χ, since the symmetric key k^* was sampled independently of ω^*. Hence, the only information about the value of m_b available to $\mathcal{A}$ is the symmetric ciphertext fragment $\chi^{(1)*}$ (and plausibly the full symmetric ciphertext χ^*, depending on how the TKEM is constructed). Furthermore, the adversary is not able to make any adaptive decryption queries under k^*. Hence there exists a machine $\mathcal{B}_2$ whose running time is essentially that of $\mathcal{A}$, such that

$$|\Pr[T_1] - 1/2| = \mathrm{Adv}_{\mathrm{DEM}}^{\mathrm{IND-PA}}(\mathcal{B}_2).$$

Summarising, we obtain the stated result.

$$\mathrm{Adv}_{\mathrm{PKE}}^{\mathrm{IND-CCA}}(\mathcal{A}) \leq 2 \cdot \mathrm{Adv}_{\mathrm{TKEM}}^{\mathrm{IND-CCA}}(\mathcal{B}_1) + \mathrm{Adv}_{\mathrm{DEM}}^{\mathrm{IND-PA}}(\mathcal{B}_2).$$

$\square$

4.3   Constructions of Tag-KEMs with Partial Ciphertext Recovery

Having established that Tag-KEMs with partial ciphertext recovery are viable in principle, it remains to suggest a practical instantiation. We generalise a construction of Dent [11] to the tag-RKEM setting. This constructions makes use of an IND-CPA encryption scheme ($PKE.Gen, PKE.Enc, PKE.Dec$) and two hash functions H and KDF. We suppose that the message space of the encryption scheme is $PKE.msglen$ and we use the notation $PKE.Enc(m, pk; r)$ to denote applying the encryption algorithm to a message m with a public key pk using random coins r. We require two security properties of the encryption scheme: that it is γ-uniform and that it is partially one-way.

Definition 11 (γ-uniform). *An encryption scheme*

$$(PKE.Gen, PKE.Enc, PKE.Dec)$$

is γ-uniform if, for all possible messages m and ciphertexts c

$$Pr[PKE.Enc(m, pk) = c] \leq \gamma$$

where the probability is taken over the random coins of the encryption algorithm.

Definition 12 (POW-CPA). *An encryption scheme (PKE.Gen, PKE.Enc, PKE.Dec) is partially one way with respect to inputs of length $\ell \leq PKE.msglen$ if no probabilistic polynomial-time attacker $\mathcal{A} = (\mathcal{A}_1, \mathcal{A}_2)$ can win the following game with non-negligible probability:*

1. *The challenger generates a key pair $(pk, sk) = PKE.Gen(1^\kappa)$.*
2. *The attacker runs $\mathcal{A}_1$ on the input pk. $\mathcal{A}_1$ terminates by outputting a tag τ of length $PKE.msglen - \ell$ and some state information s,*
3. *The challenger randomly chooses a seed value α of length ℓ, sets $m = \alpha||\tau$ and computes $c^* = PKE.Enc(m, pk)$.*
4. *The attacker runs $\mathcal{A}_2$ on the input (c^*, s). $\mathcal{A}_2$ terminates by outputting a guess α' for α.*

The attacker wins if $\alpha' = \alpha$.

Note that if the encryption scheme is IND-CPA and ℓ is super-poly-logarithmic as a function of the security parameter, then the encryption scheme is POW-CPA secure.

We assume that we can split $PKE.msglen$ into two lengths $TKEM.keylen$ and $TKEM.msglen$ such that $PKE.msglen = TKEM.keylen + TKEM.msglen$. We construct the tag-RKEM as follows:

1. $TKEM.Gen$ runs $PKE.Gen$ to generate a public/private key pair, and appends the values of $TKEM.keylen$ and $TKEM.msglen$ to the public key.
2. $TKEM.Sym$ picks a random seed α of length $TKEM.keylen$ and derives a key $k = KDF(\alpha)$. The algorithm outputs the state information $s = (pk, \alpha)$ and the key k.
3. $TKEM.Encap$ runs in several steps:
   (a) Parse s as (pk, α).
   (b) Parse τ as $\tau^{(0)}||\tau^{(1)}$ where $\tau^{(0)}$ is $TKEM.msglen$-bits long if τ contains more than $TKEM.msglen$ bits and $\tau^{(0)} = \tau$ if τ is less than $TKEM.msglen$ bits in length.
   (c) Compute $m = \alpha||\tau^{(0)}$.
   (d) Compute $r = H(\alpha, \tau)$.
   (e) Compute $\psi = PKE.Enc(m, pk; r)$.
   (f) Output ψ.
4. $TKEM.Decap$ runs in several steps:
   (a) Compute $m = PKE.Dec(\psi, sk)$.
   (b) Parse m as $\alpha||\tau^{(0)}$ where α is $TKEM.keylen$ bits in length.
   (c) If $\tau^{(0)}$ is less than $TKEM.msglen$ bits in length and $\tau^{(1)} \neq \emptyset$, then output $\perp$.
   (d) Compute $\tau = \tau^{(0)}||\tau^{(1)}$.
   (e) Compute $r = H(\alpha||\tau)$.
   (f) If $\psi \neq PKE.Enc(m, pk; r)$ then output $\perp$.
   (g) Otherwise output $k = KDF(\alpha)$.

Theorem 3. *Suppose there exists an attacker $\mathcal{A}$ against the tag-RKEM in the random oracle model that makes at most q_D decapsulation oracle queries, q_K queries to the KDF-oracle, q_H queries to the H-oracle, and breaks the IND-CCA security with advantage $Adv_{TKEM}^{IND-CCA}$. Then there exists an attacker $\mathcal{B}$ against the POW-CPA security of the public key encryption scheme (with respect to the length TKEM.keylen) with advantage*

$$Adv_{PKE}^{POW-CPA} \geq \frac{1}{q_D + q_K + q_H}\left\{ Adv_{TKEM}^{IND-CCA} - q_D/2^{TKEM.keylen} - q_D\gamma\right\}$$

Corollary 1. *If TKEM.keylen grows super-poly-logarithmically, γ is negligible and (PKE.Gen, PKE.Enc, PKE.Dec) is partially one-way with respect to the length TKEM.keylen, then the tag-KEM construction is secure.*

If we instantiate this construction using the RSA-OAEP encryption scheme and a passively secure DEM, then the result construction will encrypt a message of length n using $n + 4\kappa + 2$ bits. This saves $\kappa + 1$ more bits than the RKEM construction given in Section 3.

Acknowledgements

This work was partially supported by the European Commission through the IST Programme under Contract IST-2002-507932 ECRYPT.

The problem of studying the efficiency and provable security of partial-message-recovery systems was suggested to the authors by Daniel J. Bernstein. Bernstein, who had incorporated a partial-message-recovery system (similar to that of Shoup [16]) into the benchmarking tools for the eBATS project [8].

The authors would also like to thank Mihir Bellare for explaining some subtle points related to the RSA-OAEP scheme and to James Birkett for reviewing a draft of the paper.

References

1. Abe, A., Gennaro, R., Kurosawa, K., Shoup, V.: Tag-KEM/DEM: A new framework for hybrid encryption and a new analysis of Kurosawa-Desmedt KEM. In: Cramer, R.J.F. (ed.) EUROCRYPT 2005. LNCS, vol. 3494, pp. 128–146. Springer, Heidelberg (2005)
2. Bellare, M.: Personal correspondence (2007)
3. Bellare, M., Desai, A., Jokipii, E., Rogaway, P.: A concrete security treatment of symmetric encryption. In: Proceedings of the 38th FOCS, pp. 394–403. IEEE, Los Alamitos (1997)
4. Bellare, M., Namprempre, C.: Authenticated encryption: Relations among notions and analysis of the generic composition paradigm. In: Okamoto, T. (ed.) ASIACRYPT 2000. LNCS, vol. 1976, pp. 531–545. Springer, Heidelberg (2000)
5. Bellare, M., Rogaway, P.: Random oracles are practical: A paradigm for designing efficient protocols. In: ACM Conference on Computer and Communications Security, pp. 62–73 (1993)

6. Bellare, M., Rogaway, P.: Optimal asymmetric encryption. In: De Santis, A. (ed.) EUROCRYPT 1994. LNCS, vol. 950, pp. 92–111. Springer, Heidelberg (1995)
7. Bellare, M., Rogaway, P.: The exact security of digital signatures — how to sign with RSA and Rabin. In: Maurer, U.M. (ed.) EUROCRYPT 1996. LNCS, vol. 1070, pp. 399–416. Springer, Heidelberg (1996)
8. Bernstein, D.J., Lange, T.: eBATS. ECRYPT Benchmarking of Asymmetric Systems (2007), http://www.ecrypt.eu.org/ebats/
9. Boneh, D., Lynn, B., Shacham, H.: Short signatures from the Weil pairing. In: Boyd, C. (ed.) ASIACRYPT 2001. LNCS, vol. 2248, pp. 514–532. Springer, Heidelberg (2001)
10. Cramer, R., Shoup, V.: Design and analysis of practical public-key encryption schemes secure against adaptive chosen ciphertext attack. SIAM Journal on Computing 33(1), 167–226 (2004)
11. Dent, A.W.: A designer's guide to KEMs. In: Paterson, K.G. (ed.) Cryptography and Coding. LNCS, vol. 2898, pp. 133–151. Springer, Heidelberg (2003)
12. Fujisaki, E., Okamoto, T.: Secure integration of asymmetric and symmetric encryption schemes. In: Wiener, M.J. (ed.) CRYPTO 1999. LNCS, vol. 1666, pp. 537–554. Springer, Heidelberg (1999)
13. Hofheinz, D., Mueller-Quade, J., Steinwandt, R.: On modeling IND-CCA security in cryptographic protocols. Cryptology ePrint Archive, Report, 2003 /024 (2003), http://eprint.iacr.org/
14. National Institute of Standards and Technology. Recommendation for key management - part 1: General. Technical Report NIST Special Publication 800-57, National Institute of Standards and Technology (2006)
15. Pintsov, L.A., Vanstone, S.A.: Postal revenue collection in the digital age. In: Frankel, Y. (ed.) FC 2000. LNCS, vol. 1962, pp. 105–120. Springer, Heidelberg (2001)
16. Shoup, V.: A proposal for an ISO standard for public key encryption. Cryptology ePrint Archive, Report, 2001/112 (2001), http://eprint.iacr.org/2001/112/
17. Shoup, V.: ISO/IEC FCD,18033-2 – Information technology – Security techniques – Encryption algorithms – Part 2: Asymmetric ciphers. Technical report, International Organization for Standardization (2004) http://shoup.net/iso/std6.pdf
18. Wander, A.S., Gura, N., Eberle, H., Gupta, V., Shantz, S.C.: Energy analysis of public-key cryptography for wireless sensor networks. In: PERCOM 2005: Proceedings of the Third IEEE International Conference on Pervasive Computing and Communications, pp. 324–328. IEEE Computer Society, Los Alamitos (2005)

A   KEM-DEM Framework

A.1   Definitions

Definition 13 (KEM). *A key encapsulation mechanism is defined by KEM =* (*KEM.Gen, KEM.Encap, KEM.Decap*) *as an ordered tuple of three algorithms.*

1. *A probabilistic key generation algorithm KEM.Gen. It takes as input a security parameter* 1^κ, *and outputs a private/public keypair* (sk, pk). *As part of the public key there is a parameter KEM.keylen that specifies the length of the symmetric keys used by the DEM.*

2. *A probabilistic key encapsulation algorithm PKE.Encap. It takes as input a public key pk, and outputs a symmetric key k of length KEM.keylen, and an encapsulation ψ.*
3. *A deterministic decapsulation algorithm PKE.Decap. It takes as input a private key sk and an encapsulation ψ and outputs either a key k or the unique error symbol $\perp$.*

The KEM is sound if for almost all valid keypairs (sk, pk), whenever (k, ψ) was the output of PKE.Encap(pk), we have $k = PKE.Decap(sk, \psi)$.

Definition 14 (DEM). *We define a data encapsulation mechanism DEM = (DEM.Enc, DEM.Dec) as an ordered pair of algorithms.*

1. *A deterministic encryption algorithm DEM.Enc. It takes as input a message m and a symmetric key k of a specified length DEM.keylen, and outputs a ciphertext χ.*
2. *A deterministic decryption algorithm DEM.Dec. It takes as input a ciphertext χ and a symmetric key k of specified length, and outputs either a message m or the unique error symbol $\perp$.*

The DEM is sound as long as $m = DEM.Dec_k\big(DEM.Enc_k(m)\big)$ holds.

We assume that a DEM can take inputs of arbitrary length messages, thus the fact that the DEM can take a message of arbitrary length implies that any resulting hybrid encryption scheme can also take arbitrary length messages.

Definition 15 (KEM+DEM Construction). *Given a KEM and a DEM, where the keys output by the KEM are of correct length for use with the DEM, i.e. DEM.keylen = KEM.keylen, we construct a hybrid PKE scheme as follows.*

- *The key generation algorithm PKE.Gen is implemented using KEM.Gen.*
- *The encryption algorithm PKE.Enc is implemented as follows.*
    1. *Compute a key/encapsulation pair $(k, \psi) = KEM.Encap(pk)$.*
    2. *Encrypt the message to obtain a ciphertext $\chi = DEM.Enc_k(m)$.*
    3. *Output the ciphertext $c = (\psi, \chi)$.*
- *The decryption algorithm PKE.Dec is implemented as follows.*
    1. *Parse the ciphertext to obtain $(\psi, \chi) = c$.*
    2. *Compute the symmetric key $k = KEM.Decap(sk, \psi)$.*
    3. *If $k = \perp$, return $\perp$ and halt.*
    4. *Decrypt the message $m = DEM.Dec_k(\chi)$.*
    5. *If $m = \perp$, return $\perp$ and halt.*
    6. *Output m.*

The soundness of the KEM+DEM construction follows from the soundness of the individual KEM and DEM. A hybrid PKE scheme created from an IND-CCA secure KEM and an IND-CCA secure DEM is itself secure [10].

A.2   Security Models

For hybrid components such as KEMs and DEMs, we may adapt the indistinguishability criterion for public key schemes for each. For a KEM, the fundamental requirement is to ensure that the adversary does not learn anything about a key from its encapsulation. Indeed, given a key and an encapsulation the adversary should not be able to tell whether a given key is the one contained in an encapsulation. This is (slightly confusingly, since it is really a question of whether the key presented is "real" or "random") usually referred to as IND-CCA security.

Definition 16 (IND-CCA Game for KEM). *The IND-CCA game for a given key encapsulation mechanism KEM is played between the challenger and an adversary* $\mathcal{A} = (\mathcal{A}_1, \mathcal{A}_2)$. *For a specified security parameter* 1^κ, *the game proceeds as follows.*

1. *The challenger generates a private/public keypair* $(sk, pk) = KEM.Gen(1^\kappa)$.
2. *The adversary runs* $\mathcal{A}_1$ *on the input* pk. *During its execution,* $\mathcal{A}_1$ *may query a decapsulation oracle* $\mathcal{O}_D$ *that takes an encapsulation* ψ *as input, and outputs* $KEM.Decap(sk, \psi)$. *The algorithm terminates by outputting some state information* s.
3. *The challenger generates a real key and its encapsulation, by calling* $(k_0, \psi^*) = KEM.Encap(pk)$, *as well as a random key* k_1 *drawn uniformly from the keyspace of the KEM. It also picks a random bit* $b \xleftarrow{R} \{0, 1\}$.
4. *The adversary runs* $\mathcal{A}_2$ *on the input* (k_b, ψ^*, s). *During its execution,* $\mathcal{A}_2$ *has access to the decapsulation oracle as before, but it may not ask for the decapsulation of* ψ^*. *The algorithm terminates by outputting a guess* b' *for the value of* b.

We say that $\mathcal{A}$ *wins the IND-CCA game whenever* $b = b'$. *The advantage of* $\mathcal{A}$ *is the probability*

$$\text{Adv}_{KEM}^{IND-CCA}(\mathcal{A}) = \left| Pr[\mathcal{A} \text{ wins }] - 1/2 \right|.$$

For DEMs, we give two security notions, the first of which is an adaption of the above notion of IND-security with respect to the input message.

Definition 17 (IND-PA Game for DEM). *The IND-PA game for a given data encapsulation mechanism DEM is played between the challenger and an adversary* $\mathcal{A} = (\mathcal{A}_1, \mathcal{A}_2)$. *For a specified security parameter* 1^κ, *the game proceeds as follows.*

1. *The challenger generates a random symmetric key* k.
2. *The adversary runs* $\mathcal{A}_1$ *on the input* 1^κ. *The algorithm terminates by outputting two messages* m_0 *and* m_1 *of equal length, and some state information* s.
3. *The challenger generates a random bit* $b \xleftarrow{R} \{0, 1\}$ *and encrypts the plaintext* m_b, *by calling* $\chi^* = DEM.Enc_k(m_b)$.

4. *The adversary runs $\mathcal{A}_2$ on the input (χ^*, s). The algorithm terminates by outputting a guess b' for the value of b.*

We say that $\mathcal{A}$ wins the IND-PA game whenever $b = b'$. The advantage of $\mathcal{A}$ is the probability

$$\mathrm{Adv}_{\mathrm{DEM}}^{\mathrm{IND-PA}}(\mathcal{A}) = \left| Pr[\mathcal{A} \ wins] - 1/2 \right|.$$

Note, that a stronger notion of security exists which is IND-CCA, in this notion we give the adversary in both stages access to a decryption oracle with respect to the challengers key, subject to the constraint that in the second stage the adversary may not call the decryption oracle on the challenge ciphertext.

The other security notion we require is ciphertext integrity or INT-CTXT. This was first defined in [4], however we will only require a one-time security notion.

Definition 18 (INT-CTXT Game for DEM). *The INT-CTXT game for a given data encapsulation mechanism DEM is played between the challenger and an adversary $\mathcal{A} = (\mathcal{A}_1, \mathcal{A}_2)$. For a specified security parameter 1^κ, the game proceeds as follows.*

1. *The challenger generates a random symmetric key k.*
2. *The adversary runs $\mathcal{A}_1$ on the input 1^κ. During its execution, $\mathcal{A}_1$ may query a decryption oracle $\mathcal{O}_D$ with respect to the key k; that takes a ciphertext χ as input, and outputs $DEM.Dec_k(\chi)$. The algorithm terminates by outputting a single messages m, and some state information s.*
3. *The challenger encrypts the plaintext m, by calling $\chi^* = DEM.Enc_k(m)$.*
4. *The adversary runs $\mathcal{A}_2$ on the input (χ^*, s). As before the adversary has access to it decryption oracle $\mathcal{O}_D$, however it may not call its oracle on the target ciphertext χ^*. The algorithm terminates by outputting a ciphertext $\chi' \neq \chi^*$.*

The adversary wins the game if it the ciphertext χ' is a valid ciphertext, i.e. it can be decrypted by the decryption algorithm. The advantage of $\mathcal{A}$ is the probability

$$\mathrm{Adv}_{\mathrm{DEM}}^{\mathrm{INT-CTXT}}(\mathcal{A}) = Pr[\mathcal{A} \ wins].$$

It is proved in [4] that in the many-time setting, a scheme which is both IND-PA and INT-CTXT will be IND-CCA as well. It is straightforward to verify that this property also holds for one-time encryption schemes.

We note that a symmetric cipher which is IND-PA (such as a secure block cipher in CBC mode with fixed IV) can be made INT-CTXT by adding a secure Message Authentication Code using the basic Encrypt-then-MAC construction. This is also the "standard" way of producing an IND-CCA symmetric cipher [10].

B   Proof of Theorem 3

In this section we prove the the tag-RKEM construction given in Section 4.3 is secure. In other words, we prove the following theorem:

Theorem 3. *Suppose there exists an attacker $\mathcal{A}$ against the tag-RKEM in the random oracle model that makes at most q_D decapsulation oracle queries, q_K queries to the KDF-oracle, q_H queries to the H-oracle, and breaks the IND-CCA security with advantage $Adv_{TKEM}^{IND-CCA}$. Then there exists an attacker $\mathcal{B}$ against the POW-CPA security of the public key encryption scheme (with respect to the length TKEM.keylen) with advantage*

$$Adv_{PKE}^{POW-CPA} \geq \frac{1}{q_D + q_K + q_H} \left\{ Adv_{TKEM}^{IND-CCA} - q_D/2^{TKEM.keylen} - q_D\gamma \right\}$$

Proof. The proof is very similar to the construction of Dent [11]. We model both the hash function H and the key derivation function KDF as random oracles. The tag-RKEM attacker can then gain no advantage in determining whether the challenge key k^* is correct key or not unless the attacker queries the KDF oracle on the challenge seed value α^*. This can be done either implicitly (by making a valid decapsulation oracle query that uses the value α^* as its seed) or explicitly by querying the KDF-oracle directly. We show that it is computationally infeasible for the attacker to make a valid decapsulation oracle query using the seed α^* without querying the H-oracle on some message $\alpha^*||\tau$. Hence, the only way that the attacker can gain a non-negligible advantage is to query one of the random oracles with a value involving α^*. We can therefore recover α^*, and solve the POW-CPA problem by guessing which oracle query contains α^*.

We use non-programmable random oracles. These random oracles are simulated using two lists KDFLIST and HLIST. In both cases, when a query is made to the random oracle on an input x, then oracle searches the relevant list for a record (x, y). If such a record exists, then the oracle outputs y; otherwise, the oracle generates a random value y of the appropriate size, adds (x, y) to the appropriate list, and outputs y.

Again, we use a game-hopping proof. Let T_i be the event that the tag-RKEM attacker wins in Game i. Let Game 0 be the normal IND-CCA attack game for $\mathcal{A}$. Hence,

$$Adv_{TKEM}^{IND-CCA} = |Pr[T_0] - 1/2| .$$

Let Game 1 be identical to Game 0 except that the attacker is immediately deemed to have lost the game if, on conclusion of the game, it turns out that $\mathcal{A}$ queried the decapsulation oracle on the challenge ciphertext ψ^* *before* the challenge ciphertext was issued. (We are forced to do this as the simulated decapsulation oracle that we will define would incorrectly decapsulate this ciphertext as $\perp$). Let E_1 be the event that $\mathcal{A}_1$ submits ψ^* to the decapsulation oracle. Since $\mathcal{A}_1$ has no information about ψ^* at this point, this would require $\mathcal{A}_1$ to guess the value of ψ^*, which implicitly means that $\mathcal{A}_1$ has guessed the value of α^* as ψ^* uniquely defines the value of α^*. Hence, $Pr[E_1] \leq q_D/2^{TKEM.keylen}$ and we obtain the following relation:

$$|Pr[T_0] - Pr[T_1]| \leq Pr[E_1] \leq q_D/2^{TKEM.keylen} .$$

Game 2 will be identical to Game 1 except that we change the decapsulation oracle to the following simulated decapsulation oracle:

1. The oracle takes as input an encapsulation ψ and a tag τ. It parses τ as $\tau^{(0)}||\tau^{(1)}$.
2. The oracle searches HLIST for an entry with input values (α, τ) and output value r such that $\psi = PKE.Enc(\alpha||\tau^{(0)}, pk; r)$.
3. If such a record exists, then the oracle outputs $KDF(r)$; otherwise, it output $\perp$.

Game 2 functions identically to Game 1 unless the attacker makes a decapsulation oracle query which is valid in Game 1 but declared invalid in Game 2. Let E_2 be the event that this occurs. The only way that E_2 can occur is if $\mathcal{A}$ submits a ciphertext ψ and tag τ for decapsulation such that $H(\alpha, \tau) = r$ and $\psi = PKE.Enc(\alpha||\tau^{(0)}, pk; r)$ but $\mathcal{A}$ has not submitted $\alpha||\tau$ to the H-oracle. This means that, in the view of the attacker, the value r is completely random. Hence, the probability that ψ is an encryption of $\alpha||\tau^{(0)}$ is bounded by γ and we obtain the following relation:

$$|Pr[T_1] - Pr[T_2]| \leq q_D \gamma.$$

We may now construct an attacker $\mathcal{B} = (\mathcal{B}_1, \mathcal{B}_2)$ against the POW-CPA property of the encryption scheme and relate $\mathcal{B}$'s success probability to $\mathcal{A}$'s advantage in Game 2. $\mathcal{B}_1$ takes pk as input and runs as follows:

1. Run $\mathcal{A}_1$ on the input pk. Simulates the oracles to which $\mathcal{A}$ has access as in Game 2. $\mathcal{A}_1$ terminates by outputting some state information s_1.
2. Generates a random key k^*.
3. Run $\mathcal{A}_2$ on the input (k^*, s_1). Simulate the oracles to which $\mathcal{A}$ has access as in Game 2. $\mathcal{A}_2$ terminates by outputting some state information s_2 and a tag τ^*.
4. Parse τ^* as $\tau^{*(0)}||\tau^{*(1)}$.
5. Output the tag $\tau^{*(0)}$.

The challenger will now pick a random seed α^* and form the challenge ciphertext ψ^* by encrypting $\alpha^*||\tau^{*(0)}$. $\mathcal{B}_2$ takes ψ^* as input and runs as follows:

1. If $\mathcal{A}_1$ or $\mathcal{A}_2$ made a decapsulation oracle query on the value ψ^*, then output $\perp$ and halt.
2. Run A_3 on the input ψ^* and s_2. Simulate the oracles to which $\mathcal{A}$ has access as in Game 2. $\mathcal{A}_3$ terminates by outputting a bit b'.
3. Randomly choose an entry from the total number of records in KDFLIST and HLIST, and extract the value α from this query. Output α as the guess for α^*.

$\mathcal{B}$ completely simulates the oracles to which $\mathcal{A}$ has access up until the point that $\mathcal{A}$ makes a KDF-oracle query on α^* or an H-oracle query on $\alpha^*||\tau$. If $\mathcal{A}$ does not make such a query, then its advantage is 0; hence, the probability that $\mathcal{A}$ makes such a query is equal to $\mathcal{A}$'s advantage in Game 2.

Therefore, the probability that $\mathcal{B}$ correctly solves the POW-CPA problem is equal to the probability that $\mathcal{A}$ makes a KDF-oracle query on α^* or an H-oracle query on $\alpha^*||\tau$, and $\mathcal{B}$ correctly guesses a record that contains a reference to α^*.

Since there are at most $q_D + q_K + q_H$ records in KDFLIST and HLIST, this will happen with probability at least $1/q_D + q_K + q_H$. Hence,

$$
\begin{aligned}
Adv_{PKE}^{POW-CPA} &\geq \frac{1}{q_D + q_K + q_H} \cdot |Pr[T_2] - 1/2| \\
&\geq \frac{1}{q_D + q_K + q_H} \left\{ Adv_{TKEM}^{IND-CCA} - q_D/2^{TKEM.keylen} - q_D\gamma \right\}
\end{aligned}
$$

This proves the theorem. $\square$

Randomness Reuse:
Extensions and Improvements

Manuel Barbosa[1] and Pooya Farshim[2]

[1] Departamento de Informática, Universidade do Minho,
Campus de Gualtar, 4710-057 Braga, Portugal
`mbb@di.uminho.pt`
[2] Department of Computer Science, University of Bristol,
Merchant Venturers Building, Woodland Road, Bristol BS8 1UB, United Kingdom
`farshim@cs.bris.ac.uk`

Abstract. We extend the generic framework of reproducibility for reuse of randomness in multi-recipient encryption schemes as proposed by Bellare et al. (PKC 2003). A new notion of *weak reproducibility* captures not only encryption schemes which are (fully) reproducible under the criteria given in the previous work, but also a class of efficient schemes which can only be used in the single message setting. In particular, we are able to capture the single message schemes suggested by Kurosawa (PKC 2002), which are more efficient than the direct adaptation of the multiple message schemes studied by Bellare et al. Our study of randomness reuse in key encapsulation mechanisms provides an additional argument for the relevance of these results: by taking advantage of our weak reproducibility notion, we are able to generalise and improve multi-recipient KEM constructions found in literature. We also propose an efficient multi-recipient KEM provably secure in the standard model and conclude the paper by proposing a notion of *direct reproducibility* which enables tighter security reductions.

Keywords: Randomness Reuse. Multi-Recipient. Hybrid Encryption.

1 Introduction

Generating randomness for cryptographic applications is a costly and security-critical operation. It is often assumed in security analysis that parameters are sampled from a perfect uniform distribution and are handled securely. Moreover, various operations performed by a cryptographic algorithm depend on the random coins used within the algorithm. These operations, such as a group exponentiation, can be quite costly and prevent the use of the scheme in constrained devices. Therefore, minimising the amount of fresh randomness required in cryptographic algorithms is important for their overall efficiency and security.

One approach to minimise this problem is to reuse randomness across multiple instantiations of cryptographic algorithms, namely in the context of batch operations where (possibly different) messages are encrypted to multiple recipients.

S.D. Galbraith (Eds.): Cryptography and Coding 2007, LNCS 4887, pp. 257–276, 2007.
© Springer-Verlag Berlin Heidelberg 2007

This avenue must be pursued with caution, since randomness reuse may hinder the security of cryptographic schemes. However, when possible, this technique allows for significant savings in processing load and bandwidth, since partial results (and even ciphertext elements) can be shared between multiple instances of a cryptographic algorithm.

Examples of this method are the multi-recipient encryption schemes proposed by Kurosawa [11], the mKEM scheme by Smart in [12], the certificateless encryption scheme in [3] where randomness is shared between the identity-based and public-key components, and the cryptographic workflow scheme in [2].

Bellare et al. [4], building on the work of Kurosawa [11], systematically study the problem of reusing randomness. The authors examine multi-recipient encryption, and consider the particular case of constructing such schemes by running multiple instances of a public-key encryption (PKE) scheme, whilst sharing randomness across them. An interesting result in this work is a general method for identifying PKE schemes that are secure when used in this scenario. Schemes which satisfy the so-called *reproducibility test* are guaranteed to permit a hybrid argument proof strategy which is generally captured in a *reproducibility theorem*. Bellare et al. later leveraged on these results to propose a stateful encryption framework [5] which enables more efficient encryption operations.

In this paper we extend the above theoretical framework supporting the reuse of randomness to construct multi-recipient encryption schemes. The main contribution of this paper is a more permissive test that permits constructing a wider class of efficient *single message* multi-recipient schemes. Of particular interest are the optimised modified versions of the ElGamal and Cramer-Shoup multi-recipient encryption schemes briefly mentioned by Kurosawa in the final section of [11]. We show that these schemes do not fit the randomness reuse framework originally proposed by Bellare et al. and propose extensions to the original definitions which capture these as well as other similar schemes. The technique that we employ to prove the main Theorem (Theorem 1) deviates from that of [4] and may be of independent interest in other contexts.

We then turn our attention to the KEM/DEM paradigm and focus on key encapsulation mechanisms [7]. Adaptation of the results in [4] is straightforward if one focuses on multi-recipient KEMs generating independent keys for each recipient. The interesting case arises when one considers single key multi-recipient KEMs. To construct these schemes efficiently by reusing randomness, we define the notion of *public key independent* KEM[1]. However, we find that if such a KEM satisfies an appropriate modification of the reproducibility test of Bellare et al. it cannot be secure. To compensate for this negative result, we propose an alternative generic construction of efficient single key multi-recipient KEMs based on weakly secure and weakly reproducible PKEs. We also present a concrete efficient construction, which is secure in the standard model.

The paper is structured as follows. In Section 2 we define what we mean by secure multi-recipient PKEs and full reproducibility, and go on to define

---

[1] This closely related to the notion of *partitioned* identity-based KEMs independently proposed by Abe et al. in [1].

weak reproducibly. Concrete schemes are analysed in Section 3. In Section 4 we examine extensions of the previous results to KEMs. Finally, in Section 5 we propose a new approach to the reproducibility generalisation that captures tighter security reductions, discuss the results we obtained using this strategy, and conclude with some associated open problems.

2   A New Notion of Reproducibility

2.1   Multi-recipient PKEs

An n-Multi-Recipient PKE (n-MR-PKE) [4] is defined similarly to a standard PKE scheme, with two exceptions: (1) The key generation algorithm is parameterised with a domain parameter I which ensures compatibility between users' key pairs and various spaces[2]. We denote the randomness, message and ciphertext spaces by the letters $\mathcal{R}$, $\mathcal{M}$ and $\mathcal{C}$ respectively; and (2) The encryption algorithm takes a list of n message/public key tuples and outputs a list of n ciphertexts.

Given a PKE scheme we can build the *associated* n-MR-PKE as follows. The key generation and decryption algorithms are identical to the underlying PKE (which we call the *base* PKE). Encryption is defined naturally by running multiple parallel instances of the base PKE encryption algorithm. If the randomness tapes in all instantiations are constant, the resulting n-MR-PKE scheme is called *randomness reusing*. In case there are common parameters that may be shared by all public keys to improve overall efficiency[3], these are included in the domain parameter I. The formal security model for an n-MR-PKE, as defined in [4], considers the possibility of insider attacks by allowing the adversary to corrupt some of the users by maliciously choosing their public keys. This ensures that security is still in place between the legitimate recipients, or in other words, that there is no "cross-talk" between the ciphertexts intended for different recipients.

In this work we are interested in a special case of n-MR-PKEs where the same message is sent to all recipients. We refer to this special case as *single message* (n-SM-PKE for simplicity), and note that such a scheme can also take advantage of randomness reuse. This specific case is a recurring use-case of n-MR-PKEs in practice, and one could ask if the single message restriction makes it any easier to construct n-MR-PKE schemes. More precisely, is there a wider range of schemes that can be used to construct efficient n-SM-PKEs through randomness reuse?

Below is the simplified security model for n-SM-PKEs. There is an important difference to the n-MR-PKE model: the adversary is no longer able to corrupt users. The reason for this is that, since all recipients will be getting the same message, there is no need to enforce security across the individual ciphertexts. We will also see in Section 4 that this weaker model is particularly relevant in

---

[2] This parameter is generated once for all users and henceforth, unless specifically stated otherwise, we leave its generation as implicit to simplify notation.

[3] For example in a Diffie–Hellman based scheme, this might include a domain modulus and generator which all parties use to create key pairs.

the hybrid encryption scenario. Throughout the game, the adversary also has access to $\mathcal{O}_1$ and $\mathcal{O}_2$, which denote a set of oracles, as follows:

- If atk = CPA then $\mathcal{O}_1 = \mathcal{O}_2 = $ NULL;
- If atk = CCA[4] then $\mathcal{O}_1$ is a set of decryption oracles one for each PK_i and $\mathcal{O}_2$ is same as $\mathcal{O}_1$ except that no component C_i^* of the challenge ciphertext C^* can be submitted to an oracle corresponding to PK_i.

IND-atk
1. For $i = 1, \ldots, n$
   $(SK_i, PK_i) \leftarrow \mathbb{G}_{n-SM-PKE}(I)$
2. $(M_0, M_1, s) \leftarrow A_1^{\mathcal{O}_1}(PK_1, \ldots, PK_n)$
3. $b \leftarrow \{0, 1\}$
4. $C^* \leftarrow \mathbb{E}_{n-SM-PKE}(M_b, (PK_i)_{i=1}^n)$
5. $b' \leftarrow A_2^{\mathcal{O}_2}(C^*, s)$

$\mathrm{Adv}_{n-SM-PKE}^{IND-atk}(A) := |2\Pr[b' = b] - 1|.$

2.2 Weak Reproducibility for PKEs

Definition 1. *A PKE scheme is* fully reproducible *if there exists a probabilistic polynomial time (PPT) algorithm R such that the following experiment returns 1 with probability 1.*

1. $(PK, SK), (PK', SK') \leftarrow \mathbb{G}_{PKE}(I)$
2. $r \leftarrow \mathcal{R}_{PKE}(I); M, M' \leftarrow \mathcal{M}_{PKE}(I)$
3. $C \leftarrow \mathbb{E}_{PKE}(M, PK; r)$
4. If $R(PK, C, M', PK', SK') = \mathbb{E}_{PKE}(M', PK'; r)$ return 1, else return 0

It is shown in [4] that an IND-atk secure PKE scheme satisfying the above definition can be used to construct an efficient IND-atk secure n-MR-PKE, by reusing randomness across n PKE instances. This result is interesting in itself, as it constitutes a generalisation of a proof strategy which can be repeated, almost without change, for all schemes satisfying the reproducibility test. This is a hybrid argument where an n-MR-PKE attacker is used to construct an attacker against the base scheme. The reproducibility algorithm generalises the functionality required to extend a challenge in the single-user PKE security game, to construct a complete challenge for the n-MR-PKE security game.

The SM security model proposed in the previous section is somewhat simpler than the original model in [4], so it is conceivable that a wider range of PKE schemes can be used to construct secure n-SM-PKEs, namely efficient randomness reusing ones. Hence, we are interested in defining a less restrictive version of reproducibility that permits determining whether a PKE scheme can be safely used in the single message scenario, even if it does not satisfy the full reproducibility test above. The following definition achieves this.

---

[4] In this paper we use IND-CCA to denote a fully adaptive chosen ciphertext attack sometimes denoted by IND-CCA2.

Definition 2. *A PKE scheme is* weakly reproducible *(wREP) if there exists a PPT algorithm R such that the following experiment returns 1 with probability 1.*

1. $(SK_1, PK_1), (SK_2, PK_2), (SK_3, PK_3) \leftarrow \mathbb{G}_{PKE}(I)$
2. $r \leftarrow \mathcal{R}_{PKE}(I); M, M' \leftarrow \mathcal{M}_{PKE}(I)$
3. $C_1 \leftarrow \mathbb{E}_{PKE}(M, PK_1; r); C_2 \leftarrow \mathbb{E}_{PKE}(M, PK_2; r)$
4. *If* $R(PK_1, C_1, M, PK_2, SK_2) \neq C_2$ *return* 0
5. *If* $R(PK_1, C_1, M', PK_3, SK_3) \neq R(PK_2, C_2, M', PK_3, SK_3)$ *return* 0, *else return* 1

Similarly to the original REP definition, the wREP definition follows from the generalisation of the hybrid argument which allows reducing the security of a randomness reusing n-SM-PKE to that of its base scheme. The intuition behind the definition is as follows. We are dealing which single message schemes. Therefore we only require correct reproduction when the two messages are the same. When the messages are different, we relax the definition and require only that R is source-PK independent (condition 5). This property is easy to check.

To see why more schemes might satisfy this definition, note that R is not even required to produce a valid ciphertext when the messages are different. In Section 3 we analyse specific PKE schemes and give a formal separation argument which establishes that the wREP definition is meaningful: there are schemes which satisfy this definition and which are not fully reproducible. Conversely, it is easy to check that the following Lemma holds, and that wREP fits in the original reproducibility generalisation.

Lemma 1. *Any scheme which is fully reproducible is also weakly reproducible.*

The following theorem shows that the wREP definition is sufficient to guarantee n-SM-PKE security. The proof uses techniques which are somewhat different from that in [4] and may be of independent interest in other contexts.

Theorem 1. *The associated randomness reusing n-SM-PKE scheme of an IND-atk public-key encryption scheme is IND-atk secure if the base PKE is weakly reproducible. More precisely, any PPT attacker A with non negligible advantage against the randomness reusing n-SM-PKE scheme can be used to construct attackers B and D against the base PKE, such that:*

$$\text{Adv}_{n\text{-SM-PKE}}^{\text{IND-atk}}(A) \leq n \cdot \text{Adv}_{\text{PKE}}^{\text{IND-atk}}(B) + (n-1) \cdot \text{Adv}_{\text{PKE}}^{\text{IND-atk}}(D).$$

Proof. We present the argument for the IND-CPA case, since the IND-CCA version is a straightforward extension where simulators use their knowledge of secret keys and external oracles to answer decryption queries. We begin by defining the following experiment, parameterised with an IND-atk attacker A against the randomness reusing n-SM-PKE scheme, and indexed by a coin b and an integer l such that $0 \leq l \leq n$.

$\text{Exp}_{l,b}(A)$

1. $(\hat{PK}, \hat{SK}) \leftarrow \mathbb{G}_{\text{PKE}}(I)$
2. $(PK_i, SK_i) \leftarrow \mathbb{G}_{\text{PKE}}(I)$, for $1 \leq i \leq n$
3. $(M_0, M_1, s) \leftarrow A_1(PK_1, \ldots, PK_n)$
4. $\hat{C} \leftarrow \mathbb{E}_{\text{PKE}}(M_b, \hat{PK})$
5. $C_i \leftarrow R(\hat{PK}, \hat{C}, M_1, PK_i, SK_i)$, for $1 \leq i \leq l$
6. $C_i \leftarrow R(\hat{PK}, \hat{C}, M_0, PK_i, SK_i)$, for $l + 1 \leq i \leq n$
7. $c \leftarrow A_2(C_1, \ldots, C_n, s)$
8. Return c

Looking at this experiment, and recalling from the wREP definition that R performs perfect reproduction when the input message is the same as that inside the input ciphertext, we can write the following equation:

$$\text{Adv}_{n-\text{SM}-\text{PKE}}^{\text{IND}-\text{atk}}(A) = |\Pr[\text{Exp}_{n,1}(A) = 1] - \Pr[\text{Exp}_{0,0}(A) = 1]|.$$

This follows from the advantage definition, and fact that when $(l, b) = (n, 1)$, then $\hat{C}$ will encapsulate M_1, and all challenge ciphertexts are reproduced with M_1, which gives rise to a valid n-IND-atk ciphertext encapsulating M_1. The same happens for M_0, when $(l, b) = (0, 0)$.

We now define a probabilistic algorithm B which tries to break the base PKE scheme using A.

$B_1(\bar{PK})$

1. Select l at random such that $1 \leq l \leq n$
2. $(PK_l, SK_l) \leftarrow (\bar{PK}, \perp)$
3. $(PK_i, SK_i) \leftarrow \mathbb{G}_{\text{PKE}}(I)$, for $1 \leq i \leq n$ and $i \neq l$
4. $(M_0, M_1, s) \leftarrow A_1(PK_1, \ldots, PK_n)$
5. Return $(M_0, M_1, (M_0, M_1, l, \bar{PK}, (PK_1, SK_1), \ldots, (PK_n, SK_n), s))$

$B_2(\bar{C}, (M_0, M_1, l, \bar{PK}, (PK_1, SK_1), \ldots, (PK_n, SK_n), s))$

1. $C_l \leftarrow \bar{C}$
2. $C_i \leftarrow R(\bar{PK}, \bar{C}, M_1, PK_i, SK_i)$, for $1 \leq i \leq l - 1$
3. $C_i \leftarrow R(\bar{PK}, \bar{C}, M_0, PK_i, SK_i)$, for $l + 1 \leq i \leq n$
4. $\hat{b} \leftarrow A_2(C_1, \ldots, C_n, s)$
5. Return $\hat{b}$

To continue the proof, we will require the following two Lemmas, which we shall prove shortly.

Lemma 2. *For $1 \leq l \leq n - 1$, and for any PPT adversary A, there is an adversary D such that*

$$\text{Adv}_{\text{PKE}}^{\text{IND}-\text{atk}}(D) = |\Pr[\text{Exp}_{l,1}(A) = 1] - \Pr[\text{Exp}_{l,0}(A) = 1]|.$$

Lemma 3. *For $1 \leq i \leq n$, the output of algorithm B and that of $\text{Exp}_{l,b}(A)$ are related as follows:*

$$\Pr[\hat{b} = 1 | l = i \wedge \bar{b} = 1] = \Pr[\text{Exp}_{i,1}(A) = 1]$$

$$\Pr[\hat{b} = 1 | l = i \wedge \bar{b} = 0] = \Pr[\text{Exp}_{i-1,0}(A) = 1].$$

Here $\bar{b}$ is the hidden bit in $\bar{C}$.

Let us now analyse overall the probability that B returns 1, conditional on the value of the hidden challenge bit $\bar{b}$. Since B choses l uniformly at random, we may write:

$$\Pr[\hat{b} = 1 | \bar{b} = 1] = \frac{1}{n} \sum_{i=1}^{n} \Pr[\hat{b} = 1 | l = i \wedge \bar{b} = 1]$$

$$\Pr[\hat{b} = 1 | \bar{b} = 0] = \frac{1}{n} \sum_{i=1}^{n} \Pr[\hat{b} = 1 | l = i \wedge \bar{b} = 0].$$

Taking advantage of Lemma 3, we can rewrite these as:

$$\Pr[\hat{b} = 1 | \bar{b} = 1] = \frac{1}{n} \sum_{i=1}^{n} \Pr[\mathrm{Exp}_{i,1}(A) = 1]$$

$$\Pr[\hat{b} = 1 | \bar{b} = 0] = \frac{1}{n} \sum_{i=1}^{n} \Pr[\mathrm{Exp}_{i-1,0}(A) = 1].$$

Subtracting the previous equations and rearranging the terms, we get

$$n(\Pr[\hat{b} = 1 | \bar{b} = 1] - \Pr[\hat{b} = 1 | \bar{b} = 0]) -$$

$$\left(\sum_{i=1}^{n-1} \Pr[\mathrm{Exp}_{i,1}(A) = 1] - \sum_{i=1}^{n-1} \Pr[\mathrm{Exp}_{i,0}(A) = 1] \right)$$

$$= \Pr[\mathrm{Exp}_{n,1}(A) = 1] - \Pr[\mathrm{Exp}_{0,0}(A) = 1].$$

Considering the absolute values of both sides and using Lemma 2, we can write

$$n\mathrm{Adv}_{\mathrm{PKE}}^{\mathrm{IND-atk}}(B) + (n-1)\mathrm{Adv}_{\mathrm{PKE}}^{\mathrm{IND-atk}}(D) \geq \mathrm{Adv}_{n-\mathrm{SM-PKE}}^{\mathrm{IND-atk}}(A).$$

In other words

$$\mathrm{Adv}_{n-\mathrm{SM-PKE}}^{\mathrm{IND-atk}}(A) \leq (2n-1)\epsilon,$$

where ϵ is negligible and the theorem follows.   □

We now prove the required lemmas.

Proof. (Lemma 2) We build an algorithm $D_l = (D_{1,l}, D_{2,l})$ which runs A in exactly the same conditions as it is run in $\mathrm{Exp}_{l,b}$, and which can be used to win the IND-atk game against the base PKE with an advantage which is the same as A's capability of distinguishing between $\mathrm{Exp}_{l,0}$ and $\mathrm{Exp}_{l,1}$.

$D_{1,l}(\bar{\mathrm{PK}})$

1. $(\mathrm{PK}_i, \mathrm{SK}_i) \leftarrow \mathbb{G}_{\mathrm{PKE}}(I)$, for $1 \leq i \leq n$
2. $(M_0, M_1, s) \leftarrow A_1(\mathrm{PK}_1, \ldots, \mathrm{PK}_n)$
3. Return $(M_0, M_1, (M_0, M_1, \bar{\mathrm{PK}}, (\mathrm{PK}_1, \mathrm{SK}_1), \ldots, (\mathrm{PK}_n, \mathrm{SK}_n), s))$

$D_{2,l}(\bar{C}, (M_0, M_1, \bar{\mathrm{PK}}, (\mathrm{PK}_1, \mathrm{SK}_1), \ldots, (\mathrm{PK}_n, \mathrm{SK}_n), s))$

1. $C_i \leftarrow R(\bar{\mathrm{PK}}, \bar{C}, M_1, \mathrm{PK}_i, \mathrm{SK}_i)$, for $1 \leq i \leq l$
2. $C_i \leftarrow R(\bar{\mathrm{PK}}, \bar{C}, M_0, \mathrm{PK}_i, \mathrm{SK}_i)$, for $l+1 \leq i \leq n$
3. $\hat{b} \leftarrow A_2(C_1, \ldots, C_n, s)$
4. Return $\hat{b}$

D simply uses the challenge public key $\bar{\text{PK}}$ in place of $\hat{\text{PK}}$ in the experiment, and uses the PKE challenge $\bar{C}$ in place of $\hat{C}$. Note that the only visible difference to the definition of Exp is that D does not know $\bar{\text{SK}}$, which it does not need, and that the IND-atk hidden bit $\bar{b}$ is used in place of b. We can therefore write, for a given value of l:

$$\Pr[\hat{b} = 1 | \bar{b} = 1] = \Pr[\text{Exp}_{l,1}(A) = 1]|$$
$$\Pr[\hat{b} = 1 | \bar{b} = 0] = \Pr[\text{Exp}_{l,0}(A) = 1]|,$$

and consequently

$$\text{Adv}_{\text{PKE}}^{\text{IND-atk}}(D) = | \Pr[\text{Exp}_{l,1}(A) = 1] - \Pr[\text{Exp}_{l,0}(A) = 1]|. \qquad \square$$

Proof. (Lemma 3) We present here the proof for the first case in the Lemma, and leave the second case, which is proved using a similar argument, for Appendix A. The first case of the Lemma states that

$$\Pr[\hat{b} = 1 | l = i \wedge \bar{b} = 1] = \Pr[\text{Exp}_{i,1}(A) = 1].$$

We must show that the probability distribution of the inputs presented to A is exactly the same in the scenarios corresponding to both sides of the equation above. This is trivially true for the public keys that A_1 receives, since all of them are independently generated using the correct algorithm. Regarding the challenge ciphertext that A_2 gets, we start by expanding the values of $(C_1, \ldots, C_n)$.

In $\text{Exp}_{i,1}(A)$, we have $\hat{C} = \mathbb{E}_{\text{PKE}}(M_1, \hat{\text{PK}}; r)$ and:

$$C_j = R(\hat{\text{PK}}, \hat{C}, M_1, \text{PK}_j, \text{SK}_j) \text{ for } 1 \le j \le i$$
$$C_j = R(\hat{\text{PK}}, \hat{C}, M_0, \text{PK}_j, \text{SK}_j) \text{ for } i + 1 \le j \le n.$$

On the other hand, in $B_2(\bar{C}, \bar{s})$, given that $l = i$ and $\bar{b} = 1$ we have $\bar{C} = \mathbb{E}_{\text{PKE}}(M_1, \bar{\text{PK}}; r)$ and:

$$C_i = \bar{C}$$
$$C_j = R(\bar{\text{PK}}, \bar{C}, M_1, \text{PK}_j, \text{SK}_j) \text{ for } 1 \le j \le i - 1$$
$$C_j = R(\bar{\text{PK}}, \bar{C}, M_0, \text{PK}_j, \text{SK}_j) \text{ for } i + 1 \le j \le n.$$

To show that the distributions are identical, we split the argument in three parts and fix the values of all random variables, considering the case where the public keys provided to A in both cases are the same, and that the implicit randomness in both $\hat{C}$ and $\bar{C}$ is the same r. We show that the resulting challenge ciphertexts in both cases are exactly the same:

- $j = i$: Note that in the second scenario we have $C_i = \bar{C}$, while in the first scenario we have $C_i = R(\hat{\text{PK}}, \hat{C}, M_1, \text{PK}_i, \text{SK}_i)$. Since $\hat{C}$ encrypts M_1, the result of R is perfect and equal to $\mathbb{E}_{\text{PKE}}(M_1, \text{PK}_i; r) = \bar{C}$.
- $j < i$: In this range, challenge components are identical in both scenarios: they are perfect reproductions $\mathbb{E}_{\text{PKE}}(M_1, \text{PK}_j; r)$, since M_1 is passed to R both in encrypted and plaintext form.

- $j > i$: In this range, challenge components are outputs of R, but in this case we cannot claim that they are identical without resorting to the properties of the wREP algorithm. For different message reproduction, condition 5 of Definition 2 ensures that

$$R(\hat{PK}, \hat{C}, M_0, PK_j, SK_j) = R(\bar{PK}, \bar{C}, M_0, PK_j, SK_j)$$

as required.

This means that the first case of the Lemma follows. □

3 Kurosawa's Efficient Schemes

In this section we analyse modified versions of ElGamal and Cramer-Shoup encryption schemes briefly mentioned by Kurosawa [11] as a way to build efficient single-message multiple-recipient public key encryption schemes. These schemes permit establishing a separation between the original reproducibility notion proposed by Bellare et al. and the one we introduced in the previous section.

3.1 Modified ElGamal

The *modified ElGamal* encryption scheme is similar to the ElGamal encryption scheme and operates as follows. The key generation algorithm $\mathbb{G}_{\mathsf{PKE}}(I)$ on input $I := (p, g)$ returns the key pair $(\mathsf{SK}, \mathsf{PK}) = (1/x, g^x)$ for $x \leftarrow \mathbb{Z}_p^*$. The encryption algorithm $\mathbb{E}_{\mathsf{PKE}}(M, \mathsf{PK}; r)$ returns the ciphertext $(u, v) := ((g^x)^r, m \cdot g^r)$ for $r \leftarrow \mathbb{Z}_p^*$. The decryption algorithm $\mathbb{D}_{\mathsf{PKE}}(u, v, 1/x)$ returns the message $m := v/(u^{1/x})$.

Theorem 2 establishes the security of the modified ElGamal scheme as well as its weak reproducibility property. Theorem 3 shows that modified ElGamal establishes a separation between the notions of full and weak reproducibility.

Theorem 2. *Modified ElGamal is (1) IND-CPA secure under the decisional Diffie–Hellman assumption, and (2) weakly reproducible.*

Proof. (1) The proof is similar to that for the ElGamal encryption scheme.

(2) The weak reproducibility algorithm R on input $(g^x, u, v, m', g^{x'}, 1/x')$ returns $((v/m')^{x'}, v)$. We now check that R satisfies the two properties required by the wREP definition. If $m' = m$, then $v/m' = (m \cdot g^r)/m = g^r$ and the output is a valid encryption of m' under $g^{x'}$ using random coins r. Note also that R's output does not dependent on the public key g^x and hence the second property is also satisfied. □

Theorem 3. *The modified ElGamal encryption is not fully reproducible under the CDH assumption.*

Proof. Let $(g, g^a, g^b) \in G^3$ denote the CDH problem instance. Our goal is to compute g^{ab}. The reproduction algorithm on input $(p, g, g^x, g^{rx}, m \cdot g^r, m', g^y, 1/y)$ outputs $(g^{ry}, m' \cdot g^r)$. We pass to R the input $(p, g^a, g, g^b, 1, 1, g^a, 1)$ which could be written as $(p, h, h^{1/a}, h^{b \cdot 1/a}, 1, 1, h, 1)$ where $h = g^a$. Note that since R succeeds with probability 1, it will run correctly on the above input instance even though its distribution is far away from those that R takes. Here implicitly we have $x = 1/a$, from $rx = b/a$ we get $r = b$, and $m = h^{-b}$. Hence the first component of the output of R will be $(h^1)^b = g^{ab}$.   □

3.2   Modified Cramer-Shoup

Another construction of an efficient n-SM-PKE hinted at by Kurosawa in [11] is based on the CS1a encryption scheme of Cramer and Shoup [7], modified in an analogous manner to the ElGamal encryption scheme as presented in the previous section. In this case, the construction is secure against adaptive chosen ciphertext attacks in the standard model. Modified versions of the other schemes presented in [7] also pass the weak reproducibility test without being fully reproducible. The following scheme, however, is the most efficient as it shares $\hat{g}$ as a domain parameter.

The scheme is defined as follows. The domain parameter is $I := (p, g, \hat{g}, H)$, where g and $\hat{g}$ are generators of a group G of prime order p and H denotes a cryptographic hash function. The key generation algorithm $\mathbb{G}_{\mathsf{PKE}}(I)$ outputs (x_1, x_2, y_1, y_2, z), a random element of $(\mathbb{Z}_p)^5$, as the secret key and the public key is set to be $(g^{x_1}\hat{g}^{x_2}, g^{y_1}\hat{g}^{y_2}, g^z)$. Encryption and decryption algorithms are:

$\mathbb{E}_{\mathsf{PKE}}(m, \mathsf{PK})$
- $(e, f, h) \leftarrow \mathsf{PK}$
- $u \leftarrow \mathbb{Z}_p$
- $\hat{a} \leftarrow \hat{g}^u$
- $b \leftarrow h^u$
- $c \leftarrow m \cdot g^u$
- $v \leftarrow \mathrm{H}(\hat{a}, b, c)$
- $d \leftarrow e^u f^{uv}$
- Return $(\hat{a}, b, c, d)$

$\mathbb{D}_{\mathsf{PKE}}((\hat{a}, b, c, d), \mathsf{SK})$
- $(x_1, x_2, y_1, y_2, z) \leftarrow \mathsf{SK}$
- $v \leftarrow \mathrm{H}(\hat{a}, b, c)$
- $a \leftarrow b^{1/z}$
- If $a^{x_1+y_1 v}\hat{a}^{x_2+y_2 v} \neq d$ return $\bot$
- $m \leftarrow c/a$
- Return m

Theorem 4. *The modified Cramer-Shoup scheme is (1) IND-CCA under the DDH assumption and (2) weakly reproducible.*

The proof of the first part of theorem is essentially that of the standard Cramer-Shoup scheme in [7]. We omit the proof details due to space limitations. Regarding the second part of Theorem 4, the weak reproduction algorithm is a natural extension of the one presented for modified ElGamal, returning

$$(\hat{a}, (c/m')^{z'}, c, (c/m')^{u(x_1'+v'y_1')}\hat{a}^{u(x_2'+v'y_2')}),$$

where $v' := H(\hat{a}, (c/m')^{z'}, c)$. A very important distinction in this case, however, is that the reproduction algorithm produces an output which may not be a valid ciphertext. In fact, for different message reproduction, the encryption algorithm would never be able to produce something like the resulting ciphertext. The returned output is, however, indistinguishable from a valid ciphertext under the decisional Diffie–Hellman assumption. The fact that the outputs of R may not be identically distributed to the outputs of the encryption algorithm, but merely indistinguishable, implies that the proof strategy presented in [4] does *not* apply for this scheme. On the other hand, note that the technique presented in the proof of Theorem 1 covers this and other similar schemes.

4   Hybrid Encryption

Practical applications of public key encryption are based on the hybrid paradigm, where public key techniques are used to encapsulate symmetric encryption keys. Formally, this is captured by the KEM/DEM framework [7]. Sharing randomness across multiple instances of a KEM may be justified, as before, as a means to achieve computational savings when performing batch operations. In this section we study randomness reuse for KEMs, a problem which has not been formally addressed by previous work.

The KEM primitive takes the recipient's public key as the single parameter to the encapsulation algorithm. In particular, unlike what happens in PKEs, one does not control the value of the encapsulated key: this is internally generated inside the KEM primitive, and its value depends only on the recipient's public key and on the randomness tape of the encapsulation algorithm. Since in this work we are interested on the role of randomness inside cryptographic algorithms, this leads us to the following categorisation of KEMs.

Definition 3. *A KEM scheme is* public key independent *if the following experiment returns 1 with probability 1.*

1. $(\text{SK}, \text{PK}), (\text{SK}', \text{PK}') \leftarrow \mathbb{G}_{\text{KEM}}(I)$
2. $r \leftarrow \mathcal{R}_{\text{KEM}}(I)$
3. $(K, C) \leftarrow \mathbb{E}_{\text{KEM}}(\text{PK}; r); (K', C') \leftarrow \mathbb{E}_{\text{KEM}}(\text{PK}'; r)$
4. *If $K = K'$ return 1, else return 0*

Considering what happens when one shares randomness across several instances of an encapsulation algorithm immediately suggests two independent adaptations of KEMs to the multi-recipient setting. The first, which we generically call multi-recipient KEMs (n-MR-KEMs), are functionally equivalent to the independent execution of n KEM instances, thereby associating an independent encapsulated secret key to each recipient. The second, which we will call single-key multi-recipient KEMs (n-SK-KEMs), given that the same secret key is encapsulated to all recipients, is akin to the mKEM notion introduced in [12].

Adaptation of the results in [4] to n-MR-KEMs is straightforward. The same is not true, however, for n-SK-KEMs. To justify why this is the case, we present a reproducibility test for KEMs in Definition 4. It is a direct adaptation of the reproducibility test for PKEs, considering that there is no message input to the encapsulation algorithm and that this returns also the encapsulated secret key. It can be easily shown that any KEM satisfying this test can be used to construct an efficient n-MR-KEM with randomness reuse.

Definition 4. *A KEM is called* reproducible *if there exists a PPT algorithm R such that the following experiment returns 1 with probability 1.*

1. $(SK, PK), (SK', PK') \leftarrow \mathbb{G}_{KEM}(I)$
2. $r \leftarrow \mathcal{R}_{KEM}(I)$
3. $(K, C) \leftarrow \mathbb{E}_{KEM}(PK; r); (K', C') \leftarrow \mathbb{E}_{KEM}(PK'; r)$
4. *If* $R(PK, C, PK', SK') = (K', C')$ *return* 1, *else return* 0

An n-SK-KEM, referred to as an mKEM in [12], is a key encapsulation mechanism which translates the hybrid encryption paradigm to the multi-cast setting: it permits encapsulating the same secret key to several different receivers. The point is that encrypting a single message to all these recipients can then be done using a single DEM instantiation based on that unique session key, rather than n different ones. This provides, not only computational savings, but also bandwidth savings, and captures a common use of hybrid encryption in practice. The natural security model for n-SK-KEMs is shown below.

IND-atk
1. For $i = 1, \ldots, n$
   $(SK_i, PK_i) \leftarrow \mathbb{G}_{n-SK-KEM}(I)$
2. $s \leftarrow A_1^{\mathcal{O}_1}(PK_1, \ldots, PK_n)$
3. $b \leftarrow \{0, 1\}$
4. $(K_0, C^*) \leftarrow \mathbb{E}_{n-SK-KEM}((PK_i)_{i=1}^n)$
5. $K_1 \leftarrow \{0, 1\}^\kappa$
6. $b' \leftarrow A_2^{\mathcal{O}_2}(C^*, K_b, s)$

$\text{Adv}_{n-SK-KEM}^{IND-atk}(A) := |2 \Pr[b' = b] - 1|.$

As usual, the adversary also has access to $\mathcal{O}_1$ and $\mathcal{O}_2$, which denote a set of oracles, as follows:

- If atk = CPA then $\mathcal{O}_1 = \mathcal{O}_2 = $ NULL;
- If atk = CCA then $\mathcal{O}_1$ is a set of decapsulation oracles, one for each PK_i, and $\mathcal{O}_2$ is same as $\mathcal{O}_1$ except that no component C_i^* of the challenge ciphertext C^* can be submitted to an oracle corresponding to PK_i.

Unlike n-MR-KEMs there does not seem to be a natural way of constructing n-SK-KEMs from single-recipient KEMs. The fact that the same key should be encapsulated for all recipients makes public key independent KEMs the only possible candidates to be used as base KEMs. However, any public key independent scheme which satisfies a reproducibility test such as that in Definition 4 must be insecure, as anyone would be able to use the reproducibility algorithm to obtain the secret key in an arbitrary ciphertext. In the following we show how the weak reproducibility notion for PKEs we obtained in Theorem 1 actually fills this apparent theoretical gap, as it permits capturing the efficient n-SK-KEMs constructions we have found in literature. We conclude this section proposing a concrete construction of an efficient n-SK-KEM secure in the standard model.

4.1 Generic Construction of n-SK-KEMs

One trivial way to build secure randomness reusing n-SK-KEMs is to use a secure weakly reproducible encryption scheme, and to set a random message to be the ephemeral key. However, the underlying encryption scheme must have the same security guarantees as those required for the KEM. A more practical way to build a fully secure n-SK-KEM is to use a weaker PKE through the following generic construction, which generalises the mKEM scheme proposed in [12] and extends a construction by Dent in [8]. The domain parameters and key generation algorithm are the same as those of the underlying PKE. Encapsulation and decapsulation algorithms are:

$$\mathbb{E}_{n-\text{SK}-\text{KEM}}(\text{PK}_1, \dots, \text{PK}_n)$$
- $M \leftarrow \mathcal{M}_{\text{PKE}}(I)$
- $r \leftarrow H(M)$
- For $i = 1 \dots, n$
  - $C_i \leftarrow \mathbb{E}_{\text{PKE}}(M, \text{PK}_i; r)$
- $C \leftarrow (C_1, \dots, C_n)$
- $K \leftarrow \text{KDF}(M)$
- Return (K, C)

$$\mathbb{D}_{n-\text{SK}-\text{KEM}}(C, \text{SK})$$
- $M \leftarrow \mathbb{D}_{\text{PKE}}(M, \text{SK})$
- If $M = \perp$ return $\perp$
- $r \leftarrow H(M)$
- If $C \neq \mathbb{E}_{\text{PKE}}(M, \text{PK}; r)$ return $\perp$
- $K \leftarrow \text{KDF}(M)$
- Return K

Here H and KDF are cryptographic hash functions. The security of this scheme is captured via the following theorem, proved in Appendix B.

Theorem 5. *The above construction is an IND-CCA secure n-SK-KEM, if the underlying PKE is IND-CPA and weakly reproducible, and if we model H and KDF as random oracles. More precisely, any PPT attacker A with non negligible advantage against the generic n-SK-KEM can be used to construct an attacker B against the base PKE, such that:*

$$\text{Adv}_{n-\text{SK}-\text{KEM}}^{\text{IND}-\text{CCA}}(A) \leq 2n(q_H + q_K + q_D)\text{Adv}_{\text{PKE}}^{\text{IND}-\text{CPA}}(B) + \epsilon,$$

where q_H, q_K and q_D are the number of queries the adversary makes to H, KDF and decapsulation oracles and ϵ denotes a negligible quantity.

The security argument for this construction has two parts. The first part establishes the one-way security of the n-SK-PKE scheme associated with the base PKE. This follows directly from the weak reproducibility theorem in Section 3.2 and the fact that one-wayness is implied by indistinguishability[5]. The second part builds on the previous result to achieve IND-CCA security in the n-SK-KEM setting, using a general construction laid out by Dent in [8]. In this construction one models the hash function H and KDF as random oracles and shows that the queries placed by any adversary with non-negligible advantage in breaking the n-SK-KEM scheme can be used to invert the one-wayness of the underlying n-SM-PKE scheme.

The mKEM in [12] fits the general framework we introduced in this paper by instantiating the above construction with the ElGamal encryption scheme. The results in this work permit introducing two interesting enhancements over the mKEM in [12] if the above construction is instantiated with the modified ElGamal scheme:

---

[5] We assume that various message spaces have exponential size in the security parameter.

- Stronger security guarantees disallowing benign malleability. The security model in [12] disallows decapsulation queries on any ciphertext which decapsulates to the same key as that implicit in the challenge.
- More efficient encryption algorithm, saving $n - 1$ group operations.

4.2   An Efficient n-SK-KEM Secure in the Standard Model

In this section we propose an efficient n-SK-KEM scheme which is IND-CCA secure in the standard model. To the best of our knowledge, it is the first such construction to achieve this level of security and efficiency. The scheme is an adaptation of a KEM proposed by Cramer and Shoup in [7], which is public key dependent and therefore cannot be used as a black-box to construct an n-SK-KEM. The adapted scheme is defined as follows.

The domain parameter is $I := (p, g, \hat{g}, H, \mathrm{KDF})$, where g and $\hat{g}$ are generators of a group G of prime order p, H is a cryptographic hash function and KDF is a key derivation function. The key generation algorithm $\mathbb{G}_{n-\mathrm{SK-KEM}}(I)$ outputs $\mathrm{SK} = (x_1, x_2, y_1, y_2, z)$, a random element of $\mathbb{Z}_p^4 \times \mathbb{Z}_p^*$, as the secret key and $\mathrm{PK} = (e, f, h) = (g^{x_1}\hat{g}^{x_2}, g^{y_1}\hat{g}^{y_2}, g^z)$ as the public key. Encapsulation and decapsulation algorithms are:

$\mathbb{E}_{n-\mathrm{SK-KEM}}(\mathrm{PK}_1, \ldots, \mathrm{PK}_n)$
- $u \leftarrow \mathbb{Z}_p$
- $\hat{a} \leftarrow \hat{g}^u$
- $b \leftarrow g^u$
- $K \leftarrow \mathrm{KDF}(\hat{a}, b)$
- For $1 \leq i \leq n$
  $(e_i, f_i, h_i) \leftarrow \mathrm{PK}_i$
  $a_i \leftarrow h_i^u$
  $v_i \leftarrow H(\hat{a}, a_i)$
  $d_i \leftarrow e_i^u f_i^{uv_i}$
- Return $(K, \hat{a}, a_1, \ldots, a_n, d_1, \ldots, d_n)$

$\mathbb{D}_{n-\mathrm{SK-KEM}}((\hat{a}, a, d), \mathrm{SK})$
- $(x_1, x_2, y_1, y_2, z) \leftarrow \mathrm{SK}$
- $v \leftarrow H(\hat{a}, a)$
- $b \leftarrow a^{1/z}$
- If $b^{x_1+vy_1}\hat{a}^{x_2+vy_2} \neq d$
  return $\perp$
- $K \leftarrow \mathrm{KDF}(\hat{a}, b)$
- Return K

A proof that the n-SK-KEM scheme proposed above is IND-CCA secure under the decisional Diffie–Hellman assumption, provided that the hash function is target collision resistant, and that the KDF function is entropy smoothing will appear in the full version of this paper.

5   Tighter Reductions

In [4] the authors present tighter security reductions for the multi-recipient randomness reusing schemes associated with the ElGamal and Cramer-Shoup encryption schemes. These reductions rely on the random self-reducibility property of the DDH problem. The tighter reductions are achieved by using this property to unfold a single DDH problem instance, so that it can be embedded in the multiple challenge ciphertext components required in the multiple user setting. In these proofs, the extra public keys and challenge ciphertexts required in the reduction are chosen in a controlled manner. For instance, one public key might have a known discrete logarithm with respect to another. The following notion of reproducibility could be viewed as a generalisation of this type of proof strategy for tight reductions.

Definition 5. *A PKE scheme is called* directly reproducible *(dREP) if there exists a set of PPT algorithms $R = (R_1, R_2, R_3)$ such that the following experiment returns 1 with probability 1.*

1. $(\mathsf{SK}, \mathsf{PK}) \leftarrow \mathbb{G}_{\mathsf{PKE}}(I)$
2. $(\mathsf{PK}', s) \leftarrow R_1(\mathsf{PK})$
3. $M \leftarrow \mathcal{M}_{\mathsf{PKE}}(I); r \leftarrow \mathcal{R}_{\mathsf{PKE}}(I)$
4. $C \leftarrow \mathbb{E}_{\mathsf{PKE}}(M, \mathsf{PK}; r); C' \leftarrow \mathbb{E}_{\mathsf{PKE}}(M, \mathsf{PK}'; r)$
5. *If $C' \neq R_2(C, s)$ return 0*
6. *If $C \neq R_3(C', s)$ return 0, else return 1*

We require the distributions of PK *and* PK' *to be identical.*

Note that R_1 controls the generation of the public keys and the main reproduction algorithm (R_2) may take advantage of the state information produced by the first algorithm. The existence of the third algorithm is required for the simulation of decryption oracles for CCA secure schemes. It is easy to verify that

Theorem 6. *The associated randomness reusing n-SM-PKE scheme of a directly reproducible and IND-atk secure encryption scheme is also secure in the IND-atk sense. More precisely, any PPT attacker A against the randomness reusing n-SM-PKE scheme can be used to build an attacker B against the base scheme, such that:*

$$\mathrm{Adv}_{n\text{-}\mathsf{SM}\text{-}\mathsf{PKE}}^{\mathtt{IND}-\mathtt{atk}}(A) \leq \mathrm{Adv}_{\mathsf{PKE}}^{\mathtt{IND}-\mathtt{atk}}(B).$$

The above notion of reproducibility, not only permits deriving tighter security reductions, but also gives rise to a new test for detecting additional schemes which allow randomness reuse. In fact, it can be shown that a modified version of the escrow ElGamal encryption scheme is directly but not weakly reproducible (see Appendix C).

Furthermore, unlike weak and full reproducibility, this new notion respects the Fujisaki-Okamoto transformation [9] for building IND-CCA secure schemes, as it does not explicitly handle the encrypted message. It therefore establishes a new set of chosen-ciphertext secure single message multi-recipient schemes with tight security reductions in the random oracle model.

Direct reproducibility also poses an interesting problem, which concerns public key encryption schemes with chosen ciphertext security in the standard model. In particular, the case of the Cramer-Shoup encryption scheme remains open, as we were unable to construct the required reproduction algorithms. We leave it as an open problem to find such an algorithm, or to design an analogous reproducibility test which admits encryption schemes which are IND-CCA secure in the standard model.

References

1. Abe, M., Cui, Y., Imai, H., Kiltz, E.: Efficient Hybrid Encryption from ID-Based Encryption. Cryptology ePrint Archive, Report 2007/023 (2007)
2. Al-Riyami, S.S., Malone-Lee, J., Smart, N.P.: Escrow-Free Encryption Supporting Cryptographic Workflow. International Journal of Information Security 5, 217–230 (2006)
3. Al-Riyami, S.S., Paterson, K.G.: Certificateless Public-Key Cryptography. In: Laih, C.-S. (ed.) ASIACRYPT 2003. LNCS, vol. 2894, pp. 452–473. Springer, Heidelberg (2003)
4. Bellare, M., Boldyreva, A., Staddon, J.: Randomness Re-Use in Multi-recipient Encryption Schemes. In: Desmedt, Y.G. (ed.) PKC 2003. LNCS, vol. 2567, pp. 85–99. Springer, Heidelberg (2002)
5. Bellare, M., Kohno, T., Shoup, V.: Stateful Public-Key Cryptosystems: How to Encrypt with One 160-bit Exponentiation. In: 3th ACM Conference on Computer and Communications Security – CCS, ACM, New York (2006)
6. Boneh, D., Franklin, M.: Identity-Based Encryption from the Weil Pairing. SIAM Journal on Computing 32, 586–615 (2003)
7. Cramer, R., Shoup, V.: A Practical Public-Key Cryptosystem Provably Secure against Adaptive Chosen Ciphertext Attack. In: Krawczyk, H. (ed.) CRYPTO 1998. LNCS, vol. 1462, pp. 13–25. Springer, Heidelberg (1998)
8. Dent, A.W.: A Designer's Guide to KEMs. In: Paterson, K.G. (ed.) Cryptography and Coding. LNCS, vol. 2898, pp. 133–151. Springer, Heidelberg (2003)
9. Fujisaki, E., Okamoto, T.: Secure Integration of Asymmetric and Symmetric Encryption Schemes. In: Wiener, M.J. (ed.) CRYPTO 1999. LNCS, vol. 1666, pp. 537–554. Springer, Heidelberg (1999)
10. Hastad, J.: Solving Simultaneous Modular Equations of Low Degree. SIMA Journal on Computing 17(2) (April 1988)
11. Kurosawa, K.: Multi-Recipient Public-Key Encryption with Shortened Ciphertext. In: Naccache, D., Paillier, P. (eds.) PKC 2002. LNCS, vol. 2274, pp. 48–63. Springer, Heidelberg (2002)
12. Smart, N.P.: Efficient Key Encapsulation to Multiple Parties. In: Blundo, C., Cimato, S. (eds.) SCN 2004. LNCS, vol. 3352, pp. 208–219. Springer, Heidelberg (2005)

A Proof of the Second Case of Lemma 3

Proof. We now prove the Lemma for the case

$$\Pr[\hat{b} = 1 | l = i \wedge \bar{b} = 0] = \Pr[\mathtt{Exp}_{i-1,0}(A) = 1].$$

The argument is similar to the previous case. We must show that the probability distribution of the inputs presented to A is exactly the same in the scenarios corresponding to both sides of the equation above. This is trivially true for the public keys that A_1 receives, since all of them are independently generated using the correct algorithm. Regarding the challenge ciphertext that A_2 gets, we start by expanding the values of $(C_1, \ldots, C_n)$.

In $\mathrm{Exp}_{i-1,0}(A)$, we have $\hat{C} = \mathbb{E}_{\mathrm{PKE}}(M_0, \hat{\mathrm{PK}}; r)$ and

$$C_j = R(\hat{\mathrm{PK}}, \hat{C}, M_1, \mathrm{PK}_j, \mathrm{SK}_j) \text{ for } 1 \le j \le i-1$$
$$C_j = R(\hat{\mathrm{PK}}, \hat{C}, M_0, \mathrm{PK}_j, \mathrm{SK}_j) \text{ for } i \le j \le n.$$

On the other hand, in $B_2(\bar{C}, \hat{s})$, given that $l = i$ and $\bar{b} = 0$ we have $\bar{C} = \mathbb{E}_{\mathrm{PKE}}(M_0, \mathrm{PK}; r)$ and

$$C_i = \bar{C}$$
$$C_j = R(\mathrm{PK}, \bar{C}, M_1, \mathrm{PK}_j, \mathrm{SK}_j) \text{ for } 1 \le j \le i-1$$
$$C_j = R(\mathrm{PK}, \bar{C}, M_0, \mathrm{PK}_j, \mathrm{SK}_j) \text{ for } i+1 \le j \le n.$$

To show that the distributions are identical, we split the argument in three parts and fix the values of all random variables, considering the case where the public keys provided to A in both cases are the same, and that the implicit randomness in both $\hat{C}$ and $\bar{C}$ is the same r. We show that the resulting challenge ciphertexts in both cases are exactly the same:

- $j = i$: Note that in the second scenario we have $C_i = \bar{C}$, while in the first scenario we have $C_i = R(\hat{\mathrm{PK}}, \hat{C}, M_0, \mathrm{PK}_i, \mathrm{SK}_i)$. Since $\hat{C}$ encrypts M_0, the result of R is perfect and equal to $\mathbb{E}_{\mathrm{PKE}}(M_0, \mathrm{PK}_i; r) = \bar{C}$.
- $j < i$: In this range, challenge components are outputs of R, but in this case we cannot claim that they are identical without resorting to the properties of R described in Definition 2 for different message reproduction, which ensure that

$$R(\hat{\mathrm{PK}}, \hat{C}, M_1, \mathrm{PK}_j, \mathrm{SK}_j) = R(\bar{\mathrm{PK}}, \bar{C}, M_1, \mathrm{PK}_j, \mathrm{SK}_j)$$

  as required.
- $j > i$: In this range, challenge components are identical in both scenarios: they are perfect reproductions $\mathbb{E}_{\mathrm{PKE}}(M_0, \mathrm{PK}_j; r)$, since M_0 is passed to R both in encrypted and plaintext form.

This means that the second case of the Lemma follows. $\square$

B   Proof of Theorem 5

Proof. Let A denote an IND-CCA adversary against the generic construction with non-negligible advantage. Modelling hash functions as random oracles, we construct an algorithm B with non-negligible advantage in the OW-CPA game for the n-SM-PKE. One-way security notion can be easily adapted to multi-recipient schemes. Note that one-wayness of an n-SM-PKE is not necessarily implied by the one-wayness of its base PKE [10]. However, since indistinguishability implies one-wayness and indistinguishability property is inherited from the base scheme due to the wREP property, we do have that the n-SM-PKE is OW-CPA. The concrete reduction is:

$$\mathrm{Adv}^{\mathrm{OW-CPA}}_{n\text{-}\mathrm{SM\text{-}PKE}}(A) \le \mathrm{Adv}^{\mathrm{IND-CPA}}_{n\text{-}\mathrm{SM\text{-}PKE}}(B) + \epsilon_1 \le n\mathrm{Adv}^{\mathrm{IND-CPA}}_{\mathrm{PKE}}(C) + \epsilon_2.$$

Here ϵ_1 and ϵ_2 are negligible quantities, assuming that the message space has a size super-polynomial in the security parameter. We omit the straightforward details of the proof.

On receiving the n public keys for the OW-CPA game, B passes these values on to algorithm A. During A's first stage, algorithm B replies to A's oracle queries as follows:

- H queries: B maintains a list $L \subseteq \mathcal{M}_{n-\text{SM-PKE}}(I) \times \mathcal{R}_{n-\text{SM-PKE}}(I)$ which contains at most q_H pairs (M, r). On input of M, if $(M, r) \in L$ then B returns r, otherwise it selects r at random from the appropriate randomness space, appends (M, r) to the list and returns r.
- KDF queries: B maintains a list $L_K \subseteq \mathcal{M}_{n-\text{SM-PKE}}(I) \times \mathcal{K}_{n-\text{SK-KEM}}(I)$ which contains at most $q_K + q_D$ pairs (M, k). On input of M, if $(M, k) \in L_K$ then B returns k, otherwise it selects k at random from the appropriate key space, appends (M, k) to the list and returns k.
- Decapsulation queries: on input (C, PK), B checks for each $(M, r) \in L$ if $\mathbb{E}_{\text{PKE}}(M, \text{PK}; r) = C$; if such a pair exists, B calls the KDF simulation procedure on value M and returns the result to A. Otherwise B returns $\perp$.

At some point A will complete its first stage and return some state information. At this point, B calls the outside challenge oracle, and obtains a challenge ciphertext $(C_1, \ldots, C_n)$ on some unknown M^*. Algorithm B now checks if A has queried for decapsulation on a tuple (C_ℓ, PK_ℓ) during its first stage. If this is the case, algorithm B terminates. Otherwise it generates a random K^* and provides this to A along with the challenge ciphertext.

In the second stage, B answers A's oracle queries as in stage one. When A terminates, B randomly returns a message from L or L_K.

Now we analyse the probability that this answer is correct.

B's execution has no chance of success (event S_B) if it terminates at the end of A's first stage (event T). Therefore:

$$\Pr[S_B] = \Pr[S_B \wedge \neg T] = \Pr[S_B | \neg T] \Pr[\neg T]$$

Note that the challenge encapsulation is independent of A's view in the first stage, so that A could only have queried decapsulation for one of the challenge encapsulations by pure chance. However, the size of the valid encapsulation space for each public key is the same as the message space. This means that the probability that B continues to execute is

$$\Pr[\neg T] = 1 - \frac{q_D}{M}$$

where $M = |\mathcal{M}_{n-\text{SM-PKE}}(I)|$.

Given that termination does not take place, B's simulation could be imperfect if one of the following events occur:

- Event E_1: The adversary places a decapsulation query for a valid ciphertext, and B returns $\perp$.

– Event E_2: The adversary queries H or KDF for the unknown M^* value.

Event E_1 occurs if A finds a valid ciphertext without querying H to obtain the randomness required to properly construct it. The probability of this is

$$\Pr[E_1] \leq \frac{q_D \gamma_n}{R},$$

where $R = |\mathcal{R}_{n-\text{SM}-\text{PKE}}(I)|$ and $\gamma_n = \gamma_n(I)$ is the least upper bound such that for every n-tuple $(\text{PK}_i)_{i=1}^n$, every $M \in \mathcal{M}_{n-\text{SM}-\text{PKE}}(I)$, every $j \in \{1, \dots, n\}$ and every $C \in \mathcal{C}_{n-\text{SM}-\text{PKE}}(I)$ we have

$$|\{r \in \mathcal{R}_{n-\text{SM}-\text{PKE}}(I) : [\mathbb{E}_{n-\text{SM}-\text{PKE}}(M, (\text{PK}_i)_{i=1}^n; r)]_j = C\}| \leq \gamma_n(I).$$

This follows from the fact that, since H is modelled as a random oracle, A can only achieve this by guessing the randomness value. Moreover, the probability that a given randomness generates a valid ciphertext is at most γ_n/R and there are at most q_D such queries.

Note that we can write

$$\Pr[E_1 \vee E_2] \leq \Pr[E_1] + \Pr[E_2] \leq \frac{q_D \gamma_n}{R} + \Pr[E_2].$$

On the other hand, since A operates in the random oracle it can have no advantage if event $E_1 \vee E_2$ does not occur. Hence we can write

$$
\begin{aligned}
(1/2)\text{Adv}_{n-\text{SK}-\text{KEM}}^{\text{IND}-\text{CCA}}(A) &= \Pr[S_A] - 1/2 \\
&= \Pr[S_A \wedge (E_1 \vee E_2)] + \Pr[S_A \wedge \neg(E_1 \vee E_2)] - 1/2 \\
&\leq \Pr[E_1 \vee E_2] + 1/2 - 1/2 \\
&\leq \Pr[E_2] + \frac{q_D \gamma_n}{R}.
\end{aligned}
$$

Now:

$$
\begin{aligned}
\text{Adv}_{n-\text{SM}-\text{PKE}}^{\text{OW}-\text{CPA}}(B) &= \Pr[S_B] = \Pr[S_B|\neg T](1 - \frac{q_D}{M}) \\
&= \frac{1}{|L| + |L_K|} \Pr[E_2](1 - \frac{q_D}{M}) \\
&\geq \frac{1}{q_H + q_K + q_D}(\Pr[E_2] - \frac{q_D}{M}).
\end{aligned}
$$

and rearranging the terms

$$\Pr[E_2] \leq (q_H + q_K + q_D)\text{Adv}_{n-\text{SM}-\text{PKE}}^{\text{OW}-\text{CPA}}(B) + \frac{q_D}{M}$$

Putting the above two results together we get:

$$\text{Adv}_{n-\text{SK}-\text{KEM}}^{\text{IND}-\text{CCA}}(A) \leq 2(q_H + q_K + q_D)\text{Adv}_{n-\text{SM}-\text{PKE}}^{\text{OW}-\text{CPA}}(B) + 2q_D(\frac{1}{M} + \frac{\gamma_n}{R}).$$

$\square$

C   Direct and Weak Reproducibility Separation

Let us consider a modified version of a scheme proposed by Boneh and Franklin [6] known as escrow ElGamal. In this scheme the domain parameter is $I := (p, g, h)$ where $h = g^t$ for $t \leftarrow \mathbb{Z}_p^*$. The key generation algorithm outputs $(1/x, g^x)$ as the secret-public key pair for $x \leftarrow \mathbb{Z}_p^*$. The encryption algorithm on input a message m and a public key g^x returns $(u, v) := ((g^x)^r, m \cdot e(g, h)^r)$ where r is random in $\mathbb{Z}_p^*$. One is able to decrypt this ciphertext using the secret key $1/x$ by computing $m := v/e(u, h^{1/x})$. Here $e : G \times G \rightarrow G_T$ is a non-degenerate efficiently computable bilinear map [6].

The randomness reuse properties of this scheme are as follows.

Theorem 7. *The modified escrow ElGamal encryption scheme given above is (1) IND-CPA under the decisional bilinear Diffie–Hellman assumption; (2) directly reproducible; and (3) not weakly reproducible if the computational Diffie–Hellman assumption holds in G.*

Proof. (1) The security proof is analogous to that of escrow ElGamal.

(2) The direct reproducibility algorithm $R = (R_1, R_2, R_3)$ operates as follows. Algorithm R_1 on input a public key g^x returns $((g^x)^s, s)$ where s is a random element in $\mathbb{Z}_p^*$. The algorithm R_2 on input a ciphertext $(u, v) = (g^{xr}, m \cdot e(g, h)^r)$ and state information s returns (u^s, v). It is easily seen that R produces a valid encryption of m under $(g^x)^s$. Algorithm R_3 returns $(u^{1/s}, v)$. Note that the public key $(g^x)^s$ is identically distributed to public keys returned by the key generation algorithm.

(3) Let $(g, g^a, g^b) \in G^3$ denote the CDH problem instance. Our goal is to compute g^{ab}. The reproduction algorithm on input

$$(p, g, h, g^x, g^{rx}, m \cdot e(g, h)^r, m, g^y, 1/y)$$

outputs $(g^{ry}, m \cdot e(g, h)^r)$. To compute g^{ab} we pass to R the input

$$(p, g^a, g, g, g^b, e(g^b, g^a), 1, g^a, 1).$$

This could be written as:

$$(p, g', g'^{1/a}, g'^{1/a}, g'^{b \cdot 1/a}, e(g', g'^{1/a})^b, 1, g', 1),$$

where $g' = g^a$. Note again that since R succeeds with probability 1, it will run correctly on the above input instance. Here implicitly we have $x = 1/a$, and from $rx = b/a$ we have $r = b$, and $m = 1$. Therefore the first component of the output will be $(g'^1)^b = g^{ab}$. $\square$

On the Connection Between Signcryption and One-Pass Key Establishment

M. Choudary Gorantla, Colin Boyd, and Juan Manuel González Nieto

Information Security Institute, Queensland University of Technology
GPO Box 2434, Brisbane, QLD 4001, Australia
mc.gorantla@isi.qut.edu.au, {c.boyd,j.gonzaleznieto}@qut.edu.au

Abstract. There is an intuitive connection between signcryption and one-pass key establishment. Although this has been observed previously, up to now there has been no formal analysis of this relationship. The main purpose of this paper is to prove that, with appropriate security notions, one-pass key establishment can be used as a signcryption KEM and vice versa. In order to establish the connection we explore the definitions for signcryption (KEM) and give new and generalised definitions. By making our generic construction concrete we are able to provide new examples of a signcryption KEM and a one-pass key establishment protocol.

Keywords: Key establishment, Signcryption, Signcryption KEM.

1 Introduction

Zheng [1] introduced the notion of *signcryption* as an asymmetric cryptographic primitive that provides both privacy and authenticity at greater efficiency than the generic composition of signature and encryption schemes. A seemingly unrelated cryptographic primitive is *key establishment* which aims to allow parties to establish a shared key that can be used to cryptographically protect subsequent communications. Most key establishment protocols are interactive, but many such protocols provide simplified *one-pass* versions which only use a single message. One-pass key establishment provides the opportunity for very efficient constructions, even though they will typically provide a lower level of security than interactive protocols.

Zheng [2] later observed that a signcryption scheme can be used as a key transport protocol by simply choosing a new key and sending it in a signcrypted message. This intuitively gives the desired properties for key establishment since the signcryption gives assurance to the sender that the key is available only to the recipient, and assurance to the recipient that the key came from the sender. However, this work contains neither a security model nor a proof for this construction and there remains currently no formal treatment. Since key establishment is notoriously tricky to get right, it is important to decide exactly what security properties such a construction can provide. The main purpose

S.D. Galbraith (Eds.): Cryptography and Coding 2007, LNCS 4887, pp. 277–301, 2007.
© Springer-Verlag Berlin Heidelberg 2007

of this paper is to define the appropriate notions of security and show how signcryption and one-pass key establishment can be related under those notions.

SECURITY FOR SIGNCRYPTION. Since the introduction of signcryption, different definitions of security have emerged. An, Dodis and Rabin [3] divided security notions for signcryption into two types: *outsider security* assumes that the adversary is not one of the participants communicating while *insider security* allows the adversary to be one of the communicating parties. Insider security is a stronger notion than outsider security as it protects the authenticity of a sender from a malicious receiver and privacy of a receiver from a malicious sender. Therefore insider security implies the corresponding notion for outsider security. The notions of outsider security and insider security with respect to authenticity are similar to third-person unforgeability and receiver unforgeability defined for asymmetric authenticated encryption [4].

Because signcryption is intended to include functionality similar to that of digital signatures, it is natural that non-repudiation is a desirable property. Non-repudiation requires insider security since the sender of a signcrypted message must be prevented from showing that it could have been formed by the recipient of that message. A signcryption scheme with only outsider security cannot provide non-repudiation [3,5]. For key establishment there is no need for non-repudiation — it is never required for a single party to take responsibility for the shared key. On the other hand, a commonly required property for key establishment is *forward secrecy* which ensures that if the long-term key of a participant in the protocol is compromised then previously established keys will remain secure. We can regard forward secrecy as analogous to insider security in signcryption with respect to confidentiality. Compromise of the sender's private key should not allow an adversary to obtain previously signcrypted messages.

In addition to forward secrecy, another common security requirement for key establishment is security against compromise of ephemeral protocol data. This is not considered in the existing models for signcryption schemes and so it is not possible, in general, to convert from a signcryption scheme to a key establishment protocol with this stronger security notion. We will argue later that there is good reason that signcryption schemes should consider security against compromise of ephemeral data. In particular this observation allows us to explain a potential weakness observed by Dent in one of his own constructions [6].

SIGNCRYPTION KEMs. Cramer and Shoup [7] formalised the concept of hybrid encryption schemes which securely use public key encryption techniques to encrypt a session key, and symmetric key encryption techniques to encrypt the actual message. This hybrid construction has a key encapsulation mechanism (KEM) and a data encapsulation mechanism (DEM) as its underlying tools. A KEM is similar to a public key encryption scheme except that it is used to generate a random key and its encryption. In a series of papers, Dent [8,6,9] extended this hybrid paradigm to signcryption, resulting in the construction of signcryption KEM and signcryption DEM with different security notions.

Although we could use plain signcryption schemes to provide one-pass key establishment as suggested by Zheng [2], signcryption KEMs seem better suited to the job. This is because they do just what we need by providing a new random key, yet have the potential to be more efficient than plain signcryption. Note, however, that the remarks regarding the differences in security models between signcryption and key establishment apply equally when signcryption KEMs are considered in place of plain signcryption.

CONTRIBUTIONS. We provide new definitions of security for signcryption and subsequently for signcryption KEMs. We then show the suitability of these new notions in deriving one-pass key establishment protocols from signcryption KEMs. Generic constructions of signcryption KEMs from one-pass key establishment protocols and vice versa are proposed. These constructions are instantiated using existing schemes; in particular we use HMQV [10], the most efficient currently known key agreement protocol with a proof of security, to derive a new signcryption KEM with strong security properties. One of the main observations of our paper is that the security models for key establishment are stronger than those normally accepted for signcryption. Moreover, the stronger security seems to be just as appropriate for signcryption as it is for key establishment. Specific contributions of the paper are:

- new definitions for signcryption (KEM)s;
- generic construction from one-pass key establishment to signcryption and vice versa;
- the first secure signcryption KEM with forward secrecy;
- an attack on a signcryption KEM of Dent [6].

The remainder of this introduction briefly surveys related work and outlines the Canetti–Krawczyk model for key establishment. Section 2 then considers the current definitions of security for signcryption and how they can be strengthened. Section 3 examines the outsider secure signcryption KEMs designed by Dent [6]. The generic construction of signcryption KEM from one-pass key establishment is covered in Section 4 while the reverse construction is covered in Section 5.

1.1 Related Work

An et al. [3] defined security notions for signcryption schemes as insider and outsider security in the two-user setting. They also described how to extend these notions to the multi-user setting. Baek et al. [11] independently attempted to provide similar notion of security for signcryption, but their model was not completely adequate. Recently, the same authors [12] extended these notions to match the corresponding definitions given by An et al. However, their security notions are still not complete, as discussed in Section 2.1.

Zheng [2] informally showed how a signcryption scheme can be used as a key transport protocol. Dent [8,6] and Bjørstad and Dent [13] discussed how a signcryption KEM can be used as a one-pass key establishment protocol. Bjørstad and Dent [13] proposed the concept of signcryption tag-KEM and claimed that better key establishment mechanisms can be built with this. However, none of these papers formally defined security in a model that is suitable for key establishment protocols. Moreover, the confidentiality notion defined for all these

KEMs does not offer security against insider attacks. Insider security for confidentiality enables achieving *forward secrecy* i.e. the compromise of the sender's private key does not compromise the confidentiality of signcryptions created using that key [3]. However, this has been ignored or its absence is treated as a positive feature called "Past Message Recovery" in the earlier work [1,12].

1.2 The Canetti–Krawczyk Model

To analyse the security of key establishment protocols, we use the Canetti-Krawczyk (CK) model [14,15], which we briefly describe here. The CK model has primarily been used for multi-pass key establishment protocols, but it can also be used to analyse one-pass key establishment protocols without modification as shown by Krawczyk [10].

In the CK model a protocol π is modelled as a collection of n programs running at different parties, $P_1, \ldots, P_n$. Each invocation of π within a party is defined as a *session*, and each party may have multiple sessions running concurrently. The communications network is controlled by an adversary $\mathcal{A}^\pi$, which schedules and mediates all sessions between the parties. When first invoked within a party, π calls an initialization function that returns any information needed for the bootstrapping of the cryptographic authentication functions (e.g. private keys and authentic distribution of other parties' public keys). After this initialization stage, the party waits for activation. $\mathcal{A}^\pi$ may activate a party P_i by means of a $\mathsf{send}(\pi_{i,j}^s, \lambda)$ request where λ is an empty message[1]. This request instructs P_i to commence a session with party P_j. In response to this request P_i outputs a message m intended for party P_j. s is a session identifier unique amongst all sessions between P_i and P_j. In this paper, where we only consider one-pass key establishment, we define the session-id as the tuple (P_i, P_j, m), where m is the unique message sent by P_i to P_j. The adversary activates the receiver P_j with an incoming message using the request $\mathsf{send}(\pi_{j,i}^s, m)$.

$\mathcal{A}^\pi$ is responsible for transmitting messages between parties, and may fabricate or modify messages when desired. Upon activation, the parties perform some computations and update their internal state. Two sessions are said to be *matching sessions* if their session-ids are identical.

In addition to the activation of parties, $\mathcal{A}^\pi$ can perform the following queries.

1. $\mathsf{corrupt}(P_i)$. With this query $\mathcal{A}^\pi$ learns the entire current state of P_i including long-term secrets, session internal state and unexpired session keys. From this point on, $\mathcal{A}^\pi$ may issue any message in which P_i is specified as the sender and play the role of P_i. The adversary is allowed to replace the public key of a corrupted user by any value of its choice.
2. $\mathsf{session\text{-}key}(\pi_{i,j}^s)$. This query returns the unexpired session key (if any) accepted by P_i during a given session s with P_j.

---

[1] Here we use the notation of Bellare and Rogaway [16] rather than the original notation of CK model, which uses a query named $\mathsf{establish\text{-}session}$ to achieve the same result.

3. session-state($\pi_{i,j}^s$). This query returns all the internal state information of party P_i associated to a particular session s with P_j; the state information does not include the long term private key.
4. session-expiration($\pi_{i,j}^s$). This query can only be performed on a completed session. It is used for defining *forward secrecy* and ensures that the corresponding session key is erased from P_i's memory. The session is thereafter said to be expired;
5. test-session($\pi_{i,j}^s$). To respond to this query, a random bit b is selected. If $b = 0$ then the session key is output. Otherwise, a random key is output chosen from the probability distribution of keys generated by the protocol. This query can only be issued to a session that has not been *exposed*. A session is exposed if the adversary performs any of the following actions:
    – a session-state or session-key query to this session or to the matching session, or
    – a corrupt query to either partner before the session expires at that partner.

Security is defined based on a game played by the adversary. In this game $\mathcal{A}^\pi$ interacts with the protocol. In the first phase of the game, $\mathcal{A}^\pi$ is allowed to activate sessions and perform corrupt, session-key, session-state and session-expiration queries as described above. The adversary then performs a test-session query to a party and session of its choice. The adversary is not allowed to expose the test-session. $\mathcal{A}^\pi$ may then continue with its regular actions with the exception that no other test-session query can be issued. Eventually, $\mathcal{A}^\pi$ outputs a bit b' as its guess on whether the returned value to the test-session query was the session key or a random value, then halts. $\mathcal{A}^\pi$ wins the game if $b = b'$. The definition of security is as follows.

Definition 1. *A key establishment protocol π is called* session key (SK-) secure with *forward secrecy if the following properties are satisfied for any adversary $\mathcal{A}^\pi$.*

1. *If two uncorrupted parties complete matching sessions then they both output the same key.*
2. *The probability that $\mathcal{A}^\pi$ guesses correctly the bit b is no more than $\frac{1}{2}$ plus a negligible function in the security parameter.*

We define the advantage of $\mathcal{A}^\pi$ to be twice the probability that $\mathcal{A}^\pi$ wins, minus one. Hence the second requirement will be met if the advantage of $\mathcal{A}^\pi$ is negligible.

As discussed by Krawczyk [10] it is impossible to achieve forward secrecy in less than three rounds with any protocol authenticated via public keys and without previously established shared state between the parties. With one-pass key establishment, using public keys and with no previous shared secret state, the best that can be achieved is *sender forward secrecy*, whose definition is the same as above, except that sessions can only be expired at the sender. Canetti and Krawczyk also provide a definition of *SK-security without forward secrecy*, where the adversary is not allowed to expire sessions at all.

As mentioned above, for the one-pass key establishment protocols in this paper we define the session-id to be the concatenation of the identities of the peers and the unique message sent in that session. This formulation of session-id prevents the adversary from replaying the message from one protocol in a different session since the model insists that each session has a unique session-id. This may be seen as an artificial way of preventing replay attacks which are an inherent limitation of one-pass key establishment protocols: in reality an adversary can simply replay the single message of the protocol and it will be accepted by the recipient unless it includes some time-varying value. This situation could be addressed by including a time-stamp to uniquely identify a session, and assuming that all parties have access to a universal time oracle [17]. We do not explore that approach further in this paper.

2   Security Definitions

We now present definitions of security for signcryption schemes that complement those of Baek et al. [12]. We then extend these definitions to arrive at new notions of security for signcryption KEMs in the multi-user setting.

2.1   New Security Notions for Signcryption

A signcryption scheme $\mathcal{SC}$ is specified by five polynomial-time algorithms: common-key-gen, sender-key-gen, receiver-key-gen, signcryption and unsigncryption.

common-key-gen: is a probabilistic polynomial time (PPT) algorithm that takes the security parameter k as input and outputs the common/public parameters params used in the scheme. These parameters include description of the underlying groups and hash functions used.

sender-key-gen: is a PPT algorithm that takes params as input and outputs the sender's public-private key pair (pk_s, sk_s) used for signcryption.

receiver-key-gen: is a PPT algorithm that takes params as input and outputs the receiver's public-private key pair (pk_r, sk_r) used for unsigncryption.

signcryption: is a PPT algorithm that takes params, a sender's private key sk_s, a receiver's public key pk_r and message m to be signcrypted as input. It returns a signcryptext C.

unsigncryption: is a deterministic polynomial-time algorithm that takes params, a sender's public key pk_s, a receiver's private key sk_r and a signcryptext C as input. It outputs either a plaintext m or an error symbol $\perp$.

For $\mathcal{SC}$ to be considered valid it is required that unsigncryption(pk_s, sk_r, signcryption(sk_s, pk_r, m)) $= m$ for all sender key pairs (pk_s, sk_s) and receiver key pairs (pk_r, sk_r).

For all the security notions defined in this paper we distinguish two users Alice and Bob as sender and receiver respectively. Depending on the notion of security one of these users, or both, will be adversary's target.

We define insider and outsider security for signcryption schemes in the multi-user setting based on the discussion given by An et al. [3]. It is natural to consider the security of a signcryption scheme in the multi-user setting where Alice can signcrypt messages for any user including Bob, and any user including Alice can signcrypt messages for Bob. Hence, in this model the adversary is given the power to obtain signcryptions of Alice created for any user through a flexible signcryption oracle (FSO). Similarly, the adversary is also given access to a flexible unsigncryption oracle (FUO) that unsigncrypts a given signcryptext created for Bob by any user. Because of these additional adversarial powers, security in the two-user setting does not imply security in the multi-user setting [12].

In any signcryption scheme users employ a private key for two purposes: for signcrypting messages sent to other users and for unsigncrypting messages received from other users. The private keys used for each of these purposes, along with their corresponding public keys, may be the same or they may be different. This is much the same as the option to use the same, or different, keys for decryption and for signing. In keeping with common practice we will assume that each key pair is used for a single purpose. Therefore we specify that all users have two different key pairs for signcryption and for unsigncryption. In contrast, An et al. [3] assumed that a single key pair is used for both signcryption and unsigncryption. Their security model therefore requires giving the adversary access to additional oracles for unsigncrypting by Alice and signcrypting by Bob. To the same end, our model allows the receiver key pair of Alice and sender key pair of Bob to be given to the adversary.

Baek et al. [12] defined only outsider security for confidentiality and insider security for unforgeability in the multi-user setting. In this section we make the desired security notions for signcryption complete by defining insider security for confidentiality and outsider security for unforgeability in the multi-user setting. We assume that the challenger fixes the set of users $\{U_1, \ldots, U_n\}$ and their key pairs before its interaction with the adversary[2]. Note that the security notions defined by Baek et al. [12] implicitly assume the same. The adversary is given all the public keys of the users initially so that it can choose the public keys from the given set when accessing the oracles. One can relax this restriction in our definitions by allowing the adversary to query the FSO and/or FUO with arbitrarily chosen public keys.

Let (pk_A, sk_A) be the public-private key pair used by Alice for signcryption and let (pk_B, sk_B) be the public-private key pair used by Bob for unsigncryption. The behaviour of Alice's FSO and Bob's FUO is described below:

FSO: On the input (pk_r, m), FSO returns a signcryptext C generated using sk_A and the public key pk_r on the message m. The adversary may choose Bob's public key pk_B as the receiver public key, i.e., $pk_r = pk_B$.

FUO: On the input (pk_s, C), FUO returns a plaintext m or a $\perp$ symbol after performing unsigncryption on C using sk_B and the public key pk_s. The adversary may choose pk_s to be the public key of Alice i.e. $pk_s = pk_A$.

For the sake of completeness, we first present our new definition for insider confidentiality and then briefly describe the definition for outsider confidentiality

---

[2] This can be seen as a way of preventing the adversary from registering replaced public keys with the certifying authority for a particular user.

given by Baek et al. [12]. Similarly, we present our new definition for outsider unforgeability and then strengthen the definition for insider unforgeability given by Baek et al. [12].

Insider confidentiality. An insider adversary $\mathcal{A}^{CCA}$ against confidentiality of $\mathcal{SC}$ is assumed to have knowledge of all key pairs except Bob's private key used for unsigncryption. The goal of $\mathcal{A}^{CCA}$ is to break the confidentiality of messages signcrypted for Bob by any user. $\mathcal{A}^{CCA}$ can issue $FUO(pk_s, C)$ for any sender's public key pk_s and a signcryptext C. Alice's FSO can be simulated by $\mathcal{A}^{CCA}$ itself with its knowledge of the corresponding private key. We define this security notion as FUO-IND-CCA2 that simulates chosen ciphertext attacks against $\mathcal{SC}$.

- *Challenge Phase:* After adaptively asking the FUO queries, $\mathcal{A}^{CCA}$ outputs two equal length messages m_0, m_1 and a public key $pk_{s'}$ and submits them to the challenger. The challenger chooses $b \in_R \{0, 1\}$ and gives $\mathcal{A}^{CCA}$ a challenge signcryptext C^* created on m_b using the private key $sk_{s'}$ corresponding to $pk_{s'}$ and Bob's public key pk_B.

$\mathcal{A}^{CCA}$ can continue asking the FUO queries except the trivial $FUO(pk_{s'}, C^*)$. However, an FUO query on C^* using a public key $pk_s \neq pk_{s'}$ is still allowed.

- *Guess Phase:* Finally, $\mathcal{A}^{CCA}$ outputs a bit b' and wins the game if $b' = b$.

The advantage of $\mathcal{A}^{CCA}$ in winning the FUO-IND-CCA2 game is:

$$Adv_{\mathcal{A}^{CCA}, \mathcal{SC}} \overset{def}{=} 2 \cdot \Pr[b' = b] - 1$$

Outsider confidentiality. An adversary against outsider confidentiality of $\mathcal{SC}$ is assumed to know all private keys except Alice's private key used for signcryption and Bob's private key used for unsigncryption. The goal of the adversary in this notion is to break the confidentiality of messages signcrypted by Alice for Bob. The outsider confidentiality notion FSO/FUO-IND-CCA2 [12] gives the adversary access to both FSO and FUO. After adaptively asking the FSO and FDO queries, the challenge and guess phases are carried on as described above. The advantage of the adversary in winning the FSO/FUO-IND-CCA2 game is also defined in the same way as above.

Outsider unforgeability. An outsider adversary $\mathcal{A}^{CMA}$ against unforgeability of $\mathcal{SC}$ is assumed to know all private keys except Alice's private key used for signcryption and Bob's private used for unsigncryption. The goal of $\mathcal{A}^{CMA}$ is to forge a valid signcryptext created by Alice for Bob. $\mathcal{A}^{CMA}$ is given access to both FSO and FUO.

After querying the FSO and FUO adaptively, $\mathcal{A}^{CMA}$ outputs a forgery C^*. A weak (conventional) notion of unforgeability requires C^* to be a valid signcryption created by Alice for Bob on a new message m^* i.e. m^* was never queried to FSO. For the notion of strong unforgeability C^* has to be a valid signcryption of Alice created for Bob on a message m^* such that C^* was never an output of FSO, although $FSO(m^*, pk_r)$ might have been issued earlier, even for $pk_r = pk_B$. We call this notion FSO/FUO-sUF-CMA for strong unforgeability against chosen message attacks by $\mathcal{A}^{CMA}$. The advantage of $\mathcal{A}^{CMA}$ in winning the FSO/FUO-sUF-CMA game is the probability of $\mathcal{A}^{CMA}$ outputting such a C^*.

Insider unforgeability. An adversary against insider unforgeability of $\mathcal{SC}$ is assumed to know all the private keys except Alice's private key used for signcryption. The goal of adversary in this notion is to produce a valid forgery of a signcryptext created by Alice for any other user. The insider unforgeability notion FSO-UF-CMA [12] gives the adversary access to FSO, as FUO can be simulated by the adversary itself. After querying the FSO adaptively, the adversary outputs a forgery (C^*, pk_r^*) for any receiver's public key pk_r^*. The FSO-UF-CMA notion does not require the message m^* to be new although only FSO(m^*, pk_r), for $pk_r \neq pk_r^*$ queries are allowed. We strengthen this notion to FSO-sUF-CMA by allowing FSO(m^*, pk_r) even for $pk_r = pk_r^*$.

2.2 New Security Notions for Signcryption KEM

A signcryption KEM $\mathcal{SK}$ is specified by five polynomial-time algorithms: common-key-gen, sender-key-gen, receiver-key-gen, encapsulation and decapsulation. The algorithms common-key-gen, sender-key-gen and receiver-key-gen are the same as those defined in Section 2.1. The sender's key pair (pk_s, sk_s) and the receiver's key pair (pk_r, sk_r) are now used for encapsulation and decapsulation respectively.

encapsulation: is a PPT algorithm that takes params, a sender's private key sk_s and a receiver's public key pk_r as input. It returns the pair (K, C), where, K is a symmetric key and C is its encapsulation.

decapsulation: is a deterministic polynomial-time algorithm that takes params, a sender's public key pk_s, a receiver's private key sk_r and an encapsulation C. It outputs either a symmetric key K or an error symbol $\bot$.

For $\mathcal{SK}$ to be valid it is required that if $(K, C) =$ encapsulation(sk_s, pk_r), then decapsulation$(pk_s, sk_r, C) = K$ for all sender key pairs (pk_s, sk_s) and receiver key pairs (pk_r, sk_r).

Dent [8,6,9] defined both insider and outsider security notions for signcryption KEMs in the two-user setting. He also provided an informal description of how to define security for signcryption KEMs in the multi-user setting [8]. Recently, Yoshida and Fujiwara [18] defined security notion for signcryption Tag-KEMs in the multi-user setting. Here, we present new notions of security for signcryption KEMs in the multi-user setting building on the definitions in Section 2.1.

In the security model for a signcryption KEM in the multi-user setting the adversary is given the power to obtain encapsulations of Alice created for any user through a flexible encapsulation oracle (FEO). The adversary is also given access to a flexible decapsulation oracle (FDO) that decapsulates a given encapsulation created for Bob by any user. Let (pk_A, sk_A) be the public-private key pair used by Alice for encapsulation and let (pk_B, sk_B) be the public-private key pair used by Bob for decapsulation.

The challenger initially fixes the set of users $\{U_1, \ldots, U_n\}$ and their key pairs. The adversary is given all the public keys of the users initially so that it can choose the public keys from the given set when accessing the oracles. The behaviour of Alice's FEO and Bob's FDO is described below.

FEO: On receiving pk_r, FEO returns a pair (K, C), where C is an encapsulation of K generated using sk_A and pk_r. The adversary may choose pk_B as the receiver's public key, i.e., $pk_r = pk_B$.

FDO: On receiving (pk_s, C), FDO returns a symmetric key K or a $\perp$ symbol after performing decapsulation on C using sk_B and pk_s. The adversary may choose pk_A as the sender's public key i.e. $pk_s = pk_A$.

Insider confidentiality. An insider adversary $\mathcal{A}^{CCA}$ against confidentiality of $\mathcal{SK}$ is assumed to have knowledge of all key pairs except Bob's private key used for decapsulation. The goal of $\mathcal{A}^{CCA}$ is to break the confidentiality of encapsulations created for Bob by any user. It is given access only to FDO as the oracle FEO can be simulated with the knowledge of Alice's private key used for encapsulation. We call this notion of security FDO-IND-CCA2.

– *Challenge Phase:* After adaptively asking the FDO queries, $\mathcal{A}^{CCA}$ outputs a public key $pk_{s'}$. The challenger generates a valid symmetric key, encapsulation pair (K_0, C^*) using the private key $sk_{s'}$ corresponding to $pk_{s'}$ and Bob's public key pk_B. It selects a key K_1 randomly from the symmetric key distribution. It then chooses $b \in_R \{0,1\}$ and gives (K_b, C^*) as the challenge.

$\mathcal{A}^{CCA}$ can continue its execution except asking the FDO$(pk_{s'}, C^*)$ query that trivially decides the guess. However, an FDO query on C^* using a public key $pk_s \neq pk_{s'}$ is still allowed.

– *Guess Phase:* Finally, $\mathcal{A}^{CCA}$ outputs a bit b' and wins the game if $b' = b$.

The advantage of $\mathcal{A}^{CCA}$ in winning the FDO-IND-CCA2 game is

$$Adv_{\mathcal{A}^{CCA}, \mathcal{SK}} \overset{def}{=} 2 \cdot \Pr[b' = b] - 1$$

Outsider confidentiality. For outsider confidentiality the adversary is assumed to know all the private keys except Alice's private key used for encapsulation and Bob's private key used for decapsulation. The goal of the adversary in this notion is to break the confidentiality of encapsulations created by Alice for Bob. The adversary must be given access to both FEO and FDO. We call this notion FEO/FDO-IND-CCA2. After adaptively asking the FEO and FDO queries, the challenge and guess phases are carried on as described above. The advantage of an adversary in winning the FEO/FDO-IND-CCA2 game is also defined in the same way as above.

Outsider unforgeability. An outsider adversary $\mathcal{A}^{CMA}$ against unforgeability of $\mathcal{SK}$ is assumed to know all private keys except Alice's private key used for encapsulation and Bob's private used for decapsulation. The goal of $\mathcal{A}^{CMA}$ is to forge a valid symmetric key and encapsulation pair (K^*, C^*) such that C^* is an encapsulation of K^* created by Alice for Bob. It is given access to both FEO and FDO.

After querying FEO and FDO adaptively, $\mathcal{A}^{CMA}$ produces a forgery (K^*, C^*). It wins the game if $\mathsf{decapsulation}(pk_A, sk_B, C^*) = K^* \neq \perp$. The trivial restriction for (K^*, C^*) to be considered valid is that (K^*, C^*) was never an output of FEO. We call this notion FEO/FDO-sUF-CMA for strong unforgeability against outsider attacks. The advantage of $\mathcal{A}^{CMA}$ in winning the FEO/FDO-sUF-CMA game is the probability of $\mathcal{A}^{CMA}$ outputting such a (K^*, C^*).

Insider unforgeability. An insider adversary against unforgeability of $\mathcal{SK}$ is assumed to know all the private keys except Alice's private key used for encapsulation. The goal of the adversary in this notion is to forge a valid encapsulation created by Alice to any other user. It is given access only to FEO as all keys for decapsulation are known to the adversary. We call this notion FEO-sUF-CMA for strong unforgeability against insider attacks. After adaptively querying the FEO, the adversary outputs a forgery (K^*, C^*, pk_r^*). It wins the FEO-sUF-CMA game if $\mathsf{decapsulation}(pk_A, sk_r^*, C^*) = K^* \neq \perp$. The trivial restriction for (K^*, C^*, pk_r^*) to be considered valid is that (K^*, C^*) was never an output of FEO. The advantage of the adversary in winning the FEO-sUF-CMA game is the probability of outputting such a (K^*, C^*, pk_r^*).

2.3   On the Unforgeability Notion for Signcryption KEMs

Dent [6] defined a different notion, called Left-or-Right (LoR) security, for outsider unforgeability of signcryption KEMs. He showed that an adversary that can output a valid forgery (K^*, C^*) under the notion described in the previous section can be efficiently turned into another adversary that can win the LoR game. Dent pointed out that LoR security is a strong requirement for outsider secure signcryption KEMs. An outsider secure signcryption KEM under our definition can be combined with an outsider secure signcryption DEM in the notion defined by Dent [6] to achieve an outsider secure hybrid signcryption scheme. However, this composition may not yield a tight reduction when compared to the hybrid signcryption scheme that is composed of an LoR secure signcryption KEM and an outsider secure signcryption DEM. In our generic constructions we show that our notion FEO/FDO-sUF-CMA is enough when relating one-pass key establishment protocols with signcryption KEMs. One may still use an LoR secure signcryption KEM to derive a one-pass key establishment protocol, as LoR security guarantees security under the FEO/FDO-sUF-CMA notion.

We emphasise that an insider secure hybrid signcryption scheme can never be guaranteed using the definition of insider unforgeability for signcryption KEM described in the previous section. Dent [9] described the impossibility of achieving an insider secure hybrid signcryption scheme by generic composition of such a signcryption KEM and an insider secure signcryption DEM. The difficulty is that a signcryption KEM that generates symmetric keys and encapsulations independent of the message to be signcrypted cannot provide the non-repudiation service and thus cannot be insider secure. But our definitions of security for insider security of signcryption KEMs are still useful when one observes their connection with one-pass key establishment protocols. A signcryption KEM that is insider confidential in our definition can be used to derive a one-pass key establishment protocol that provides sender forward secrecy. Similarly, we speculate that a signcryption KEM that is insider unforgeable in our definition can be turned into a one-pass key establishment protocol that provides resistance to key compromise impersonation (KCI) attacks even when the receiver's private key is compromised. The latter observation is yet to be explored formally.

3   Outsider Secure Signcryption KEMs

Dent [6] proposed an outsider secure signcryption KEM called elliptic curve integrated signcryption scheme KEM (ECISS-KEM1) based on the ECIES-KEM [19]. A potential problem with the ECISS-KEM1 was identified by Dent who then proposed an improvement that is claimed to overcome this problem, but without a proof of security. Both the schemes are described below.

3.1   ECISS-KEM1

Let (G, P, q) be the system parameters, where G is a large additive cyclic group of prime order q and P is an arbitrary generator of G. Let $(P_s = sP, s)$ and $(P_r = rP, r)$ be the public-private key pairs of the sender and receiver respectively, where $s, r \in_R Z_q^*$. ECISS-KEM1 is described in Figure 1. The scheme uses a hash function that outputs a key of desired length. ECISS-KEM1 is proven secure by Dent in the two-user setting against outsider security (for both confidentiality and integrity) assuming the hardness of CDH problem.

- Encapsulation
    1. Choose an element $t \in_R Z_q^*$
    2. Set $K = \mathrm{Hash}(sP_r + tP)$
    3. Set $C = tP$
    4. Output (K, C)

- Decapsulation
    1. Set $K = \mathrm{Hash}(rP_s + C)$
    2. Output K

Fig. 1. ECISS-KEM1

3.2   Potential Problems with Ephemeral Data

Dent discussed a potential weakness with the scheme ECISS-KEM1 as follows. If an attacker is ever to obtain $sP_r + tP$ (through a temporary break-in), the component sP_r can be recovered easily. This means that the adversary can indefinitely impersonate the sender.

It is interesting that Dent identified this as a problem even though it is not recognised as one by any of the current security models for signcryption. On the other hand, we can see that this capability of the adversary to obtain ephemeral protocol data has already been known in the key establishment models for many years. If the KEM is used to build a one-pass key agreement protocol, then in the Canetti–Krawczyk model described in Section 1.2 the session-state query allows the adversary to obtain $sP_r + tP$ and hence break the protocol.

We would argue that the plausibility of such an attack, as well as its consequences, are equally valid for a signcryption KEM as for a key establishment protocol. Therefore we suggest that session-state queries can be usefully added to the signcryption security model to give a useful stronger notion of security. The feasibility of session-state queries will certainly vary according to the application scenario. Factors that might influence the feasibility include the security of storage during processing and the quality of practical random number generators. It may be argued that applications such as signcryption or one-pass key establishment are of limited vulnerability to session-state queries since local values can be erased immediately once they are used. In contrast, two-pass protocols often require some ephemeral values to be stored until interaction with a protocol peer are completed.

3.3   ECISS-KEM2

Having recognised the problem with ECISS-KEM1, Dent proposed another signcryption KEM (ECISS-KEM2). The system parameters, key pairs and hash function are the same as those in ECISS-KEM1. The symmetric key in the encapsulation algorithm of ECISS-KEM2 is computed as $K = \mathrm{Hash}(sP_r + tP_r)$ and its encapsulation is $C = tP$. Given an encapsulation, the symmetric key can be recovered using the deterministic decapsulation algorithm as $K = \mathrm{Hash}(rP_s + rC)$.

Dent argued that even if an attacker discovers the value $sP_r + tP_r$, it would help in recovering only a single message for which the hashed material is used to produce the symmetric key. This is because it is not easy to compute sP_r from the discovered value and C. Although the security of the scheme is stated informally, Dent claimed that a proof can be given with a non-standard security assumption. However, the attack below enables an active adversary to impersonate the sender to any receiver indefinitely.

Attack on ECISS-KEM2. An active adversary calculates C^* as $P - P_s$ and sends it to the receiver as a message from a sender with the public key P_s. This forces the receiver to compute shared key as $K = \mathrm{Hash}(rP_s + rC^*) = \mathrm{Hash}(rsP + r(P - sP)) = \mathrm{Hash}(rsP + rP - rsP) = \mathrm{Hash}(P_r)$, which can easily be computed by the adversary. Now, the adversary can use a DEM along with ECISS-KEM2 and *signcrypt* messages as having come from the original sender. The attack is possible because the random element chosen in the encapsulation algorithm and the static private key of the sender are not combined in a way that makes eliminating them from the hashed material difficult. This attack directly violates the Left or Right security defined by Dent for outsider unforgeability.

3.4 ECISS-KEM1 in Multi-user Setting

We slightly modify ECISS-KEM1 to work in a multi-user environment as shown in Figure 2. This new version has the same potential problems described in Section 3.2. As suggested by An et al. [3], the identities of the users are now embedded in the encapsulation and decapsulation processes.

- Encapsulation
  1. Choose $C \in_R G$
  2. Set $K = \text{Hash}(\hat{S}, \hat{R}, sP_r + C)$
  3. Output (K, C)

- Decapsulation
  1. Set $K = \text{Hash}(\hat{S}, \hat{R}, rP_s + C)$
  2. Output K

$\hat{S}$, $\hat{R}$ represent the identities of the sender S and receiver R respectively.

Fig. 2. ECISS-KEM1 in the multi-user setting

Theorem 1. *ECISS-KEM1 in the multi-user setting is secure in the outsider unforgeability notion in the random oracle model assuming hardness of the Gap Diffie Hellman (GDH) problem in G.*

Theorem 2. *ECISS-KEM1 in the multi-user setting is secure in the outsider confidentiality notion in the random oracle model assuming hardness of the GDH problem in G.*

The proof of Theorem 1 is provided in Appendix A. The proof of Theorem 2 is very similar to that of Theorem 1 and we omit the details.

4 Signcryption KEM from One-Pass Key Establishment

This section first discusses how a one-pass key establishment protocol π can be used as a signcryption KEM $\mathcal{SK}$. A proof of the generic construction is also provided.

The session key computed by a sender in π can be used as the symmetric key of $\mathcal{SK}$. The outgoing message of π becomes the encapsulation of the key. The session key computation process at the receiver end in π can be used as the decapsulation algorithm to retrieve the symmetric key.

Theorem 3. *If π is a one-pass key establishment protocol SK-secure with sender forward secrecy in the CK model, then it can be used as a signcryption KEM that is secure in the insider confidentiality and outsider unforgeability notions.*

Proof. To prove the theorem it is enough if we show that if $\mathcal{SK}$ is not secure in the insider confidentiality **or** outsider unforgeability notion, then π is also not secure in the CK model. Given an adversary $\mathcal{A}^{CCA}$ against insider confidentiality or $\mathcal{A}^{CMA}$ against outsider unforgeability with non-negligible advantage, we construct an adversary $\mathcal{A}^{\pi}$ against SK-security of π in the CK model that can distinguish a real session key from a random number in polynomial time.

Constructing $\mathcal{A}^\pi$ from $\mathcal{A}^{CMA}$: We start by assuming the existence of $\mathcal{A}^{CMA}$ against outsider unforgeability, with a non-negligible advantage ϵ_u. Then, we prove that $\mathcal{A}^\pi$ can distinguish a real session key from a random value with the same advantage using $\mathcal{A}^{CMA}$ as subroutine. The running time of $\mathcal{A}^\pi$ is $t_1 \leq t_u + (n_{feo} + n_{fdo})(t_s + t_k)$, where t_u is the time required for $\mathcal{A}^{CMA}$ to forge $\mathcal{SK}$, n_{feo} and n_{fdo} are the number of FEO and FDO queries issued by $\mathcal{A}^{CMA}$ respectively and t_s and t_k are the response times for send and session-key queries respectively. The view of $\mathcal{A}^{CMA}$ is simulated as below:

$\mathcal{A}^\pi$ allows $\mathcal{A}^{CMA}$ to choose two users U_A and U_B from a set of users $\{U_1, \ldots, U_n\}$. It then corrupts all the parties except U_A and U_B and gives their key pairs to $\mathcal{A}^{CMA}$. This enables $\mathcal{A}^{CMA}$ to choose public keys from the given set when accessing U_A's FEO and U_B's FDO. The queries asked by $\mathcal{A}^{CMA}$ are answered as below:

- *FEO Queries:* For an FEO query asked by $\mathcal{A}^{CMA}$ with input pk_j, the adversary $\mathcal{A}^\pi$ initiates a session by issuing a send$(\pi^s_{A,j}, \lambda)$ query and obtains the parameter C computed by the oracle $\pi^s_{A,j}$. It then issues a session-key$(\pi^s_{A,j})$ query to get the session key K accepted in that session. The pair (K, C) is returned to $\mathcal{A}^{CMA}$.

- *FDO Queries:* On an FDO query with the input (pk_i, C), $\mathcal{A}^\pi$ issues a send$(\pi^s_{B,i}, C)$. It then issues a session-key$(\pi^s_{B,i})$ query and returns the accepted session key K to $\mathcal{A}^{CMA}$. It returns a $\perp$ symbol if there is no key accepted in the session.

Answering the challenger: $\mathcal{A}^{CMA}$ finally outputs a forgery (K^*, C^*) such that C^* is a valid encapsulation of K^* created by U_A for U_B. $\mathcal{A}^\pi$ now establishes a fresh session between U_A and U_B, by issuing send$(\pi^t_{B,A}, C^*)$ and chooses it as the test session. The challenger computes the session key K_0 of the test session and selects a random value K_1 from session key distribution. It then chooses $b \in_R \{0, 1\}$ and gives K_b to $\mathcal{A}^\pi$. $\mathcal{A}^\pi$ outputs its guess as 0 if $K_b = K^*$ or 1 otherwise.

For $\mathcal{A}^{CMA}$ to be successful it has to forge a valid encapsulation of U_A created for U_B i.e. between any two users that were chosen initially. As explained above, $\mathcal{A}^\pi$ always wins whenever $\mathcal{A}^{CMA}$ outputs such a forgery by establishing a test session between those two users. Hence, the advantage of $\mathcal{A}^\pi$ constructed from $\mathcal{A}^{CMA}$ is

$$Adv_1^\pi(k) = \epsilon_u \tag{1}$$

For each FEO or FDO query, $\mathcal{A}^\pi$ has to establish a session through a send query and retrieve the session key through a session-key query. Hence, the running time of $\mathcal{A}^\pi$ is bounded by $t_1 \leq t_u + (n_{feo} + n_{fdo})(t_s + t_k)$.

Constructing $\mathcal{A}^\pi$ from $\mathcal{A}^{CCA}$: Now, we assume that there exists $\mathcal{A}^{CCA}$ against insider confidentiality with a non-negligible advantage ϵ_c. Using $\mathcal{A}^{CCA}$ as subroutine, we construct an adversary $\mathcal{A}^\pi$ that can distinguish real session key from a random value with an advantage of at least $\frac{\epsilon_c}{(n-1)}$. The running time of $\mathcal{A}^\pi$ is $t_2 \leq t_c + n_{fdo}(t_s + t_k)$, where t_c is the running time of $\mathcal{A}^{CCA}$, n_{fdo} is the number of FDO queries issued by $\mathcal{A}^{CCA}$ and t_s and t_k are the response times for send and session-key queries respectively. $\mathcal{A}^\pi$ allows $\mathcal{A}^{CCA}$ to select a user U_B from the set of users $\{U_1, \ldots, U_n\}$. The aim of $\mathcal{A}^{CCA}$ is to break the confidentiality of an encapsulation created for U_B by any other user.

$\mathcal{A}^\pi$ now initiates a session $\pi_{A,B}^t$ between U_B and any other user U_A, by issuing a send$(\pi_{A,B}^t, \lambda)$ query. It obtains the outgoing parameter C^* and establishes a matching session by issuing a send$(\pi_{B,A}^t, C^*)$ query. $\mathcal{A}^\pi$ chooses either $\pi_{A,B}^t$ or $\pi_{B,A}^t$ as the test session. The challenger selects $b \in_R \{0,1\}$ and gives real session key computed in the test session if $b = 0$ or a random value chosen from session key distribution otherwise. Let K_b be the value returned to $\mathcal{A}^\pi$. $\mathcal{A}^\pi$ now issues a session-expiration$(\pi_{A,B}^t)$ query, which ensures that the key computed in that session is erased.

$\mathcal{A}^\pi$ corrupts all the users (including U_A) except U_B and gives the key pairs to $\mathcal{A}^{CCA}$. It is now ready to answer the queries asked by $\mathcal{A}^{CCA}$:

- FDO Queries: When a decapsulation query is asked with the input (pk_i, C), $\mathcal{A}^\pi$ initiates a session through send$(\pi_{B,i}^s, C)$ query. It then issues a session-key$(\pi_{B,i}^s)$ and obtains the session key K generated in that session and returns it to $\mathcal{A}^{CCA}$. It returns a $\perp$ symbol if there is no key accepted in the session.

Answering the challenger: After adaptively asking the FDO queries $\mathcal{A}^{CCA}$ outputs a public key $pk_{s'}$. If $pk_{s'} \neq pk_A$, $\mathcal{A}^\pi$ aborts its execution. Otherwise, it gives (K_b, C^*) as the challenge to $\mathcal{A}^{CCA}$. $\mathcal{A}^{CCA}$ may continue to ask the FDO queries except the trivial one with input (pk_A, C^*). It finally returns a bit θ as its guess with an advantage ϵ_c. Incase $\theta = 0$, $\mathcal{A}^\pi$ outputs $b = 0$, which implies C^* is a valid encapsulation of K_b and thus K_b is a real session key. $\mathcal{A}^\pi$ outputs $b = 1$ otherwise.

$\mathcal{A}^{CCA}$ becomes successful if it can break the confidentiality of an encapsulation created for the initially chosen U_B by any other user. $\mathcal{A}^\pi$ wins its game with non-negligible advantage only if $\mathcal{A}^{CCA}$ outputs $pk_{s'} = pk_A$ in the challenge phase i.e. the public key of the user U_A selected by $\mathcal{A}^\pi$. This occurs with the probability $\frac{1}{(n-1)}$. Hence, the advantage of $\mathcal{A}^\pi$ when constructed from $\mathcal{A}^{CCA}$ is

$$Adv_2^\pi(k) \geq \frac{\epsilon_c}{(n-1)} \tag{2}$$

For each FDO query asked by $\mathcal{A}^{CCA}$, $\mathcal{A}^\pi$ has to establish a session through a send query and retrieve the session key through a session-key query. Hence, the running time of $\mathcal{A}^\pi$ is bounded by $t_2 \leq t_c + n_{fdo}(t_s + t_k)$.

From (1) and (2), the advantage of $\mathcal{A}^\pi$ when constructed from $\mathcal{A}^{CMA}$ or $\mathcal{A}^{CCA}$ is $Adv^\pi(k) \geq \min\{Adv_1^\pi(k), Adv_2^\pi(k)\}$, which is non-negligible. The running time of such $\mathcal{A}^\pi$ with the advantage $Adv^\pi(k)$ is $t_\pi \leq \max\{t_1, t_2\}$. But, as the protocol π is secure in the CK model $Adv^\pi(k)$ must be negligible. This is a contradiction to the construction of $\mathcal{A}^\pi$ from $\mathcal{A}^{CMA}$ or $\mathcal{A}^{CCA}$. Hence, there exists no such $\mathcal{A}^{CMA}$ or $\mathcal{A}^{CCA}$ that has non-negligible advantage against $\mathcal{SK}$ □

Note that, if π does not provide sender forward secrecy, then the resulting $\mathcal{SK}$ will be outsider secure for both confidentiality and unforgeability notions.

4.1 New Signcryption KEM from the One-Pass HMQV

The one-pass HMQV protocol proposed by Krawczyk [10] can be used as a signcryption KEM secure in the insider confidentiality and outsider unforgeability notions. This new signcryption KEM between the parties A and B in the multi-user setting is presented in Figure 3. Apart from the system parameters used for ECISS-KEM1 described in Section 3.1, a new hash function $\mathcal{H}$ defined as $\mathcal{H} : G \times \{0,1\}^* \to Z_q^*$ is used.

- Encapsulation
    1. Choose $t \in_R Z_q^*$
    2. Set $C = tP$
    3. Set $h = \mathcal{H}(C, (\hat{A}||\hat{B}))$
    4. Set $K = \mathrm{Hash}((t + sh)P_r)$
    5. Output (K, C)

- Decapsulation
    1. Set $h = \mathcal{H}(C, (\hat{A}||\hat{B}))$
    2. Set $K = \mathrm{Hash}\,(r(C + hP_s))$
    3. Output K

Fig. 3. New Signcryption KEM

4.2 Security of the New KEM

Krawczyk [10] proved the one-pass HMQV secure in the CK model. Its security is based on the XCR signature, whose security was also proven by Krawczyk in the random oracle model assuming the hardness of the CDH problem. By combining this result with Theorem 3, it follows that the new signcryption KEM is secure in the insider confidentiality and outsider unforgeability notions.

Table 1 compares the new signcryption KEM with existing signcryption KEMs in terms of security and efficiency. The security notions considered are insider and outsider security for both confidentiality and unforgeability. The efficiency is measured by number of group exponentiations required in encapsulation and decapsulation algorithms. The new signcryption KEM is the only one that has

Table 1. Security and efficiency comparisons with existing signcryption KEMs

	Confidentiality		Unforgeability		Efficiency	
	Outsider	Insider	Outsider	Insider	Encap.	Decap.
ECISS-KEM1 [6]	Y	N	Y	N	2 Exp	1 Exp
ECISS-KEM2 [6]	broken					
Dent [9]	Y	N	Y	Y	1 Exp	2 Exp
Bjørstad and Dent [13]	Y	N	Y	Y	1 Exp	2 Exp
Our new KEM	Y	Y	Y	N	2 Exp	1.5 Exp

insider security for confidentiality. It achieves this forward secrecy with an additional half-length exponentiation[3] compared to the ECISS-KEM1 in the decapsulation algorithm. Unlike the ECISS-KEM1, the discovery of ephemeral data by an adversary in the new signcryption KEM leads to compromise of only one particular communication. Moreover, the notions of security considered for all other signcryption KEMs are in the two-user setting, whereas the security of the new signcryption KEM is treated in the multi-user setting.

5   One-Pass Key Establishment from Signcryption KEM

We now consider the generic construction in the other direction. We first discuss how a signcryption KEM $\mathcal{SK}$ can be used as a one-pass key establishment protocol π. The security requirements of $\mathcal{SK}$ that can be used are first stated and a formal construction of π from $\mathcal{SK}$ is then presented.

When $\mathcal{SK}$ is used as π, the encapsulation algorithm of $\mathcal{SK}$ becomes the session key computation process by the sender in π. The generated symmetric key serves as the session key and the encapsulation of the symmetric key as the outgoing message to the receiver. The receiver can compute the same session key by executing the decapsulation algorithm on the incoming message.

For $\mathcal{SK}$ to be suitable to be used as a one-pass key establishment protocol it should be secure in the insider confidentiality and outsider unforgeability notions. Security in these notions enables the resulting protocol to have SK-security with sender forward secrecy in the CK model. For the reasons discussed in Section 3.2 security against compromise of ephemeral data is not guaranteed for π. Therefore the adversary is not allowed to have access to the session-state query.

Theorem 4. *If a signcryption KEM is secure in the insider confidentiality and outsider unforgeability notions, then it can be used as a one-pass key establishment protocol π that is SK-secure with sender forward secrecy in the CK model (without session-state queries).*

Proof. The truth value of the above theorem is the same as the statement: if π is not secure in the CK model, then $\mathcal{SK}$ is not secure in either insider confidentiality **or** outsider unforgeability notion. Hence, it is enough to show that given an adversary $\mathcal{A}^\pi$ against π that can distinguish a real session key from a random number with advantage ϵ, then either $\mathcal{A}^{CMA}$ or $\mathcal{A}^{CCA}$ against $\mathcal{SK}$ can be constructed with advantage $\epsilon' \geq \epsilon$ in polynomial time.

The proof is divided into two parts. In the first part $\mathcal{A}^{CMA}$ is constructed with non-negligible advantage only if an event Forgery (explained later) occurs. In the second part $\mathcal{A}^{CCA}$ is constructed from $\mathcal{A}^\pi$ with non-negligible advantage if the event Forgery does not occur.

Let $\{U_1, U_2, ..., U_n\}$ be set of n users and assume each user is activated at most m times by $\mathcal{A}^\pi$, where n and m are polynomials in the security parameter.

---

[3] Krawczyk [10] showed that the length of $h = \frac{q}{2}$ provides the right performance-security trade-off.

Constructing $\mathcal{A}^{CMA}$ from $\mathcal{A}^{\pi}$: We assume the existence of $\mathcal{A}^{\pi}$ that can distinguish a real session key from a random value in time t_f. We then construct $\mathcal{A}^{CMA}$ within time $t_1 \leq t_f + m(n-1)(t_{feo} + 2 \cdot t_{fdo})$, where t_{feo} and t_{fdo} are the response times for the FEO and FDO queries respectively.

The input to $\mathcal{A}^{CMA}$ consists of the sender and receiver public keys pk_A and pk_B of two users U_A and U_B from the set of n users $\{U_1, \ldots, U_n\}$ respectively. Its aim is to produce (K^*, C^*) where C^* is a valid encapsulation of K^* under sk_A and pk_B, using $\mathcal{A}^{\pi}$ as subroutine. The input of $\mathcal{A}^{CMA}$ also contains key pairs of each of the n parties in the protocol, except the sender's private key sk_A of U_A and receiver's private key sk_B of U_B for whom only the corresponding public keys are given. $\mathcal{A}^{CMA}$ wins its game only if the target session chosen by $\mathcal{A}^{\pi}$ is a session between U_A and U_B. All the queries from $\mathcal{A}^{\pi}$ that do not concern U_A and U_B can be answered directly by $\mathcal{A}^{CMA}$ which knows the private keys of the users. For queries that require the knowledge of sk_A and sk_B, $\mathcal{A}^{CMA}$ uses its own oracles and returns the messages produced to $\mathcal{A}^{\pi}$ as described below.

- send: When a $\mathsf{send}(\pi_{A,j}^s, \lambda)$ query is asked, $\mathcal{A}^{CMA}$ queries its FEO with the input pk_j and obtains (K, C). It then returns C to $\mathcal{A}^{\pi}$ as the outgoing message and keeps (s, A, j, K, C) in its encapsulation list $L_{\mathcal{E}}$. If $\mathcal{A}^{\pi}$ issues $\mathsf{send}(\pi_{B,i}^s, C)$, $\mathcal{A}^{CMA}$ queries its FDO with the input (pk_i, C). If it obtains a symmetric key K from the challenger, $\mathcal{A}^{CMA}$ marks the oracle $\pi_{B,i}^s$ as accepted and stores the value (s, B, i, K, C) in $L_{\mathcal{E}}$. If the output of the FDO is $\perp$ then the session is not accepted and the entry $(s, B, i, \perp, C)$ is stored in $L_{\mathcal{E}}$. The result of whether the session is accepted or not is made known to $\mathcal{A}^{\pi}$.

- session-key: For a $\mathsf{session\text{-}key}(\pi_{i,j}^s)$ query, it returns the key held in the session with identifier s as follows: Since a session-key key reveal query is issued only on a session that has an accepted session key, the session id s must have an entry in $L_{\mathcal{E}}$. $\mathcal{A}^{CMA}$ checks to see if there is an entry for (s, i, j) in $L_{\mathcal{E}}$ and returns the corresponding key K incase of a match. If there is no key stored in $L_{\mathcal{E}}$ along with s, the session-key is not a valid query.

- session-expiration: On the input $\pi_{i,j}^s$, $\mathcal{A}^{CMA}$ deletes the entry with s from $L_{\mathcal{E}}$. There must have been an entry in $L_{\mathcal{E}}$ because session-expiration can be issued only on a completed session.

- corrupt: When $\mathcal{A}^{\pi}$ wishes to corrupt a party U_i (for $i \neq A, B$), $\mathcal{A}^{CMA}$ fetches session keys from $L_{\mathcal{E}}$ that are generated by the oracles at U_i. It returns all these keys (if any) along with U_i's long term private key. $\mathcal{A}^{CMA}$ outputs "fail" for a corrupt query on U_A or U_B. Note that $\mathcal{A}^{CMA}$ cannot return the internal state information for a corrupt query for reasons discussed in Section 3.2.

Whenever, a $\mathsf{send}(\pi_{B,A}^s, C)$ is issued, $\mathcal{A}^{CMA}$ first checks to see if there exists an entry (s, A, B, K, C) in $L_{\mathcal{E}}$ for some s and K. If there is an entry it just returns the message that the session is accepted. Otherwise, it queries its FDO with the input (pk_A, C). If the output of FDO is $\perp$ it returns the message that the session is not accepted. If $FDO(pk_A, C) = K^* \neq \bot$, $\mathcal{A}^{CMA}$ outputs (K^*, C^*) as its forgery with $C^* = C$.

Let Forgery be the event that $\mathcal{A}^\pi$ issues a $send(\pi^s_{B,A}, C^*)$ such that C^* is a valid encapsulation under sk_A and pk_B such that it was not the response of an earlier $send(\pi^s_{A,B}, \lambda)$ query. Clearly, $\mathcal{A}^{CMA}$ wins its game only if the event Forgery occurs. If $\mathcal{A}^\pi$ ends its run without choosing a test session between U_A and U_B or if the event Forgery does not occur $\mathcal{A}^{CMA}$ outputs "fail". The probability of $\mathcal{A}^\pi$ choosing a test session that has U_A as initiator and U_B as responder is $\frac{1}{n(n-1)}$. Thus, the non-negligible advantage of $\mathcal{A}^{CMA}$ is given as:

$$Adv_{\mathcal{A}}^{CMA}(k) \geq \frac{\Pr[\text{Forgery}]}{n(n-1)} \tag{3}$$

For each send query to the oracle $\pi^s_{A,j}$, $\mathcal{A}^{CMA}$ has to query the FEO and FDO oracles. The maximum number of such queries involving U_A can be $m(n-1)$. Similarly, for each send query to $\pi^s_{B,i}$ a query to FDO is made. The maximum possible number of such queries involving U_B is $m(n-1)$. Hence, $\mathcal{A}^{CMA}$ can forge $\mathcal{SK}$ with the above advantage in time $t_1 \leq t_f + m(n-1)(t_{feo} + 2 \cdot t_{fdo})$.

Constructing $\mathcal{A}^{CCA}$ from $\mathcal{A}^\pi$: Now, we assume the existence of $\mathcal{A}^\pi$ that can distinguish a real session key from a random value in time t_d when the event Forgery does not occur. Using $\mathcal{A}^\pi$ as subroutine, we construct $\mathcal{A}^{CCA}$ within time $t_2 \leq t_d + (m(n-1)-1)t_{fdo}$, where t_{fdo} is the time required to get a response from FDO.

The input to $\mathcal{A}^{CCA}$ consists of a receiver's public key pk_B of a user U_B and the key pairs of rest of users from the set $\{U_1, \dots, U_n\}$. The aim of $\mathcal{A}^{CCA}$ is to break the confidentiality of encapsulations created for U_B by any other user using $\mathcal{A}^\pi$ as subroutine.

$\mathcal{A}^{CCA}$ returns a sender's public key pk_A of a user U_A to its challenger. The challenger gives (K_b, C^*) to $\mathcal{A}^{CCA}$ as the challenge computed as described in Section 2.2. $\mathcal{A}^{CCA}$ chooses $t \in_R \{1, \dots, m\}$. With these choices $\mathcal{A}^{CCA}$ is trying to guess $\mathcal{A}^\pi$'s choice of the target session. It is now ready to simulate the view of $\mathcal{A}^\pi$.

Except for U_B, the actions of rest of the uncorrupted users are performed by $\mathcal{A}^{CCA}$ with its knowledge of the corresponding private keys. For queries that require the knowledge of receiver's private key of U_B, $\mathcal{A}^{CCA}$ uses its own oracle.

- send: When $\mathcal{A}^\pi$ issues a $send(\pi^s_{i,B}, \lambda)$ query, $\mathcal{A}^{CCA}$ generates (K, C), where C is encapsulation of K created by U_i as it knows the private key sk_i. It then returns C to $\mathcal{A}^\pi$ as the outgoing value and keeps (s, i, B, K, C) in its encapsulation list $L_{\mathcal{E}}$. If $\mathcal{A}$ issues $send(\pi^s_{B,i}, C)$, $\mathcal{A}^{CCA}$ queries its FDO with the input (pk_i, C). If it obtains a symmetric key K from the challenger, the session is accepted and the entry (s, B, i, K, C) is stored in $L_{\mathcal{E}}$. If the output of FDO is $\bot$, the session is not accepted and the entry $(s, B, i, \bot, C)$ is stored. The result of whether the session is accepted or not is made known to $\mathcal{A}^\pi$. The t-th instantiation between U_A and U_B is handled in a special way as explained later.

- session-key: When a session-key($\pi_{i,j}^s$) is issued, $\mathcal{A}^{CCA}$ first checks to see if there is an entry for (s, i, j) in $L_{\mathcal{E}}$ and returns the corresponding key incase of a match. Otherwise, the session-key is not a valid query.
- session-expiration: On the input $\pi_{i,j}^s$, $\mathcal{A}^{CCA}$ deletes the entry with s from $L_{\mathcal{E}}$.
- corrupt: When $\mathcal{A}^\pi$ wishes to corrupt a party U_i (for $i \neq A, B$), $\mathcal{A}^{CCA}$ fetches session keys from $L_{\mathcal{E}}$ that are generated by the oracles at U_i. It returns all these keys (if any) along with U_i's long term private key. It aborts the simulation on a corrupt query on U_A or U_B.

If a send($\pi_{A,B}^t, \lambda$) query is issued, $\mathcal{A}^{CCA}$ returns C^* as the outgoing parameter. Now, $\mathcal{A}^\pi$ can choose the t-th session between U_A and U_B as its target session in one of following ways

- The session $\pi_{A,B}^t$ itself or
- A matching session established by issuing send($\pi_{B,A}^t, C^*$) query.

$\mathcal{A}^\pi$ can issue a corrupt(U_A) only after a session-expiration($\pi_{A,B}^t$). If it chooses the t-th session between A and B as the test session as expected by $\mathcal{A}^{CCA}$, K_b is returned. Eventually, $\mathcal{A}^\pi$ halts with its guess θ. If $\theta = 0$, $\mathcal{A}^{CCA}$ outputs $b = 0$ implying that K_b is a real session key and thus C^* is an encapsulation of K_b. Otherwise $b = 1$ is returned.

If Forgery occurs $\mathcal{A}^\pi$ may win its game without choosing the session in which the challenge encapsulation C^* is injected, as the test session. In this case $\mathcal{A}^{CCA}$ gets no advantage. Hence, if the event Forgery occurs or if $\mathcal{A}^\pi$ chooses a different session other than the one expected by $\mathcal{A}^{CCA}$ as test session, $\mathcal{A}^{CCA}$ outputs a random bit b with a probability $\frac{1}{2}$.

The probability of $\mathcal{A}^\pi$ choosing a session t that has U_A as initiator and U_B as responder is $\frac{1}{mn(n-1)}$. The non-negligible advantage of $\mathcal{A}^{CCA}$ is given as

$$Adv_{\mathcal{A}}^{CCA}(k) \geq \frac{(Adv_{\mathcal{A}}^\pi(k)|\overline{\text{Forgery}})}{mn(n-1)} \tag{4}$$

For each send($\pi_{B,i}^s$) query, $\mathcal{A}^{CCA}$ has to issue an FDO query. The maximum possible number of such queries involving the user U_B is $m(n-1) - 1$; excluding one in the test session. Hence, the running time of $\mathcal{A}^{CCA}$ with the above advantage is $t_2 \leq t_d + (m(n-1) - 1)t_{fdo}$.

By the theorem of total probability, the advantage of $\mathcal{A}^\pi$ is given by

$$Adv_{\mathcal{A}}^\pi = (Adv_{\mathcal{A}}^\pi|\text{Forgery}) \times \Pr(\text{Forgery}) + (Adv_{\mathcal{A}}^\pi|\overline{\text{Forgery}}) \times \Pr(\overline{\text{Forgery}})$$
$$\leq \Pr(\text{Forgery}) + (Adv_{\mathcal{A}}^\pi|\overline{\text{Forgery}})$$

However, from Equations 3 and 4, $\Pr(\text{Forgery})$ and $(Adv_{\mathcal{A}}^\pi|\overline{\text{Forgery}})$ are negligible when $\mathcal{SK}$ is secure in the insider confidentiality and outsider unforgeability notions. Hence, the advantage of an adversary $\mathcal{A}^\pi$ against one-pass key establishment protocol constructed from such an $\mathcal{SK}$ is also negligible. $\square$

New key establishment protocol: The ECISS-KEM1 in the multi user setting described in Figure 2 can be used as a one-pass key establishment protocol. However, as the ECISS-KEM1 is secure only in the outsider security model it does not provide sender forward secrecy. Moreover, as discussed in Section 3.2 it does not have security against **session-state reveal** queries. One advantage it does have over the one-pass HMQV is that its overall efficiency is better.

6   Conclusion

We have shown that there exists a duality between signcryption KEMs and one-pass key establishment protocols. However, the models typically used for defining security of key establishment are stronger than those for signcryption. Hence it has turned out that starting from signcryption KEMs we can only derive one-pass key establishment protocols with weaker security than those already known (such as HMQV).

In the other direction we have been able to use a strong one-pass key establishment protocol (HMQV) to derive a signcryption KEM with stronger properties than those known before. However, even though our signcryption KEM is stronger in terms of confidentiality, it does not provide insider secure authentication (non-repudiation). It might be possible to obtain a signcryption KEM that is insider secure with respect to both confidentiality and authentication from a one-pass key establishment protocol that is sender forward secure and resilient to KCI attacks. The feasibility of doing this is still not clear.

It remains an open question to derive hybrid signcryption schemes with insider security for both confidentiality and authentication even without using our generic constructions. Providing more signcryption schemes secure when the adversary has access to a **session-state** query also remains an interesting challenge.

Acknowledgements

The authors thank Alex Dent for his extensive comments and expert advice. This work was supported by the Australian Research Council under grant DP0773348.

References

1. Zheng, Y.: Digital Signcryption or How to Achieve Cost(Signature & Encryption) << Cost(Signature) + Cost(Encryption). In: Kaliski Jr., B.S. (ed.) CRYPTO 1997. LNCS, vol. 1294, pp. 165–179. Springer, Heidelberg (1997)
2. Zheng, Y.: Shortened Digital Signature, Signcryption and Compact and Unforgeable Key Agreement Schemes. Technical report, A submission to IEEE P1363 Standard Specifications for Public Key Cryptography (1998),
   http://grouper.ieee.org/groups/1363/StudyGroup/Hybrid.html
3. An, J., Dodis, Y., Rabin, T.: On the Security of Joint Signature and Encryption. In: Knudsen, L.R. (ed.) EUROCRYPT 2002. LNCS, vol. 2332, pp. 83–107. Springer, Heidelberg (2002)

4. An, J.: Authenticated Encryption in the Public-Key Setting: Security Notions and Analyses. Cryptology ePrint Archive, Report, 2001/079 (2001), http://eprint.iacr.org/2001/079

5. Dodis, Y.: Signcryption (Short Survey). Encyclopedia of Cryptography and Security, (2005), http://theory.lcs.mit.edu/~yevgen/surveys.html

6. Dent, A.: Hybrid Signcryption Schemes with Outsider Security. In: Zhou, J., Lopez, J., Deng, R.H., Bao, F. (eds.) ISC 2005. LNCS, vol. 3650, pp. 203–217. Springer, Heidelberg (2005)

7. Cramer, R., Shoup, V.: Design and analysis of practical public-key encryption schemes secure against adaptive chosen ciphertext attack. Technical report (2002), http://shoup.net/

8. Dent, A.: Hybrid Cryptography. Cryptology ePrint Archive, Report, 2004/210 (2004), http://eprint.iacr.org/2004/210

9. Dent, A.: Hybrid Signcryption Schemes with Insider Security. In: Boyd, C., González Nieto, J.M. (eds.) ACISP 2005. LNCS, vol. 3574, pp. 253–266. Springer, Heidelberg (2005)

10. Krawczyk, H., HMQV,: A High-Performance Secure Diffie-Hellman Protocol. In: Shoup, V. (ed.) CRYPTO 2005. LNCS, vol. 3621, pp. 546–566. Springer, Heidelberg (2005)

11. Baek, J., Steinfeld, R., Zheng, Y.: Formal Proofs for the Security of Signcryption. In: Naccache, D., Paillier, P. (eds.) PKC 2002. LNCS, vol. 2274, Springer, Heidelberg (2002)

12. Baek, J., Steinfeld, R., Zheng, Y.: Formal Proofs for the Security of Signcryption. Journal of Cryptology 20, 203–235 (2007)

13. Bjørstad, T., Dent, A.: Building Better Signcryption Schemes with Tag-KEMs. In: Yung, M., Dodis, Y., Kiayias, A., Malkin, T.G. (eds.) PKC 2006. LNCS, vol. 3958, pp. 491–507. Springer, Heidelberg (2006)

14. Bellare, M., Canetti, R., Krawczyk, H.: A Modular Approach to the Design and Analysis of Authentication and Key Exchange Protocols (Extended Abstract). In: STOC1998. Proc. of the 30th Annual ACM Symposium on Theory of Computing, pp. 419–428 (1998)

15. Canetti, R., Krawczyk, H.: Analysis of Key-Exchange Protocols and Their Use for Building Secure Channels. In: Pfitzmann, B. (ed.) EUROCRYPT 2001. LNCS, vol. 2045, pp. 453–474. Springer, Heidelberg (2001)

16. Bellare, M., Rogaway, P.: Entity Authentication and Key Distribution. In: Stinson, D.R. (ed.) CRYPTO 1993. LNCS, vol. 773, pp. 232–249. Springer, Heidelberg (1994)

17. Tin, Y.S.T., Vasanta, H., Boyd, C., González-Nieto, J.M.: Protocols with Security Proofs for Mobile Applications. In: Wang, H., Pieprzyk, J., Varadharajan, V. (eds.) ACISP 2004. LNCS, vol. 3108, pp. 358–369. Springer, Heidelberg (2004)

18. Yoshida, M., Fujiwara, T.: On the Security of Tag-KEM for Signcryption. Electr. Notes Theor. Comput. Sci. 171, 83–91 (2007)

19. International Organization for Standardization: ISO/IEC CD 18033-2, Information technology - Security techniques - Encryption Algorithms - Part 2: Asymmetric Ciphers (2003)

20. Okamoto, T., Pointcheval, D.: The Gap-Problems: A New Class of Problems for the Security of Cryptographic Schemes. In: Kim, K.-c. (ed.) PKC 2001. LNCS, vol. 1992, pp. 104–118. Springer, Heidelberg (2001)

A   Proof of Theorem 1

Proof. To prove this theorem we show that if there exists a polynomial time adversary $\mathcal{A}^{CMA}$ against the unforgeability of the KEM with non-negligible advantage ϵ, then a polynomial time algorithm $\mathcal{A}^{GDH}$ can be constructed that solves the Gap Diffie-Hellman (GDH) problem with the same advantage as $\mathcal{A}^{CMA}$. Recall that the GDH problem entails solving the Computational Diffie-Hellman (CDH) with the assistance of a decisional Diffie-Hellman oracle $\mathcal{O}_{DDH}$ [20].

Let $A = aP$, $B = bP$ be the problem instance given to $\mathcal{A}^{GDH}$ with the goal to find the value abP. $\mathcal{A}^{GDH}$ runs $\mathcal{A}^{CMA}$ and simulates the answers to the queries made by $\mathcal{A}^{CMA}$ as shown below.

- Hash: For Hash queries , $\mathcal{A}^{GDH}$ initially starts with an empty list $L_{\mathcal{H}}$. On input $(\hat{S}, \hat{R}, X)$, $\mathcal{A}^{GDH}$ first checks to see if there is an existing entry $(\hat{S}, \hat{R}, X, K)$ for some K in $L_{\mathcal{H}}$ that stores the past returned hash values. If so, it returns the corresponding K; otherwise it accesses the global encapsulation list $L_{\mathcal{E}}$ and does the following:

  if $(\hat{S}, \hat{R}, C, K) \in L_{\mathcal{E}}$ *for some K and C values* **then**
  compute $Y = (X - C)$
  if $\mathcal{O}_{DDH}(P_s, P_r, Y) = True$ **then**
     if $P_s = A$ *and* $P_r = B$ **then**
        return Y as solution to the GDH challenger and exit
     else
        return K to $\mathcal{A}^{CMA}$
        update $L_{\mathcal{H}} = L_{\mathcal{H}} \| (\hat{S}, \hat{R}, X, K)$
     end
  else
     Select K randomly from the key distribution and return it to $\mathcal{A}^{CMA}$
     update $L_{\mathcal{H}} = L_{\mathcal{H}} \| (\hat{S}, \hat{R}, X, K)$
  end
     else
     Select K randomly from the key distribution and return it to $\mathcal{A}^{CMA}$
     update$L_{\mathcal{H}} = L_{\mathcal{H}} \| (\hat{S}, \hat{R}, X, K)$
     end

- FEO: Initially $\mathcal{A}^{GDH}$ has an empty encapsulation list $L_{\mathcal{E}}$. On input (P_s, P_r), $\mathcal{A}^{GDH}$ first selects $C \in_R G$. It then checks each entry $(\hat{S}, \hat{R}, X, K)$ in $L_{\mathcal{H}}$ to see if $\mathcal{O}_{DDH}(P_s, P_r, X - C) = True$ for the same (P_s, P_r) as in the input to FEO. If so, it fetches the corresponding K from $L_{\mathcal{H}}$; otherwise it selects K randomly from the symmetric key distribution. It returns (K, C) to $\mathcal{A}^{CMA}$. Finally, $L_{\mathcal{E}}$ is updated to $L_{\mathcal{E}} = L_{\mathcal{E}} \| (\hat{S}, \hat{R}, K, C)$.

- FDO: On input (P_s, P_r, C), $\mathcal{A}^{GDH}$ first checks to see if there is an entry$(\hat{S}, \hat{R}, C, K) \in L_{\mathcal{E}}$.

In case of a match it returns the corresponding symmetric key K. Otherwise, it does the following:

if $(\hat{S}, \hat{R}, X, K) \in L_{\mathcal{H}}$ *for some* X **then**
compute $Y = (X - C)$
if $\mathcal{O}_{DDH}(P_s, P_r, Y) = True$ **then**
  if $P_s = A$ *and* $P_r = B$ **then**
    return Y as solution to the GDH challenger and exit
  else
    fetch corresponding K from $L_{\mathcal{H}}$ and return it to $\mathcal{A}^{CMA}$
    update $L_{\mathcal{E}} = L_{\mathcal{E}} \| (\hat{S}, \hat{R}, K, C)$
  end
else
  Select K randomly from the key distribution and return it to $\mathcal{A}^{CMA}$
  update $L_{\mathcal{E}} = L_{\mathcal{E}} \| (\hat{S}, \hat{R}, K, C)$
end
else
Select K randomly from the key distribution and return it to $\mathcal{A}^{CMA}$
update $L_{\mathcal{E}} = L_{\mathcal{E}} \| (\hat{S}, \hat{R}, K, C)$
end

Answering the GDH challenger: Eventually, $\mathcal{A}^{CMA}$ outputs a forgery (K^*, C^*) as an encapsulation created by S from R. For the forgery to be valid under the outsider unforgeability notion FEO/FDO-sUF-CMA, C^* must be a valid encapsulation of K^*. If C^* is a valid encapsulation of K^* then $\mathcal{A}^{CMA}$ must have queried the Hash with corresponding keying material, in which case $\mathcal{A}^{GDH}$ would have answered the GDH challenger already. Hence, the advantage of $\mathcal{A}^{GDH}$, ϵ' in solving the GDH problem is the same as the advantage of $\mathcal{A}^{CMA}$. □

Optimised Versions of the Ate and Twisted Ate Pairings

Seiichi Matsuda[1], Naoki Kanayama[2], Florian Hess[3], and Eiji Okamoto[2]

[1] University of Tsukuba, Japan
seiichi@cipher.risk.tsukuba.ac.jp
[2] University of Tsukuba, Japan
{kanayama, okamoto}@risk.tsukuba.ac.jp
[3] Technische Universität Berlin, Germany
hess@math.tu-berlin.de

Abstract. We observe a natural generalisation of the ate and twisted ate pairings, which allow for performance improvements in non standard applications of pairings to cryptography like composite group orders. We also give a performance comparison of our pairings and the Tate, ate and twisted ate pairings for certain polynomial families based on operation count estimations and on an implementation, showing that our pairings can achieve a speedup of a factor of up to two over the other pairings.

1 Introduction

Initiated by the pioneering works [18,13,5] on identity based key agreement, one-round tripartite Diffie-Hellman key exchange and identity based encryption respectively, the investigation of pairings has become one of the most attractive areas in contemporary cryptographic research. A host of pairing based protocols has been developed since 2001, offering superior efficiency or greater, novel functionality over classical protocols.

The currently only known instantiations of pairings suitable for cryptography are the Weil and Tate pairings on elliptic curves or on Jacobians of more general algebraic curves. In view of the applications, efficient algorithms for computing these pairings are of great importance.

The Tate pairing on elliptic curves is usually the most efficient choice. It is generally computed using Miller's algorithm [14,15] or a much improved version of Miller's algorithm presented in Barreto *et al* [2]. Duursma and Lee [9] subsequently introduced a very special algorithm on a class of supersingular hyperelliptic curves over finite fields. Barreto *et al* [1] generalised this algorithm to efficiently compute a particular form of the Tate pairing, called η_T pairing, on supersingular elliptic and hyperelliptic curves. The main improvement here is that the loop length in Miller's algorithm for computing the Tate pairing can usually be reduced to at most half the length when computing the η_T pairing. Hess *et al* [12] have generalised the η_T pairing in two ways to ordinary elliptic curves, retaining the efficiency advantage of the η_T pairing over the Tate pairing

S.D. Galbraith (Eds.): Cryptography and Coding 2007, LNCS 4887, pp. 302–312, 2007.

and at the same time enabling larger embedding degrees. The pairings from [12] are called ate pairing and twisted ate pairing.

The dramatic efficiency improvements of the η_T pairing and the ate and twisted ate pairings over the Tate pairing are not always possible. Let E be an elliptic curve over $\mathbb{F}_q$ with $\#E(\mathbb{F}_q) = q + 1 - t$ and let r be a large prime factor of $\#E(\mathbb{F}_q)$. The loop length in Miller's algorithm for the Tate pairing is roughly $\log_2(r)$, while the loop length of the η_T and ate pairings is roughly $\log_2(|t|)$. Since $|t| \leq 2\sqrt{q}$ by the theorem of Hasse, we have roughly $\log_2(|t|) \leq (1/2)\log_2(q)$. Now, in standard situations one has approximately $\log_2(r) = \log_2(q)$, so $\log_2(|t|) \leq (1/2)\log_2(r)$ and the statement on the loop lengths follows. But as soon as $r \leq (1/2)\log_2(q)$, the Tate pairing may actually become faster than the η_T or ate pairings.

In the present paper we observe a generalisation of the ate and twisted ate pairings which we call optimised ate and optimised twisted ate pairings. The loop length of these pairings is roughly equal to $\log_2(|S|)$, where S is any integer such that $S \equiv q \bmod r$ (we choose S to be of minimal absolute value). Note that $S \equiv t - 1 \bmod r$ because of $r|\#E(\mathbb{F}_q)$ and that roughly $|S| \leq r/2$. With this choice of S we thus obtain a pairing that is always at least as fast as the Tate pairing and the ate pairings. We also provide a performance comparison of our optimised pairings and the Tate, ate and twisted ate pairings for certain polynomial families, showing that our pairings can achieve a speedup of a factor of up to two over the other pairings.

The significance of our result is twofold. First, our pairings are very natural generalisations of the ate and twisted ate pairings. Second, while our pairings do not offer a performance improvement for standard applications of pairings in cryptography, they may prove useful for special embedding degrees or composite group orders. The use of composite group orders in pairing based protocols has recently attracted much interest, see for example [6,7]. If pairing values are to be computed in prime order subgroups with known subgroup orders (we are currently not aware of any protocols based on this situation) then our pairings can offer performance improvements.

This paper is organised as follows. Section 2 gives a brief mathematical description of the Tate pairing, η_T pairing and the ate and twisted ate pairings. Section 3 contains our main theorem about the optimised ate and optimised twisted ate pairings. Section 4 contains a performance comparison of our pairings against the Tate, ate and twisted ate pairings, using operation count estimations and our implementation of these pairings. We draw conclusions in Section 5.

2 Background

2.1 Tate Pairing

Let $\mathbb{F}_q$ be a finite field of characteristic p and E an elliptic curve defined over $\mathbb{F}_q$. Let r be a large prime coprime to q such that $r|\#E(\mathbb{F}_q)$. The embedding degree

k with respect to q and r is the smallest positive integer k such that $r|(q^k - 1)$. We also require $r^2 \nmid \#E(\mathbb{F}_{q^k})$. The point at infinity of E is denoted by O.

For every $P \in E(\mathbb{F}_{q^k})$ and integer s let $f_{s,P}$ be an $\mathbb{F}_{q^k}$-rational function with divisor $\mathrm{div}(f_{s,P}) = s(P) - (sP) - (s-1)(O)$. Note that $f_{s,P}$ is uniquely defined up to non-zero scalar multiples from $\mathbb{F}_{q^k}$.

Let $P \in E(\mathbb{F}_{q^k})[r]$ and $Q \in E(\mathbb{F}_{q^k})$. Choose an arbitrary (random) $R \in E(\mathbb{F}_{q^k})$ such that $\#\{P, O, Q + R, R\} = 4$, and let $D = (Q + R) - (R)$. Then the Tate pairing is a non-degenerate bilinear pairing defined by

$$\langle \cdot, \cdot \rangle_r : E(\mathbb{F}_{q^k})[r] \times E(\mathbb{F}_{q^k})/rE(\mathbb{F}_{q^k}) \to \mathbb{F}_{q^k}^\times/(\mathbb{F}_{q^k}^\times)^r,$$

$$\langle P, Q \rangle_r = f_{r,P}(D) \cdot \mathbb{F}_{q^k}^\times.$$

Pairing-based protocols require unique elements (and not classes) in the domain and range of the Tate pairing. Using the isomorphisms $\phi_r : E(\mathbb{F}_{q^k})[r] \to E(\mathbb{F}_{q^k})/rE(\mathbb{F}_{q^k}), Q \mapsto Q + rE(\mathbb{F}_{q^k})$ and $\chi_r : \mathbb{F}_{q^k}^\times/(\mathbb{F}_{q^k}^\times)^r \to \mu_r, x \cdot \mathbb{F}_{q^k}^\times \mapsto x^{(q^k-1)/r}$ with $\mu_r = \mathbb{F}_{q^k}^\times[r]$, the group of r-th roots of unity, we obtain the reduced Tate pairing as

$$t : E(\mathbb{F}_{q^k})[r] \times E(\mathbb{F}_{q^k})[r] \to \mu_r,$$

$$t(P, Q) = \chi_r(\langle P, \phi_r(Q) \rangle_r) = \langle P, Q \rangle_r^{(q^k-1)/r}.$$

Galbraith et al [11] shows that r can be replaced by any integer N such that $r|N|(q^k - 1)$, i.e. $t(P, Q) = \langle P, Q \rangle_N^{(q^k-1)/N}$.

One can compute $f_{r,P}(Q)$ for $Q \in E(\mathbb{F}_{q^k})$ using Miller's algorithm. For a description of Miller's algorithm and numerous optimisations see Barreto et al [2]. A particularly noteworthy optimisation from [2] is

$$t(P, Q) = f_{r,P}(Q)^{(q^k-1)/r},$$

which holds for $P \in E(\mathbb{F}_q)$.

The η_T pairing and ate and twisted ate pairings, which are discussed in the next sections, are all restrictions of some power of the reduced Tate pairing to suitable subgroups of $E(\mathbb{F}_{q^k})$.

2.2 η_T Pairing

Let E be a supersingular elliptic curve with distortion map $\psi : E(\mathbb{F}_q) \to E(\mathbb{F}_{q^k})$ and let $\#E(\mathbb{F}_q) = q + 1 - t$. Barreto et al [1] have introduced the (reduced) η_T pairing which is defined by

$$\eta_T(P, Q) = f_{T,P}(\psi(Q))^{(q^k-1)/r}$$

for $T = q - \#E(\mathbb{F}_q) = t - 1$ and $P, Q \in E(\mathbb{F}_q)[r]$. It is a bilinear and non-degenerate pairing if certain conditions are met.

The main improvement of the η_T pairing over the Tate pairing is that the loop length in Miller's algorithm for the evaluation of $f_{T,P}$ at a point is at most only half the loop length required for the evaluation of $f_{r,P}$ at a point, if $r \approx \#E(\mathbb{F}_q)$ holds true. The reason for this is that T has at most only half the bit length of $\#E(\mathbb{F}_q)$ according to the theorem of Hasse.

2.3   Ate Pairing and Twisted Ate Pairing

The ate and twisted ate pairings have been introduced by Hess *et al* [12]. These pairings can be regarded as variations or generalisations of the η_T pairing for ordinary elliptic curves.

Let E be an ordinary elliptic curve. As in the case of the η_T pairing, let $\#E(\mathbb{F}_q) = q+1-t$ and $T = t-1$. Also write $N = \gcd(T^k-1, q^k-1) > 0$ and $T^k - 1 = LN$. Let $\pi_q : (x,y) \mapsto (x^q, y^q)$ be the q-power Frobenius endomorphism on E and define two groups $\mathbb{G}_1 = E(\mathbb{F}_q)[r] = E[r] \cap \mathrm{Ker}(\pi_q - 1)$, $\mathbb{G}_2 = E[r] \cap \mathrm{Ker}(\pi_q - q)$. Moreover, define $f_{T,Q}^{\mathrm{norm}} = f_{T,Q}/(z^r f_{T,Q})(O)$, where z is a local uniformiser at O. Finally, let $c_T = \sum_{i=0}^{k-1} T^{k-1-i} q^i \equiv kq^{k-1} \bmod r$ (these definitions hold equally for positive or negative T).

The ate pairing is defined as

$$a_T(Q,P) = f_{T,Q}^{\mathrm{norm}}(P)^{c_T(q^k-1)/N}$$

for $Q \in \mathbb{G}_2$ and $P \in \mathbb{G}_1$. If $k|\#\mathrm{Aut}(E)$, then the twisted ate pairing is defined as

$$a_T^{\mathrm{twist}}(P,Q) = f_{T,P}(Q)^{c_T(q^k-1)/N}$$

for $P \in \mathbb{G}_1$ and $Q \in \mathbb{G}_2$. The ate pairing and twisted ate pairing are bilinear and non-degenerate if and only if $r \nmid L$.

Like the η_T pairing, the ate and twisted ate pairing can be computed using at most half the loop length in Miller's algorithm in comparison with the Tate pairing, if $r \approx \#E(\mathbb{F}_q)$.

2.4   Twists

The fairly restrictive condition $k|\#\mathrm{Aut}(E)$ for the twisted ate pairing (note $\#\mathrm{Aut}(E) \le 6$ for ordinary E) is related to the existence of twists of E.

Let E and E' be two ordinary elliptic curves over $\mathbb{F}_q$. The curve E' is called a twist of degree d of E if there exists an isomorphism $\psi : E' \to E$ defined over $\mathbb{F}_{q^d}$ and d is minimal with this property. Then the condition $d \,|\, \#\mathrm{Aut}(E)$ holds true if and only if E admits a twist of degree d.

Table 1 contains information about the various twists of elliptic curves in characteristic ≥ 5 together with the twisting isomorphisms. The element D has to be chosen from $\mathbb{F}_q$ such that the twisting isomorphisms are properly defined over $\mathbb{F}_{q^d}$.

Twists can be used in conjunction with the ate and twisted ate pairing to achieve point compression and a protocol depending speed up. If E' is a twist of E of degree d and $de = k$, then it is possible to choose a twisting isomorphism $\psi : E' \to E$ such that $\psi(E'(\mathbb{F}_{q^e})) = \mathbb{G}_2$. This allows to work with $Q' = \psi^{-1}(Q)$ instead of Q. Note that in the supersingular case we can have $E' = E$ and the twisted ate pairing then coincides with the η_T pairing.

Table 1. Twists of Elliptic Curves in Characteristic ≥ 5

$$d = 2 \qquad E : y^2 = x^3 + Ax + B,$$
$$E' : y^2 = x^3 + A/D^2 x + B/D^3,$$
$$\psi : E' \to E : (x, y) \mapsto (Dx, D^{3/2}y),$$
$$d = 4 \qquad E : y^2 = x^3 + Ax,$$
$$E' : y^2 = x^3 + A/Dx,$$
$$\psi : E' \to E : (x, y) \mapsto (D^{1/2}x, D^{3/4}y),$$
$$d = 3, 6 \qquad E : y^2 = x^3 + B,$$
$$E' : y^2 = x^3 + B/D,$$
$$\psi : E' \to E : (x, y) \mapsto (D^{1/3}x, D^{1/2}y).$$

3 Optimised Versions of the Ate and Twisted Ate Pairings

Let E be an ordinary elliptic curve over $\mathbb{F}_q$. The next theorem provides a generalisation of the ate and twisted ate pairings by replacing $T = t - 1$ with any integer S such that $S \equiv q \bmod r$.

Theorem 1. *Let S be any integer with $S \equiv q \bmod r$. Define $N = \gcd(S^k - 1, q^k - 1) > 0$ and $L = (S^k - 1)/N$. Let $c_S = \sum_{i=0}^{k-1} S^{k-1-i} q^i \bmod N$. Then*

$$a_S : \mathbb{G}_2 \times \mathbb{G}_1 \to \mu_r, \quad (Q, P) \mapsto f_{S,Q}^{\mathrm{norm}}(P)^{c_S(q^k - 1)/N}$$

defines a bilinear pairing. If $k \mid \#\mathrm{Aut}(E)$ then

$$a_S^{\mathrm{twist}} : \mathbb{G}_1 \times \mathbb{G}_2 \to \mu_r, \quad (P, Q) \mapsto f_{S,P}(Q)^{c_S(q^k - 1)/N}$$

also defines a bilinear pairing. Both pairings a_S and a_S^{twist} are non-degenerate if and only if $r \nmid L$.

The relation with the reduced Tate pairing is

$$a_S(Q, P) = t(Q, P)^L \quad \text{and} \quad a_S^{\mathrm{twist}}(P, Q) = t(P, Q)^L.$$

We remark that if $P = O$ or $Q = O$ then the pairing values are defined to be equal to 1. Also, if $P = Q$ (only possible for $k = 1$) then P needs to be replaced by any divisor $(P + R) - (R)$ coprime to $(Q) - (O)$ for the first pairing and by $(Q + R) - (R)$ coprime to $(P) - (O)$ for the second pairing, with $R \in E(\overline{\mathbb{F}}_q)$.

We will show that under certain conditions a suitable choice of S yields pairings a_S and a_S^{twist} which are more efficient than the ate pairing a_T and the twisted ate pairing a_T^{twist} for $T = t - 1$. For these choices of S, we call a_S and a_S^{twist} optimised ate and optimised twisted ate pairing.

Theorem 1 can also be applied to the base extensions E_e of E over $\mathbb{F}_{q^e}$ (i.e. E regarded as an elliptic curve over $\mathbb{F}_{q^e}$) for $1 \leq e \leq k - 1$. The subgroups $\mathbb{G}_1$

and $\mathbb{G}_2$ of $E[r]$ remain invariant under such base extensions and the embedding degree of E_e with respect to r is $k_e = k/\gcd(k, e)$. Hence Theorem 1 also holds true after replacing q by q^e and k by k_e for any e with $1 \leq e \leq k - 1$. This observation allows for example to apply the twisted ate pairing in the case $k \nmid \#\mathrm{Aut}(E)$ and $\gcd(k, \#\mathrm{Aut}(E)) \neq 1$, where we choose $e = k/\gcd(k, \#\mathrm{Aut}(E))$ so that $k_e = \gcd(k, \#\mathrm{Aut}(E))$ and $k_e \mid \#\mathrm{Aut}(E_e)$. Another application is to further minimise the absolute value of S. This has been observed in [22], where a number of interesting examples are given.

Proof (of Theorem 1). The proof is essentially the same as in [12], but slightly more general. In the following, we only adapt the main arguments of [12] to our setting.

We let $\psi = \pi_q$ for the ate pairing case and $\psi = \gamma\pi_q$ for the twisted ate pairing case, where $\gamma \in \mathrm{Aut}(E)$ is an automorphism of order k such that $(\gamma\pi_q)(Q) = Q$ and $(\gamma\pi_q)(P) = qP$. If we interchange P and Q for the twisted ate pairing we have $\psi(P) = P$, $\psi(Q) = qQ = SQ$ and need to consider $f_{S,Q}(P)^{cs(q^k-1)/N}$ like for the ate pairing. This allows us to deal with both cases simultaneously.

From Lemma 1 of [12] we obtain

$$t(Q, P) = f_{r,Q}(P)^{(q^k-1)/r} = f_{N,Q}(P)^{(q^k-1)/N}$$

and

$$\begin{aligned}
t(Q, P)^L &= f_{N,Q}(P)^{L(q^k-1)/N} = f_{LN,Q}(P)^{(q^k-1)/N} \\
&= f_{S^k-1,Q}(P)^{(q^k-1)/N} \\
&= f_{S^k,Q}(P)^{(q^k-1)/N}.
\end{aligned} \tag{1}$$

Lemma 2 of [1] yields

$$f_{S^k,Q} = f_{S,Q}^{S^{k-1}} f_{S,SQ}^{S^{k-2}} \cdots f_{S,S^{k-1}Q}. \tag{2}$$

Since ψ is purely inseparable of degree q, we obtain from Lemma 4 in [12]

$$f_{S,\psi^i(Q)} \circ \psi^i = f_{S,Q}^{q^i}. \tag{3}$$

We have $\psi^i(Q) = S^i Q$ and $\psi^i(P) = P$. Combining this with (2) and (3) gives

$$f_{S^k,Q}(P) = f_{S,Q}(P)^{\sum_{i=0}^{k-1} S^{k-1-i} q^i}. \tag{4}$$

Substituting (4) into (1) gives

$$t(Q, P)^L = f_{S,Q}(P)^{cs(q^k-1)/N}. \tag{5}$$

Now (5) shows that a_S and a_S^{twist} are bilinear pairings, which are non-degenerate if and only if $r \nmid L$. □

4   Performance Evaluation

We provide some families of elliptic curves admitting a twist of degree 4 and 6, and compare the costs of our optimised pairings with the standard pairings.

4.1   Polynomial Families

Assume $q = p$. If $\Delta = 1, 2, 3$ in the CM equation $4p - t^2 = \Delta V^2$, then the corresponding elliptic curves can be generated without the full CM algorithm [10], for example by randomly choosing α and β in the equation below until the correct curve is found (but see also [17]).

$$E_1 : y^2 = x^3 + \alpha x \qquad\qquad (\Delta = 1)$$
$$E_2 : y^2 = x^3 - 30\alpha x^2 + 56\alpha^3 \qquad\qquad (\Delta = 2)$$
$$E_3 : y^2 = x^3 + \beta \qquad\qquad (\Delta = 3)$$

The endomorphism rings are isomorphic to $\mathbb{Z}[\sqrt{-\Delta}]$ and E_1, E_2 and E_3 admit twists of degree 4, 2 and 6 respectively.

Let $\rho \equiv \log p / \log r$ be the ratio between the bit lengths of the finite field and the order of the subgroup. Some polynomial families such that $4p - t^2$ is a square polynomial have been presented

– in [8], for $k = 4$ and $\rho \sim 2$,
– in [19], for $k = 6$ and $\rho \sim 2$,
– in [10], for $k = 8$ and $\rho \sim 3/2$,
– in [4], for $k = 12$ and $\rho \sim 1$.

The details of these polynomial families are given in Appendix 1.

4.2   Efficiency Comparison

We follow the analysis of [16] and compare the Tate pairing $f_{r,P}(Q)$, ate pairing $f_{T,Q}(P)$, twisted ate pairing $f_{T^e,P}(Q)$, optimised ate pairing $f_{S,Q}(P)$ and optimised twisted ate pairing $f_{S_e,P}(Q)$ on ordinary elliptic curves admitting a twist of degree 6 when $k = 6, 12$ and of degree 4 when $k = 4, 8$. We refer to $f_{N,P}(Q)$ as a Miller-Lite operation and $f_{N,Q}(P)$ as a Miller-Full operation. We denote the cost of the Miller-Lite operation by C_{Lite} and the cost of the Miller-Full operation by C_{Full}. Assume both operations use projective coordinates. On the form $Y^2 = X^3 + AX + B$, the costs for Miller-operations are estimated as follows [12]. When $A = -3$:

$$C_{\text{Lite}} = (4S_1 + (2e + 7)M_1 + S_k + M_k) \log_2 N$$
$$C_{\text{Full}} = (4S_e + 6M_e + 2eM_1 + S_k + M_k) \log_2 N$$

When $A = 0$:

$$C_{\text{Lite}} = (5S_1 + (2e + 6)M_1 + S_k + M_k) \log_2 N$$
$$C_{\text{Full}} = (5S_e + 6M_e + 2eM_1 + S_k + M_k) \log_2 N$$

where $s = 2^i 3^j$, $M_s = 3^i 5^j M_1$, $S_s = M_s$ with respect to multiplication in $\mathbb{F}_{q^s}^{\times}$.

Table 2. The Costs Required for the Different Pairings

Security Level	Method	Cost (average size t)	
		Standard	Optimised
$k = 4, d = 4$	Tate	4960	
$\log_2 p \sim 320$	ate	4800	2400
$\log_2 r \sim 160$	twisted ate	4960	2480
$k = 6, d = 6$	Tate	11008	
$\log_2 p \sim 512$	ate	5504	5504
$\log_2 r \sim 256$	twisted ate	5504	5504
$k = 8, d = 4$	Tate	17664	
$\log_2 p \sim 384$	ate	16896	16896
$\log_2 r \sim 256$	twisted ate	26496	13248
$k = 12, d = 6$	Tate	26880	
$\log_2 p \sim 256$	ate	16256	16256
$\log_2 r \sim 256$	twisted ate	26880	20160

Using the parameters in Appendix 1, we estimate the loop length for each pairing. The results are given in Table 2.

When $k = 4$ the optimised ate and optimised twisted ate pairing are twice as fast as the Tate, ate and twisted ate pairing. When $k = 8$ the optimised twisted ate pairing is more efficient than the optimised ate pairing. We conclude that our optimised pairings always run at least as fast as the Tate pairing, and the loop length of the optimised (twisted) ate pairing can be at least reduced to $\frac{\deg(r)-1}{\deg(r)}$ of the loop length of the Tate pairing when $t - 1 \geq r$ for the optimised ate pairing and $(t - 1)^e \geq r$ for the optimised twisted ate pairing.

4.3   Implementation Evaluation

We have implemented all pairings on $\mathbb{G}_1 \times \mathbb{G}_2$ for $k = 4$ using the GNU MP library in C++ to demonstrate the effectiveness of our proposal. Table 3 presents the running times for Tate pairing, twisted ate pairing and optimised twisted ate pairing excluding the final powering. The detailed parameters are given in Appendix 2.

Table 3. The Running Time for the Different Pairings

Security Level	Method	Running time
$k = 4$	Tate	11.3
$\log_2 q \sim 320, \log_2 r \sim 160$	Twisted ate	11.1
MOV security ~ 1280	Optimised twisted ate	5.7

5   Conclusion

We have described very natural optimised variants of the ate and twisted ate pairing which are simultaneous improvements over the Tate, ate and twisted ate pairings. We have provided some sample polynomial families for which the loop length in Miller's algorithm for our optimised pairings is shorter by a factor of $\frac{\deg(r)-1}{\deg(r)}$ in comparison to the loop length for the Tate pairing when $t - 1 \geq r$ for the optimised ate pairing and $(t - 1)^e \geq r$ for the optimised twisted ate pairing.

Acknowledgement

We thank Xavier Boyen for pointing us to [6,7].

References

1. Barreto, P.S.L.M., Galbraith, S., O'hEigeartaigh, C., Scott, M.: Efficient pairing computation on supersingular abelian varieties. Designs, Codes and Cryptography 42(3), 239–271 (2007)
2. Barreto, P.S.L.M., Kim, H.Y., Lynn, B., Scott, M.: Efficient algorithms for pairing-based cryptosystems. In: Yung, M. (ed.) CRYPTO 2002. LNCS, vol. 2442, pp. 354–368. Springer, Heidelberg (2002)
3. Barreto, P.S.L.M., Lynn, B., Scott, M.: On the Selection of Pairing-Friendly Groups. In: Matsui, M., Zuccherato, R.J. (eds.) SAC 2003. LNCS, vol. 3006, pp. 17–25. Springer, Heidelberg (2004)
4. Barreto, P.S.L.M., Naehrig, M.: Pairing-Friendly Elliptic Curve of Prime Order. In: Preneel, B., Tavares, S. (eds.) SAC 2005. LNCS, vol. 3897, pp. 319–331. Springer, Heidelberg (2006)
5. Boneh, D., Franklin, M.: Identity-based Encryption from the Weil pairing. In: Kilian, J. (ed.) CRYPTO 2001. LNCS, vol. 2139, pp. 213–229. Springer, Heidelberg (2001)
6. Boyen, X., Waters, B.: Compact Group Signatures Without Random Oracles. In: Vaudenay, S. (ed.) EUROCRYPT 2006. LNCS, vol. 4004, pp. 427–444. Springer, Heidelberg (2006)
7. Boyen, X., Waters, B.: Full-Domain Subgroup Hiding and Constant-Size Group Signatures. In: Public Key Cryptography—PKC 2007. LNCS, vol. 4450, pp. 1–15. Springer, Heidelberg (2007)
8. Duan, P., Cui, S., Chan, C.W.: Effective Polynomial Families for Generating More Pairing-friendly Elliptic Curve. Cryptology ePrint Archive, Report, 2005/236 (2005), http://eprint.iacr.org/2005/236
9. Duursma, I., Lee, H.S.: Tate Pairing Implementation for Hyperelliptic Curves $y^2 = x^p - x + d$. In: Laih, C.-S. (ed.) ASIACRYPT 2003. LNCS, vol. 2894, pp. 111–123. Springer, Heidelberg (2003)
10. Freeman, D., Scott, M., Teske, E.: A taxonomy of pairing-friendly elliptic curves. Cryptology ePrint Archive, Report, 2006/372 (2006), http://eprint.iacr.org/2006/372
11. Galbraith, S., Harrison, K., Soldera, S.: Implementing the Tate pairing. In: Fieker, C., Kohel, D.R. (eds.) Algorithmic Number Theory Symposium–ANTS V. LNCS, vol. 2369, pp. 324–337. Springer, Heidelberg (2002)
12. Hess, F., Smart, N.P., Vercauteren, F.: The Eta Pairing Revisited. IEEE Transaction on Information Theory 52(10), 4595–4602 (2006)

13. Joux, A.: A One Round Protocol for Tripartite Diffie-Hellman. In: Bosma, W. (ed.) Algorithmic Number Theory Symposium–ANTS IV. LNCS, vol. 1838, pp. 385–394. Springer, Heidelberg (2000)
14. Miller, V.S.: Short Programs for functions on Curves (1986), http://crypto.stanford.edu/miller/miller.pdf
15. Miller, V.S.: The Weil pairing and its efficient calculation. Journal of Cryptology 17(4), 235–261 (2004)
16. Koblitz, N., Menezes, A.: Pairing-based cryptography at high security level. In: Smart, N.P. (ed.) Cryptography and Coding: 10th IMA International Conference. LNCS, vol. 3796, pp. 13–36. Springer, Heidelberg (2005)
17. Rubin, K., Silverberg, A.: Choosing the correct elliptic curve in the CM method. Cryptology ePrint Archive, Report, 2007/253 (2007), http://eprint.iacr.org/2007/253
18. Sakai, R., Ohgishi, K., Kasahara, M.: Cryptosystems based on pairing. In: Symposium on Cryptography and Information Security–SCIS 2000 (2000)
19. Scott, M.: Private communication
20. Scott, M.: Scaling security in pairing-based protocols, Cryptology ePrint Archive, Report 2005/139 (2005), http://eprint.iacr.org/2005/139
21. Scott, M., Costigan, N., Abdulwahab, W.: Implementing Cryptographic Pairings on Smartcards. In: Goubin, L., Matsui, M. (eds.) CHES 2006. LNCS, vol. 4249, pp. 134–147. Springer, Heidelberg (2006)
22. Zhao, C.-A., Zhang, F., Huang, J.: A Note on the Ate Pairing. Cryptology ePrint Archive, Report, 2007/247 (2007), http://eprint.iacr.org/2007/247

Appendix 1

Polynomial families for $k = 4, 6, 8, 12$ from [8,4,10] and the values of $e = k/\gcd(k, \#\mathrm{Aut}(E))$, $T = t - 1$, $S \equiv p \bmod r$ and $S_e \equiv p^e \bmod r$.

$k = 4$
$p = 8z^4 + 6z^2 + 2z + 1$
$r = 4z^2 + 1$
$t = 4z^2 + 2z + 2$
$\Delta V^2 = 4z^2(2z - 1)^2$
$\Delta = 1$
$e = 1$
$T = 4z^2 + 2z + 1$
$S = 2z$

$k = 6$
$p = 27z^4 + 9z^3 + 3z^2 + 3z + 1$
$r = 9z^2 + 3z + 1$
$t = 3z + 2$
$\Delta V^2 = 3z^2(6z + 1)^2$
$\Delta = 3$
$e = 1$
$T = 3z + 1$
$S = T$

$k = 8$
$p = \frac{1}{4}(81z^6 + 54z^5 + 45z^4 + 12z^3 + 13z^2 + 6z + 1)$
$r = 9z^4 + 12z^3 + 8z^2 + 4z + 1$
$t = -9z^3 - 3z^2 - 2z$
$\Delta V^2 = (3z + 1)^2$
$\Delta = 1$
$e = 2$
$T = -9z^3 - 3z^2 - 2z - 1$
$T^2 = 81z^6 + 54z^5 + 45z^4 + 30z^3 + 10z^2 + 4z + 1$
$S = T$
$S_e = p^2 \bmod r = -18z^3 - 15z^2 - 10z - 4$

$k = 12$
$p = 36z^4 + 36z^3 + 24z^2 + 6z + 1$
$r = 36z^4 + 36z^3 + 18z^2 + 6z + 1$
$t = 6z^2 + 1$
$\Delta V^2 = 3(6z^2 + 4z + 1)^2$
$\Delta = 3$
$e = 2$
$T = 6z^2$
$T^2 = 36z^4$
$S = T$
$S_e = p^2 \bmod r = -36z^3 - 18z^2 - 6z - 1$

Appendix 2

The parameters for the pairing implementation in Section 4.3.

$k = 4$
$p =$ 6802412203485154774779492598941919023699396553915045681512070169946616890505876170525361872297 49 (319 bit)
$E : y^2 = x^3 + 3x$
$E' : y^2 = x^3 + (3/D)x$, where $1/D = v^2$ and $v^2 - 2 = 0$
$\#E(\mathbb{F}_p) =$ 680241220348515477477949259894191902369939655390338170945836123217606411022317222264735061564936 (319 bit)
$\#E'(\mathbb{F}_p) =$ 680241220348515477477949259894191902369939655392670965356577910771716964918860599461430061667370 (319 bit)
$r =$ 1166397205370893777055276948271688598347500051217 (160 bit)
$t =$ 1166397205370893777055278028270394787801125664814 (160 bit)
$T =$ 1166397205370893777055278028270394787801125664813 (160 bit)
$S =$ 1079998706189453625613596 (80 bit)

Extractors for Jacobian of Hyperelliptic Curves of Genus 2 in Odd Characteristic

Reza Rezaeian Farashahi[1,2]

[1] Dept. of Mathematics and Computer Science, TU Eindhoven,
P.O. Box 513, 5600 MB Eindhoven, The Netherlands
[2] Dept. of Mathematical Sciences, Isfahan University of Technology,
P.O. Box 85145 Isfahan, Iran

Abstract. We propose two simple and efficient deterministic extractors for $J(\mathbb{F}_q)$, the Jacobian of a genus 2 hyperelliptic curve H defined over $\mathbb{F}_q$, for some odd q. Our first extractor, SEJ, called *sum extractor*, for a given point D on $J(\mathbb{F}_q)$, outputs the sum of abscissas of rational points on H in the support of D, considering D as a reduced divisor. Similarly the second extractor, PEJ, called *product extractor*, for a given point D on the $J(\mathbb{F}_q)$, outputs the product of abscissas of rational points in the support of D. Provided that the point D is chosen uniformly at random in $J(\mathbb{F}_q)$, the element extracted from the point D is indistinguishable from a uniformly random variable in $\mathbb{F}_q$. Thanks to the Kummer surface $\mathcal{K}$, that is associated to the Jacobian of H over $\mathbb{F}_q$, we propose the *sum* and *product* extractors, SEK and PEK, for $\mathcal{K}(\mathbb{F}_q)$. These extractors are the modified versions of the extractors SEJ and PEJ. Provided a point K is chosen uniformly at random in $\mathcal{K}$, the element extracted from the point K is statistically close to a uniformly random variable in $\mathbb{F}_q$.

Keywords: Jacobian, Hyperelliptic curve, Kummer surface, Deterministic extractor.

1 Introduction

A deterministic extractor for a set S is a function that converts a random point on S to a bit-string of fixed length that is statistically close to uniformly random. In this paper, we propose two simple and efficient deterministic extractors for $J(\mathbb{F}_q)$, the Jacobian of a hyperelliptic curve H of genus 2 defined over $\mathbb{F}_q$, for some odd q. Our first extractor, SEJ, called *sum extractor*, for a given point D on $J(\mathbb{F}_q)$, outputs the sum of abscissas of rational points on H in the support of D, considering D as a reduced divisor. Similarly the second extractor, PEJ, called *product extractor*, for a given point D on the $J(\mathbb{F}_q)$, outputs the product of abscissas of rational points in the support of D. Provided that the point D is chosen uniformly at random in $J(\mathbb{F}_q)$, the element extracted from the point D is indistinguishable from a uniformly random variable in $\mathbb{F}_q$.

Let $\mathcal{K}$ be the Kummer surface associated to the Jacobian of H over $\mathbb{F}_q$. Then there is a map κ from $J(\mathbb{F}_q)$ to $\mathcal{K}(\mathbb{F}_q)$, so that a point and it's opposite in $J(\mathbb{F}_q)$ are mapped to the same value. Using this map, we propose two simple

S.D. Galbraith (Eds.): Cryptography and Coding 2007, LNCS 4887, pp. 313–335, 2007.

and efficient deterministic extractors, SEK and PEK, for the Kummer surface $\mathcal{K}$. If a point K is chosen uniformly at random in $\mathcal{K}$, the element extracted from the point K is statistically close to a uniformly random variable in $\mathbb{F}_q$.

The use of hyperelliptic curves in public key cryptography was first introduced by Koblitz in [15]. The security of hyperelliptic cryptosystems is based on the difficulty of discrete logarithm problem in the Jacobian of these curves. Hyperelliptic curves of genus 2 are undergoing intensive study. They were shown to be competitive with elliptic curves in speed and security. Various researchers have been optimizing genus 2 arithmetic (see [2,16,17]). The security of genus 2 hyperelliptic curves is assumed to be similar to that of elliptic curves of the same group size (e.g see [10]).

The use of Kummer surface associated to the Jacobian of a genus 2 curve is proposed for faster arithmetic (see [7,11,16]). The scalar multiplication on the Jacobian can be used to define a scalar multiplication on the Kummer surface. It could be used to construct a Diffie-Hellman protocol (see [21]). In addition, it is shown in [21], solving the discrete logarithm problem on the Jacobian is polynomial time equivalent to solving the discrete logarithm problem on the kummer surface.

The problem of converting random points of a variety (e.g a curve or Jacobian of a curve) into random bits has several cryptographic applications. Such applications are key derivation functions, key exchange protocols and design of cryptographically secure pseudorandom number generators. As examples we can mention the well-known Elliptic Curve Diffie-Hellman protocol and Diffie-Hellman protocol in genus 2. By the end of Diffie-Hellman protocol, the parties agree on a common secret element of the group, which is indistinguishable from a uniformly random element under the decisional Diffie-Hellman assumption (denoted by DDH). However the binary representation of the common secret element is *distinguishable* from a uniformly random bit-string of the same length. Hence one has to convert this group element into a random-looking bit-string. This can be done using a deterministic extractor.

At the moment, several deterministic randomness extractors for elliptic curves are known. Kaliski [14] shows that if a point is taken uniformly at random from the union of an elliptic curve and its quadratic twist then the abscissa of this point is uniformly distributed in the finite field. Then Chevassut et al. [5], proposed the TAU technique. This technique allows to extract almost all the bits of the abscissa of a point of the union of an elliptic curve and its quadratic twist. Gürel [12] proposed an extractor for an elliptic curve defined over a quadratic extension of a prime field. It extracts almost half of the bits of the abscissa of a point on the curve. Then, Farashahi and Pellikaan proposed the similar extractor, yet more general, for hyperelliptic curves defined over a quadratic extension of a finite filed in odd characteristic [8]. Furthermore, their result for elliptic curves improves the result of [12]. Two deterministic extractors for a family of binary elliptic curves are proposed by Farashahi et al. [9]. It is shown that half of the bits of the abscissa of a point on the curve can be extracted. They also proposed two deterministic extractors for the main subgroup of an ordinary elliptic curve that has minimal 2-torsion. In our knowledge, up to now, no extractor is defined for the Jacobian of a hyperelliptic curve.

We organize the paper as follows. In the next section we introduce some notations and recall some basic definitions. In Section 3, we propose extractors SEJ and PEJ for $J(\mathbb{F}_q)$, the Jacobian of a genus 2 hyperelliptic curve H over $\mathbb{F}_q$. We show that the outputs of these extractors, for a given uniformly random point of $J(\mathbb{F}_q)$, are statistically close to a uniformly random variable in $\mathbb{F}_q$. For the analysis of these extractors, we need some bounds on the cardinalities of $\mathtt{SEJ}^{-1}(a)$ and $\mathtt{PEJ}^{-1}(b)$, for all $a, b \in \mathbb{F}_q$. We give our estimates for them in Theorems 2 and 3. Then, in Section 4, we give the proofs of the main Theorems 2 and 3. In Section 5, we propose two extractors SEK and PEK for $\mathcal{K}(\mathbb{F}_q)$, the Kummer surface related to $J(\mathbb{F}_q)$. These extractors are modified versions of the previous extractors, using the map κ from $J(\mathbb{F}_q)$ to $\mathcal{K}(\mathbb{F}_q)$. We conclude our result in Section 6. Furthermore, in appendix, we introduce some corresponding problems for the proof of the main Theorem 2.

2 Preliminaries

Let us define the notations and recall the basic definitions that are used throughout the paper.

Notation. Denote by $\mathbb{Z}_n$ the set of nonnegative integers less than n. A field is denoted by $\mathbb{F}$ and its algebraic closure by $\overline{\mathbb{F}}$. Denote by $\mathbb{F}^*$ the set of nonzero elements of $\mathbb{F}$. The finite field with q elements is denoted by $\mathbb{F}_q$, and its algebraic closure by $\overline{\mathbb{F}}_q$. Let C be a curve defined over $\mathbb{F}_q$, then the set of $\mathbb{F}_q$-rational points on C is denoted by $C(\mathbb{F}_q)$. The x-coordinate of a point P on a curve is denoted by x_P. The cardinality of a finite set S is denoted by $\#S$. We make a distinction between a variable $\mathbf{x}$ and a specific value x in $\mathbb{F}$.

2.1 Finite Field Notation

Consider the finite fields $\mathbb{F}_q$ and $\mathbb{F}_{q^2}$, where $q = p^k$, for some odd prime number p and positive integer k. Fix a polynomial representation $\mathbb{F}_{q^2} \cong \mathbb{F}_q[t]/(t^2 - \alpha)$, where α is not a quadratic residue in $\mathbb{F}_q$. Then $\mathbb{F}_{q^2}$ is a vector space over $\mathbb{F}_q$ which is generated by the basis $\{1, t\}$. That means every element x in $\mathbb{F}_{q^2}$ can be represented in the form $x = x_0 + x_1 t$, where x_0 and x_1 are in $\mathbb{F}_q$.

Let $\phi : \overline{\mathbb{F}}_q \longrightarrow \overline{\mathbb{F}}_q$ be the Frobenius map defined by $\phi(x) = x^q$.

2.2 Hyperelliptic Curves

Definition 1. *An absolutely irreducible nonsingular curve $\mathcal{H}$ of genus at least 2 is called* hyperelliptic *if there exists a morphism of degree 2 from $\mathcal{H}$ to the projective line.*

Theorem 1. *Let $\mathcal{H}$ be a hyperelliptic curve of genus g over $\mathbb{F}_q$, where q is odd. Then $\mathcal{H}$ has a plane model of the form*

$$\mathbf{y}^2 = f(\mathbf{x}),$$

where f is a square-free polynomial and $2g + 1 \leq \deg(f) \leq 2g + 2$. The plane model is singular at infinity. If $\deg(f) = 2g+1$ then the point at infinity ramifies and $\mathcal{H}$ has only one point at infinity. If $\deg(f) = 2g+2$ then $\mathcal{H}$ has zero or two $\mathbb{F}_q$-rational points at infinity.

Proof. See [1,6]. □

In this paper we consider a hyperelliptic curve $\mathcal{H}$ that has only one point at infinity. One calls $\mathcal{H}$ an *imaginary* hyperelliptic curve.

2.3   Jacobian of a Hyperelliptic Curve

Let $\mathcal{H}$ be an imaginary hyperelliptic curve of genus g over $\mathbb{F}_q$, where q is odd. Then $\mathcal{H}$ has a plane model of the form $\mathbf{y}^2 = f(\mathbf{x})$, where f is a square-free polynomial and $\deg(f) = 2g+1$. For any subfield $\mathbb{K}$ of $\overline{\mathbb{F}}_q$ containing $\mathbb{F}_q$, the set

$$\mathcal{H}(\mathbb{K}) = \{(x,y) : x, y \in \mathbb{K}, \ y^2 = f(x)\} \cup \{P_\infty\},$$

is called the set of $\mathbb{K}$-*rational points* on $\mathcal{H}$. The point P_∞ is called the *point at infinity* for $\mathcal{H}$. A point P on $\mathcal{H}$, also written $P \in \mathcal{H}$, is a point $P \in \mathcal{H}(\overline{\mathbb{F}}_q)$. The negative of a point $P = (x, y)$ on $\mathcal{H}$ is defined as $-P = (x, -y)$ and $-P_\infty = P_\infty$.

Definition 2. *A divisor D on $\mathcal{H}$ is a formal sum of points on $\mathcal{H}$*

$$D = \sum_{P \in \mathcal{H}} m_P P,$$

where $m_P \in \mathbb{Z}$, and only a finite number of the m_P are nonzero. The degree of D is defined by $\deg D = \sum_{P \in \mathcal{H}} m_P P$. The divisor D is said to be defined over $\mathbb{K}$, if for all automorphisms φ in the Galois group of $\mathbb{K}$, $\varphi(D) = \sum_{P \in \mathcal{H}} m_P \varphi(P) = D$, where $\varphi(P) = (\varphi(x), \varphi(y))$ if $P = (x, y)$ and $\varphi(P_\infty) = P_\infty$.

The set of all divisors on $\mathcal{H}$ defined over $\mathbb{K}$, denoted by $Div_{\mathcal{H}}(\mathbb{K})$, forms an additive abelian group under the addition rule

$$\sum_{P \in \mathcal{H}} m_P P + \sum_{P \in \mathcal{H}} n_P P = \sum_{P \in \mathcal{H}} (m_P + n_P) P.$$

The set $Div_{\mathcal{H}}^0(\mathbb{K})$ of all divisors on $\mathcal{H}$ of degree zero defined over $\mathbb{K}$ is a subgroup of $Div_{\mathcal{H}}(\mathbb{K})$. In particular, $Div_{\mathcal{H}}^0 = Div_{\mathcal{H}}^0(\overline{\mathbb{K}})$.

Let $\mathbb{K}[\mathcal{H}]$ be the *coordinate ring* of the plain model of $\mathcal{H}$ over $\mathbb{K}$. Then the *function field* of $\mathcal{H}$ over $\mathbb{K}$ is the field of fractions $\mathbb{K}(\mathcal{H})$ of $\mathbb{K}[\mathcal{H}]$. For a polynomial R in $\mathbb{K}[\mathcal{H}]$, the divisor of R is defined by $\mathrm{div}(R) = \sum_{P \in \mathcal{H}} \mathrm{ord}_P(R) P$, where $\mathrm{ord}_P(R)$ is the order of vanishing of R at P. For a rational function $R = F/G$, where $F, G \in \mathbb{K}[\mathcal{H}]$, the divisor of R is defined by $\mathrm{div}(R) = \mathrm{div}(F) - \mathrm{div}(G)$ and is called a *principal divisor*. The *group of principal divisors* on $\mathcal{H}$ over $\mathbb{K}$ is denoted by $\mathcal{P}_\mathcal{H}(\mathbb{K}) = \{\mathrm{div}(R) : R \in \mathbb{K}(\mathcal{H})\}$. Specially $\mathcal{P}_\mathcal{H} = \mathcal{P}_\mathcal{H}(\overline{\mathbb{K}})$ is called the *group of principal divisors* on $\mathcal{H}$.

Definition 3. *The* Jacobian *of* $\mathcal{H}$ *over* $\mathbb{K}$ *is defined by*

$$J_\mathcal{H}(\mathbb{K}) = Div_\mathcal{H}^0(\mathbb{K})/\mathcal{P}_\mathcal{H}(\mathbb{K}).$$

Similarly, the Jacobian *of* $\mathcal{H}$ *is defined by* $J_\mathcal{H} = Div_\mathcal{H}^0/\mathcal{P}_\mathcal{H}$.

For each nontrivial class of divisors in $J_\mathcal{H}(\mathbb{K})$, there exist a unique divisor D on $\mathcal{H}$ over $\mathbb{K}$ of the form

$$D = \sum_{i=1}^{r} P_i - rP_\infty,$$

where $P_i = (x_i, y_i) \neq P_\infty$, $P_i \neq -P_j$, for $i \neq j$, and $r \leq g$. Such a divisor is called a *reduced* divisor on $\mathcal{H}$ over $\mathbb{K}$. By using Mumford's representation [19], each reduced divisor D on $\mathcal{H}$ over $\mathbb{K}$ can be uniquely represented by a pair of polynomials $[u(x), v(x)]$, $u, v \in \mathbb{K}[x]$, where u is monic, $\deg(v) < \deg(u) \leq g$, and $u \mid (v^2 - f)$. Precisely $u(x) = \prod_{i=1}^{r}(x - x_i)$ and $v(x_i) = y_i$. The neutral element of $J_\mathcal{H}(\mathbb{K})$, denoted by $\mathcal{O}$, is represented by $[1, 0]$. Cantor's algorithm, [3], efficiently computes the sum of two reduced divisors in $J_\mathcal{H}(\mathbb{K})$ and expresses it in reduced form.

2.4   Kummer Surface

Let H be an imaginary hyperelliptic curve of genus 2 defined over $\mathbb{F}_q$, for odd q. Then H has a plane model of the form

$$\mathbf{y}^2 = f(\mathbf{x}) = x^5 + f_4x^4 + f_3x^3 + f_2x^2 + f_1x + f_0, \tag{1}$$

where $f_i \in \mathbb{F}_q$ and f is a square-free polynomial. Then for the curve H, there exist a quartic surface $\mathcal{K}$ in $\mathbb{P}^3$, called the Kummer surface, which is given by the equation

$$A(k_1, k_2, k_3)k_4^2 + B(k_1, k_2, k_3)k_4 + C(k_1, k_2, k_3) = 0,$$

where

$$A(k_1, k_2, k_3) = k_2^2 - 4k_1k_3,$$
$$B(k_1, k_2, k_3) = -2(2f_0k_1^3 + f_1k_1^2k_2 + 2f_2k_1^2k_3 + f_3k_1k_2k_3 + 2f_4k_1k_3^2 + k_2k_3^2),$$
$$\begin{aligned} C(k_1, k_2, k_3) = &-4f_0f_2k_1^4 + f_1^2k_1^4 - 4f_0f_3k_1^3k_2 - 2f_1f_3k_1^3k_3 - 4f_0f_4k_1^2k_2^2 \\ &+ 4f_0k_1^2k_2k_3 - 4f_1f_4k_1^2k_2k_3 + 2f_1k_1^2k_3^2 - 4f_2f_4k_1^2k_3^2 + f_3^2k_1^2k_3^2 \\ &- 4f_0k_1k_2^3 - 4f_1k_1k_2^2k_3 - 4f_2k_1k_2k_3^2 - 2f_3k_1k_3^3 + k_3^4. \end{aligned}$$

Let $J(\mathbb{F}_q)$ be the Jacobian of H over $\mathbb{F}_q$ (see Subsection 2.3). Then there is a map

$$\kappa : J(\mathbb{F}_q) \longrightarrow \mathcal{K}(\mathbb{F}_q),$$

where $\kappa(D) = \kappa(-D)$, for all $D \in J(\mathbb{F}_q)$ and $\kappa(\mathcal{O}) = (0,0,0,1)$. This map does not preserve the group structure, however, endows a pseudo-group structure on $\mathcal{K}$ (see [4]). In particular, a scalar multiplication on the image of κ is defined by

$$m\kappa(D) = \kappa(mD),$$

for $m \in \mathbb{Z}$ and $D \in J(\mathbb{F}_q)$. It could be used for a Diffie-Hellman protocol (see [21]). Furthermore, the above definition can be extended to have a scalar multiplication on $\mathcal{K}$. Since each point on $\mathcal{K}$ can be pulled back to the Jacobian of H or to the Jacobian of the quadratic twist of H.

2.5   Deterministic Extractor

In our analysis we use the notion of a deterministic extractor, so let us recall it briefly. For general definition of extractors we refer to [20,22].

Definition 4. *Let X and Y be S-valued random variables, where S is a finite set. Then the statistical distance $\Delta(X,Y)$ of X and Y is*

$$\Delta(X,Y) = \tfrac{1}{2} \sum_{s \in S} |\Pr[X = s] - \Pr[Y = s]|.$$

Let U_S denote a random variable uniformly distributed on S. We say that a random variable X on S is δ-uniform, if $\Delta(X, U_S) \leq \delta$.

Note that if the random variable X is δ-uniform, then no algorithm can distinguish X from U_S with advantage larger than δ, that is, for all algorithms $D : S \longrightarrow \{0,1\}$

$$|\Pr[D(X) = 1] - \Pr[D(U_S) = 1]| \leq \delta.$$

See [18].

Definition 5. *Let S, T be finite sets. Consider the function $\mathrm{Ext} : S \longrightarrow T$. We say that Ext is a deterministic (T, δ)-extractor for S if $\mathrm{Ext}(U_S)$ is δ-uniform on T. That means*

$$\Delta(\mathrm{Ext}(U_S), U_T) \leq \delta.$$

In the case that $T = \{0,1\}^k$, we say Ext is a δ-deterministic extractor for S.

In this paper we consider deterministic $(\mathbb{F}_q, \delta)$-extractors. Observe that, converting random elements of $\mathbb{F}_q$ into random bit strings is a relatively easy problem. For instance, one can represent an element of $\mathbb{F}_q$ by a number in $\mathbb{Z}_q$ and convert this number to a bit-string of a length equal or very close to the bit length of q (e.g. see [13]). Furthermore, if q is close to a power of 2, that is, $0 \leq (2^n - q)/2^n \leq \delta$ for a small δ, then the uniform element $U_{\mathbb{F}_q}$ is statistically close to n uniformly random bits. The following simple lemma is a well-known result (the proof can be found, for instance, in [5]).

Lemma 1. *Under the condition that $0 \leq (2^n - q)/2^n \leq \delta$, the statistical distance between $U_{\mathbb{F}_q}$ and U_{2^n} is bounded from above by δ.*

3   Extractors for Jacobian

In this section we propose two extractors for the Jacobian of a hyperelliptic curve of genus 2 in odd characteristic. Then we analyse them.

We recall that H is an imaginary hyperelliptic curve of genus 2 defined over $\mathbb{F}_q$, for odd q, and $J(\mathbb{F}_q)$ is the Jacobian of H over $\mathbb{F}_q$. The hyperelliptic curve H has a plane model of the form $\mathbf{y}^2 = f(\mathbf{x})$, where f is a monic square-free polynomial of degree 5 (see equation (1)).

3.1   Sum Extractor for Jacobian

Definition 6. *The sum extractor* SEJ *for the Jacobian of* H *over* $\mathbb{F}_q$ *is defined as the function* SEJ $: J(\mathbb{F}_q) \longrightarrow \mathbb{F}_q$, *by*

$$\mathrm{SEJ}(D) = \begin{cases} \sum_{i=1}^{r} x_{P_i} & \text{if } D = \sum_{i=1}^{r} P_i - rP_\infty, \ 1 \le r \le 2 \\ 0 & \text{if } D = \mathcal{O}. \end{cases}$$

Remark 1. By using Mumford's representation for the points of $J(\mathbb{F}_q)$, the function SEJ is defined as

$$\mathrm{SEJ}(D) = \begin{cases} -u_1 & \text{if } D = [x^2 + u_1 x + u_0, v_1 x + v_0], \\ -u_0 & \text{if } D = [x + u_0, v_0], \\ 0 & \text{if } D = [1, 0]. \end{cases}$$

The following theorem gives the estimates for $\#\mathrm{SEJ}^{-1}(a)$, for all a in $\mathbb{F}_q$. In Subsection 3.3, we use the result of this theorem to analyse the extractor SEJ. We give a proof of Theorem 2 in Section 4.

Theorem 2. *For all* $a \in \mathbb{F}_q^*$,

$$\left| \#\mathrm{SEJ}^{-1}(a) - q \right| \le 8\sqrt{q} + 1$$

and

$$\left| \#\mathrm{SEJ}^{-1}(0) - (q+1) \right| \le 8\sqrt{q} + 1.$$

3.2   Product Extractor for Jacobian

Definition 7. *The product extractor* PEJ *for the Jacobian of* H *over* $\mathbb{F}_q$ *is defined as the function* PEJ $: J(\mathbb{F}_q) \longrightarrow \mathbb{F}_q$, *by*

$$\mathrm{PEJ}(D) = \begin{cases} \prod_{i=1}^{r} x_{P_i} & \text{if } D = \sum_{i=1}^{r} P_i - rP_\infty, \ 1 \le r \le 2 \\ 0 & \text{if } D = \mathcal{O}. \end{cases}$$

Remark 2. By using Mumford's representation for the points of $J(\mathbb{F}_q)$, the function PEJ is defined as

$$\mathrm{PEJ}(D) = \begin{cases} u_0 & \text{if } D = [x^2 + u_1 x + u_0, v_1 x + v_0], \\ -u_0 & \text{if } D = [x + u_0, v_0], \\ 0 & \text{if } D = [1, 0]. \end{cases}$$

The next theorem shows the estimates for $\#\text{PEJ}^{-1}(b)$, for all b in $\mathbb{F}_q$.

Theorem 3. *Let $b \in \mathbb{F}_q^*$. Let $I_f = \{z \in \mathbb{F}_q^* : f_1 = z^2, f_2 = zf_4\}$. Then*

$$\left|\#\text{PEJ}^{-1}(b) - q\right| \leq \begin{cases} 8\sqrt{q} + 3 & \text{if } f_0 \neq 0, \\ 6\sqrt{q} + 3 & \text{if } f_0 = 0 \text{ and } b \notin I_f, \\ q + 4\sqrt{q} & \text{if } f_0 = 0 \text{ and } b \in I_f. \end{cases}$$

For $b = 0$,

$$\left|\#\text{PEJ}^{-1}(0) - (eq + 1)\right| \leq 4e\sqrt{q},$$

where $e = \#\{(x,y) \in H(\mathbb{F}_q) : x = 0\}$.

3.3   Analysis of the Extractors

In this subsection we show that provided the divisor D is chosen uniformly at random in $J(\mathbb{F}_q)$, the element extracted from the divisor D by SEJ or PEJ is indistinguishable from a uniformly random element in $\mathbb{F}_q$.

Let A be a $\mathbb{F}_q$-valued random variable that is defined as

$$A = \text{SEJ}(D), \quad \text{for } D \in_R J(\mathbb{F}_q).$$

Proposition 1. *The random variable A is statistically close to the uniform random variable $U_{\mathbb{F}_q}$.*

$$\Delta(A, U_{\mathbb{F}_q}) = O(\frac{1}{\sqrt{q}}).$$

Proof. Let $a \in \mathbb{F}_q$. For the uniform random variable $U_{\mathbb{F}_q}$, $\Pr[U_{\mathbb{F}_q} = a] = 1/q$. Also for the $\mathbb{F}_q$-valued random variable A,

$$\Pr[A = a] = \frac{\#\text{SEJ}^{-1}(a)}{\#J(\mathbb{F}_q)}.$$

The genus of H is 2, so by Hasse-Weil's Theorem we have

$$(\sqrt{q} - 1)^4 \leq \#J(\mathbb{F}_q) \leq (\sqrt{q} + 1)^4.$$

Theorem 2 gives the bound for $\#\text{SEJ}^{-1}(a)$, for all $a \in \mathbb{F}_q$. Hence

$$\Delta(A, U_{\mathbb{F}_q}) = \frac{1}{2} \sum_{a \in \mathbb{F}_q} \left|\Pr[A = a] - \Pr[U_{\mathbb{F}_q} = a]\right|$$

$$= \frac{1}{2} \sum_{a \in \mathbb{F}_q} \left|\frac{\#\text{SEJ}^{-1}(a)}{\#J(\mathbb{F}_q)} - \frac{1}{q}\right|$$

$$= \frac{\left|q\#\text{SEJ}^{-1}(0) - \#J(\mathbb{F}_q)\right|}{2q\#J(\mathbb{F}_q)} + \sum_{a \in \mathbb{F}_q^*} \frac{\left|q\#\text{SEJ}^{-1}(a) - \#J(\mathbb{F}_q)\right|}{2q\#J(\mathbb{F}_q)}.$$

Then

$$\Delta(A, U_{\mathbb{F}_q}) \leq \frac{(12q\sqrt{q} - 4q + 4\sqrt{q} - 1) + (q - 1)(12q\sqrt{q} - 5q + 4\sqrt{q} - 1)}{2q(\sqrt{q} - 1)^4}$$

$$= \frac{12q\sqrt{q} - 5q + 4\sqrt{q}}{2(\sqrt{q} - 1)^4} = \frac{6 + \epsilon(q)}{\sqrt{q}},$$

where $\epsilon(q) = \frac{43q\sqrt{q} - 68q + 48\sqrt{q} - 12}{2(\sqrt{q} - 1)^4}$. If $q \geq 570$, then $\epsilon(q) < 1$. □

Corollary 1. SEJ *is a deterministic* $(\mathbb{F}_q, O(\frac{1}{\sqrt{q}}))$-*extractor for* $J(\mathbb{F}_q)$.

Proof. Proposition 1 concludes the proof of this corollary. □

Corollary 2. PEJ *is a deterministic* $(\mathbb{F}_q, O(\frac{1}{\sqrt{q}}))$-*extractor for* $J(\mathbb{F}_q)$.

Proof. The result of Theorem 3 implies the proof of this corollary. □

4 Proofs of Theorems 2 and 3

In this section we give the proofs of Theorems 2 and 3. In other words, we are going to count the cardinalities of $\#\text{SEJ}^{-1}(a)$, $\#\text{PEJ}^{-1}(b)$, for all $a, b \in \mathbb{F}_q$. In Subsection 4.1, we recall some notes on the Jacobian of H over $\mathbb{F}_q$. We give the proof of Theorem 2 in Subsection 4.2. Then, we sketch the proof of Theorem 3 in Subsection 4.3.

4.1 Notes on the Jacobian of H over $\mathbb{F}_q$

We recall from Section 3 that $J(\mathbb{F}_q)$ is the Jacobian of H over $\mathbb{F}_q$. We partition $J(\mathbb{F}_q)$ as $J(\mathbb{F}_q) = J_0 \cup J_1 \cup J_2$, where $J_0 = \{\mathcal{O}\}$ and J_r, for $r = 1, 2$ is defined as

$$J_r = \{D \in J(\mathbb{F}_q) : D = [u(x), v(x)], \deg(u) = r\}.$$

Recall that $\mathcal{O}$ is represented by $[1, 0]$.

Note that D is defined over $\mathbb{F}_q$, that means for all automorphisms φ in the Galois group of $\mathbb{F}_q$, $\varphi(D) = D$.

Let $D \in J_1$, then $D = P - P_\infty$, where $P = (x_P, y_P) \in H(\mathbb{F}_q)$. The Mumford's representation for D is $[x - x_P, y_P]$.

Let $D \in J_2$, then $D = P + Q - 2P_\infty$, where $P, Q \neq P_\infty$ and $P \neq -Q$. The divisor D is represented by $[u(x), v(x)]$, such that $u(x) = (x - x_P)(x - x_Q)$ and v is the line through P and Q. Since D is defined over $\mathbb{F}_q$, then $\phi(D) = \phi(P) + \phi(Q) - 2\phi(P_\infty) = D$, where ϕ is the Frobenius map. There are two cases for D.

- Suppose $\phi(P) = P$. Since $\phi(D) = D$, then $\phi(Q) = Q$. Thus $P, Q \in H(\mathbb{F}_q)$. That means

$$D = P + Q - 2P_\infty, \; P, Q \in H(\mathbb{F}_q), \; P, Q \neq P_\infty, P \neq -Q.$$

In this case the polynomial u is reducible over $\mathbb{F}_q$.

– Suppose $\phi(P) \neq P$. Since $\phi(D) = D$, so $\phi(P) = Q$ and $\phi(Q) = P$. Then $\phi(\phi(P)) = P$. Hence $P \in H(\mathbb{F}_{q^2})$. That means

$$D = P + \phi(P) - 2P_\infty, \ P \in H(\mathbb{F}_{q^2}), \ P \neq P_\infty, \phi(P) \neq \pm P.$$

In this case the polynomial u is irreducible over $\mathbb{F}_q$.

Let

$$\mathcal{J} = \{(P,Q) : P, Q \in H(\mathbb{F}_q), \ P, Q \neq P_\infty, Q \neq -P\},$$

$$\mathcal{J}^\phi = \{(P, \phi(P)) : P \in H(\mathbb{F}_{q^2}), P \neq P_\infty, \ \phi(P) \neq -P\}.$$

Lemma 2. *Let* $\sigma : \mathcal{J} \longrightarrow J_2$ *be the map defined by*

$$\sigma(P,Q) = P + Q - 2P_\infty,$$

and let $\sigma_\phi : \mathcal{J}^\phi \longrightarrow J_2$ *be the map defined by*

$$\sigma_\phi(P, \phi(P)) = P + \phi(P) - 2P_\infty.$$

Then $\#\sigma^{-1}(D) + \#\sigma_\phi^{-1}(D) = 2$, *for all* $D \in J_2$.

Proof. Let $D \in J_2$. Then we have the following cases.

1. Assume $D = P + Q - 2P_\infty$, such that $P, Q \in H(\mathbb{F}_q)$, $P, Q \neq P_\infty$ and $Q \neq P$. Clearly $\sigma^{-1}(D) = \{(P,Q), (Q,P)\}$ and $\sigma_\phi^{-1}(D) = \emptyset$.
2. Assume $D = P + \phi(P) - 2P_\infty$, such that $P \in H(\mathbb{F}_{q^2})$, $P \neq P_\infty$ and $\phi(P) \neq P$. Clearly $\sigma^{-1}(D) = \emptyset$ and $\sigma_\phi^{-1}(D) = \{(P, \phi(P)), (\phi(P), P)\}$.
3. Assume $D = 2P - 2P_\infty$, where $P \in H(\mathbb{F}_q)$, $P \neq P_\infty$. It is easy to see that $\sigma^{-1}(D) = \sigma_\phi^{-1}(D) = \{(P,P)\}$. ☐

4.2 Proof of Theorem 2

For the proof of Theorem 2, we need several propositions. First, by Proposition 2, we transform our problem to the problem of computing sum of the cardinalities of corresponding sets in Definition 8. Second, in proposition 3, we give a formula for this sum in terms of the cardinalities of some curves. Finally, by using Hasse-Weil Theorem, we obtain tight estimates for $\#\mathrm{SEJ}^{-1}(a)$, for all $a \in \mathbb{F}_q$.

Definition 8. *Let* $a \in \mathbb{F}_q$. *Define*

$$\Sigma_a = \{(P,Q) : P, Q \in H(\mathbb{F}_q), \ x_P + x_Q = a\},$$

$$\Sigma_a^\phi = \{(P, \phi(P)) : P \in H(\mathbb{F}_{q^2}), \ x_P + x_{\phi(P)} = a\}.$$

Proposition 2. *For all* $a \in \mathbb{F}_q$,

$$\#(\mathrm{SEJ}^{-1}(a) \cap J_2) = \frac{\#\Sigma_a + \#\Sigma_a^\phi}{2} - 1.$$

Proof. Let $a \in \mathbb{F}_q$. Let $\mathcal{S}_a = \sigma^{-1}(\text{SEJ}^{-1}(a) \cap J_2)$ and $\mathcal{S}_a^\phi = \sigma_\phi^{-1}(\text{SEJ}^{-1}(a) \cap J_2)$ (see Lemma 2). Then $\Sigma_a = \mathcal{S}_a \cup \mathcal{E}_a$ and $\Sigma_a^\phi = \mathcal{S}_a^\phi \cup \mathcal{E}_a^\phi$, where $\mathcal{E}_a = \{(P, Q) : (P, Q) \in \Sigma_a, Q = -P\}$ and $\mathcal{E}_a^\phi = \{(P, \phi(P)) : (P, \phi(P)) \in \Sigma_a^\phi, \phi(P) = -P\}$. Since $\mathcal{S}_a$ and $\mathcal{E}_a$ are disjoint, so $\#\Sigma_a = \#\mathcal{S}_a + \#\mathcal{E}_a$. Similarly, $\#\Sigma_a^\phi = \#\mathcal{S}_a^\phi + \#\mathcal{E}_a^\phi$.

Assume $(P, -P)$ is a point of $\mathcal{E}_a$ or $\mathcal{E}_a^\phi$, then $x_P = \frac{a}{2}$. Obviously P is a point of $H(\mathbb{F}_q)$ or $H(\mathbb{F}_{q^2})$. Suppose $f(\frac{a}{2}) = 0$. Then $P \in H(\mathbb{F}_q)$ and $P = -P$. That means $\mathcal{E}_a = \mathcal{E}_a^\phi = \{(P, P)\}$. Now, suppose $f(\frac{a}{2}) \neq 0$. So $P \neq -P$. If $P \in H(\mathbb{F}_q)$, then $\mathcal{E}_a = \{(P, -P), (-P, P)\}$ and $\mathcal{E}_a^\phi = \emptyset$. Otherwise, P is a point of $H(\mathbb{F}_{q^2})$. Thus $\phi(P) = -P$. Hence $\mathcal{E}_a = \emptyset$ and $\mathcal{E}_a^\phi = \{(P, -P), (-P, P)\}$. In other words $\#\mathcal{E}_a + \#\mathcal{E}_a^\phi = 2$.

Lemma 2 implies that $\#\mathcal{S}_a + \#\mathcal{S}_a^\phi = 2\#(\text{SEJ}^{-1}(a) \cap J_2)$. That concludes the proof of this proposition. □

Proposition 2 gives the estimate for the cardinality of $\text{SEJ}^{-1}(a)$, for $a \in \mathbb{F}_q$, in terms of the sum of the cardinalities of Σ_a and Σ_a^ϕ. Now, we are dealing to have a tight estimate for $\#\Sigma_a + \#\Sigma_a^\phi$, for all $a \in \mathbb{F}_q$. In order to do that, we define a curve $\mathcal{X}_a$, for $a \in \mathbb{F}_q$. Then, in Proposition 3, we give a formula for $\#\Sigma_a + \#\Sigma_a^\phi$ in terms of the cardinalities of $H(\mathbb{F}_q)$ and $\mathcal{X}_a(\mathbb{F}_q)$. After that, using the Hasse-Weil's Theorem, we obtain a tight estimate for $\#\Sigma_a + \#\Sigma_a^\phi$.

The hyperellitic curve H has the plane model defined by

$$\mathbf{y}^2 = f(\mathbf{x}) = \prod_{i=1}^{5}(\mathbf{x} - \lambda_i), \tag{2}$$

where λ_i are pairwise distinct elements of $\overline{\mathbb{F}}_q$. (see equation (1)). Define the two-variable polynomial $\Phi \in \mathbb{F}_q[\mathbf{x}_0, \mathbf{x}_1]$ as $\Phi(\mathbf{x}_0, \mathbf{x}_1) = f(\mathbf{x}_0)f(\mathbf{x}_1)$. Clearly Φ is a symmetric polynomial. Let $\mathbf{a} = \mathbf{x}_0 + \mathbf{x}_1$ and $\mathbf{b} = \mathbf{x}_0 \mathbf{x}_1$. Then from equation (2), we obtain

$$\Phi(\mathbf{x}_0, \mathbf{x}_1) = \prod_{i=1}^{5}((\mathbf{x}_0 - \lambda_i)(\mathbf{x}_1 - \lambda_i)) = \prod_{i=1}^{5}(\mathbf{x}_0 \mathbf{x}_1 - \lambda_i(\mathbf{x}_0 + \mathbf{x}_1) + \lambda_i^2)$$

Define the two-variable polynomial Ψ in $\mathbb{F}_q[\mathbf{a}, \mathbf{b}]$ by

$$\Psi(\mathbf{a}, \mathbf{b}) = \prod_{i=1}^{5}(\mathbf{b} - \lambda_i \mathbf{a} + \lambda_i^2). \tag{3}$$

For $a \in \mathbb{F}_q$, let $\mathcal{X}_a$ be the affine curve defined over $\mathbb{F}_q$, by the equation

$$\mathbf{y}^2 = \Psi_a(\mathbf{b}) = \Psi(a, \mathbf{b}). \tag{4}$$

Proposition 3. *Let* $a \in \mathbb{F}_q$. *Then*

$$\#\Sigma_a + \#\Sigma_a^\phi = 2(\#H(\mathbb{F}_q) + \#\mathcal{X}_a(\mathbb{F}_q) - q - 1).$$

Proof. See Proposition 12. □

Clearly the affine curve $\mathcal{X}_a$ is absolutely irreducible, for all $a \in \mathbb{F}_q$. The curve $\mathcal{X}_a$ is nonsingular for almost all $a \in \mathbb{F}_q$. Furthermore, the genus of the nonsingular model of $\mathcal{X}_a$ is at most 2. By using the Hasse-Weil's bound for the nonsingular model of $\mathcal{X}_a$, we obtain an estimate for $\#\mathcal{X}_a(\mathbb{F}_q)$.

Proposition 4. *For all $a \in \mathbb{F}_q$,*

$$|\#\mathcal{X}_a(\mathbb{F}_q) - q| \le 4\sqrt{q}.$$

Proof. See Subsection B.1.   □

Proof (Theorem 2). Let $a \in \mathbb{F}_q$. Proposition 2 shows that

$$\#(\mathrm{SEJ}^{-1}(a) \cap J_2) = \frac{\#\Sigma_a + \#\Sigma_a^\phi}{2} - 1.$$

From Proposition 3, we have

$$\#\Sigma_a + \#\Sigma_a^\phi = 2(\#H(\mathbb{F}_q) + \#\mathcal{X}_a(\mathbb{F}_q) - q - 1).$$

Then by using Hasse-Weil's bound for H we obtain

$$|\#H(\mathbb{F}_q) - q - 1| \le 4\sqrt{q}.$$

Furthermore, from Proposition 4 we have

$$|\#\mathcal{X}_a(\mathbb{F}_q) - q| \le 4\sqrt{q}.$$

Hence

$$\left|\#(\mathrm{SEJ}^{-1}(a) \cap J_2) - q\right| \le 8\sqrt{q}.$$

Clearly $\#(\mathrm{SEJ}^{-1}(a) \cap J_1)$ equals $0, 1$ or 2. If $a = 0$, then $\#(\mathrm{SEJ}^{-1}(a) \cap J_0)$ equals 1, otherwise equals 0. So the proof of Theorem 2 is completed.   □

4.3   Proof of Theorem 3

The proof of Theorem 3 is similar to the proof of Theorem 2. First, in Proposition 5, we give the estimate for the cardinality of $\mathrm{PEJ}^{-1}(b)$, for $b \in \mathbb{F}_q^*$, in terms of the sum of the cardinalities of Π_b and Π_b^ϕ. Second, in Proposition 6, we give a relation between $\#\Sigma_a + \#\Sigma_a^\phi$ and the cardinalities of $\mathcal{H}(\mathbb{F}_q)$ and $\mathcal{X}_a(\mathbb{F}_q)$. Finally, Hasse-Weil Theorem concludes the proof of Theorem 3.

Definition 9. *Let $b \in \mathbb{F}_q^*$. Define*

$$\Pi_b = \{(P, Q) : P, Q \in H(\mathbb{F}_q),\ x_P x_Q = b\},$$
$$\Pi_b^\phi = \{(P, \phi(P)) : P \in H(\mathbb{F}_{q^2}),\ x_P x_{\phi(P)} = b\}.$$

Proposition 5. *For all $b \in \mathbb{F}_q^*$,*

$$\#(\mathrm{PEJ}^{-1}(b) \cap J_2) = \frac{\#\Pi_b + \#\Pi_b^\phi}{2} - r_b,$$

where r_b equals the number of square roots of b in $\mathbb{F}_q^*$.

Proof. The proof of this proposition is similar to the proof of Proposition 2. So we leave it for the interested reader. □

Consider the polynomial $\Psi \in \mathbb{F}_q[\mathbf{a}, \mathbf{b}]$ defined by the equation (3). Let $\mathcal{X}_b$ be the affine curve defined over $\mathbb{F}_q$, by the equation

$$\mathbf{y}^2 = \Psi_b(\mathbf{a}) = \prod_{i=1}^{5}(b - \lambda_i \mathbf{a} + \lambda_i^2), \tag{5}$$

for $b \in \mathbb{F}_q^*$.

Proposition 6. *Let* $b \in \mathbb{F}_q^*$. *Then*

$$\#\Pi_b + \#\Pi_b^\phi = 2(\#H(\mathbb{F}_q) + \#\mathcal{X}_b(\mathbb{F}_q) - q - e),$$

where $e = \#\{(x,y) \in H(\mathbb{F}_q) : x = 0\}$.

Proof. The proof of this proposition is similar to the proof of Proposition 3. □

The affine curve $\mathcal{X}_b$ is absolutely irreducible and nonsingular, for almost all $b \in \mathbb{F}_q$. In fact the curve $\mathcal{X}_b$ is reducible if and only if $\lambda_i = 0$, for some i, and $b \in I_f$, where $I_f = \{z \in \mathbb{F}_q^* : f_1 = z^2, f_2 = z f_4\}$. Provided the curve $\mathcal{X}_b$ is absolutely irreducible, the genus of the nonsingular model of $\mathcal{X}_b$ is at most 2. Then Hasse-Weil's Theorem gives the estimates for $\#\mathcal{X}_b(\mathbb{F}_q)$.

Proposition 7. *Let* $b \in \mathbb{F}_q$. *Then*

$$|\#\mathcal{X}_b(\mathbb{F}_q) - q| \leq \begin{cases} 4\sqrt{q} & \text{if } f_0 \neq 0, \\ 2\sqrt{q} & \text{if } f_0 = 0 \text{ and } b \notin I_f, \\ q & \text{if } f_0 = 0 \text{ and } b \in I_f. \end{cases}$$

Proof. See Subsection B.2. □

Proof (Theorem 3). Let $b \in \mathbb{F}_q^*$. Proposition 5 shows that

$$\#(\text{PEJ}^{-1}(b) \cap J_2) = \frac{\#\Pi_b + \#\Pi_b^\phi}{2} - r_b,$$

where r_b equals the number of square roots of b in $\mathbb{F}_q$. It is easy to see that $0 \leq \#(\text{PEJ}^{-1}(b) \cap J_1) \leq 2$ and $\#(\text{PEJ}^{-1}(b) \cap J_0) = 0$. So

$$|\#\text{PEJ}^{-1}(b) - q| \leq \frac{\left|\#\Pi_b + \#\Pi_b^\phi - 2q\right|}{2} + 2.$$

From Proposition 6, we have

$$\#\Pi_b + \#\Pi_b^\phi = 2(\#H(\mathbb{F}_q) + \#\mathcal{X}_b(\mathbb{F}_q) - q - e),$$

where e is the number of points on $H(\mathbb{F}_q)$ whose abscissa equals zero. Note that $0 \leq e \leq 2$. Hence

$$\left|\#\Pi_b + \#\Pi_b^\phi - 2q\right| \leq 2\left|\#H(\mathbb{F}_q) + \#\mathcal{X}_b(\mathbb{F}_q) - 2q - 1\right| + 2.$$

Hasse-Weil's Theorem gives the bound for $\#H(\mathbb{F}_q)$. Then Proposition 7 concludes the proof of Theorem 3 for all $b \in \mathbb{F}_q^*$.

Now assume that $b = 0$. It is easy to see $\#\mathrm{PEJ}^{-1}(0) = e\#H(\mathbb{F}_q) - e + 1$, where e equals the number of points of $H(\mathbb{F}_q)$ whose abscissa equals zero. So the proof of Theorem 3 is completed.   □

5   Extractors for Kummer Surface

Consider the hyperelliptic curve H that is defined in equation (1). Let $\mathcal{K}$ be the Kummer surface related to $J(\mathbb{F}_q)$ (Jacobian of H over $\mathbb{F}_q$). We recall that each point of $J(\mathbb{F}_q)$ can be uniquely represented by at most 2 points on H. Then there is a map

$$\kappa : J(\mathbb{F}_q) \longrightarrow \mathcal{K}(\mathbb{F}_q)$$
$$P + Q - 2P_\infty \longmapsto (1 : a : b : c)$$
$$P - P_\infty \longmapsto (0 : 1 : x_P : x_P^2)$$
$$\mathcal{O} \longmapsto (0 : 0 : 0 : 1),$$

where $a = x_P + x_Q$, $b = x_P x_Q$ and

$$c = \begin{cases} \dfrac{\widetilde{B}(a,b) - 2y_P y_Q}{(x_P - x_Q)^2} & \text{if } P \neq Q \\ \dfrac{\widetilde{C}(a,b)}{4y_P^2} & \text{if } P = Q, \end{cases}$$

with

$$\widetilde{B}(a,b) = ab^2 + f_3 ab + f_1 a + 2f_4 b^2 + 2f_2 b + 2f_0,$$
$$\widetilde{C}(a,b) = C(1,a,b).$$

5.1   Sum Extractor for Kummer Surface

In this subsection we define the *sum extractor* SEK for the Kummer surface $\mathcal{K}$. Then we define the *sum extractor* SEKJ as the restriction of SEK to the image of κ. We briefly mention the analysis of these extractors.

Definition 10. *The sum extractor* SEK *for the Kummer surface* $\mathcal{K}$ *is defined as the function* SEK $: \mathcal{K}(\mathbb{F}_q) \longrightarrow \mathbb{F}_q$, *by*

$$\mathrm{SEK}(k_1 : k_2 : k_3 : k_4) = \begin{cases} \dfrac{k_2}{k_1} & \text{if } k_1 \neq 0, \\ \dfrac{k_3}{k_2} & \text{if } k_1 = 0, k_2 \neq 0, \\ 0 & \text{otherwise.} \end{cases}$$

The following theorem gives the estimates for $\#\mathrm{SEK}^{-1}(a)$, for all a in $\mathbb{F}_q$. By using the result of this theorem, one can show that SEK is a deterministic $(\mathbb{F}_q, O(\frac{1}{\sqrt{q}}))$-extractor for $\mathcal{K}(\mathbb{F}_q)$.

Theorem 4. *For all $a \in \mathbb{F}_q^*$,*

$$\left| \#\text{SEK}^{-1}(a) - q \right| \leq 4\sqrt{q}$$

and

$$\left| \#\text{SEK}^{-1}(0) - (q+1) \right| \leq 4\sqrt{q}.$$

Proof. Note that each point on $\mathcal{K}$ can be pulled back to the Jacobian of H or to the Jacobian of the quadratic twist of H. Furthermore, the map κ is $2:1$ on all points except the points of order 2 in the Jacobian of H where it is $1:1$. Then, the proof of Theorem 2 and the application of that proof for the sum extractor for the Jacobian of the quadratic twist of H conclude the proof of this Theorem.

$\square$

The scalar multiplication on $\kappa(J(\mathbb{F}_q))$ could be used for a variant of Diffie-Hellman protocol on this set. For instance, consider the case that $J(\mathbb{F}_q)$ is a cyclic group with generator D_g. Then $\kappa(D_g)$ is the generator of $\kappa(J(\mathbb{F}_q))$. That brings us to define the following extractor for this set.

Definition 11. *The sum extractor* SEKJ *for* $\kappa(J(\mathbb{F}_q))$, *is defined as the restriction of the extractor* SEK *to* $\kappa(J(\mathbb{F}_q))$.

The following theorem shows that $\#\text{SEJ}^{-1}(a) = 2\#\text{SEKJ}^{-1}(a)$, for almost all $a \in \mathbb{F}_q$. One can show that SEKJ is a deterministic $(\mathbb{F}_q, O(\frac{1}{\sqrt{q}}))$-extractor for $\kappa(J(\mathbb{F}_q))$ (see Subsection 3.3).

Proposition 8. *For all $a \in \mathbb{F}_q$,*

$$\#\text{SEKJ}^{-1}(a) = \frac{\#\text{SEJ}^{-1}(a) + d_a}{2},$$

where d_a is the number of two torsion points of $J(\mathbb{F}_q)$ in $\text{SEJ}^{-1}(a)$.

Proof. The fact that the map κ is $2:1$ on all points except the points of order 2 in the Jacobian of H where it is $1:1$, concludes the proof of this proposition. $\square$

Remark 3. It is easy to see that $0 \leq d_a \leq 3$ and $\sum_{a \in \mathbb{F}_q} d_a$ equals the number of two torsion points of $J(\mathbb{F}_q)$, which is bounded by 16.

5.2   Product Extractor for Kummer Surface

In this subsection we define the *product extractor* PEK for the $\mathcal{K}$. We briefly mention the analysis of this extractor.

Definition 12. *The product extractor* PEK *for the Kummer surface $\mathcal{K}$ is defined as the function* PEK $: \mathcal{K}(\mathbb{F}_q) \longrightarrow \mathbb{F}_q$, *by*

$$\text{PEK}(k_1 : k_2 : k_3 : k_4) = \begin{cases} \dfrac{k_3}{k_1} & \text{if } k_1 \neq 0, \\[2mm] \dfrac{k_3}{k_2} & \text{if } k_1 = 0, k_2 \neq 0, \\[2mm] 0 & \text{otherwise.} \end{cases}$$

The next theorem gives the estimates for $\#\text{PEK}^{-1}(b)$, for all b in $\mathbb{F}_q$. The result of this theorem implies that PEK is a deterministic $(\mathbb{F}_q, O(\frac{1}{\sqrt{q}}))$-extractor for $\mathcal{K}(\mathbb{F}_q)$.

Theorem 5. *Let $b \in \mathbb{F}_q$. Let $I_f = \{z \in \mathbb{F}_q^* : f_1 = z^2, f_2 = zf_4\}$. Then*

$$\left|\#\text{PEK}^{-1}(b) - q\right| \leq \begin{cases} 4\sqrt{q} + 1 & \text{if } f_0 \neq 0, \\ 2\sqrt{q} + 1 & \text{if } f_0 = 0 \text{ and } b \notin I_f, \\ q - 1 & \text{if } f_0 = 0 \text{ and } b \in I_f. \end{cases}$$

Furthermore, one can define the *product extractor* PEKJ for $\kappa(J(\mathbb{F}_q))$ as the restriction of the extractor PEK to $\kappa(J(\mathbb{F}_q))$.

6 Conclusion

We propose the *sum* and *product* extractors, SEJ and PEJ, for $J(\mathbb{F}_q)$, the Jacobian of a genus 2 hyperelliptic curve H over $\mathbb{F}_q$. We show that the outputs of these extractors, for a given uniformly random point of $J(\mathbb{F}_q)$, are statistically close to a uniformly random variable in $\mathbb{F}_q$. To show the latter we need some bounds on the cardinalities of $\text{SEJ}^{-1}(a)$ and $\text{PEJ}^{-1}(b)$, for all $a, b \in \mathbb{F}_q$. To have these estimates, we introduce some corresponding problems. In new problems, we are looking for bounds on the cardinality of some curves. We give our estimates in Theorems 2 and 3 using Hasse-Weil Theorem.

Thanks to the Kummer surface $\mathcal{K}$, that is associated to the Jacobian of H over $\mathbb{F}_q$, we propose the *sum* and *product* extractors, SEK and PEK, for $\mathcal{K}(\mathbb{F}_q)$. These extractors are the modified versions of the extractors SEJ and PEJ. Provided a point K is chosen uniformly at random in $\mathcal{K}$, the element extracted from the point K is statistically close to a uniformly random variable in $\mathbb{F}_q$.

Our proposed extractors can be generalized for the Jacobian of hyperelliptic curves of higher genus.

Acknowledgment. The author thanks to the anonymous referees for several useful suggestions.

References

1. Artin, E.: Algebraic Numbers and Algebraic Functions. Gordon and Breach, New York (1967)
2. Avanzi, R.M.: Aspects of Hyperelliptic Curves over Large Prime Fields in Software Implementations. In: Joye, M., Quisquater, J.-J. (eds.) CHES 2004. LNCS, vol. 3156, pp. 148–162. Springer, Heidelberg (2004)
3. Cantor, D.: Computing in the Jacobian of a Hyperelliptic Curve. Mathematics of Computation 48(177), 95–101 (1987)
4. Cassels, J.W.S., Flynn, E.V.: Prolegomena to a Middlebrow Arithmetic of Curves of Genus 2. Cambridge University Press, Cambridge (1996)
5. Chevassut, O., Fouque, P., Gaudry, P., Pointcheval, D.: The Twist-Augmented Technique for Key Exchange. In: Yung, M., Dodis, Y., Kiayias, A., Malkin, T.G. (eds.) PKC 2006. LNCS, vol. 3958, pp. 410–426. Springer, Heidelberg (2006)

6. Cohen, H., Frey, G.: Handbook of Elliptic and Hyperelliptic Curve Cryptography, Chapman & Hall/CRC, New York (2006)
7. Duquesne, S.: Montgomery Scalar Multiplication for Genus 2 Curves. In: Buell, D.A. (ed.) ANTS 2004. LNCS, vol. 3076, pp. 153–168. Springer, Heidelberg (2004)
8. Farashahi, R.R., Pellikaan, R.: The Quadratic Extension Extractor for (Hyper)Elliptic Curves in Odd Characteristic. In: WAIFI 2007. LNCS, vol. 4547, pp. 219–236. Springer, Heidelberg (2007)
9. Farashahi, R.R., Pellikaan, R., Sidorenko, A.: Extractors for Binary Elliptic Curves. In: WCC 2007. Workshop on Coding and Cryptography, pp. 127–136 (2007)
10. Gaudry, P.: An Algorithm for Solving the Discrete Log Problem on Hyperelliptic Curves. In: Preneel, B. (ed.) EUROCRYPT 2000. LNCS, vol. 1807, pp. 3419–3448. Springer, Heidelberg (2000)
11. Gaudry, P.: Fast genus 2 arithmetic based on Theta functions, Cryptology ePrint Archive, Report 2005/314 (2005), http://eprint.iacr.org/
12. Gürel, N.: Extracting bits from coordinates of a point of an elliptic curve, Cryptology ePrint Archive, Report 2005/324 (2005), http://eprint.iacr.org/
13. Juels, A., Jakobsson, M., Shriver, E., Hillyer, B.K.: How to turn loaded dice into fair coins. IEEE Transactions on Information Theory 46(3), 911–921 (2000)
14. Kaliski, B.S.: A Pseudo-Random Bit Generator Based on Elliptic Logarithms. In: Odlyzko, A.M. (ed.) CRYPTO 1986. LNCS, vol. 263, pp. 84–103. Springer, Heidelberg (1987)
15. Koblitz, N.: Hyperelliptic Cryptosystem. J. of Cryptology 1, 139–150 (1989)
16. Lange, T.: Montgomery Addition for Genus Two Curves. In: Buell, D.A. (ed.) ANTS 2004. LNCS, vol. 3076, pp. 307–309. Springer, Heidelberg (2004)
17. Lange, T.: Formulae for Arithmetic on Genus 2 Hyperelliptic Curves. aaecc 15(1), 295–328 (2005)
18. Luby, M.: Pseudorandomness and Cryptographic Applications. Princeton University Press, USA (1994)
19. Mumford, D.: Tata Lectures on Theta II. In: Progress in Mathematics, vol. 43 (1984)
20. Shaltiel, R.: Recent Developments in Explicit Constructions of Extractors. Bulletin of the EATCS 77, 67–95 (2002)
21. Smart, N.P., Siksek, S.: A Fast Diffie-Hellman Protocol in Genus 2. Journal of Cryptology 12, 67–73 (1999)
22. Trevisan, L., Vadhan, S.: Extracting Randomness from Samplable Distributions. In: IEEE Symposium on Foundations of Computer Science, pp. 32–42 (2000)

Appendix

A   Corresponding Problems

In this section we are dealing with computing the bounds for the cardinalities of Σ_a and Σ_a^ϕ, for $a \in \mathbb{F}_q$ (see Definition 8). We reconsider Definition 8 related to an affine curve with an arbitrary genus. In particular, the sum of Σ_a and Σ_a^ϕ are related to subsets of points of the Jacobian of a genus 2 hyperelliptic (see Proposition 2).

Let $\mathcal{C}$ be an affine curve that is defined over $\mathbb{F}_q$ by the equation

$$\mathbf{y}^2 = f(\mathbf{x}),$$

where $f(\mathbf{x}) \in \mathbb{F}_q[x]$ is a monic polynomial of a positive degree d. Let $a \in \mathbb{F}_q$. We recall that

$$\Sigma_a = \{(P, Q) : P, Q \in \mathcal{C}(\mathbb{F}_q), \ x_P + x_Q = a\},$$
$$\Sigma_a^\phi = \{(P, \phi(P)) : P \in \mathcal{C}(\mathbb{F}_{q^2}), \ x_P + x_{\phi(P)} = a\}.$$

Note that we reconsider Definition 8 that is now related to the affine curve $\mathcal{C}$.

A.1   Cardinality of Σ_a

For an element $a \in \mathbb{F}_q$, the set Σ_a includes the ordered pairs of points on $\mathcal{C}(\mathbb{F}_q)$, such that the sum of their abscissas equals a.

Let $\mathcal{C}_a$ be the affine curve defined over $\mathbb{F}_q$ by the equation

$$\mathbf{z}^2 = f_a(\mathbf{x}) = f(a - \mathbf{x}).$$

Let $\mathcal{C}_a^\star$ be the affine curve over $\mathbb{F}_q$, that is defined by the following equation.

$$\mathbf{w}^2 = f_a^\star(\mathbf{x}) = f(\mathbf{x})f(a - \mathbf{x}).$$

The next proposition gives a formula for the cardinality of Σ_a in terms of the numbers of $\mathbb{F}_q$-rational points of curves $\mathcal{C}$ and $\mathcal{C}_a^\star$.

Lemma 3. *Define*

$$T_a = \{(P, Q) : P \in \mathcal{C}(\mathbb{F}_q), Q \in \mathcal{C}_a(\mathbb{F}_q), \ x_P = x_Q\}.$$

Then $\#T_a = \#\Sigma_a$.

Proof. Clearly $((x, y), (x', y')) \in T$ if and only if $((x, y), (a - x', y')) \in \Sigma_a$.   □

Lemma 4. *Define the function* $\pi_{T_a} : T_a \longrightarrow \mathbb{F}_q$ *by* $\pi_{T_a}(P, Q) = x_P$. *Define the projection map* $\pi_\mathcal{C} : \mathcal{C}(\mathbb{F}_q) \longrightarrow \mathbb{F}_q$ *by* $\pi_\mathcal{C}(P) = x_P$. *Similarly define the projection maps* $\pi_{\mathcal{C}_a}$ *and* $\pi_{\mathcal{C}_a^\star}$, *for the curves* $\mathcal{C}_a$, $\mathcal{C}_a^\star$. *Then*

$$\#\pi_\mathcal{C}^{-1}(x) + \#\pi_{\mathcal{C}_a}^{-1}(x) + \#\pi_{\mathcal{C}_a^\star}^{-1}(x) = 2 + \#\pi_{T_a}^{-1}(x),$$

for all $x \in \mathbb{F}_q$.

Proof. Define $m(x) = \#\pi_{T_a}^{-1}(x)$ and $r(x) = \#\pi_\mathcal{C}^{-1}(x) + \#\pi_{\mathcal{C}_a}^{-1}(x) + \#\pi_{\mathcal{C}_a^\star}^{-1}(x)$, for $x \in \mathbb{F}_q$. We shall prove that $r(x) = 2 + m(x)$, for all $x \in \mathbb{F}_q$.

Let $x \in \mathbb{F}_q$. Let $X_{T_a} = \pi_{T_a}(T_a)$. First we assume that $x \in X_{T_a}$ and $f_a^\star(x) \neq 0$. Then there exist points $P = (x, y) \in \mathcal{C}(\mathbb{F}_q)$ and $Q = (x, z) \in \mathcal{C}_a(\mathbb{F}_q)$. Let $R = (x, w)$, where $w = yz$. So R is a point on $\mathcal{C}_a^\star(\mathbb{F}_q)$. Note that y, z and w are nonzero elements in $\mathbb{F}_q$. So $-P = (x, -y) \neq P$, also $-Q \neq Q$ and $-R \neq R$. Then it is easy to see that $\pi_\mathcal{C}^{-1}(x) = \{P, -P\}$, $\pi_{\mathcal{C}_a}^{-1}(x) = \{Q, -Q\}$ and $\pi_{\mathcal{C}_a^\star}^{-1}(x) = \{R, -R\}$. So $r(x) = 6$. Also $\pi_T^{-1}(x) = \{(P, Q), (P, -Q), (-P, Q), (-P, -Q)\}$. That means $m(x) = 4$.

Second we assume that $x \in \mathbb{F}_q \setminus X_{T_a}$ and $f_a^\star(x) \neq 0$. Since $x \notin X_{T_a}$, then $\pi_T^{-1}(x) = \emptyset$ and $m(x) = 0$. If there exist a point $P = (x, y) \in \mathcal{C}(\mathbb{F}_q)$ then

$\pi_{\mathcal{C}}^{-1}(x) = \{P, -P\}$ and $\pi_{\mathcal{C}_a}^{-1}(x) = \emptyset$, since $x \notin X_{T_a}$. Also $\pi_{\mathcal{C}_a^\star}^{-1}(x) = \emptyset$, since if there exist a point $R = (x, w) \in \mathcal{C}_a^\star(\mathbb{F}_q)$, then $(x, w/y) \in \mathcal{C}_a(\mathbb{F}_q)$, which contradicts the assumption that $x \notin X_{T_a}$. Hence $r(x) = 2$. Similarly if there exist a point $Q = (x, z) \in \mathcal{C}_a(\mathbb{F}_q)$, then $\pi_{\mathcal{C}_a}^{-1}(x) = \{Q, -Q\}$ and $\pi_{\mathcal{C}}^{-1}(x) = \pi_{\mathcal{C}_a^\star}^{-1}(x) = \emptyset$. That means $r(x) = 2$. Therefore assume that there do not exist points on $\mathcal{C}(\mathbb{F}_q)$ or $\mathcal{C}_a(\mathbb{F}_q)$, with the abscissa equals x. So $f(x)$ and $f_a(x)$ are not squared in $\mathbb{F}_q$. Hence $f_a^\star(x)$ is a squared in $\mathbb{F}_q$. Let w be the square root of $f_a^\star(x)$. Then $R = (x, z) \in \mathcal{C}_a^\star(\mathbb{F}_q)$. Therefore $\pi_{\mathcal{C}_a^\star}^{-1}(x) = \{R, -R\}$ and $\pi_{\mathcal{C}}^{-1}(x) = \pi_{\mathcal{C}_a}^{-1}(x) = \emptyset$. Thus $r(x) = 2$.

Third we assume that $x \in X_{T_a}$ and $f_a^\star(x) = 0$. So $\pi_{\mathcal{C}_a^\star}^{-1}(x) = \{P_0\}$, where $P_0 = (x, 0)$. Since $f_a^\star(x) = 0$, then $f(x) = 0$ or $f_a(x) = 0$. If both of $f(x)$ and $f_a(x)$ are zero, then $\pi_{\mathcal{C}}^{-1}(x) = \pi_{\mathcal{C}_a}^{-1}(x) = \{P_0\}$. Also $\pi_T^{-1}(x) = \{(P_0, P_0)\}$. Hence in this case $r(x) = 3$ and $m(x) = 1$. If $f(x) = 0$, but $f_a(x) \neq 0$, then there exist a point $Q = (x, z) \in \mathcal{C}_a(\mathbb{F}_q)$, where $z \neq 0$. Hence $\pi_{\mathcal{C}}^{-1}(x) = \{P_0\}$ and $\pi_{\mathcal{C}_a}^{-1}(x) = \{Q, -Q\}$. Also $\pi_T^{-1}(x) = \{(P_0, Q), (P_0, -Q)\}$. Therefore $r(x) = 4$ and $m(x) = 2$. Similarly in the case that $f(x) \neq 0$ and $f_a(x) = 0$, $r(x) = 4$ and $m(x) = 2$.

Finally we assume that $x \in \mathbb{F}_q \setminus X_{T_a}$ and $f_a^\star(x) = 0$. So $\pi_{\mathcal{C}_a^\star}^{-1}(x) = \{P_0\}$. If $f(x) = 0$, then $\pi_{\mathcal{C}}^{-1}(x) = \{P_0\}$ but $\pi_{\mathcal{C}_a}^{-1}(x) = \emptyset$, since $x \notin X_{T_a}$. Hence $r(x) = 2$ and $m(x) = 0$. If $f_a(x) = 0$, then $\pi_{\mathcal{C}}^{-1}(x) = \emptyset$ and $\pi_{\mathcal{C}_a}^{-1}(x) = \{P_0\}$. Therefore $r(x) = m(x) + 2$, for all $x \in \mathbb{F}_q$. $\quad\square$

Proposition 9. *For all* $a \in \mathbb{F}_q$,

$$\#\Sigma_a = 2\#\mathcal{C}(\mathbb{F}_q) + \#\mathcal{C}_a^\star(\mathbb{F}_q) - 2q.$$

Proof. From Lemma 4, we have

$$\#\mathcal{C}(\mathbb{F}_q) + \#\mathcal{C}_a(\mathbb{F}_q) + \#\mathcal{C}_a^\star(\mathbb{F}_q) = \sum_{x \in \mathbb{F}_q} (\#\pi_{\mathcal{C}}^{-1}(x) + \#\pi_{\mathcal{C}_a}^{-1}(x) + \#\pi_{\mathcal{C}_a^\star}^{-1}(x))$$

$$= \sum_{x \in \mathbb{F}_q} (2 + \#\pi_{T_a}^{-1}(x)) = 2q + \#T_a.$$

From Lemma 3, we have $\#T_a = \#\Sigma_a$. Since $\#\mathcal{C}(\mathbb{F}_q) = \#\mathcal{C}_a(\mathbb{F}_q)$, so the proof of this proposition is finished. $\quad\square$

A.2   Cardinality of Σ_a^ϕ

For $a \in \mathbb{F}_q$, let $\mathcal{C}_a'$ be the affine curve that is defined by the equation

$$\mathbf{y}^2 = F_a(\mathbf{x}) = f(a + \mathbf{x}t)f(a - \mathbf{x}t).$$

Remark 4. The affine curve $\mathcal{C}_a'$, for $a \in \mathbb{F}_q$, is defined over $\mathbb{F}_q$ (see [8]). Furthermore,

$$\#\mathcal{C}_a'(\mathbb{F}_q) = \#\{P \in \mathcal{C}(\mathbb{F}_{q^2}) : x_P = a + x_1 t, \ x_1 \in \mathbb{F}_q\}.$$

Theorem 3 in [8] gives the bound for $\#\mathcal{C}_a'(\mathbb{F}_q)$.

Proposition 10. $\#\Sigma_a^\phi = \#C'_{\frac{a}{2}}(\mathbb{F}_q)$, *for all $a \in \mathbb{F}_q$.*

Proof. Let $P \in \mathcal{C}(\mathbb{F}_{q^2})$, where $x_P = x_0 + x_1 t$ and $x_0, x_1 \in \mathbb{F}_q$. Since $t^q = -t$, so $x_P + x_{\phi(P)} = 2x_0$. That means $(P, \phi(P)) \in \Sigma_a^\phi$ if and only if $x_0 = \frac{a}{2}$. Then Remark 4 concludes the proof of this proposition. $\qquad\square$

A.3   On the Sum of $\#\Sigma_a$ and $\#\Sigma_a^\phi$

In the proof of Theorem 2 (Subsection 4.2), we are dealing to have a tight estimate for $\#\Sigma_a + \#\Sigma_a^\phi$, for all $a \in \mathbb{F}_q$. Following the result of Propositions 9 and 10, one can obtain separate estimates for $\#\Sigma_a$ and $\#\Sigma_a^\phi$. Then add them together to have an estimate for $\#\Sigma_a + \#\Sigma_a^\phi$, for $a \in \mathbb{F}_q$. But this estimate is not tight. Using the result of Proposition 12, we give a tight estimate for it. For the proof of Proposition 12, we need several lemmas.

We recall some details from Subsection 4.2. The two-variable polynomial Φ in $\mathbb{F}_q[\mathbf{x}_0, \mathbf{x}_1]$ is defined as $\Phi(\mathbf{x}_0, \mathbf{x}_1) = f(\mathbf{x}_0)f(\mathbf{x}_1)$. Furthermore, the two-variable polynomial Ψ in $\mathbb{F}_q[\mathbf{a}, \mathbf{b}]$ is defined by

$$\Psi(\mathbf{a}, \mathbf{b}) = \prod_{i=1}^{d} (\mathbf{b} - \lambda_i \mathbf{a} + \lambda_i^2),$$

where λ_i are roots of f in $\overline{\mathbb{F}}_q$. For $a \in \mathbb{F}_q$, the affine curve $\mathcal{X}_a$ is defined over $\mathbb{F}_q$, by the equation

$$\mathbf{y}^2 = \Psi_a(\mathbf{b}) = \Psi(a, \mathbf{b}).$$

Lemma 5. *Define the map $\rho : C_a^\star(\mathbb{F}_q) \longrightarrow \mathbb{F}_q$ by*

$$\rho(x, y) = x(a - x).$$

Let $b \in \mathbb{F}_q$. Assume $\rho^{-1}(b) \neq \emptyset$. Let $(x, y) \in \rho^{-1}(b)$. Then

$$\#\rho^{-1}(b) = \begin{cases} 1, & \text{if } x = \frac{a}{2} \text{ and } y = 0, \\ 2, & \text{if } x = \frac{a}{2} \text{ and } y \neq 0 \text{ or } x \neq \frac{a}{2} \text{ and } y = 0, \\ 4, & \text{otherwise.} \end{cases}$$

Proof. Let $(x, y) \in \rho^{-1}(b)$. It is obvious that $(x, y) \in \rho^{-1}(b)$ if and only if $(x, -y) \in \rho^{-1}(b)$. Furthermore x is a root of polynomial $\tau(\mathbf{x}) = \mathbf{x}^2 - a\mathbf{x} + b$. $\square$

Lemma 6. *Define the map $\varrho : C'_{\frac{a}{2}}(\mathbb{F}_q) \longrightarrow \mathbb{F}_q$ by*

$$\varrho(x, y) = \frac{a^2}{4} - \alpha x^2.$$

Let $b \in \mathbb{F}_q$. Assume $\varrho^{-1}(b) \neq \emptyset$. Let $(x, y) \in \varrho^{-1}(b)$. Then

$$\#\varrho^{-1}(b) = \begin{cases} 1, & \text{if } x = 0 \text{ and } y = 0, \\ 2, & \text{if } x = 0 \text{ and } y \neq 0 \text{ or } x \neq 0 \text{ and } y = 0, \\ 4, & \text{otherwise.} \end{cases}$$

Proof. Let $(x, y) \in \varrho^{-1}(b)$. It is obvious that $(x, y) \in \varrho^{-1}(b)$ if and only if $(x, -y) \in \varrho^{-1}(b)$. Furthermore x is a root of polynomial $\widetilde{\tau}(\mathbf{x}) = \alpha \mathbf{x}^2 - \frac{a^2}{4} + b$. Thus $(x, y) \in \varrho^{-1}(b)$ if and only if $(-x, y) \in \varrho^{-1}(b)$. $\qquad\square$

Lemma 7. *Define the projection map* $\pi : \mathcal{X}_a(\mathbb{F}_q) \longrightarrow \mathbb{F}_q$ *by* $\pi(b, y) = b$. *Then*

$$\#\rho^{-1}(b) + \#\varrho^{-1}(b) = 2\#\pi^{-1}(b),$$

for all $b \in \mathbb{F}_q$.

Proof. Let $b \in \mathbb{F}_q$, such that $\pi^{-1}(b) \neq \emptyset$. So there exist a point $(b, y) \in \mathcal{X}_a(\mathbb{F}_q)$. Hence $y^2 = \Psi_a(b) = \Psi(a, b)$. If $y = 0$, then $\pi^{-1}(b) = \{(b, 0)\}$. So $\#\pi^{-1}(b) = 1$. If $y \neq 0$, then $\pi^{-1}(b) = \{(b, y), (b, -y)\}$. Hence $\#\pi^{-1}(b) = 2$. Consider the polynomials $\tau, \widetilde{\tau} \in \mathbb{F}_q[\mathbf{x}]$, that are defined as $\tau(\mathbf{x}) = \mathbf{x}^2 - a\mathbf{x} + b$ and $\widetilde{\tau}(\mathbf{x}) = \alpha \mathbf{x}^2 - \frac{a^2}{4} + b$. Let $\mathcal{D}$ be the discriminant of τ, that is $\mathcal{D} = a^2 - 4b$. Then $\alpha \mathcal{D}$ is the discriminant of $\widetilde{\tau}$. We explain in three cases for $\mathcal{D}$.

First, assume $\mathcal{D} = 0$. Hence $\frac{a}{2}$ is the multiple root of τ. Since $y^2 = \Psi(a, b)$, then $y^2 = \Phi(\frac{a}{2}, \frac{a}{2}) = (f(\frac{a}{2}))^2$. Thus $(\frac{a}{2}, y) \in \mathcal{C}_a^\star(\mathbb{F}_q)$ and $(0, y) \in \mathcal{C}'_{\frac{a}{2}}(\mathbb{F}_q)$. Since $\mathcal{D} = 0$, then $b = \frac{a^2}{4}$, so $(\frac{a}{2}, y) \in \rho^{-1}(b)$ and $(0, y) \in \varrho^{-1}(b)$. From Lemmas 5 and 6, if $y = 0$, then $\#\rho^{-1}(b) = \#\varrho^{-1}(b) = 1$, else $\#\rho^{-1}(b) = \#\varrho^{-1}(b) = 2$.

Second, assume $\mathcal{D}$ is a square in $\mathbb{F}_q^\star$. So τ is reducible in $\mathbb{F}_q[\mathbf{x}]$. Let x_0, x_1 be the distinct roots of τ in $\mathbb{F}_q$. Then $x_0 + x_1 = a$ and $x_0 x_1 = b$. Since $y^2 = \Psi(a, b)$, then $y^2 = \Phi(x_0, x_1) = f(x_0)f(x_1)$. Thus (x_0, y) and (x_1, y) are points of $\mathcal{C}_a^\star(\mathbb{F}_q)$ and $\rho^{-1}(b)$. From Lemma 5, if $y = 0$, then $\#\rho^{-1}(b) = 2$, else $\rho^{-1}(b) = 4$, since x_0 and x_1 do not equal $\frac{a}{2}$. Since $\mathcal{D}$ is a square in $\mathbb{F}_q^\star$ and α is a non-square in $\mathbb{F}_q$, then $\alpha \mathcal{D}$, the discriminant of $\widetilde{\tau}$, is a non-square in $\mathbb{F}_q^\star$. That means $\widetilde{\tau}(\mathbf{x})$ has no root in $\mathbb{F}_q$. So $\varrho^{-1}(b) = \emptyset$.

Third, assume $\mathcal{D}$ is a non-square in $\mathbb{F}_q$. Hence $\tau(\mathbf{x})$ has no root in $\mathbb{F}_q$. So $\rho^{-1}(b) = \emptyset$. Also $\alpha \mathcal{D}$ is a square in $\mathbb{F}_q^\star$. Thus $\widetilde{\tau}$ is reducible in $\mathbb{F}_q[\mathbf{x}]$. Let x_0, x_1 be the distinct roots of $\widetilde{\tau}$ in $\mathbb{F}_q$. Clearly $x_0 = -x_1$ and $x_0 x_1 = -\frac{\mathcal{D}}{4\alpha}$. Let $z_0 = \frac{a}{2} + x_0 t$ and $z_1 = \frac{a}{2} + x_1 t$. Then $z_0 + z_1 = a$ and $z_0 z_1 = b$. Since $y^2 = \Psi(a, b)$, then $y^2 = \Phi(z_0, z_1) = f(z_0)f(z_1)$. So $y^2 = F_{\frac{a}{2}}(x_0) = F_{\frac{a}{2}}(x_1)$. Thus (x_0, y) and (x_1, y) are points of $\mathcal{C}'_{\frac{a}{2}}(\mathbb{F}_q)$ and $\varrho^{-1}(b)$. From Lemma 6, if $y = 0$, then $\#\varrho^{-1}(b) = 2$, else $\varrho^{-1}(b) = 4$, since x_0 and x_1 do not equal 0.

Now, let $b \in \mathbb{F}_q$, such that $\pi^{-1}(b) = \emptyset$. Then $\rho^{-1}(b) = \varrho^{-1}(b) = \emptyset$. Since if $(x, y) \in \rho^{-1}(b)$, then $x(a - x) = b$ and $(x, y) \in \mathcal{C}_a^\star(\mathbb{F}_q)$. So $y^2 = f(x)f(a - x)$. Then $y^2 = \Phi(x, a - x) = \Psi(a, b) = \Psi_a(b)$. Thus $(b, y) \in \mathcal{X}_a(\mathbb{F}_q)$, which is a contradiction. Also if $(x, y) \in \varrho^{-1}(b)$, then $\frac{a^2}{4} - \alpha x^2 = b$ and $(x, y) \in \mathcal{C}'_{\frac{a}{2}}(\mathbb{F}_q)$. Hence $y^2 = f(\frac{a}{2} + xt)f(\frac{a}{2} - xt)$. Then $y^2 = \Phi(\frac{a}{2} + xt, \frac{a}{2} - xt) = \Psi(a, b) = \Psi_a(b)$. Thus $(b, y) \in \mathcal{X}_a(\mathbb{F}_q)$, which is a contradiction. $\qquad\square$

Proposition 11. $\#\mathcal{C}_a^\star(\mathbb{F}_q) + \#\mathcal{C}'_{\frac{a}{2}}(\mathbb{F}_q) = 2\#\mathcal{X}_a(\mathbb{F}_q)$, *for all* $a \in \mathbb{F}_q$.

Proof. Let $a \in \mathbb{F}_q$. From Lemma 7, $\#\rho^{-1}(b) + \#\varrho^{-1}(b) = 2\#\pi^{-1}(b)$, for all $b \in \mathbb{F}_q$. Then

$$\#\mathcal{C}_a^{\star}(\mathbb{F}_q) + \#\mathcal{C}_{\frac{a}{2}}'(\mathbb{F}_q) = \sum_{b \in \mathbb{F}_q} \#\rho^{-1}(b) + \sum_{b \in \mathbb{F}_q} \#\varrho^{-1}(b)$$

$$= \sum_{b \in \mathbb{F}_q} 2\#\pi^{-1}(b) = 2\#\mathcal{X}_a(\mathbb{F}_q).$$

$\square$

Proposition 12. *Let $a \in \mathbb{F}_q$. Then*

$$\#\Sigma_a + \#\Sigma_a^{\phi} = 2(\#\mathcal{C}(\mathbb{F}_q) + \#\mathcal{X}_a(\mathbb{F}_q) - q).$$

Proof. Propositions 9, 10 and 11 conclude the proof of this proposition. $\square$

B   Proofs of Propositions

In this section we prove Propositions 4 and 7.

B.1   Proof of Proposition 4

Proof (Proposition 4). Clearly the affine curve $\mathcal{X}_a$ is absolutely irreducible for all $a \in \mathbb{F}_q$. The affine curve $\mathcal{X}_a$ may be singular. Let $\sigma_{i,j} = \lambda_i + \lambda_j$, for all integers i, j such that $1 \le i < j \le 5$. Let s_a be the number of $\sigma_{i,j}$ that are equal to a. Then the polynomial $\Psi_a(\mathbf{b})$ has s_a double roots, since λ_i are pairwise distinct. That means $\mathcal{X}_a$ has s_a singular points. Note that $0 \le s_a \le 2$. If $s_a = 0$, then $\mathcal{X}_a$ is is an absolutely nonsingular affine curve of genus 2. In fact, the genus of the nonsingular model of $\mathcal{X}_a$ equals $2 - s_a$. By using Hasse-Weil bound for the nonsingular model of $\mathcal{X}_a$, we obtain

$$|\#\mathcal{X}_a(\mathbb{F}_q) - q| \le 2(2 - s_a)\sqrt{q} + s_a \le 4\sqrt{q}.$$

So the proof of this proposition is completed. $\square$

B.2   Proof of Proposition 7

Proof (Proposition 7). Let $b \in \mathbb{F}_q$. Let $\delta_{i,j} = \lambda_i \lambda_j$, for all integers i, j such that $1 \le i < j \le 5$. Let s_b be the number of $\delta_{i,j}$ that are equal to b. Then the polynomial $\Psi_b(\mathbf{a})$ has s_b double roots, since λ_i are pairwise distinct.

If $f(0) \ne 0$, then $\lambda_i \ne 0$, for all integer $0 \le i \le 5$. Then the degree of $\Psi_b(\mathbf{a})$ equals 5. So the affine curve $\mathcal{X}_b$ is absolutely irreducible for all $b \in \mathbb{F}_q$. Since $\Psi_b(\mathbf{a})$ has s_b double root, thus $\mathcal{X}_b$ has s_b singular points. In fact, the genus of the nonsingular model of $\mathcal{X}_b$ equals $2 - s_b$. By using Hasse-Weil bound for the the number of $\mathbb{F}_q$-rational points of the nonsingular model of $\mathcal{X}_b$, we obtain

$$|\#\mathcal{X}_b(\mathbb{F}_q) - q| \le 2(2 - s_b)\sqrt{q} + s_b \le 4\sqrt{q}.$$

If $f(0) = 0$, then there exists an integer i such that $\lambda_i = 0$. If $b = 0$, clearly $\#\mathcal{X}_b(\mathbb{F}_q) = q$. Now assume that $b \neq 0$. Then the degree of $\Psi_b(\mathbf{a})$ equals 4. In this case, one could show that, $s_b = 2$ if and only if $b \in I_f$. If $s_b = 2$, then $\Psi_b(\mathbf{a})$ is square, so the affine curve $\mathcal{X}_b$ is reducible. Hence we have only the trivial bound for $\#\mathcal{X}_b(\mathbb{F}_q)$, that is

$$|\#\mathcal{X}_b(\mathbb{F}_q) - q| \leq q.$$

Otherwise $s_b \leq 1$. So $\Psi_b(\mathbf{a})$ is a non-square. Hence the affine curve $\mathcal{X}_b$ is absolutely irreducible. Furthermore $\mathcal{X}_b$ has s_b singular points and the genus of the nonsingular model of $\mathcal{X}_b$ equals $1 - s_b$. By using Hasse-Weil bound we obtain

$$|\#\mathcal{X}_b(\mathbb{F}_q) - q| \leq 2(1 - s_b)\sqrt{q} + s_b \leq 2\sqrt{q}.$$

So the proof of this proposition is finished.   □

Constructing Pairing-Friendly Elliptic Curves Using Gröbner Basis Reduction*

Waldyr D. Benits Junior** and Steven D. Galbraith***

Mathematics Department,
Royal Holloway University of London,
Egham, Surrey TW20 0EX, UK
[w.benits-junior,steven.galbraith]@rhul.ac.uk

Abstract. The problem of constructing elliptic curves suitable for pairing applications has received a lot of attention. One of the most general methods to solve this problem is due to Scott. We propose a variant of this method which replaces an exhaustive search with a Gröbner basis calculation. This makes the method potentially able to generate a larger set of families of parameters. We present some new families of parameters for pairing-friendly elliptic curves.

Keywords: Pairing-friendly elliptic curves, Gröbner bases.

1 Introduction

Pairing-Based Cryptography has been a research area of great interest since the work of Sakai, Oghishi and Kasahara [17], Joux [15] and Boneh and Franklin [7].

However, to make pairing computation feasible, we have to work with some special curves, called "PAIRING-FRIENDLY" elliptic curves, in that the embedding degree is relatively small and the curve has a large prime order subgroup. For a survey of methods to construct pairing-friendly elliptic curves, we refer to [13].

One option is to use supersingular curves, but in this case, we are restricted to an embedding degree $\leqslant 6$. Hence, the use of ordinary elliptic curves with small embedding degree is very attractive, although these curves are very rare [3].

Now we give some notation that will be used throughout the paper. Let p be a large prime, let $E : y^2 = x^3 + ax + b$ be an elliptic curve over $\mathbb{F}_p$ and let t be the trace of Frobenius so that $n = \#E(\mathbb{F}_p) = p + 1 - t$ is the curve order. Suppose r is a large prime factor of n and $h = n/r$ a "small" integer called the cofactor. Let k be the embedding degree with respect to r (i.e., $r \mid (p^k - 1)$) and let $\rho = \log(p)/\log(r)$.

* The work described in this paper has been supported in part by the European Commission through the IST Programme under Contract IST-2002-507932 ECRYPT. The information in this document reflects only the author's views, is provided as is and no guarantee or warranty is given that the information is fit for any particular purpose. The user thereof uses the information at its sole risk and liability.
** This author thanks the Brazilian Navy for support.
*** This author thanks the EPSRC for support.

S.D. Galbraith (Eds.): Cryptography and Coding 2007, LNCS 4887, pp. 336–345, 2007.
© Springer-Verlag Berlin Heidelberg 2007

To obtain a suitable elliptic curve, we should choose parameters that not only satisfy the required conditions for security and efficiency (see [13]), but also satisfy the CM equation (1) below, so that we can construct an elliptic curve using the Complex Multiplication (CM) method (see [2]). The problem is therefore to find suitable parameters p, r, k, a, b such that pairing computation is feasible and the system is secure.

A very common approach is to obtain polynomial families of parameters for given embedding degree k and CM discriminant D. More precisely, one finds polynomials $P(x), R(x), T(x)$ with integer coefficients such that $P(x)$ and $R(x)$ can represent primes and such that if x_0 is such that $p = P(x_0)$ is a prime then there is an elliptic curve E over $\mathbb{F}_p$ with complex multiplication of discriminant D and with number of points divisible by $R(x_0)$ and the corresponding subgroup has (at least, generically) embedding degree k.

The paper is organized as follows: Section 2 reviews the existing methods for finding polynomial families of parameters, showing their characteristics and restrictions. Section 3 presents Scott's method [18] in detail. Section 4 presents our improvement of the exhaustive search part of Scott's method. Section 5 gives a comparison between our method and Scott's method. Section 6 concludes the paper and proposes further research. Finally, Appendix A gives a step-by-step example of our method obtained using MAGMA [8].

2   Some Existing Methods

The first polynomial families of parameters for ordinary elliptic curves with small embedding degree were proposed by Miyaji, Nakabayashi and Takano [16] in 2001. In their work, they give families of ordinary elliptic curves with embedding degree 3, 4 and 6. The restriction to $k \leq 6$ is undesirable for higher security levels. Freeman [12] extended this method to $k = 10$.

In 2002, Cocks and Pinch proposed a method to generate elliptic curves with larger values of k using the CM method (see Chapter IX of [6]). In this method, r is an input to the algorithm that generates p and t, so one can freely choose r with very low Hamming weight, for example. One of the key ideas of their method is to use equation (4) below and to obtain t from considering primitive k-th roots modulo r. The main restriction in this method is that $\rho \approx 2$, which is considered too big for certain applications.

Brezing and Weng [9] gave a method to generate polynomial families by exploiting polynomials $f(x)$ such that $\mathbb{Q}(\zeta_k, \sqrt{D}) \cong \mathbb{Q}(x)/(f(x))$. They also exploit equation (4) to choose polynomials $T(x)$ corresponding to elements of order k in $\mathbb{Q}(x)/(f(x))$. This method uses small values for D but it only works well for certain values for k.

Barreto, Lynn and Scott [4] wrote $n = \frac{hr}{d}$ for $h > 1$ and some integer d and arranged that the RHS of the CM equation (1) becomes of the form $(D'u)^2$ by using adequate values for d, t and h (that can be easily found by simple search algorithms). They have solutions for families of curves with many values of k.

The best results from [9,4] have ρ approximately 1.25. As these solutions are produced by using small values of D, the curves can be quickly found by the

CM method. Both methods use $r = \Phi_k(t - 1)$, where Φ_k is the k-th cyclotomic polynomial, which restricts the number of curves that can be found.

Barreto and Naehrig [5], building on work of Galbraith, McKee and Valença [14], found a particular solution for a curve of prime order (which is the optimum case, when we have $h = 1$ and $\rho \approx 1$) with $k = 12$, using a proper factor of $\Phi_k(T(x) - 1)$, when $T(x) = 6x^2 + 1$. Curves in the resulting family are known as "BN curves".

Scott [18] extended the ideas of [9,4], by using $R(x)$ as any proper factor of $\Phi_k(T(x) - 1)$ (and not just $R(x) = \Phi_k(T(x) - 1)$). In this method (which is described in greater detail in Section 3 below) one first checks, for given $T(x)$, $H(x)$ and d, if $4H(x)R(x) - d(T(x) - 2)^2$ is a perfect square. If so then one can compute $U(x)$ and $P(x)$ and then construct a curve using the CM method. Similar ideas were also used by Duan, Cui and Chan [11]; their best results is $\rho = 1.06$, but only when $k = 96$.

The main drawback in these cases is that you have to limit the coefficient sizes and the degree of the polynomial $H(x)$ since it is found by an exhaustive search. Because of that, if there are solutions for $H(x)$ with relatively high degree or large coefficients then those solutions will not be easily found by such methods. Our contribution is to replace the exhaustive search by solving a system of multivariate polynomial equations using Gröbner basis reduction.

3   Scott's Method

In this section, we will present in detail the method proposed by Scott [18] (also see Section 6 of [13]). The starting point is the CM equation:

$$t^2 - 4p = Du^2. \tag{1}$$

We also have:

$$n = \#E(\mathbb{F}_p) = p + 1 - t \equiv 0 \bmod r \tag{2}$$
$$p^k - 1 \equiv 0 \bmod r \tag{3}$$

Substituting (2) into (3) gives

$$(t - 1)^k \equiv 1 \bmod r. \tag{4}$$

Since k is the embedding degree, it follows that $t - 1$ is a primitive kth root of unity modulo r, which is equivalent to $r \mid \Phi_k(t - 1)$.

Substituting $p = n + t - 1$ into (1), we have:

$$(t - 2)^2 - 4n = Du^2. \tag{5}$$

To have the RHS of the equation (5) to be a perfect square one can multiply equation (5) by an integer d, such that $dD = (D')^2$. Setting $n = hr$ for a "small" cofactor h and defining $h' = hd$ gives

$$d(t - 2)^2 - 4h'r = (D'u)^2. \tag{6}$$

The problem is to find integers (d, t, h', r) such that the LHS of equation (6) is a perfect square. Algorithm 1 presents the method to achieve this.

Step 2 of Algorithm 1 uses an exhaustive search over possible values for the cofactor $H'(x)$. In practice, this means that we have to impose limits on the

---

Algorithm 1. Scott's method

System Parameters: $\Phi_k(x)$.
Input: embedding degree k.
Output: polynomials $P(x), T(x), H(x)$ and $R(x)$.
1. Given k (embedding degree), choose a polynomial $T(x)$, such that $\Phi_k(T(x)-1)$ has a suitable factorisation as $R(x)R'(x)$;
2. Write $H'(x) = h'_0 + h'_1 x + h'_2 x^2 + ... + h'_n x^n$;
   Use an exhaustive search (see below) over the variables $(h'_0, h'_1, ..., h'_n, d)$ until $d(T(x) - 2)^2 - 4H'(x)R(x)$ is a perfect square;
3. Compute $H(x) = H'(x)/d$, $D \in \mathbb{Z}$ the square-free part of d (i.e., $d = Dw^2$ for some $w \in \mathbb{Q}$), $P(x) = H(x)R(x)+T(x)-1$ and $U(x) = \frac{\sqrt{d(T(x)-2)^2 - 4H'(x)R(x)}}{D}$ and check if $P(x)$ is irreducible;
5. If there exists an $x_0 \in \mathbb{Z}$ such that $P(x_0)$ and $R(x_0)$ are both primes, then return $P(x), T(x), H(x)$ and $R(x)$

---

coefficient sizes and on the degree of $H'(x)$. Choose an integer B as a bound for the size of coefficients in $H'(x)$ and choose a bound n on the degree of $H'(x)$. Note that the degree n should be relatively small (since we do not want the cofactor h to be large), but it is not clear that the coefficients of $H(x)$ are necessarily very small. The algorithm needs B^{n+1} steps to find the coefficients of $H'(x)$. If we allow negative coefficients for $H'(x)$, then we have the coefficients of H' bounded between $[-B, B]$ and therefore, $(2B)^{n+1}$ steps. Typically we might also have $-B \leqslant D < 0$ giving $B(2B)^{n+1}$ steps. Therefore, there is a practical limit on the polynomials $H(x)$ which can be found by Scott's method.

4   Our Refinement of Scott's Method

As discussed above, if there are families of elliptic curves such that $H(x)$ has large coefficients and/or high degree then Scott's method will take a long time to find them. Our idea is to replace the exhaustive search by solving a system of multivariate polynomial equations.

Our starting point is the equation (5), derived from the CM equation. Putting $n = hr$ and rearranging terms gives

$$(t - 2)^2 - Du^2 = 4hr. \tag{7}$$

This can be rewritten as

$$(t - 2)^2 - Du^2 \equiv 0 \bmod (4r). \tag{8}$$

As we seek a polynomial family, we have polynomials $T(x)$ and $R(x)$ and we want to solve the equation

$$(T(x) - 2)^2 - DU(x)^2 \equiv 0 \bmod (4R(x)) \tag{9}$$

for D and $U(x)$. We will solve this equation by expressing the problem in terms of solving a system of multivariate equations. We discard the trivial solution $D = 1$ and $U(x) = T(x) - 2$.

Precisely, write $U(x) = u_0 + u_1 x + \cdots + u_n x^n \in \mathbb{Q}[x]$ where the u_i are indeterminates and where n is some guess of the degree. Also, let D be an indeterminate[1]. Let $m = \deg(R(x))$. One can compute the LHS of (9) and reduce modulo $R(x)$ to obtain a polynomial of degree $\leq (m-1)$ in the indeterminates u_i and D. The congruence to 0 therefore gives m non-linear equations over $\mathbb{Q}$ in the $n+2$ variables $u_0, \ldots, u_n, D$. If $n + 2 < m$ then we have a potentially overdetermined system and one can hope to solve it using Gröbner basis techniques (see [1], [10]). Our implementation uses MAGMA; in Appendix A we give a step-by-step example.

Once $U(x)$ has been obtained one can then solve for $H(x)$ by polynomial division as

$$H(x) = \frac{(T(x) - 2)^2 - DU(x)^2}{4R(x)}.$$

We now explain that the condition $n + 2 < m$ is not a serious problem when one is trying to produce families of parameters with a good value of ρ. In practice we usually have $\deg(R(x)) \geq 2\deg(T(x))$ and so, by the equation $4H(x)R(x) = (T(x) - 2)^2 - DU(x)^2$ it follows that $\deg(H(x)) + \deg(R(x)) = 2\deg(U(x))$. In other words, $\deg(H(x)) + m = 2n$. Since we do not want to have large cofactors, it follows that $\deg(H(x))$ should be much smaller than m, and so $n < m$. In any case, since we are working modulo $R(x)$ it is unnecessary to consider $\deg(U(x)) \geq \deg(R(x))$. Furthermore, even if $n + 2 \geq m$ then one can guess one or more of the coefficients u_i (an example of this is given in Appendix A).

Finally, we can compute $P(x) = H(x)R(x) + T(x) - 1$ and check if $P(x)$ is irreducible and if there is an integer x_0 such that $P(x_0)$ is prime.

Algorithm 2. Our refinement of Scott's method

System Parameters: $\Phi_k(x)$.

Input: embedding degree k.

Output: polynomials $P(x), T(x), H(x)$ and $R(x)$.

1. Given k (embedding degree), find a polynomial $T(x)$ such that $\Phi_k(T(x) - 1)$ has a suitable factorisation as $R(x)R'(x)$; Let $m = \deg(R(x))$;
2. Write $U(x) = u_0 + u_1 x + u_2 x^2 + \ldots + u_n x^n$ for suitable n;
   Reduce $[(T(x) - 2)^2 - DU(x)^2] \bmod (4R(x))$ to get m multivariate polynomials in the $n + 2$ variables u_i and D;
3. Determine the set of solutions over $\mathbb{Q}$ to the system of multivariate polynomial equations using Gröbner basis methods;
4. Compute $H(x) = \frac{(T(x)-2)^2 - DU(x)^2}{4R(x)}$ and $P(x) = H(x)R(x) + T(x) - 1$ and check if $P(x)$ is irreducible;
5. If there exists an $x_0 \in \mathbb{Z}$ such that $P(x_0)$ and $R(x_0)$ are both primes, then return $P(x), T(x), H(x)$ and $R(x)$

Algorithm 2 gives a brief description of our method. In Step 1 we just choose $T(x)$ by a simple exhaustive search algorithm; the choice of $T(x)$ is equally crucial to both Scott's method and our refinement. Note that we focused attention on

---

[1] A further extension is to take $D(x)$ to be a polynomial, but this quickly gives CM discriminants which are too large.

the case where $R(x)$ is a "proper factor" of $\Phi_k(T(x) - 1)$, but our method still works if $\Phi_k(T(x) - 1)$ is irreducible.

5 Discussion and Examples

In principle, Scott's method can find every family of pairing-friendly curves. However, in practice the search required to find the coefficients of the polynomial $H(x)$ could be prohibitive. Further, it is impossible to tell when using Scott's method whether one has found all solutions or whether there exist polynomials whose coefficients are outside the search range.

Our refinement also finds all families, and the dependence of the running time on the size of coefficients seems to be much less severe. Also, with our approach one computes all families at once and so one knows that every possible family has been found for a given choice of $T(x), R(x)$ and degree of $U(x)$.

Note however that the complexity of Gröbner basis methods depends badly on the number of variables. Hence, we are restricted to using polynomials $H(x)$ of relatively low degree. As mentioned above, to get low values for ρ one needs $\deg H(x)$ to be relatively small, so this is not a serious issue.

We now present some examples of curves with polynomials $H(x)$ which are not very quickly found by Scott's method. This illustrates the potential advantage of our approach over Scott's method. We give two examples below of polynomial families of pairing-friendly curves which, to the best of our knowledge, have not previously appeared in the literature. In both examples the size of $H(x)$ is such that a relatively large computation would have been required to find it using Scott's method, whereas our solution takes less than a second. Note that both examples have $\rho \geq 1.5$ which means that these families are not necessarily suitable for some applications.

We were very surprised that our efforts yielded only these two families. It seems to be the case that almost all examples of polynomial families of pairing-friendly curves have polynomials $H(x)$ with coefficients of relatively low height. We have no explanation for this fact.

Example 1: $k = 14$
$T(x) = x + 1$
$D = -7$
$R(x) = x^6 - x^5 + x^4 - x^3 + x^2 - x + 1$
$U(x) = \frac{1}{7}(-2x^5 + 2x^4 - 2x^3 + 4x^2 - 3x + 1)$
$P(x) = \frac{1}{7}(x^{10} - 2x^9 + 3x^8 - 6x^7 + 8x^6 - 8x^5 + 8x^4 - 7x^3 + 6x^2 + 2x + 2)$
$H(x) = \frac{1}{7}(x^2 + x + 2)(x - 1)^2 = \frac{1}{7}(x^4 - x^3 + x^2 - 3x + 2)$
$\rho = 1.65$

In this case, $R(x) = \Phi_{14}(T(x) - 1)$ is irreducible. For $x_0 = 134222987$ we find a 267-bit prime p and a 162-bit prime r. It took about 0.5 seconds for Magma to compute the Gröbner basis for this example.

Example 2: $k = 12$

$T(x) = x^3 + x^2 + 4x + 2$

$D = -4$

$R(x) = x^4 + 2x^3 + 5x^2 + 4x + 1$

$U(x) = -2x^3 - 3x^2 - \frac{17}{2}x - 4$

$P(x) = \frac{1}{4}(17x^6 + 50x^5 + 181x^4 + 280x^3 + 405x^2 + 288x + 68)$

$H(x) = \frac{1}{4}(17x^2 + 16x + 64)$

$\rho = 1.5$

For $x_0 = 1099511631452$ we find a 242-bit prime p and a 160-bit prime r. The Gröbner basis computation in Magma for this example took about 0.4 seconds. Note that the coefficients of $H(x)$ are larger than in Example 1, so Scott's method would typically take longer to find this example, whereas the running time of our method is about the same in both cases. The other advantage is that we know we have found all solutions, whereas with Scott's method one doesn't know if there is another solution just outside the search range.

6   Conclusion

In this paper we proposed a refinement to the method of constructing elliptic curves proposed by Michael Scott, in that we avoid the exhaustive search used to find the cofactor $H(x)$.

We remark that both methods have the choice of $T(x)$ as a bottleneck, since that polynomial is found by exhaustive search. Galbraith, McKee and Valença [14] gave examples of quadratic families of polynomials $Q(x)$ such that $\Phi_k(Q(x))$ splits, but for higher degree families it still remains an open problem to classify such polynomials.

Acknowledgements

We thank Mike Scott and Ben Smith for helpful comments.

References

1. Adams, W., Loustaunau, P.: An introduction to Grobner bases. In: Graduate Studies in Math, vol. 3, Oxford University Press, New York (1994)
2. Avanzi, R., Cohen, H., Doche, C., Frey, G., Lange, T., Nguyen, K., Vercauteren, F.: Handbook of elliptic and hyperelliptic curve cryptography. In: Discrete Mathematics and its Applications, Chapman & Hall/CRC (2006)
3. Balasubramanian, R., Koblitz, N.: The improbability that an elliptic curve has subexponential discrete log problem under the Menezes-Okamoto-Vanstone algorithm. Journal of Cryptology 11(2), 141–145 (1998)
4. Barreto, P., Lynn, B., Scott, M.: Constructing elliptic curves with prescribed embedding degrees. In: Cimato, S., Galdi, C., Persiano, G. (eds.) SCN 2002. LNCS, vol. 2576, pp. 263–273. Springer, Heidelberg (2003)

5. Barreto, P., Naehrig, M.: Pairing-friendly elliptic curves of prime order. In: Preneel, B., Tavares, S. (eds.) SAC 2005. LNCS, vol. 3897, pp. 319–331. Springer, Heidelberg (2006)
6. Blake, I., Seroussi, G., Smart, N.: Advances in elliptic curve cryptography. Cambridge University Press, Cambridge (2005)
7. Boneh, D., Franklin, M.: Identity based encryption from the Weil pairing. In: Kilian, J. (ed.) CRYPTO 2001. LNCS, vol. 2139, pp. 213–229. Springer, Heidelberg (2001)
8. Bosma, W., Cannon, J., Playoust, C.: The Magma algebra system I: The user language. Journal of Symbolic Computation 24, 235–265 (1997)
9. Brezing, F., Weng, A.: Elliptic curves suitable for pairing based cryptography. Codes and Cryptography 37, 133–141 (2005)
10. Cox, D.A., Little, J.B., O'Shea, D.: Ideals, varieties and algorithms: an introduction to computational algebraic geometry and commutative algebra. Springer, Heidelberg (1992)
11. Duan, P., Cui, S., Chan, C.W.: Effective polynomial families for generating more pairing-friendly elliptic curves. Cryptology ePrint Archive, Report 2005/236 (2005), http://eprint.iacr.org/2005/236
12. Freeman, D.: Constructing pairing-friendly elliptic curves with embedding degree 10. In: Hess, F., Pauli, S., Pohst, M. (eds.) ANTS-VII. LNCS, vol. 4076, pp. 452–465. Springer, Heidelberg (2006)
13. Freeman, D., Scott, M., Teske, E.: A taxonomy of pairing-friendly elliptic curves, Cryptology ePrint Archive, Report 2006/372 (2006), http://eprint.iacr.org/2006/372
14. Galbraith, S.D., McKee, J.F., Valenca, P.C.: Ordinary abelian varieties having small embedding degree, Finite Fields and Applications (to appear), http://eprint.iacr.org/2004/365
15. Joux, A.: A one-round protocol for tripartite Diffie-Hellman. In: Bosma, W. (ed.) ANTS IV. LNCS, vol. 1838, pp. 385–394. Springer, Heidelberg (2000)
16. Miyaji, A., Nakabayashi, M., Takano, S.: New explicit conditions of elliptic curve traces for FR-reduction. IEICE Transactions on Fundamentals E84-A(5), 1234–1243 (2001)
17. Sakai, R., Ohgishi, K., Kasahara, M.: Cryptosystems based on pairing. In: 2000 Symposium on Cryptography and Information Security (SCIS2000), Okinawa, Japan (2000)
18. Scott, M.: Generating families of pairing-friendly elliptic curves (Preprint, 2006)

A   A Step-by-Step Example

We demonstrate our method with example 2 of Section 5.

Let $T(x) = x^3 + x^2 + 4x + 2$, $k = 12$. We have

$$\Phi_{12}(x) = x^4 - x^2 + 1$$

and so

$$\Phi_{12}(T(x) - 1) = x^{12} + 4x^{11} + 22x^{10} + 56x^9 + 157x^8 + 268x^7 + 457x^6 + 506x^5$$
$$+ 493x^4 + 298x^3 + 82x^2 + 8x + 1$$
$$= (x^4 + 2x^3 + 5x^2 + 4x + 1)(x^8 + 2x^7 + 13x^6 + 16x^5 + 51x^4$$
$$+ 32x^3 + 61x^2 + 4x + 1)$$

So, we take $R(x) = x^4 + 2x^3 + 5x^2 + 4x + 1$. One can also use the other factor $R'(x)$; this leads to a family with $\deg P(x) = 12$ and $\deg H(x) = 4$ so we have the same value ρ but a more sparse family (since we expect a degree 12 polynomial to represent fewer primes of a given size than a degree 4 polynomial does).

Now, we write $U(x) = u_3 x^3 + u_2 x^2 + u_1 x + u_0$ and reduce $[(T(x) - 2)^2 - dU(x)^2] \bmod (4R(x))$ to obtain

$$(-2du_0u_3 - 2du_1u_2 + 4du_1u_3 + 2du_2^2 + 2du_2u_3 - 8du_3^2 - 4)x^3$$
$$+(-2du_0u_2 - du_1^2 + 10du_1u_3 + 5du_2^2 - 12du_2u_3 - 12du_3^2 - 5)x^2$$
$$+(-2du_0u_1 + 8du_1u_3 + 4du_2^2 - 14du_2u_3 - 6du_3^2 - 16)x$$
$$+(-du_0^2 + 2du_1u_3 + du_2^2 - 4du_2u_3 - du_3^2 - 4)$$

We want $[(T(x) - 2)^2 - 4U(x)^2] \equiv 0 \bmod (4R(x))$, so we have the following system of equations:

$$2du_0u_3 + 2du_1u_2 - 4du_1u_3 - 2du_2^2 - 2du_2u_3 + 8du_3^2 + 4 = 0$$
$$2du_0u_2 + du_1^2 - 10du_1u_3 - 5du_2^2 + 12du_2u_3 + 12du_3^2 + 5 = 0$$
$$2du_0u_1 - 8du_1u_3 - 4du_2^2 + 14du_2u_3 + 6du_3^2 + 16 = 0$$
$$du_0^2 - 2du_1u_3 - du_2^2 + 4du_2u_3 + du_3^2 + 4 = 0$$

Finding solutions to this system of equations is the same as finding rational points on the affine algebraic set defined by them. Recall that if I is a set of polynomials in $K[x_1, ..., x_n]$ then the subset V of K^n consisting of all $(a_1, ..., a_n) \in K^n$ such that $f(a_1, ..., a_n) = 0$ for all $f \in I$ is an algebraic set. If V is irreducible then it is called a variety.

We use MAGMA to find solutions to this system (using Gröbner basis reduction). At this point, we have four non-linear equations and five variables (u_0, u_1, u_2, u_3, d). We need first to fix one of the variables, so that we have a zero-dimensional algebraic set.[2] We take $u_0 = 1$ (which incidently prevents the trivial solution).

```
Magma V2.12-19    Thu Jan 25 2007 09:38:55    [Seed = 2105557683]
Type ? for help.  Type <Ctrl>-D to quit.
> R<d,u1,u2,u3> := PolynomialRing(RationalField(),4);
> u0:= 1;
> I:= ideal<R|-2*d*u0*u3 - 2*d*u1*u2 + 4*d*u1*u3 + 2*d*u2^2 +
>       2*d*u2*u3 - 8*d*u3^2 - 4, - 2*d*u0*u2 - d*u1^2 + 10*d*u1*u3 +
>       5*d*u2^2 - 12*d*u2*u3 - 12*d*u3^2 - 5, - 2*d*u0*u1 +
>       8*d*u1*u3 + 4*d*u2^2 - 14*d*u2*u3 - 6*d*u3^2 - 16, - d*u0^2 +
>       2*d*u1*u3 + d*u2^2 - 4*d*u2*u3 - d*u3^2 - 4>;
> print Variety(I);
[ <-196/3, 15/7, 11/14, 1/2>, <-64, 17/8, 3/4, 1/2> ]
>
```

---

[2] This does not lead to an exhaustive search, since our equations are of the form $df_i(u_0, \dots, u_n) = c_i$ where f_i is homogeneous of degree 2 and c_i is a constant. Hence, if $(d, u_0, \dots, u_n)$ is a solution with $u_0 \neq 0$ then so is $(du_0^2, 1, u_1/u_0, \dots, u_n/u_0)$.

To simplify the expressions we take $u_0 = -4$ (this reduces d to a fundamental discriminant for the second solution), and then
Variety $= [< -49/12, -60/7, -22/7, -2 >, < -4, -17/2, -3, -2 >]$.

Taking the point $(-4, -17/2, -3, -2)$ gives the polynomial $U(x) = -2x^3 - 3x^2 - \frac{17}{2}x - 4$ and $D = -4$. We then compute

$$H(x) = \frac{(T(x) - 2)^2 - DU(x)^2}{4R(x)} = \frac{17x^2 + 16x + 64}{4}$$

and

$$P(x) = H(x)R(x) + T(x) - 1 = \frac{17x^6 + 50x^5 + 181x^4 + 280x^3 + 405x^2 + 288x + 68}{4}$$

which is an irreducible polynomial over $\mathbb{Q}$.

If $x \simeq 2^{40}$ we get 242-bit values p and 160-bit values r, giving $\rho = \frac{\log(p)}{\log(r)} \simeq 1.5$. As mentioned above, taking $x_0 = 1099511631452$ gives both $P(x_0)$ and $R(x_0)$ prime as required.

Taking the other rational point leads to a family with $D = -3$ and $P(x) = (13x^6 + 40x^5 + 142x^4 + 223x^3 + 317x^2 + 222x + 52)/3$.

Efficient 15,360-bit RSA Using Woop-Optimised Montgomery Arithmetic*

Kamel Bentahar and Nigel P. Smart

Department of Computer Science, University of Bristol,
Merchant Venturers Building, Woodland Road,
Bristol, BS8 1UB, United Kingdom
{bentahar, nigel}@cs.bris.ac.uk

Abstract. The US government has published recommended RSA key sizes to go with AES-256 bit keys. These are huge and it is not clear what is the best strategy to implement modular arithmetic. This paper aims to investigate a set of possibilities from straight Montgomery and Barrett arithmetic through to combining them with Karatsuba and Toom-Cook style techniques.

1 Introduction

Modular Arithmetic is a fundamental component in many public-key cryptosystems such as RSA, ElGamal and ECC. A common feature of most of these schemes is that the *modulus is fixed* for most, if not all, calculations. This fact can be cleverly exploited to reduce the cost of the modular arithmetic operations and, in fact, there are two widely used methods which do so, namely the Montgomery and Barrett reduction algorithms.

Precomputed values related to the fixed modulus are used in the Montgomery and Barrett methods in order to reduce the cost of the modular reduction. In fact, the main effort in these methods is spent in evaluating two half-multiplications as opposed to an expensive long division. Given that the sizes of the moduli used in current systems are small or moderate, these multiplications are usually computed using the classical schoolbook multiplication and there has never been a need for using asymptotically faster multiplication methods.

However, NIST has recently recommended using moduli sizes as big as 15360 bits to match the security level of AES-256 [13, p. 63]. With this in mind, it now becomes worthwhile to explore the improvements that can be made by using asymptotically faster multiplication methods in combination with any "tricks" that may render them practical even for moderate sizes. We will, in fact, see that an error correction technique called *wooping* [3] allows us to overcome the

---

* The work described in this paper has been supported in part by the European Commission through the IST Programme under Contract IST-2002-507932 ECRYPT. The information in this document reflects only the author's views, is provided as is and no guarantee or warranty is given that the information is fit for any particular purpose. The user thereof uses the information at its sole risk and liability.

S.D. Galbraith (Eds.): Cryptography and Coding 2007, LNCS 4887, pp. 346–363, 2007.

difficulties that arise when trying to go beyond the obvious simple substitution of multiplication methods. These difficulties are due to carry-propagation when computing upper-half products with recursive methods, a problem that does not arise when using traditional combinations such as the Karatsuba-Comba-Montgomery (KCM) method [7,17].

Using a formal computational cost model, we estimate the exact cost of the Montgomery and Barrett modular reduction algorithms. We then introduce some variants using the Karatsuba and Toom-3 multiplication methods, and analyse the savings that can be theoretically achieved. These variants have been implemented in C using the GMP library (GNU Multiple Precision arithmetic library) [5], and the relevant results are reported here and compared with the theoretical estimates.

The authors would like to thank Pooya Farshim and Dan Page for their insightful discussions and suggestions, and also the anonymous reviewers whose comments have helped improve the quality of this paper greatly.

1.1   Notation and Assumptions

We assume that we have a machine that can do arithmetic operations on *word* sized operands, which we will refer to as *base operations*, and that it has access to an unlimited *random access memory*. The first assumption is true for most modern machines whereas, strictly speaking, the second is not true as memory is always limited in practice and there is some cost associated with fetching or moving data – a cost that depends on the size and location of the data and also on the speed and size of the RAM and Cache. If enough care is taken then a good implementation should be able to bring this extra cost to a minimum. Also, in order to simplify the task of analysing algorithms, we will limit ourselves to the study of *sequential* machines and do not consider any aspect of parallelism.

We represent large integers as arrays of machine words, with the basic arithmetic operations done with the usual classical schoolbook methods, unless otherwise mentioned. A cost expression of the form $x\mathcal{M} + y\mathcal{A}$ denotes the cost of performing x base multiplications and y base additions. In order to make comparison feasible, we introduce a parameter μ such that $1\mathcal{M} = \mu\mathcal{A}$. This parameter depends on the machine's architecture and implementation details. To keep our notation light, we will omit the unit $\mathcal{A}$ in formulae of the form $a\mathcal{M} + b\mathcal{A} = (a\mu + b)\mathcal{A}$ and would simply write $a\mu + b$.

Let us now estimate the cost of schoolbook addition and multiplication in our model. We have $\mathcal{A}(n) = n$ for the cost of adding two n-word integers, and $\mathcal{M}(n) = n^2\mathcal{M} + 2n(n-1)\mathcal{A}$ for the cost of multiplying two n-word integers.

$$\mathcal{A}(n) = n \quad \text{and} \quad \mathcal{M}(n) = (\mu + 2)n^2 - 2n. \tag{1}$$

We let $\mathcal{M}_u(n)$ and $\mathcal{M}_\ell(n)$ denote the cost of computing the *upper* and *lower* halves of the product of two n-word integers, respectively. The cost of computing the lower half product is $\mathcal{M}_\ell(n) = \frac{1}{2}n(n+1)\mathcal{M} + n(n-1)\mathcal{A}$, which leads to

$$\mathcal{M}_\ell(n) = (\frac{\mu}{2} + 1)n^2 + (\frac{\mu}{2} - 1)n. \tag{2}$$

In principal, we have $\mathcal{M}_u(n) = \mathcal{M}_\ell(n)$ but there is a small extra cost due to the fact that we need to keep track of carries from the lower half of the product, a fact which will be crucial in the sequel.

We also set R to be the least power of the basis that is greater than n-words i.e. if a word holds β bits then the basis is

$$b = 2^\beta \quad \text{and} \quad R = (2^\beta)^n = 2^{\beta n}.$$

Then, the subscripts ℓ and u respectively denote the lower and upper part of a number in the sense that

$$x_\ell = x \bmod R \quad \text{and} \quad x_u = \lfloor x/R \rfloor.$$

We will assume that the word size is $\beta = 32$ bits, which is the standard word-size in most desktop computers. So, for a 15360-bit integer we will need $n = 480$ words. If the word size is 64 bits then n drops to 240.

2 The Karatsuba and Toom-Cook Multiplication Algorithms

The next two subsections will review the Karatsuba and Toom-Cook fast integer multiplication algorithms and analyse their cost according to the cost model presented in the introduction. A more comprehensive treatment of these and other methods can be found in [9, p. 294–311]. We will also consider the computation of upper and lower halves of products as these will save us on the overall cost, [8].

Recall that, according to our computational cost model, we will not take the cost of memory operations into account and we will assume that they are for free.

2.1 Karatsuba Integer Multiplication

This is a popular divide-and-conquer algorithm for faster multiplication introduced by Karatsuba and published by Ofman [14]. It achieves an asymptotic complexity of $\mathcal{O}(n^{\lg 3}) = \mathcal{O}(n^{1.585})$, as opposed to $\mathcal{O}(n^2)$ for the schoolbook method.

Let $u, v \in \mathbb{N}$ be represented as n-word integers in base $b = 2^\beta$, where $n = 2t$. Write $u = u_1 b^t + u_0$ and $v = v_1 b^t + v_0$, where u_0, u_1, v_0, v_1 are t-word integers. Then

$$uv = w_2 b^{2t} + w_1 b^t + w_0,$$

where
$$w_2 = u_1 v_1$$
$$w_1 = (u_0 + u_1)(v_0 + v_1) - w_0 - w_2$$
$$w_0 = u_0 v_0$$

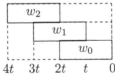

In practice, computing $u_0 + u_1$ and $v_0 + v_1$ may result in an overflow, so extra care has to be taken when computing these values. Alternatively, we can compute

$$w_1 = w_0 + w_2 - (u_0 - u_1)(v_0 - v_1),$$

which uses subtraction instead of addition.

If we use the Karatsuba method recursively to multiply operands greater than or equal to a fixed threshold value T and switch to schoolbook multiplication thereafter then the cost function is

$$\mathcal{K}(n) = \begin{cases} 3\mathcal{K}(n/2) + 4n & \text{for } n \geq T \\ \mathcal{M}(n) & \text{for } n < T \end{cases} \tag{3}$$

The method of solving such recurrence equations is outlined in Appendix A, and applying it to this equation we get (for $n \geq T$)

$$\mathcal{K}(n) = \underbrace{[(\mu + 2)\frac{T}{2^{\{\lg(T/n)\}}} + 6]\left(\frac{3}{4}\right)^{\{\lg(T/n)\}}}_{\text{Bounded by a constant } (\mu, T \text{ are fixed})} T \cdot \left(\frac{n}{T}\right)^{\lg 3} - 8n = \mathcal{O}(n^{\lg 3}).$$

The case where n is odd can be dealt with by letting $t = \lceil n/2 \rceil$, but it is more efficient to set $t = \lfloor n/2 \rfloor$ allowing u_1, v_1 to be $(t+1)$-word integers while keeping u_0, v_0 as t-word integers and treating the extra bits explicitly. In this case, we have

$$K(n) = 2\mathcal{K}((n+1)/2) + K((n-1)/2) + 4n, \quad \text{for odd } n.$$

With this optimisation, it becomes very difficult to write a closed form for the solution, if it is possible at all. So, we will be satisfied with a sample plot. The graph in Figure 1 shows the ratio $\mathcal{M}(n)/\mathcal{K}(n)$ and illustrates the savings that can be made by using the Karatsuba multiplication method instead of the schoolbook method.

2.2   Toom-Cook Multiplication

This method also uses a divide-and-conquer strategy and can be considered as a generalisation of the Karatsuba method. The general framework here is to exploit polynomial arithmetic. We first write the two integers u, v that we want to multiply as degree r polynomials $u(x), v(x)$ whose coefficients are the base b digits of u and v. We then evaluate the polynomials at as many points as needed to uniquely define their product $w(x) = u(x)v(x)$ through interpolation: $2r + 1$ points. Now, multiplying the values of the two polynomials $u(x), v(x)$ at the chosen points, we get the values of the product $w(x)$ at the same points. Given these $2r + 1$ values, we can now recover $w(x)$ by interpolation; and to get the product of the original integers we simply evaluate $w(x)$ at the base b (release the carries). This yields a multiplication method having complexity $\mathcal{O}(n^{\log(2r+1)/\log(r+1)})$. Note that the Karatsuba method can viewed as a special case of this framework when $r = 1$ (linear polynomials).

We will describe the popular instance known as Toom-3 multiplication. Toom-3 achieves a complexity $\mathcal{O}(n^{\log_3 5}) = \mathcal{O}(n^{1.465})$ by taking the polynomials $u(x)$ and $v(x)$ to be quadratic. Suppose we want to multiply two n-word integers u and v, where $n = 3t$. First, represent them as polynomials:

$$u = u(x)|_{x=b^t} = u_0 + u_1 b^t + u_2 b^{2t},$$
$$v = v(x)|_{x=b^t} = v_0 + v_1 b^t + v_2 b^{2t}.$$

Now, to evaluate $w = uv$, we first evaluate $w(x) = u(x)v(x)$ at $x = 0, 1, -1, 2, \infty$. Then, knowing the values of $w(x) = w_4 x^4 + w_3 x^3 + w_2 x^2 + w_1 x + w_0$ at five points,

we interpolate the coefficients of w. We have

$$w_4 = u_2 v_2$$
$$w_3 = u_2 v_1 + u_1 v_2$$
$$w_2 = u_2 v_0 + u_1 v_1 + u_0 v_2$$
$$w_1 = u_0 v_1 + u_1 v_0$$
$$w_0 = u_0 v_0$$

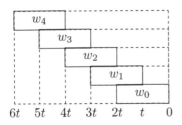

$$
\begin{aligned}
w(x)|_{x=0} &= u_0 v_0 & &= w_0 \\
w(x)|_{x=+1} &= (u_2 + u_1 + u_0)(v_2 + v_1 + v_0) & &=: \alpha \\
w(x)|_{x=-1} &= (u_2 - u_1 + u_0)(v_2 - v_1 + v_0) & &=: \beta \\
w(x)|_{x=2} &= (4u_2 + 2u_1 + u_0)(4v_2 + 2v_1 + v_0) &=: \gamma \\
w(x)|_{x=\infty} &\overset{\text{def}}{=} \lim_{x\to\infty} u(x)v(x)/x^4 = u_2 v_2 & &= w_4
\end{aligned}
$$

So we get w_0 and w_4 right away, and what remains is to find w_1, w_2, w_3. Solving the previous system of equations we get

$$w_2 = (\alpha + \beta)/2 - w_4 - w_0$$
$$w_3 = +w_0/2 - 2w_4 + (\gamma - \beta)/6 - \alpha/2$$
$$w_1 = -w_0/2 + 2w_4 - (\gamma + 2\beta)/6 + \alpha$$

Hence, the cost function for Toom-3 is $\mathcal{T}(n) = 5\mathcal{T}(n/3) + [3\mathcal{A}(t) + 4\mathcal{A}(t) + 3\mathcal{A}(t)] + 4\mathcal{A}(t)$, i.e.

$$\mathcal{T}(n) = 5\mathcal{T}(n/3) + 14n/3. \tag{4}$$

When n is not a multiple of 3, we set $t = \lceil n/3 \rceil$ and allow u_2 and v_2 to be shorter than t words. This makes implementation easier, as was done in the GMP library.

We introduce a second threshold value $T' > T$ such that if $n < T$ then we use schoolbook multiplication, if $T \leq n < T'$ then we use Karatsuba multiplication, and if $n > T'$ then we use Toom-3 multiplication.

Figure 1 shows the plots of the ratios $\mathcal{M}(n)/\mathcal{K}(n)$ and $\mathcal{M}(n)/\mathcal{T}(n)$ for $\mu = 1.2$, $T = 23$ and $T' = 133$, hence showing the expected speedup that can be made over the schoolbook multiplication method for these particular parameter values. These specific values seem to be generic and will be used throughout this paper for the purpose of illustration.

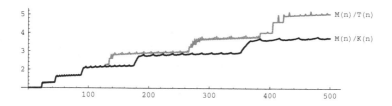

Fig. 1. Plots for $\mathcal{M}(n)/\mathcal{K}(n)$ and $\mathcal{M}(n)/\mathcal{T}(n)$

The threshold values $T = 23$ and $T' = 133$ are those of the Pentium-4 machines that were used for testing (2.80GHz, 512KB cache, model 2), as estimated by GMP's tuning program, and can easily be estimated for other architectures.

Note, however, that the exact value of μ is hard to pin down because execution times depend on the ordering of instructions and data, which may lead to significant savings through pipelining. Luckily, it turns out that small variations in μ have little theoretical impact on the cost ratios considered here, as μ essentially only affects the leading coefficient which varies slowly as a function of μ. The value 1.2 for μ was experimentally chosen from a set of possible values in the range $(1, 1.5)$, which were obtained using loops to measure the average times for word operations on a few different computers and then fitting the collected data to estimate the value of μ for each architecture. Values for μ can also be estimated theoretically through the tables presented in [6].

2.3   Short Products

We will make use of methods for computing the lower and upper half products (short products), so we will study their costs next. We start with a general method that applies to all multiplication algorithms [12,8] then present some specific solutions suited to the Karatsuba method.

A General Method. First, we will introduce a visual aid that will make explaining this method easier and more intuitive. When multiplying two numbers using schoolbook multiplication we stack the partial products in a shape similar to the one on the left in Figure 2 prior to adding them up; and to find the lower half product, for example, we only need to compute the results in the shaded triangle.

Fig. 2. Calculation of short products

Let $S(n)$ be the cost of computing a short product of two n-word integers. If we take a portion ρn, where $0.5 \leq \rho < 1$, of both operands and compute their full product, corresponding to the darker shade on the right in Figure 2, and then compute the remaining terms using short products again, corresponding to the two lighter shaded triangles, then we find that this method would cost

$$S(n) = \mathcal{M}(\rho n) + 2S((1 - \rho)n).$$

Since the multiplication methods we are considering all cost $\mathcal{M}(n) = n^\alpha$, for some $\alpha \in (1, 2]$, we find that

$$S(n) \leq \frac{\rho^\alpha}{1 - 2(1 - \rho)^\alpha} \mathcal{M}(n) \quad =: C_\rho \mathcal{M}(n).$$

The factor C_ρ is minimal at $\hat{\rho} = 1 - 2^{-1/(\alpha-1)}$, and the following table summarises the results for the methods that we are interested in. It should be noted that these are the best asymptotically and thus there may be better choices for ρ when n is small or moderate.

Method	α	$\hat{\rho}$	$C_{\hat{\rho}}$
Schoolbook	2	0.5	0.5
Karatsuba	$\lg 3$	0.694	0.808
Toom-3	$\log_3 5$	0.775	0.888

Note that if we fix n and look for the best value of $\hat{\rho}$ we may get a slightly different value. For the case where $n = 480$, this value turns out to be about 0.80 for Karatsuba and 0.88 for Toom-3.

The next Karatsuba-specific methods are actually special cases of this general setup with $\rho = 0.5$. They are easier to implement and may be faster in practice. Note, however, that doing the same for Toom-3 produces a slower method and hence it has not been considered.

Lower Half Products Using the Karatsuba Method. For this we need to compute

$$(w_0 + w_1 b^t + w_2 b^{2t}) \bmod b^n = (u_0 v_0 + [(u_0 v_1 + u_1 v_0) \bmod b^t]b^t) \bmod b^n,$$

which costs $\mathcal{K}_\ell(n) = \mathcal{K}(t) + 2\mathcal{K}_\ell(t) + 2\mathcal{A}(t)$, i.e.

$$\mathcal{K}_\ell(n) = \mathcal{K}(n/2) + 2\mathcal{K}_\ell(n/2) + n. \tag{5}$$

Upper Half Products Using the Karatsuba Method. This time, we have to compute

$$\left\lfloor \frac{w_2 b^{2t} + w_1 b^t + w_0}{b^{2t}} \right\rfloor = carry + u_1 v_1 + \left\lfloor \frac{u_0 v_1 + u_1 v_0}{b^t} \right\rfloor .$$

The carry results from adding $u_0 v_0$ to $(u_0 v_1 + u_1 v_0) b^t$, in the full multiplication, and hence is either 0 or 1.

If we ignore the carry and use the "faulty" recursive method suggested by this formula then the maximum error $\epsilon(n)$ will satisfy the recurrence equation

$$\epsilon(n) = 2\epsilon(n/2) + 1 \quad \text{and} \quad \epsilon(n) = 0 \text{ for } n < T.$$

By the result in Appendix A, we deduce that $\epsilon(n) = 2^{\lceil \lg(n/T) \rceil} - 1 \leq 2n/T - 1$. So, computing upper-half products, up-to an error of order $\mathcal{O}(n)$, can be done at the cost of $\mathcal{K}_u(n) = \mathcal{K}(t) + 2\mathcal{K}_u(t) + 2\mathcal{A}(t)$, i.e.

$$\mathcal{K}_u(n) = \mathcal{K}(n/2) + 2\mathcal{K}_u(n/2) + n \quad = \mathcal{K}_\ell(n). \tag{6}$$

It turns out that, when the faulty result of this method is used in the reduction algorithms, we can correct the computation by using a nice technique, known as *wooping*, which is due to Bos [3, p. 281–284]. The next subsection will introduce this technique while the correction steps are detailed in section 3.1.

To see how much faster these methods are, we plot $\mathcal{M}_\ell(n)/\mathcal{K}_\ell(n)$ (using both the general and the specific method) and $\mathcal{M}_\ell(n)/\mathcal{T}_\ell(n)$ – see Figure 3. The same speed-ups apply to the upper-half product methods too as they essentially have the same cost.

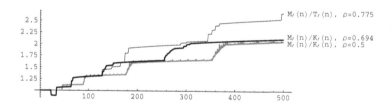

Fig. 3. Plots for $\mathcal{K}(n)/\mathcal{K}_\ell(n)$ and $\mathcal{T}(n)/\mathcal{T}_\ell(n)$

2.4 Wooping

The wooping technique allows us to verify the outcome of a set of integer operations. The idea is to perform the same operations modulo a *small* prime number and then compare the results. More specifically, we reduce the operands modulo the prime number first then operate on them with the corresponding "small" modular operations. For example, if the operation is $z \leftarrow x \cdot y$ then we randomly choose a small prime p and compute $\tilde{x} \leftarrow x \bmod p$ and $\tilde{y} \leftarrow y \bmod p$ first, then we compute $\tilde{z} \leftarrow \tilde{x} \cdot \tilde{y} \bmod p$; and for the comparison we reduce z modulo p and compare the result with $\tilde{z}$. If the compared small results do not agree then there certainly is an error in the full integer computation, but if they agree then there is a low chance that an error has occurred.

If p is a *randomly chosen prime number* then the probability that it will fail to reveal the error is $1/p$, so one can choose other prime numbers for the wooping test to increase confidence.

Note, however, that the manner in which we use this technique is slightly different because we already know that there is an error in our computation and we just want to correct it. Furthermore, since we are not in an adversarial setup, this correction scheme is *deterministic* and *always successful*. What we do is choose a woop modulus that is bigger than the largest possible error, and then correct the integer computation by adding the difference between the two reduced values.

As a toy example of how to use the wooping technique for correction, let us consider a device that can multiply integers but sometimes introduces an error of $+1$ in the result. Suppose that we wanted to compute 4×5 but we got 21 as the answer. First, note that we can choose the woop modulus to be 2 as that is enough to reveal the error. Now, now we check that $(4 \bmod 2) \times (5 \bmod 2) = 0 \times 1 = 0$ whereas $21 \bmod 2 = 1$, so we correct the computation by subtracting 1 from 21 to get the correct answer of 20. For the exact details of how to use this technique in our work, see section 3.1.

On a side note, as an alternative to wooping one may consider computing enough extra words to the right of the truncated upper-product in order to ensure a small probability of a carry being missed. This is in fact suggested in [8] and the extra words are referred to as "guard digits." This alternative is more complicated to implement because of the extra storage and will most likely be more expensive, especially if two or more guard digits are needed. Wooping on the other hand requires negligible storage and computational overhead.

3   The Montgomery and Barrett Reductions

Given a fixed n-word modulus m, we want to reduce $2n$-word integers modulo m as fast as possible. We will now describe two practical algorithms used for this purpose, namely the Montgomery and Barrett reduction methods.

3.1   Montgomery Reduction

The Montgomery reduction algorithm is described in Algorithm 1. We note that the practical version used with schoolbook multiplication does not require direct calculation of lower or upper half products, but the quoted cost remains the same.

We can see that the cost of Algorithm 1 is $\mathcal{M}_\ell(n) + \mathcal{M}_u(n) + 2\mathcal{A}(n)$. So, its cost using schoolbook multiplication is (using $\mathcal{M}_\ell = \mathcal{M}_u$)

$$\mathcal{C}_{mr,cl}(n) = (\mu + 2)n^2 + \mu n. \tag{7}$$

Algorithm 1. Montgomery reduction
Input: n-word integer m, $-m^{-1} \bmod R$ where $R = b^n$, and $z < mR$.
Output: $zR^{-1} \bmod m$.

1: $u \leftarrow (-m^{-1})z \bmod R$	$\langle \mathcal{M}_\ell(n) \rangle$
2: $x \leftarrow (z + um)/R$	$\langle \mathcal{M}_u(n) + \mathcal{A}(n) \rangle$
3: **if** $x \geq m$ **then**	
4: $x \leftarrow x - m.$	$\langle \mathcal{A}(n) \rangle$
5: **end if**	
6: Return x	

The Karatsuba Variant with Wooping. Recall, from section 2.1, that we can compute upper-half products using Karatsuba multiplication with an error of $\mathcal{O}(n)$. We will now explain how to use the wooping correction idea in our case. Let $\lambda \in \mathbb{N}$ be a modulus greater than the magnitude of the maximum possible error resulting from ignoring the carry in the "faulty" upper-half Karatsuba method.

We first compute the product $u \leftarrow (-m^{-1})z \bmod R$ with a low-half Karatsuba multiplication. Now, for $x \leftarrow (z + um)/R$, note that a good approximation to this value is given by $x_u + (um)_u$, which will be off by at most 1 (carry). An extra error will come from the fact that we are using a faulty Karatsuba multiplication for the upper-half product. To correct the approximate answer, we now compute $(z + um)/R$ modulo λ and compare it with the reduction of the approximate value: Given that the error magnitude is less than λ then we will be able to deduce the offset from the correct answer by comparing these reduced values, and therefore correct our answer. This is the "trick" that allows us to be satisfied with an approximation to $(um)_u$ and save on its computation.

If we further choose $\lambda = b^l - 1$, for some $l \in \mathbb{N}$, then reduction modulo λ becomes rather efficient. In fact, to reduce an n-word number modulo $b^l - 1$, we only need $\lceil n/l \rceil$ additions on numbers of size l words, costing a total of $n\mathcal{A}$. In practice, for $b = 2^{32}$, we take $l = 1$ as this is enough to correct errors for $n < b$. Also, note that with this choice of λ and $R = b^n$ we have $R \equiv 1 \pmod{\lambda}$, so the computation of $(z + um)/R \bmod \lambda$ requires no inversion.

With this choice of $\lambda = b - 1$ and $R = b^n$, the correction steps involve computing $z + um \bmod \lambda$, costing about $(2n + n + n)\mathcal{A} + 1\mathcal{M} + 2\mathcal{A}$, and $x \bmod \lambda$, costing about $n\mathcal{A}$, where x is the result of step 2 of the algorithm. Then, computing the offset and correction will cost $2\mathcal{A}$. So the cost of the Karatsuba variant of Montgomery reduction is about

$$\mathcal{C}_{mr,2}(n) = \mathcal{K}_\ell(n) + \mathcal{K}_u(n) + 7n + \mu + 4. \tag{8}$$

The Toom-3 Variant (with Wooping). We proceed exactly the same as in the Karatsuba variant, and the cost is then found to be

$$\mathcal{C}_{mr,3}(n) = \mathcal{T}_\ell(n) + \mathcal{T}_u(n) + 7n + \mu + 4. \tag{9}$$

Comparison. Figure 4 shows the graphs of $\mathcal{C}_{mr,cl}/\mathcal{C}_{mr,2}$ and $\mathcal{C}_{mr,cl}/\mathcal{C}_{mr,3}$, and serves to illustrate the improvements that can be made with these two variants of Montgomery reduction.

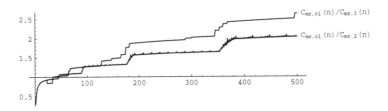

Fig. 4. Plots for $\mathcal{C}_{mr,cl}/\mathcal{C}_{mr,2}$ and $\mathcal{C}_{mr,cl}/\mathcal{C}_{mr,3}$

Crossing Point. From the graph we see that the crossing point is at about 40 words, so we should expect that the Karatsuba variant of Montgomery reduction will start to be effective from moduli sizes of about 1280 bits.

3.2   Montgomery Multiplication

Montgomery multiplication aims to achieve fast multiplication and reduction in one go. There exists an efficient *interleaved* version where multiplications and division by R are interleaved and performed word-by-word. This approach keeps the memory costs minimal and makes implementation easier. There does not seem to be an easy way in which this can be done with the faster multiplication methods. We do however consider using an iterative version of Karatsuba in section 4.1. The interleaved version is described in Algorithm 2.

Let $\mathcal{M}_1(n)$ denote the cost multiplying an n-word integer by a single word integer. Then
$$\mathcal{M}_1(n) = n\mathcal{M} + (n-1)\mathcal{A} = (\mu+1)n - 1.$$

So, the cost of the interleaved Montgomery multiplication is $n[2\mathcal{M}+1\mathcal{A}+2(n\mathcal{M}+(n-1)\mathcal{A})] + \mathcal{A}(n)$.

$$\mathcal{C}_{mm,cl}(n) = 2(\mu+1)n^2 + 2\mu n. \tag{10}$$

The Karatsuba Variant (with Wooping). To compute the Montgomery multiplication of X and Y: $XYR^{-1} \bmod m$, we first multiply X by Y using the Karatsuba method then we Montgomery-reduce the result as described in subsection 3.1. This will then cost

$$\mathcal{C}_{mm,2}(n) = \mathcal{K}(n) + \mathcal{C}_{mr,2}(n). \tag{11}$$

Algorithm 2. Montgomery multiplication

Input: $X = xR \bmod m$ and $Y = ym \bmod R$ as n-word integers, $R = b^n$, $m' = -m^{-1} \bmod b$.

Output: $XYR^{-1} \bmod m$.

1: $z \leftarrow 0$
2: **for** $i = 0, \ldots, n-1$ **do**
3:     $u \leftarrow (z_0 + x_i y_0)m' \bmod b$ $\qquad\qquad\qquad\qquad\qquad\qquad\qquad$ $\langle 2\mathcal{M} + 1\mathcal{A} \rangle$
4:     $z \leftarrow (z + x_i y + um)/b$ $\qquad\qquad\qquad\qquad\qquad\qquad$ $\langle 2\mathcal{M}_1(n) + 2\mathcal{A}(n) \rangle$
5: **end for**
6: **if** $z \geq m$ **then**
7:     $z \leftarrow z - m$ $\qquad\qquad\qquad\qquad\qquad\qquad\qquad\qquad\qquad\qquad$ $\langle \mathcal{A}(n) \rangle$
8: **end if**
9: Return z

The Toom-3 Variant. Here we also proceed exactly the same as in the Karatsuba variant. The cost this time is found to be

$$\mathcal{C}_{mm,3}(n) = \mathcal{T}(n) + \mathcal{C}_{mr,3}(n). \qquad (12)$$

Comparison. Figure 5 shows the plots of $\mathcal{C}_{mm,cl}/\mathcal{C}_{mm,2}$ and $\mathcal{C}_{mm,cl}/\mathcal{C}_{mm,3}$, which illustrate the gain that is theoretically achievable with these variant of Montgomery multiplication.

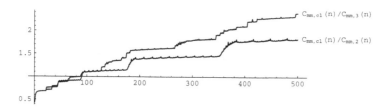

Fig. 5. Plots for $\mathcal{C}_{mm,cl}/\mathcal{C}_{mm,2}$ and $\mathcal{C}_{mm,cl}/\mathcal{C}_{mm,3}$

Crossing Point. From the previous graph we find that the crossing point is at about 90 words, implying that 2880 bits is the point from which our variant starts to be advantageous.

3.3 Barrett Reduction

Let z be a $2n$-word integer and m be a fixed n-word modulus. Recall that we can reduce z modulo m using Euclidean division of z by m: $z = qm + (z \bmod m)$.

Barrett's idea is to avoid division by computing a good estimate for the quotient $q = \lfloor z/m \rfloor$ as follows

$$q = \left\lfloor \frac{z_u b^n + z_\ell}{m} \right\rfloor \approx \frac{b^n z_u}{m} = \frac{b^{2n}}{m} \cdot \frac{z_u}{b^n} \approx \left\lfloor \frac{\mu z_u}{b^n} \right\rfloor =: \tilde{q} \quad \text{where } \mu = \left\lfloor \frac{b^{2n}}{m} \right\rfloor.$$

Note that $\tilde{q} \leftarrow \lfloor (\mu z_u)/b^n \rfloor = (\mu z_u)_u$, and so it can be computed as an upper-half product (using wooping to correct the small error that results from ignoring carry-propagation from the lower half).

It can be shown that if $z < m^2$ then $q - 2 \leq \tilde{q} \leq q$. So a good estimate for the remainder is $z - \tilde{q}m$ which we can correct by subtracting m from it at most twice. Algorithm 3 describes this method in detail, [11, p. 604].

Algorithm 3. Barrett reduction
Input: n-word modulus m, $\mu = \lfloor b^{2n}/m \rfloor$ and $z < m^2$.
Output: $z \bmod m$.

1: $z' \leftarrow \lfloor z/b^{n-1} \rfloor$, $\tilde{q} \leftarrow \lfloor z'\mu/b^{n+1} \rfloor$	$\langle \approx \mathcal{M}_u(n) \rangle$
2: $r \leftarrow (z \bmod b^{n+1}) - (\tilde{q}m \bmod b^{n+1})$	$\langle \approx \mathcal{M}_\ell(n) + \mathcal{A}(n) \rangle$
3: **if** $r < 0$ **then**	
4: $r \leftarrow r + b^{n+1}$	$\langle \mathcal{A}(n) \rangle$
5: **end if**	
6: **while** $r \geq m$ **do**	
7: $r \leftarrow r - m$	$\langle \leq 2\mathcal{A}(n) \rangle$
8: **end while**	
9: Return r	

From this description, we see that the approximate cost of the Barrett reduction method is $\mathcal{M}_u(n) + \mathcal{M}_\ell(n) + 4\mathcal{A}(n)$. So if schoolbook multiplication is used then this reduction method will cost

$$\mathcal{C}_{br,cl}(n) = (\mu + 2)n^2 + (\mu + 2)n = (\mu + 2)n(n + 1). \tag{13}$$

If Karatsuba multiplication is used then the reduction will cost

$$\mathcal{C}_{br,2}(n) = \mathcal{K}_\ell(n) + \mathcal{K}_u(n) + 4n \tag{14}$$

and similarly for Toom-3 we get

$$\mathcal{C}_{br,3}(n) = \mathcal{T}_\ell(n) + \mathcal{T}_u(n) + 4n. \tag{15}$$

Comparison. Figure 6 represents $\mathcal{C}_{br,cl}/\mathcal{C}_{br,2}$ and $\mathcal{C}_{br,cl}/\mathcal{C}_{br,3}$, and we can see from it that the cutoff point is at around 40-50 words.

By comparing (14) with (8) we come to the interesting conclusion that $\mathcal{C}_{br,2}(n) < \mathcal{C}_{mr,2}(n)$ which means that the theory expects the Barrett reduction with wooping to be slightly faster than Montgomery reduction with wooping – as opposed to the schoolbook versions where the opposite is the case. The same remark holds for the Toom-3 variants too.

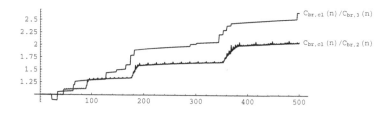

Fig. 6. Plots for $\mathcal{C}_{br,cl}/\mathcal{C}_{br,2}$ and $\mathcal{C}_{br,cl}/\mathcal{C}_{br,3}$

4   Further Potential Improvements

4.1   Cache Oblivious Version

If the cache size is too small then the recursive nature of the used multiplication methods and the large sizes of the operands may cause cache misses and hence slow the computation considerably.

One possible way of circumventing this problem is to use an iterative version of the multiplication algorithms. For a description of an iterative version of the Karatsuba method see [10].

4.2   FFT-Transform Based Multiplication

FFT multiplication methods are asymptotically faster than the other methods described in this paper but the cutoff point after which it becomes faster in practice is very high. It is argued in [2] that the Schönhage method becomes as efficient as the Karatsuba and Toom-3 methods at about $2^{17} = 131,072$ bits, which is close to the value of the generic FFT multiplication threshold used in the GMP library namely 30 times Toom-3's threshold $30 \times 128 = 3840$ words (122,880 bits). These sizes are too high for our purpose. The reader may be interested in having a look at [18] to a see a report on concrete implementation of a wide range of multiplication methods (but run on an old machine).

The authors of [15] suggest using *cyclic convolutions* instead of half products and achieve, in [16], a complexity of $\Theta(2.5n \log n)$ for a reduction algorithm with the use of *negacyclic convolutions*.

5   Experimental Results

We implemented Montgomery Multiplication in three flavours: The classical interleaved version, the new Karatsuba and Toom-3 with wooping variants and, finally, a naive version where we first multiply using the fastest available multiplication method then Montgomery-reduce the resulting product using the efficient word-level version of Algorithm 1 (GMP's `redc` function). These were

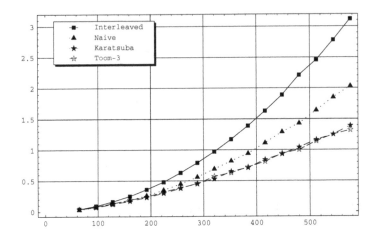

Fig. 7. Montgomery Multiplication times in milliseconds

implemented in C using the GMP library [5] with the low-level mpn set of functions for speed (as they are SSE2-optimized). We also implemented the RSA exponentiation by adapting GMP's mpz_powm function which uses the efficient sliding window method for exponentiation [4,1].[1]

The times needed to perform each of these two computations were averaged for operand sizes from 64 words (2,048 bits) up to 576 words (18,432 bits), with a step size of 32 words, and then plotted to ease comparison of the different methods. Figure 7 shows the times for Montgomery Multiplication and Figure 8 summarises the times obtained for RSA exponentiation (Times are given in milliseconds).

The experiments were done on Intel Pentium 4 machines (2.80GHz, 512KB cache, model 2). The threshold values that were used are $(T, T') = (23, 133)$ as estimated by GMP's tuning program tuneup. We bring the reader's attention to the fact that GMP uses slightly different threshold values for squaring, for which a more optimised code is used. (For our machines, they are 57 and 131 respectively but there is a large margin of error in them). Note also that, in our implementation of the short products algorithms, we used halves of T and T' for the thresholds.

---

[1] A careful analysis of the sliding window method for bit-size η done by H. Cohen in [1] shows that there exists $\rho > 1$ such that this method requires $\eta - \frac{1}{2}(k^2 + k + 2)/(k + 1) + \mathcal{O}(\rho^{-\eta})$ squarings and $\eta/(k + 1) - \frac{1}{2}k(k + 3)/(k + 1)^2 + \mathcal{O}(\rho^{-\eta})$ multiplications. We note that GMP optimises the window size k depending on the exponent's bit-size η by finding the least k such that $2\eta > 2^k(k^2 + 3k + 2) = 2^k(k + 1)(k + 2)$. The following table shows when a window of size k is first used for $n < 1000$ ($\eta = 32n$).

k	3	4	5	6	7	8	9	10
n	1	3	8	22	57	145	361	881

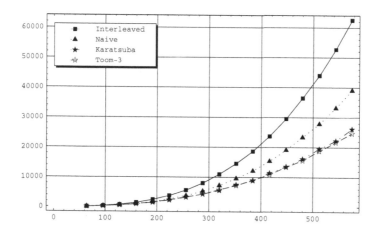

Fig. 8. RSA exponentiation times in milliseconds

We get, in particular, that an execution of a 15,360-bit RSA exponentiation on these Pentium 4 machines takes 16.1 seconds with the Karatsuba variant and 15.6s with the Toom-3 variant on average, compared to about 23.3s with the naive version and 36.45s for the traditional interleaved Montgomery multiplication.

Finally, we note that although the experimental cutoff points do not fit very accurately with the theory, because of the parallelism present in modern processors (pipelining), these are not far from the expected theoretical values, and the general trends are indeed as expected.

References

1. Cohen, H.: Analysis of the Sliding Window Powering Algorithm. Journal of Cryptology 18(1), 63–76 (2005)
2. Coronado, L.C.: García. Can Schönhage multiplication speed up the RSA decryption or encryption? MoraviaCrypt (2007),
   http://www.cdc.informatik.tu-darmstadt.de/mitarbeiter/coronado.html
   (preprint available)
3. Ferguson, N., Schneier, B.: Practical Cryptography, New York, NY, USA. John Wiley & Sons, Chichester (2003)
4. Gordon, D.M.: A Survey of Fast Exponentiation Methods. Journal of Algorithms 27(1), 129–146 (1998)
5. Granlund, T.: GNU multiple precision arithmetic library 4.1.2.,
   http://swox.com/gmp
6. Granlund, T.: Instruction latencies and through put for AMD and Intel x86 processors. (September 2, 2007), http://swox.com/doc/x86-timing.pdf
7. Großschädl, J., Avanzi, R.M., Savaş, E., Tillich, S.: Energy-efficient software implementation of long integer modular arithmetic. In: Rao, J.R., Sunar, B. (eds.) CHES 2005. LNCS, vol. 3659, pp. 75–90. Springer, Heidelberg (2005)

8. Hars, L.: Fast truncated multiplication for cryptographic applications. In: Rao, J.R., Sunar, B. (eds.) CHES 2005. LNCS, vol. 3659, pp. 211–225. Springer, Heidelberg (2005)
9. Knuth, D.E.: Seminumerical Algorithms. In: The Art of Computer Programming, 3rd edn., Addison-Wesley Longman, Reading (1998)
10. Lei, C.-L., Liu, C.-B., Huang, C.-H.: Design and implementation of long-digit karatsuba's multiplication algorithm using tensor product formulation. In: The Ninth Workshop on Compiler Techniques for High-Performance Computing, pp. 23–30 (2003)
11. Menezes, A.J., van Oorschot, P.C., Vanstone, S.A.: Handbook Of Applied Cryptography. CRC Press, Boca Raton (1997)
12. Mulders, T.: On computing short products. Technical Report 276, Dept of CS, ETH Zurich (November 1997),
    `ftp://ftp.inf.ethz.ch/pub/publications/tech-reports/2xx/276.pdf`
13. National Institute of Standards and Technology (NIST). Recommendation for key management - part 1: General. Technical Report NIST Special Publication 800-57, National Institute of Standards and Technology (2006),
    `http://csrc.nist.gov/publications/nistpubs/800-57/SP800-57-Part1.pdf`
14. Ofman, Y., Karatsuba, A.: Multiplication of multidigit numbers on automata. Soviet Physics - Doklady 7, 595–596 (1963)
15. Phatak, D.S., Goff, T.: Fast modular reduction for large wordlengths via one linear and one cyclic convolution. In: Computer Arithmetic, 2005. ARITH-17 2005. 17th IEEE Symposium, pp. 179–186 (2005)
16. Phatak, D.S., Goff, T.: Low complexity algorithms for fast modular reduction: New results and a unified framework. Technical report, Computer Science and Electrical Engineering Department. University of Maryland, Baltimore County, Baltimore, MD 21250 (2006)
17. Scott, M.P.: Comparison of methods for modular exponentiation on 32-bit intel 80x86 processors, `ftp://ftp.computing.dcu.ie/pub/crypto/timings.ps`
18. Zuras, D.: More on squaring and multiplying large integers. IEEE Transactions on Computers 43(8), 899–908 (1994)

A   Recurrence Equations of the Form $\mathcal{R}(n) = a\mathcal{R}(n/b) + cn + d$

We are interested in solving the recurrence equation $\mathcal{R}(n) = a\mathcal{R}(n/b) + cn + d$ subject to the *threshold condition*

$$\mathcal{R}(n) = f(n) \quad \text{for } n < T = b^\tau,$$

where T is a fixed threshold value and f is a given function. We distinguish two case according to whether a and b are equal or not.

– Let us examine the case where $a \neq b$ first. Set $k = \log_b n$, then by induction we get

$$\mathcal{R}(b^k) = a^\ell \mathcal{R}(b^{k-\ell}) + \frac{(a/b)^\ell - 1}{a/b - 1} \cdot cb^k + \frac{a^\ell - 1}{a - 1}d \quad \text{for any } \ell \in \mathbb{N}.$$

We want ℓ to be the least number such that $b^{k-\ell}$ is just below the threshold T i.e. $b^{k-\ell} < b^\tau \le b^{k-(\ell-1)}$, so we set

$$\ell = \lceil k - \tau \rceil = \lceil \log_b(n/T) \rceil \quad =: \ell_b(n/T). \tag{16}$$

We then get that (Using $\lceil x \rceil = x + \{-x\}$)

$$k - \ell = k - \lceil k - \tau \rceil = \tau - \{\tau - k\},$$

where $\{x\}$ denotes the fractional part of x.

So, for $n \ge T$ and $a \ne b$, we have the following solution

$$\mathcal{R}(n) = a^{\ell_b(n/T)} f\left(\frac{T}{b^{\{\log_b(T/n)\}}}\right) + \frac{(a/b)^{\ell_b(n/T)} - 1}{a/b - 1} \cdot cn + \frac{b^{\ell_b(n/T)} - 1}{b - 1} d. \tag{17}$$

— If $a = b$ then induction yields

$$\mathcal{R}(b^k) = b^\ell \mathcal{R}(b^{k-\ell}) + \ell c b^k + \frac{b^\ell - 1}{b - 1} d \quad \text{for any } \ell \in \mathbb{N}.$$

With the same choice of ℓ as before, we get for $n \ge T$ and $a = b$

$$\mathcal{R}(n) = b^{\ell_b(n/T)} f\left(\frac{T}{b^{\{\log_b(T/n)\}}}\right) + cn\ell_b(n/T) + \frac{b^{\ell_b(n/T)} - 1}{b - 1} d. \tag{18}$$

Toward Acceleration of RSA Using 3D Graphics Hardware[*,**]

Andrew Moss, Daniel Page, and Nigel P. Smart

Department of Computer Science,
Merchant Venturers Building,
Woodland Road,
Bristol, BS8 1UB,
United Kingdom
{moss,page,nigel}@cs.bris.ac.uk

Abstract. Demand in the consumer market for graphics hardware that accelerates rendering of 3D images has resulted in commodity devices capable of astonishing levels of performance. These results were achieved by specifically tailoring the hardware for the target domain. As graphics accelerators become increasingly programmable however, this performance has made them an attractive target for other domains. Specifically, they have motivated the transformation of costly algorithms from a general purpose computational model into a form that executes on said graphics hardware. We investigate the implementation and performance of modular exponentiation using a graphics accelerator, with the view of using it to execute operations required in the RSA public key cryptosystem.

1 Introduction

Efficient arithmetic operations modulo a large prime (or composite) number are core to the performance of public key cryptosystems. RSA [22] is based on arithmetic in the ring $\mathbb{Z}_N$, where $N = pq$ for large prime p and q, while Elliptic Curve Cryptography (ECC) [11] can be parameterised over the finite field $\mathbb{F}_p$ for large prime p. With a general modulus m taking the value N or p respectively, on processors with a w-bit word size, one commonly represents $0 \leq x < m$ using a vector of $n = \lceil m/2^w \rceil$ radix-2^w digits. Unless specialist co-processor hardware is used, modular operations on such numbers are performed in software using well known techniques [15,2] that operate using native integer machine operations. Given the significant computational load, it is desirable to accelerate said operations using instruction sets that harness Single Instruction

* The work described in this paper has been supported in part by the European Commission through the IST Programme under Contract IST-2002-507932 ECRYPT. The information in this document reflects only the author's views, is provided as is and no guarantee or warranty is given that the information is fit for any particular purpose. The user thereof uses the information at its sole risk and liability.

** The work described in this paper has been supported in part by EPSRC grant EP/C522869/1.

S.D. Galbraith (Eds.): Cryptography and Coding 2007, LNCS 4887, pp. 364–383, 2007.

Multiple Data (SIMD) parallelism; in the context of ECC, a good overview is given by Hankerson et al. [11, Chapter 5]. Although dedicated vector processors have been proposed for cryptography [9] these are not commodity items.

In an alternative approach, researchers have investigated cryptosystems based on arithmetic in fields modulo a small prime m or extension thereof. Since ideally we have $m < 2^w$, the representation of $0 \leq x < m$ is simply one word; low-weight primes [7] offer an efficient method for modular reduction. Examples that use such arithmetic include Optimal Extension Fields (OEF) [1] which can provide an efficient underpinning for ECC; torus based constructions such as T_{30} [8]; and the use of Residue Number Systems (RNS) [16, Chapter 3] to implement RSA. Issues of security aside, the use of such systems is attractive as operations modulo m may be more efficiently realised by integer based machine operations. This fact is reinforced by the aforementioned potential for parallelism; for example, addition operations in an OEF can be computed in a component-wise manner which directly maps onto SIMD instruction sets [11, Chapter 5].

However, the focus on use of integer operations in implementation of operations modulo large and small numbers ignores the capability for efficient floating point computation within commodity desktop class processors. This feature is often ignored and the related resources are left idle: from the perspective of efficiency we would like to utilise the potential for floating point arithmetic to accelerate our implementations. Examples of this approach are provided in work by Bernstein which outline high-performance floating point based implementations of primitives such as Poly1305 [3] and Curve25519 [4]. Beyond algorithmic optimisation, use of floating point hardware in general purpose processors such as the Intel Pentium 4 offered Bernstein some significant advantages. Specifically, floating point operations can often be executed in parallel with integer operations; there is often a larger and more orthogonally accessible floating point register file available; good scheduling of floating point operations can often yield a throughput close to one operation per-cycle.

Further motivation for use of this type of approach is provided by the recent availability of programmable, highly SIMD-parallel floating point co-processors in the form of Graphics Processing Units (GPU). Driven by market forces these devices have developed at a rate that has outpaced Moore's Law: for example, the Nvidia 7800-GTX uses 300 million transistors to deliver roughly 185 Gflop/s in contrast with the 55 million transistor Intel Pentium 4 which delivers roughly 7 Gflop/s. Although general purpose use of the GPU is an emerging research area [10], until recently the only published prior usage for cryptography was by Cook et al. [5] who implemented block and stream ciphers using the OpenGL command-set; we are aware of no previous work accelerating computationally expensive public key primitives. Further, quoted performance results in previous work are somewhat underwhelming, with the GPU executing AES at only 75% the speed of a general purpose processor. This was recently improved, using modern GPU hardware, by Harrison and Waldron [12] who also highlight the problems of overhead in communication with the card and miss reporting of host processor utilisation while performing GPU computation.

This paper seeks to gather together all three strands of work described above. Our overall aim is arithmetic modulo a large number so we can execute operations required in the RSA public key cryptosystem; we implement this arithmetic with an RNS based approach which performs arithmetic modulo small floating point values. The end result is an implementation which firstly fits the GPU programming model, and secondly makes effective use of SIMD-parallel floating point operations on which GPU performance relies. We demonstrate that with some caveats, this implementation makes it possible to improve performance using the GPU versus that achieved using a general purpose processor (or CPU). An alternative approach is recent work implementing a similar primitive on the IBM Cell [6], another media-biased vector processor. However, the radically different special purpose architecture of the GPU makes the task much more difficult than on the general purpose IBM Cell, hence our differing approach.

We organise the paper as follows. In Section 2 we give an overview of GPU architecture and capabilities. We use Section 3 to describe the algorithms used to implement modular exponentiation in RNS before describing the GPU implementation in Section 4. The experimental results in Section 4.3 compare the GPU implementation with one on a standard CPU, with conclusions in Section 5.

2   An Overview of GPU Architecture

Original graphics accelerator cards were special purpose hardware accelerators for the OpenGL and DirectX Application Programming Interfaces (APIs). Programs used the API to describe a 3D scene using polygons. The polygons have surfaces filled with a 2D pattern called a texture. The API produced an image for display to the user. Images are arrays of Picture Elements, or *pixels*, formed by the perspective-correct projection of the primitives onto a 2D plane. Each pixel describes the colour and intensity of a point on the display.

Graphics cards developed to allow the fixed functionality to be reprogrammed. Vector shaders are programs that transform 3D vectors within the graphics pipeline by custom projections and calculations. Pixel shaders allow the value of pixels to be specified by the programmer, as part of this specification a set of textures can be indexed by the program. Results can be directed into textures held in memory rather than to the display. We ignore 3D functionality and render a single scaled 2D rectangle parallel to the display plane, enforcing a 1:1 relation between input and output pixels, thereby turning the GPU into a vector processor. Each pixel is a 4-vector of single-precision floating-point values.

In the GPU programming model a single pixel shader is executed over a 2D rectangle of pixels. Each output pixel is computed by a separate instance of the shader, with no communication between program instances. Control-flow proceeds in lockstep between the instances to implement a SIMD vector processor. The program instance can use its 2D position within the array to parameterise computations, furthermore we can provide *uniform* variables which are constant over a single rendering step; each program instance has read-only access to such variables which are used to communicate parameters from the host to the shader

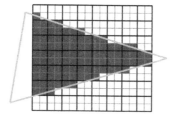

Fig. 1. Rasterisation of a graphics primitive into pixels

programs. The output of a value parametrised only by the coordinates of the pixel characterises the GPU programming model as *stream-processing*.

In this paper we specifically consider the Nvidia 7800-GTX as an archetype of the GPU. From here on, references to the GPU can be read as meaning the GPU within the Nvidia 7800-GTX accelerator. The pixel shading step within the GPU happens when the 3D geometry is *rasterised* onto the 2D image array. This process is shown in Figure 1, each rendered group of 2×2 pixels is termed a *quad*. The GPU contains 24 pipelines, arranged as six groups of quad-processors; each quad-processor operates on 4 pixels containing 16 values. Each pixel pipeline contains two execution units that can dispatch two independent 4-vector operations per clock cycle. If there are enough independent operations within the pixel-shader then each pipeline can dispatch a total of four vector operations per clock cycle. This gives a theoretical peak performance of $24 \times 4 = 96$ vector operations, or 384 single-precision floating point operations, per clock cycle.

The GPU contains a number of ports that are connected to textures stored in local memory. Each port is uni-directional and allows either pixels to be read from a texture, or results to be written to a texture. The location of the pixels output is fixed by the rendering operation and cannot be altered within a pixel shader instance. In stream processing terminology *gather* operations, e.g. accumulation, are possible but *scatter* operations are not. Read and write operations cannot be mixed on the same texture within a rendering step.

The lack of concurrent read and write operations, and communication between execution units, limits the class of programs that can be executed in a shader program, in particular modifying intermediate results is not directly possible. A solution to this problem called *ping-ponging* has been developed by the general purpose community [10]. The technique shown in Figure 2 uses multiple shader programs executing in sequence, with the textures holding intermediate results being written in one pass, and then read in a subsequent pass. We have identified the implementation of modular exponentiation using RNS as possible using this technique. There is a constant overhead associated with the setup of each pixel shader program, split between the OpenGL API, the graphics card driver and the latency filling the GPU pipelines. To achieve high-performance, this constant cost must amortised over a large texture. Increasing the number of ping-ponging steps increases the size of data-set required to break-even.

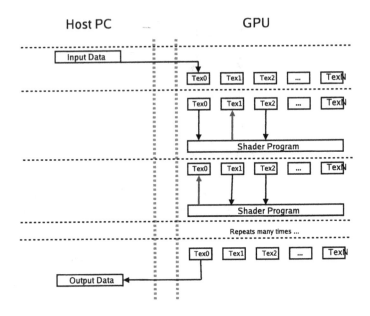

Fig. 2. Ping-pong operation in the GPU to produce intermediate results

Texture lookup within a pixel shader allows any pixel in a texture bound to an input port to be read. Texture lookups are not free operations; at a core frequency of 430 MHz, latency is in the order of 100 clock cycles. Exposing this latency is not feasible so it is hidden by multi-threading the quads processed on each quad-processor. Quads are issued in large batches, and all reads from instances within those batches are issued to a cache. This produces the two dimensional cache locality that is observed in GPU execution. Once a pixel has passed through the pipeline it is retired into the texture buffer. Recombination of the output stream is performed by 16 Raster Operators (ROP) which introduce a bottleneck of 25.6GB/s bandwidth shared between all shader instances in every pass. This hard limit creates a lower bound of six cycles on the execution of each shader instance; lacking the operations to fill these "free" cycles is a bottleneck.

3   Realising Modular Exponentiation With RNS

The RSA [22] public key cryptosystem uses a public key pair (N, e) where N is chosen to be the product of two large primes p and q that are kept secret. A private key d is selected such that $e \cdot d \equiv 1 \pmod{\phi(N)}$. To encrypt a plaintext message m, the ciphertext c is computed as $c = m^e \pmod{N}$ while to reverse the operation and decrypt the ciphertext one computes $m = c^d \pmod{N}$. As such, the core computational requirement is efficient arithmetic modulo N, in particular modular exponentiation. Efficient realisation of this primitive relies heavily on efficient modular multiplication as a building block.

We focus on implementing modular multiplication using a Residue Number System (RNS) with standard values of N; that is, taking N as a 1024-bit number. The advantage of using RNS is that operations such as addition and multiplication can be computed independently for the digits that represent a given value. As such, RNS operations can be highly efficient on a vector processor, such as the GPU, since one can operate component-wise in parallel across a vector of digits. Beyond this, selection of the exponentiation algorithm itself requires some analysis of the trade-off between time and space; see Menezes et al. [14, Chapter 14] for an overview.

In the specific case of RSA, the performance of a single modular exponentiation can be improved by employing small encryption exponent variants and the Chinese Remainder Theorem (CRT) [20] to accelerate the public and private key operations respectively. Since these cases are essentially specialisations of general modular exponentiation, we do not consider them. Instead we focus on using the GPU to implement the general case; one can expect performance improvements by specialising our techniques to suit, but this is a somewhat trivial step.

More relevant performance improvements can be achieved over multiple invocations of the modular exponentiation primitive. Consider an application where one is required to decrypt k ciphertexts c_i with the same decryption exponent, i.e.

$$m_i = c_i^d \pmod{N_i}, \quad i \in \{1, 2, \ldots k\}. \tag{1}$$

These separate computations can be viewed as a single SIMD-parallel program; control flow is uniform over the k operations, while the data values differ. As justification consider a server communicating with k clients: each client encrypts and communicates information using a single public key pair; batches of ciphertexts are decrypted with a common exponent. Considering this form of operation is important since it enables us to capitalised on both fine-grained parallelism at the RNS arithmetic level, and course-grained parallelism in the exponentiation level.

3.1   Standard Arithmetic in an RNS

Numbers are stored in a conventional radix number representation by selecting coefficients of the various powers of the radix that sum to the number. On processors with a w-bit word size, one typically selects the radix $b = 2^w$ so that each coefficient can be stored in a separate word. An RNS operates by selecting a *basis*, a set of co-prime integers that are fixed for all of the numbers being operated upon. An example basis B might be defined as

$$B = \langle\, B[1], B[2], \ldots, B[n] \,\rangle \text{ with } \gcd(B[i], B[j]) = 1 \text{ whenever } i \neq j$$

For efficiency, each of the moduli $B[i]$ should fit within a word. The size of a basis is defined as the product of the moduli, that is

$$|B| = \prod_{i=1}^{n} B[i]$$

such that the largest uniquely representable number is less than $|B|$. Arithmetic operations performed in B implicitly produce results modulo $|B|$.

When the integer x is encoded into an RNS representation in the basis B, it is stored as a vector of words; there are n words, one for each $B[i] \in B$. Each component of the vector holds the residue of x modulo the respective $B[i]$. Thus, using standard notation, the encoding of x under basis B is the vector

$$x_B = < \; x \bmod B[1], \; x \bmod B[2], \; \ldots, \; x \bmod B[n] \; >$$

such that $x_B[i]$ denotes the i-th component of the vector. The Chinese Remainder Theorem (CRT) defines a bijection between the integers modulo $|B|$ and the set of representations stored in the basis B. To decode x_B, the CRT is applied

$$x = \sum_{i=1}^{n} \left(\hat{B}[i] \cdot \frac{x_B[i]}{\hat{B}[i]} \bmod B[i] \right) \bmod |B|$$

where $\hat{B}[i] = \frac{|B|}{B[i]}$. It is useful to rewrite the CRT as shown in Equation 2. In this form the reduction factor k is directly expressed, which is used by the base extension algorithms in Section 3.3. Note that unless values are initially stored or transmitted in RNS representation, the conversion process represents an overhead.

$$x = \sum_{i=1}^{n} \hat{B}[i] \cdot \frac{x_B[i]}{\hat{B}[i]} - k|B| \tag{2}$$

Multiplication and addition of two integers x_B and y_B, encoded using an RNS representation under the basis B, is performed component-wise on the vectors. For example, multiplication is given by the vector

$$x_B \cdot y_B = \; \langle \; x[1] \cdot y[1] \bmod B[1], \; x[2] \cdot y[2] \bmod B[2], \; \ldots, \; x[n] \cdot y[n] \bmod B[n] \; \rangle$$

Each component is independent, and so on a vector architecture individual terms in the computation can be evaluated in parallel. Assuming m execution units on the GPU, and n moduli in the RNS basis, then a single operation will only take $\lceil \frac{n}{m} \rceil$ clock cycles. Eliminating the communication, and hence synchronisation, between the execution units makes this speed possible. It is imperative to emphasise this advantage of RNS; propagation of carries between words within the GPU is very difficult to achieve and thus SIMD-parallel methods for realising arithmetic modulo N on general purpose processors are not viable.

3.2   Modular Arithmetic in an RNS

Although we have described standard multiplication using RNS, implementation of modular exponentiation depends on modular multiplication. In a positional number system this operation is commonly performed using Montgomery representation [15]. To define the Montgomery representation of x, denoted x_M,

---

Algorithm 1. Cox-Rower Algorithm [21] for Montgomery multiplication in RNS

Input : x and y, both encoded into A and B.
Output: $x \cdot y \cdot A^{-1}$, encoded into both A and B.

In Base A	In Base B
$s_A \leftarrow x_A \cdot y_A$	$s_B \leftarrow x_B \cdot y_B$
	$t_B \leftarrow s_B \cdot \left(-N^{-1} \bmod |B| \right)$

$$\text{base extend } t_A \leftarrow t_B$$

$u_A \leftarrow t_A \cdot N_A$
$v_A \leftarrow s_A + u_A$
$w_A \leftarrow v_A \cdot \left(|B|^{-1} \bmod A \right)$

$$\text{base extend } w_A \rightarrow w_B$$

return w_A, w_B

---

one selects an $R = b^t > N$ for some integer t; the representation then specifies that $x_M \equiv xR \pmod{N}$. To compute the product of x_M and y_M, termed Montgomery multiplication, one interleaves a standard integer multiplication with an efficient reduction by R

$$x_M \star y_M = x_M y_M R^{-1}$$

The same basic approach can be applied to integers represented in an RNS [18,21]. As the suitable input range of the Posch algorithm [18] is tighter than the output produced, a conditional subtraction may be required. To avoid this conditional control flow we have used the Cox-Rower algorithm of Kawamura et al. [21]. Roughly, a basis $|A|$ is chosen as the R by which intermediate reductions are performed. Operations performed within the basis A are implicitly reduced by $|A|$ for free. Montgomery multiplication requires the use of integers up to RN, or $|A|N$ in size. In order to represent numbers larger than $|A|$ a second basis B is chosen.

The combination of the values from A and B allow representation of integers less than $|A| \cdot |B|$. Consideration of the residues of the integer stored in either basis allows a free reduction by $|A|$ or by $|B|$. To take advantage of this free reduction, we require a base extension operation. When a number represented in RNS basis A, say x_A, is extended to basis B to form $x_B = x_A \bmod B$, the residue of $x \bmod |A|$ is computed modulo each of the moduli in B.

Algorithm 1 details the Cox-Rower algorithm for Montgomery multiplication in an RNS. The algorithm computes the product in both bases, using the product modulo the first basis to compute the reduction. Each of the basic arithmetic operations on encoded numbers is highly efficient on a vector architecture as are computed component-wise in parallel. Efficient implementation of the Montgomery Multiplication requires both the reduction of x_A modulo $|A|$, and the base extension to be inexpensive. Note that in the RNS representation A, reduction by $|A|$ is free, as it is a side-effect of representation in that basis.

3.3   Base Extension

There are several well-known algorithms for RNS base extension in the literature [24,19,21]. In each case the convention is to measure time complexity (or circuit depth) by the number of operations to produce a residue modulo a single prime. For use in Montgomery multiplication as described above, each algorithm must be executed n times to produce residues for each target moduli.

The earliest result is the Szabo-Tanaka algorithm [24] with $O(n)$ time complexity. The RNS number is first converted into a Mixed Residue System (MRS). The MRS is a positional format where the coefficients of the digits are products of the primes in the RNS system, rather than powers of a radix. So the RNS value x_A would be represented in MRS as an n-vector $m(x_A)$ such that

$$ x_A = \sum_{i=1}^{n} \left(m(x_A)[i] \cdot \prod_{j=0}^{i-1} A[j] \right) \quad \text{where } A[0] = 1 $$

This conversion is achieved in an n-step process. In each step the residue of the smallest remaining prime is used as the next MRS digit. This MRS digit is subtracted from the remaining residues, and the resultant number is an exact product of the prime; the number can be divided efficiently by the prime through multiplication by the inverse. Each step consume a single digit of the RNS representation and produces a single digit in the MRS representation. The Szabo-Tanaka algorithm then proceeds to accumulate the product of each MRS digit and its digit coefficient modulo each of the target moduli to construct the RNS representation. To allow single word operations the residues of the coefficients can be precomputed. The algorithm is highly suitable for a vector architecture as it uses uniform control flow over each vector component.

Posch et al. [19] and Kawamura et al. [21] both achieve $O(\log n)$. The approaches are similar and use the CRT as formulated in Equation 2, computing a partial approximation of k. The computation is iterated until the result has converged to within a suitable error bound. The iterative process requires complex control-flow creating inefficiency in the GPU programming model.

Shenoy and Kumaresan [23] claim that an extra redundant residue can be carried through arithmetic operations, and used to speed up the conversion process. Unfortunately their approach does not appear to work for modulo arithmetic. Assuming that our system operates in basis B, all arithmetic operations are implicitly modulo $|B|$. The redundant channel r is co-prime to $|B|$ and thus results mod r cannot be "carried through" from one operation to another. Whenever the result of an operation is greater than $|B|$, the result in the redundant channel would need to be reduced by $|B|$ *before* the reduction by r. To make this approach work for systems using modulo arithmetic, each operation requires an expensive reduction into base r before the redundant channel can be used to speed up the reduction for the other bases in the system.

Algorithm 2. Cox-Rower Montgomery multiplier with Szabo-Tanaka Extension

---

Input   : x_A, x_B and y_A, y_B.
Output: $x \cdot y \cdot A^{-1}$, encoded into w_A, w_B.

Stage1 $s_A \leftarrow x_A \cdot y_B$; $s_B \leftarrow x_B \cdot y_B$; $t_B \leftarrow s_B \cdot -N^{-1} \pmod{|B|}$

Stage2   **for** $i = 1$ **upto** n **do**
  $m[i] \leftarrow t_B[i]$;
  $t_B[i+1\,..\,n] \leftarrow t_B[i+1\,..\,n] - t_B[i]$;
  $t_B[i+1\,..\,n] \leftarrow t_B[i+1\,..\,n] \cdot B^{-1}[i,n]$

Stage3   **for** $i = 1$ **upto** n **do**
  $t_A[i] \leftarrow \sum_j^n m[j] \cdot C_A[i,j] \pmod{|A|}$

Stage4 $u_A = t_A \cdot N$; $v_A = s_A + u_A$; $w_A = v_A \cdot |B|^{-1} \pmod{|A|}$

Stage5   **for** $i = 0$ **upto** n **do**
  $m[i] \leftarrow w_A[i]$;
  $w_A[i+1\,..\,n] \leftarrow w_A[i+1\,..\,n] - w_A[i]$;
  $w_A[i+1\,..\,n] \leftarrow w_A[i+1\,..\,n] \cdot A^{-1}[i,n]$

Stage6   **for** $i = 1$ **upto** n **do**
  $w_B[i] \leftarrow \sum_j^n m[j] \cdot C_B[i,j] \pmod{|B|}$
return w_A, w_B

---

4   Mapping RNS Arithmetic to the GPU

Algorithm 2 is an overview of our GPU implementation of Montgomery multiplication. In the array-language syntax each variable refers to a vector, and an indexed variable refers to a single component. The algorithm describes operations on a vector architecture without specifying an explicit representation. On the GPU each stage will be encoded as a separate shader program written in the OpenGL Shading Language (GLSL [17]). One instance of the shader will execute for each vector component (each vector is the same length). Operations performed over vectors refer to a component-wise application; vector operations do not require communication between shaders.

The horizontal bars that separate each stage represent parallel barriers in the algorithm; either a communication is required between independent shaders, or a change in the *shape* of the control-flow that executes. As the control-flow of each shader is lock-stepped this requires a new rendering operation. Executing single instances of the shader over the different components of a vector provides a single implicit loop; the explicit loops must either be unfolded within the shader or implemented in multiple rendering steps. When communication is required between individual shaders in the loop then a new rendering step will be required. For example the loop in Stage 2 performs an update on the elements of t_B beyond the i-th position. Each iteration $i + 1$ requires the updated value from the i-th position. Hence each iteration of the loop requires a new rendering step, both as a barrier and to allow the computed value to be retrieve from the ping-pong.

Stages one and four constitute the Cox-Rower algorithm [21] for Montgomery multiplication in RNS. Variable names are retained from Algorithm 1 for clarity. Stages two and three together form a single base extension; converting the RNS value in t_B into MRS form in the vector m, then reducing these digits by the moduli of basis A to produce $t_A = t_B \bmod A$. The matrix B^{-1} stores the inverse of each prime in B under every other prime. The MRS coefficients reduced by each prime in A are stored in the matrix C_A. The operation is repeated in stages five and six, extending w_A into w_B to provide the result in both bases.

The most important issue in implementing Algorithm 1 is how the vector components are mapped into the GPU. Memory on the GPU is organised as two dimensional arrays called textures. Each element (pixel) is a tuple of four floating-point values. The floating-point format uses a 24-bit mantissa that allows the representation of integers up to $2^{24} - 1$. Within this range arithmetic operations produce the correct values. When this range is exceeded precision is lost as the floating-point format stores the most significant bits. This poses a significant problem for cryptographic application where the least significant bits are required.

There are two aspects to the issue of choosing a mapping from the vectors in Algorithm 1 that both impact performance. The issue of how to represent words within each floating point value is covered in Section 4.1. How the arrangement of the tuples of floating point values affects the execution of shaders is covered in Section 4.2. Neither topic can be explained completely independently of the other and unfortunately some material is split across both sections. In Section 4.3 we describe the performance results of the different choices of representation.

4.1 Floating-Point Arithmetic

The most problematic operation is multiplication of two words, and then reducing the result by a modulus in the system. The simplest method of avoiding overflow is to use half-word values; bounding the moduli by 2^{12}. This ensures that products are correctly represented in a single word and the GLSL mod operation can be used for reduction. Twice as many moduli are required when they are restricted to half-words, this causes two major problems: the necessary memory bandwidth is doubled, the complexity of the base extension is $O(n^2)$ so doubling the number of digits will quadruple the amount of computation required. The main advantages are the simplicity of the implementation and the fast speed of word-sized modulo multiplications. The mod compiles into 3 instructions that are executed in 1.5 cycles producing a modular multiplication of 2 cycles. In the results section we refer to the code using this representaton as Implementation-A.

An alternative approach is to use primes less than 2^{24} as moduli, and perform multi-precision arithmetic to handle the double-word product. Techniques [13] for multi-precision arithmetic in floating point are well known. Dekker-style splits can be used to compute the low-order bits by subtracting partial products from the high-order bits in a floating-point product. The 15 operations required are executed in 7.5 cycles but the technique requires guard bits in the floating-point unit to guarantee correctness. Experiments to determine the validity of this approach on the floating-point implementation in the 7800-GTX were unsuccesful.

The reductions in complexity and memory traffic from full-word moduli are desirable, and so we have investigated alternatives to the multi-precision approach. One possibility is to pack two half-words in every word. This creates the reduction in memory traffic, and once a word is retrieved it can be unpacked into two separate values. The modular multiplication is then the 2 cycle operation over half-words. The number of digits in the system is not reduced, but there is still a performance advantage. Unfortunately the Nvidia compiler necessary to use GLSL on the GPU is still immature, and has difficulty performing register colouring correctly when this much data is active in a shader. Examination of the output of the compiler shows that twice as many registers as required are allocated, and that the actual performance of the shader suffers. As there is no way to load binary code directly onto the device it is not possible to check the performance of correctly scheduled code with proper register allocation. Experiments with similar code fragments suggest that when the compiler matures enough to compile this code correctly there will be a significant increase in performance for this method. Similar experiments [25] have shown dramatic decreases in performance for each power of two that the number of registers exceeds, this suggests that scheduling shader invocations with a shared register bank causes the architectural problem, and that the compiler is unable to avoid this case. This representation is used in Implementation-C along with the specialisation techniques described below.

Another representation is provided by the condition that RNS moduli are not required to be prime, merely coprime to each other. Consider a modulus m chosen with two prime factors p and q, such that $m < 2^{24}$. For $a, b \in (0 .. m - 1)$ the product $a \cdot b \bmod m$ can be computed without intermediate values that exceed 2^{24}. To do so, compute:

$$a_p \leftarrow a \bmod p, \; a_q \leftarrow a \bmod q,$$

$$b_p \leftarrow b \bmod p, \; b_q \leftarrow b \bmod q,$$

$$t_1 \leftarrow a_p \cdot b_p \bmod p, \; t_2 \leftarrow t_1 \cdot q^{-1} \bmod p, \; t_3 \leftarrow t_2 \cdot q,$$

$$t_4 \leftarrow a_q \cdot b_q \bmod q, \; t_5 \leftarrow t_4 \cdot p^{-1} \bmod q, \; t_6 \leftarrow t_5 \cdot p,$$

$$(t_3 \cdot s(t_3) \cdot m) + t_6$$

The modulo operations in each term are necessary to keep the intermediate values within the word range and prevent overflow. The final term is the product reduced by m. The function $s(x)$ denotes the sign of x and is a primitive function in GLSL. This unusual reduction trades multiplications for inexpensive modulo operations. On the GPU the cost of a word-sized mod is only 1.5 cycles, whereas the cost of a multiplication that overflows the word is large[1]. By avoiding overflow in the floating-point registers with mod operations we are treating them

---

[1] Observed cost per operation increased with the number of operations due to register spilling in the Nvidia compiler.

as mantissa sized integer registers. The entire operation takes 16 cycles. Although this is much higher than the half-word approach the total cost of the shaders is comparable because register spilling is reduced and the GPU appears to dispatch the shader faster. In our results this representation is used in Implementation-D.

The reduction described can be seen as an unfolding of the CRT over two moduli. This leads to our final approach to modular multiplication in the GPU. The independent results under p and q do not have to be recombined on each operation. Each digit in the RNS that we are using is represented by a smaller scale RNS comprised of two primes. This digit can be operated on $\bmod p$ and $\bmod q$ for fast operation, or the entire digit can be used as part of the larger RNS. This choice of representation with a fast map between the two allows us to compute the chain of operations in a shader independently in each base, but to execute the base conversion over a lower number of digits and so reduce the complexity. When this reduction is combined with the specialisation technique described below, many of the mappings from $x \bmod m \mapsto \langle x_p, x_q \rangle$ are paired with mappings from $\langle x_p, x_q \rangle \mapsto x \bmod m$ and so can be removed from the computation. Each of the individual modular multiplications are primitive GLSL `mod` operations. This results in a substantial increase in overall performance. The code using this technique is refered to as Implementation-E.

4.2   Memory Layout and Shader Specialisation

Each instance of a shader executes on a single pixel in a texture. Four floating-point values are computed by the execution of every shader, and are stored in a fixed location within the texture. When a shader is rendered over a rectangle in a texture the output location is fixed for each instance. The execution of each shader then proceeds in lockstep, with the same instructions being dispatched in each processor pipeline.

Initially the only distinction between the execution of the programs is the co-ordinate of the pixel, available to the program as a 2D vector `gl_TexCoord[0]`. One major difficulty in writing GPU shaders is distinguishing the computation in that particular instance, from that performed in every other instance. Although shaders can make differing texture lookups from other instances, sufficient information to decide which data to lookup must be encoded in the pixel coordinates.

Our strategy for memory layout is to encode which instance, and which vector components within that instance a particular pixel holds by the spatial location of the pixel. Each row of the texture holds c instances. The pixels containing the first instance are located in columns $0, c, 2c, \ldots, nc$ where $4n$ is the number of moduli in the RSA system. The moduli associated with the 4 residues in a pixel are consistent across c columns, and the entire height of the texture. The advantage of this striding pattern is that contiguous rectangles describe pixels associated with particular moduli.

Of the four textures available for input and output, we use two to perform a ping-pong operation. Essentially these textures are used to hold the *new* and *previous* states of mutable variables. For the case of 1024-bit instances, 88 moduli less than 2^{12} describe the system, or 44 moduli less than 2^{24}. Because values are required in bases A and B we split the textures in two vertically and have two independent sets of columns to hold values. We use the largest textures possible[2] on the drivers and hardware which are 1024×1024 pixels in size. For the half-word case this allows 23 instances per row for a total of 23552 parallel instances. In the full-word case a texture stores 47104 parallel instances. For the binary operations in the Cox-Rower Algorithm we use a third texture as a data source. The texture is updated between modular multiplication algorithms as part of the exponentiation. The fourth texture contains auxiliary read-only data such as the values of inverses, primes and other data.

Our initial experimental results used this layout scheme, but profiling of the performance suggested that the memory traffic reading the auxiliary texture dominated the runtime. Current work in the GPGPU community suggests that this approach of treating textures as arrays is common. One reason for this is that the 'arrays' offered in GLSL are not true arrays because of hardware limitations. All array indices must be compile-time constants, which limits the potential for storing auxiliary data in local lookup tables and indexing it by the current pixel coordinate.

To circumvent this restriction we have employed a program transformation technique called specialisation. Some of the inputs to a specialised program are fixed to constants known at compile-time. This allows constant propagation to remove operations in the compiled program. Constants are created by fissioning program templates into separate shaders that run over c columns in the texture. For each pixel rendered by a particular shader the set of moduli is constant. This propagates through the program and makes the other auxiliary data constant by virtue of the fact that the lookup table indices become constant. After specialisation the program template generates either 11 or 22 separate shaders, depending on if half-word or full-word arithmetic is used. Within each shader the constant auxiliary data is coded as a local array of constants. In GLSL for Nvidia GPUs constant data is encoded within the instruction stream, removing all memory transfers to fetch the data. As a side-effect this frees the fourth texture which may allow performance improvements in the future. An example of this technique is shown in Listing 1.1. The code is a program template for the MRS reduction step in Implementation-A. The <1> syntax refers to terms that are compile-time constants, our implementation generates multiple shaders from each template replacing the parameters with constant values in each instance — these generated programs can then be compiled as normal GLSL code. The implementations that use half-word values store a 1024-bit number in 22 pixels, using 4 floating-point numbers in each. Each template generates 22 separate shaders in this case. The full-word versions store the same value as 11 pixels, and generate 11 separate shader programs.

Despite the increased costs of running shaders over smaller regions, and flushing the program cache on the GPU because of the large number of programs being executed — the technique still produces a significant increase in performance.

---

[2] The 7800-GTX supports 4096×4096 textures but we were unable to get stable results with the current drivers.

Performance of Version E on 1024-bit exponentiation

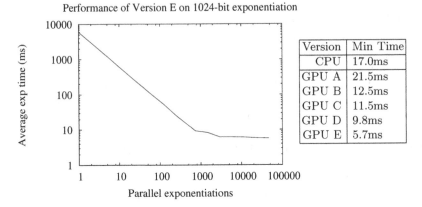

Version	Min Time
CPU	17.0ms
GPU A	21.5ms
GPU B	12.5ms
GPU C	11.5ms
GPU D	9.8ms
GPU E	5.7ms

Fig. 3. Performance of 1024-bit Modular Exponentiation

4.3 Experimental Results

Our goal in evaluating the performance of our GPU based implementation is comparison with a roughly equivalent implementation on a general purpose processor (CPU), namely an 2.2 GHz AMD-64 3200+. This processor was mounted in a computer that also acted as the host for the graphics accelerator; all experiments were conducted on this common platform.

To explore the optimisation space for the algorithm on the 7800-GTX and gauge the importance of the trade-offs described in Section 4 we implemented several versions. In all cases the modular multiplication was used within a binary exponentiation method [14, Chapter 14] to compute 1024-bit modular exponentiations. The exponents were all chosen to have hamming weights of 512. A standard C implementation of Montgomery multiplication was written for our CPU; this included small assembly language inserts to expose critical features in the processor (i.e. retrieving double word products). The CPU based Montgomery multiplication used a 4-bit sliding window method [14, Chapter 14] to compute the same 1024-bit modular exponentiations. The CPU implementation achieved 17ms on our hardware, using sliding windows which we have not implemented on the GPU due to difficulties in memory management. The performance of each implementation is shown in Figure 3 along with the details of how the increase in parallelism is reflected in the average operation times (throughput). The latency for the data in this graph is the number of operations × the average time.

Implementation-A. This is an unoptimised version that uses the half-word representation and no specialisation. This was a first attempt to naively treat the GPU as a set of fast memory arrays and ALUs. The result was somewhat disappointing given the 35 : 1 performance ratio between the two platforms.

Implementation-B. This version is specialised over 22 shaders for each template; expanding the templates produces $2 \cdot 22 + 8 = 52$ shaders in total.

The half-word representation is still used but some low level optimisations including changes to the internal texture format used by OpenGL have been applied. The increase in performance is dramatic, although somewhat hampered by the switching cost of 52 separate shaders being executed.

Implementation-C. This version alters Implementation-B by adapting the full-word data representation; there are still 52 shaders but with an overall increase in performance because of the lower memory bandwidth requirements.

Implementation-D. This version alters Implementation-B by adapting smooth word-size representation. Although the reduction is expensive compared to the previous methods, the combination of reduction and only 11 shaders increases the performance again. The increase was not uniform, the shaders for Stage2 benefited from increased performance, while the shaders for Stage3 decreased in performance by 50%. The relative expense of the reduction (compared to GLSL mod methods) means that performance is dependent on the memory bandwidth to computation ratio in the particular shader.

Implementation-E. Finally, this version alters Implementation-D, still using the smooth word-size representation, by substituting the reductions for GLSL mod operations working in each of the paired bases. The improvement in performance is dramatic, resulting in a 3 fold increase over the CPU implementation. This result is despite the lack of windowing in the GPU implementation.

Each of these figures was produced over many parallel instances of exponentiations which results in a large latency. This is necessary because the drivers are optimised for graphics workloads, and the rapid switching of shader programs is causing problems. Some of the latency issues could be rectified if Nvidia optimised their drivers for non-graphics workloads, but this would require an uptake of GPGPU in the marketplace.

Our performance figures for exponentiation on the CPU should be seen only as a broad guideline. The implementation uses a common choice of algorithms but has not had extensive low-level optimisation. Reports from other researchers and the figures in the *Crypto++* benchmarks suggest that aggressive low-level optimisation can reduce this figure on our reference platform to about 5ms. The GPU implementation cannot be aggressively optimised in the same way, but gives results comparable to the fastest CPU implementation. Our experiments suggest that given low-level access to the GPU (bypassing the GLSL compiler) there are similar potential performance increases in the GPU platform.

5   Conclusion

We have presented an investigation into the implementation and performance of modular exponentiation, the core computational operation in cryptosystems such as RSA, on a commodity GPU graphics accelerator. In a sense, this work represents an interesting aside from implementation on devices normally constrained by a lack of computational or memory resources: here the constraint is architecture targeted at a different application domain. Previous results in GPGPU have shown that such a limited architecture can be successfully targeted at problems outside of the intended domain. We believe that integer modular exponentiation is the most radically different problem domain so far reported.

In an attempt to mitigate this difference and fit the algorithm into the GPU programming model, we employed vector arithmetic in an RNS which allowed us to capitalise on fine-grained SIMD-parallel floating point computation. Our experimental results show that there is a significant latency associated with invoking operations on the GPU, due to overhead imposed by OpenGL and transfer of data to and from the accelerator. Even so, if a large number of similar modular exponentiations are required, the GPU can capitalise on course-grained parallelism at the exponentiation level and out-perform a CPU. Although this comparison is uneven in a number of respects (the CPU uses windowed exponentiation while the GPU can not, the GPU uses an unreasonably large number of parallel exponentiations) it is crucial to see that exploiting a commodity resource for cryptographic primitives offers interesting possibilities for the future. Although current drivers miss-report 100% utilisation of the CPU during GPU operation, this is not a requirement since synchronisation between CPU and GPU is essentially implemented as a spin-lock. Most of the GPU computation time does not require CPU intervention and manually scheduling code before calling the glFinish() synchronisation allows use of the CPU for other tasks. This raises the possibility of utilising an otherwise idle resource for cryptographic processing.

5.1   Further Work

The advantages and disadvantages highlighted by our approach detail a number of issues worthy of further research; we expect that this will significantly improve GPU performance beyond the presented proof of concept.

– Due to the difficulty of managing multiple textures (that might represent pre-computed input, for example) on the GPU, in this work we opted to use a very basic exponentiation algorithm. Clearly huge advantages can be made if some form of pre-computation and windowing can be used [14, Chapter 14]; we aim to investigate this issue as a means of optimisation in further work. In particular the specialisation technique frees up a texture that can be used as a second output in *multi-texturing*.
– A major limiting factor in deploying these techniques in real systems is the large number of parallel exponentiations that need to be executed in order to break-even on various overheads imposed by the GPU. In particular, the driver software for the GPU is optimised to deal with repeated execution of a limited set of pixel shaders on large data-sets; this contrasts with our demands for larger numbers of pixel-shaders and comparatively smaller data-sets.

– Program transformation and specialisation techniques to generating code for these exotic and quickly evolving architectures is paramount to success. Our target architecture forced us to use somewhat complex instruction flows (consisting of 52 generated shaders); we expect that by specialising, for a specific RSA exponent for example, the associated overhead can be significantly reduced. An automated system for achieving this task efficiently is work in progress.

5.2   Evolving Architectures

The CPU market is rapidly evolving. A new class of architecture is released every year, with roughly twice the performance of the previous generation. This rapid progress makes the GPU a desirable target for cryptographic code, however it also creates a problem that results may not stay relevant for very long. While this paper was being written a new generation of Nvidia GPU, the 8800-GTX, has been released. The strict separation between programmer and hardware that is imposed by the driver allows GPUs to be completely redesigned in each generation.

The new 8800-GTX uses a higher core clock speed and faster memory to change the memory / computation trade-offs in GPU programming. The vector units are finer-grained, with 128 scalar ALUs available to the programmer. The cache is lockable through software and primitive are available to control the mapping of software instances onto the vector array. Finally, the memory architecture allows scatter operations which enlarges the set of programs that can be written.

All of these changes would make interesting future research for implementing cryptographic code. However the results in this paper are still relevant as the *cost* in terms of transistor count, or power consumption for these more fully featured ALUs is higher than the simple ALUs used in the 7800-GTX. Hence these results are applicable to future resource constrained systems with a similar architecture.

References

1. Bailey, D.V., Paar, C.: Efficient Arithmetic in Finite Field Extensions with Application in Elliptic Curve Cryptography. Journal of Cryptology 14(3), 153–176 (2001)
2. Barrett, P.D.: Implementing the Rivest, Shamir and Adleman Public Key Encryption Algorithm on a Standard Digital Signal Processor. In: Odlyzko, A.M. (ed.) CRYPTO 1986. LNCS, vol. 263, pp. 311–323. Springer, Heidelberg (1987)
3. Bernstein, D.J.: The Poly1305-AES Message-Authentication Code. In: Gilbert, H., Handschuh, H. (eds.) FSE 2005. LNCS, vol. 3557, pp. 32–49. Springer, Heidelberg (2005)
4. Bernstein, D.J.: Curve25519: New Diffie-Hellman Speed Records. In: Yung, M., Dodis, Y., Kiayias, A., Malkin, T.G. (eds.) PKC 2006. LNCS, vol. 3958, pp. 207–228. Springer, Heidelberg (2006)

5. Cook, D.L., Keromytis, A.D., Ioannidis, J., Luck, J.: CryptoGraphics: Secret Key Cryptography Using Graphics Cards. In: Menezes, A.J. (ed.) CT-RSA 2005. LNCS, vol. 3376, pp. 334–350. Springer, Heidelberg (2005)

6. Costigan, N., Scott, M.: Accelerating SSL using the Vector processors in IBM's Cell Broadband Engine for Sony's Playstation 3. Cryptology ePrint Archive, Report 2007/061 (2007)

7. Crandall, R.E.: Method and Apparatus for Public Key Exchange in a Cryptographic System. U.S. Patent Number 5,159,632 (1992)

8. van Dijk, M., Granger, R., Page, D., Rubin, K., Silverberg, A., Stam, M., Woodruff, D.: Practical Cryptography in High Dimensional Tori. In: Cramer, R.J.F. (ed.) EUROCRYPT 2005. LNCS, vol. 3494, pp. 234–250. Springer, Heidelberg (2005)

9. Fournier, J., Moore, S.: A Vectorial Approach to Cryptographic Implementation. In: International Conference on Digital Rights Management (2005)

10. GPGPU: General-Purpose Computation Using Graphics Hardware. http://www.gpgpu.org/

11. Hankerson, D., Menezes, A., Vanstone, S.: Guide to Elliptic Curve Cryptography. Springer, Heidelberg (2004)

12. Harrison, O., Waldron, J.: AES Encryption Implementation and Analysis on Commodity Graphics Processing Units. In: Cryptographic Hardware and Embedded Systems (CHES). LNCS, vol. 4727, pp. 209–226. Springer, Heidelberg (2007)

13. Knuth, D.E.: The Art of Computer Programming, 3rd edn., vol. 1-3. Addison-Wesley, Reading (1997), Additions to v.2: http://www-cs-faculty.stanford.edu/~knuth/err2-2e.ps.gz

14. Menezes, A.J., van Oorschot, P.C., Vanstone, S.A.: Handbook of Applied Cryptography. CRC Press, Boca Raton (1997)

15. Montgomery, P.L.: Modular Multiplication Without Trial Division. Mathematics of Computation 44, 519–521 (1985)

16. Parhami, B.: Computer Arithmetic: Algorithms and Hardware Designs. Oxford University Press, Oxford (2000)

17. Randi, R.J.: OpenGL Shading Language. Addison Wesley, Reading (2004)

18. Posch, K.C., Posch, R.: Modulo Reduction in Residue Number Systems. IEEE Transactions on Parallel and Distributed Systems 6(5), 449–454 (1995)

19. Posch, K.C., Posch, R.: Base Extension Using a Convolution Sum in Residue Number Systems. Computing 50, 93–104 (1993)

20. Quisquater, J-J., Couvreur, C.: Fast Decipherment Algorithm for RSA Public-key Cryptosystem. IEE Electronics Letters 18(21), 905–907 (1982)

21. Kawamura, S., Koike, M., Sano, F., Shimbo, A.: Cox-Rower Architecture for Fast Parallel Montgomery Multiplication. In: Preneel, B. (ed.) EUROCRYPT 2000. LNCS, vol. 1807, pp. 523–538. Springer, Heidelberg (2000)

22. Rivest, R., Shamir, A., Adleman, L.M.: A Method for Obtaining Digital Signatures and Public-key Cryptosystems. Communications of the ACM 21(2), 120–126 (1978)

23. Shenoy, P.P., Kumaresan, R.: Fast Base Extension Using a Redundant Modulus in RNS. IEEE Transactions on Computers 38(2), 292–297 (1989)

24. Szabo, N.S., Tanaka, R.I.: Residue Arithmetic and its Applications to Computer Technology. McGraw-Hill, New York (1967)

25. Bucks, I.: Invited Talk at Eurographics/SIGGRAPH Workshop on Graphics Hardware (2003), http://graphics.stanford.edu/~ianbuck/GH03-Brook.ppt

Appendix: Program Listing

This source code is a program template. At run-time our implementation substitutes a constant value for the template parameters indicated by <0> and <1>. These values are determined by the batch size and control the number of exponentiation instances encoded into each row, and the size of the texture respectively.

Listing 1.1. An example program template

```
1   uniform sampler2D T;    // The matrix with intermediate results
2                           // from the previous stage of ping-pong
3   uniform sampler2D P;    // The matrix with auxillery precomputed data
4                           // such as A,B,B^-1 etc
5   uniform float i;        // The loop index - shader executes onces per iteration
6   void main(void)
7   {
8       vec2   where    = gl_TexCoord[0].xy;     // Retrieve the xy parameters for
9                                                // this instance.
10      // Which of the moduli covers this target pixel
11      float base      = floor( where.x / <0> );
12
13      // Which exponentiation instance we are within the row
14      float inst      = mod( where.x, <0> );
15
16      // The location within P of the inverse, the scaling accounts for
17      // coordinate normalisation
18      vec2  invWhere = vec2( i, base ) / <1>;
19      vec4  bInvs = texture2D( P, invWhere );
20
21      // The moduli in base A for this pixel (4 components of tB)
22      vec4   primes   = texture2D( P, vec2(88+base,0) / <1> );
23
24      // Retrieve the "current" values from the ping-pong texture
25      vec4   v    = texture2D( T, where / <1> );
26      // Retreive the values of the current subtraction digit from P (v')
27      vec4   v2   = texture2D( T, vec2(i * <0> +inst,where.y) / <1> );
28      vec4   t2   = mod( (v-v2) * bInvs, primes );
29
30      // Switch between passing through v or t2. This guarded form is
31      // recognised by the compiler and produces straight-line code
32      float c = (where.x<=i) ? 1 : 0;
33      gl_FragColor = v*vec4(c) + t2*(1-(vec4)c);
34  }
```

Multi-key Hierarchical Identity-Based Signatures*

Hoon Wei Lim and Kenneth G. Paterson

Information Security Group
Royal Holloway, University of London
Egham, Surrey TW20 0EX, UK
{h.lim, kenny.paterson}@rhul.ac.uk

Abstract. We motivate and investigate a new cryptographic primitive that we call *multi-key hierarchical identity-based signatures* (multi-key HIBS). Using this primitive, a user is able to prove possession of a set of identity-based private keys associated with nodes at arbitrary levels of a hierarchy when signing a message. Our primitive is related to, but distinct from, the notions of identity-based multi-signatures and aggregate signatures. We develop a security model for multi-key HIBS. We then present and prove secure an efficient multi-key HIBS scheme that is based on the Gentry-Silverberg hierarchical identity-based signature scheme.

1 Introduction

Research in identity-based cryptography (IBC) [32] has proliferated in recent years since the discovery of novel pairing-based key agreement protocols due to Sakai *et al.* [30] and Joux [23], and the seminal work of Boneh and Franklin [8] giving the first secure and practical identity-based encryption (IBE) scheme.

Several proposals for identity-based signature (IBS) schemes, such as [12,21,26], followed quickly after Boneh and Franklin's publication [8]. More recent proposals for IBS can be found in [4,5,17,27], for example. However, IBS schemes are arguably less interesting than IBE schemes. They suffer from the key escrow property that is inherent in IBC, which makes non-repudiation of these signatures more difficult to achieve. Moreover, there is a generic construction that creates an IBS scheme from two instantiations of any (normal) public key signature scheme [5]. This is related to the fact that, with a normal,

---

* The first author's research was supported by the EPSRC under grant EP/D051878/1.The work described in this paper has been supported in part by the European Commission through the IST Programme under Contract IST-2002-507932 ECRYPT. The information in this document reflects only the authors' views, is provided as is and no guarantee or warranty is given that the information is fit for any particular purpose. The user thereof uses the information at its sole risk and liability.

S.D. Galbraith (Eds.): Cryptography and Coding 2007, LNCS 4887, pp. 384–402, 2007.

certificate-based signature, all information necessary for verification can be included along with the basic signature to create a self-contained package. Nevertheless, many real world applications of identity-based signatures have been proposed in the literature, such as lightweight email authentication [1], and biometric authentication [11]. Attractive variants of identity-based signatures, for example identity-based multi-signatures (IBMS) [6,18], identity-based aggregate signatures (IBAS) [18] and identity-based threshold signatures (IBTS) [2] have also been proposed. These applications and primitives offer other advantages over the use of certificate-based signatures, such as significant savings in communication bandwidth. Moreover, IBS schemes can be deployed in identity-based infrastructures so as to provide a complement to IBE schemes whilst making use of the same infrastructural components and code base.

In hierarchical identity-based cryptography (HIBC) [7,10,19,22], multiple levels of private key generators (PKGs) and users form a tree-like structure which mimics the existing hierarchical PKI model. A user at any level of the tree can encrypt or sign a message targeting any intended recipient at any level, using only a set of shared cryptographic system parameters published by the root PKG. A hierarchical identity-based signature (HIBS, or HIDS in [19]) scheme is then the analogue of an IBS in the hierarchical setting. One interesting application of the hierarchical approach is that it can be used for delegation in a natural way: an act of delegation can be carried out by the issuance of a private key by a node in the tree to its child. A delegatee's delegated credential can be checked by performing only one signature verification, regardless of the length of the delegation chain [14].

In this paper, we introduce, motivate and develop a new cryptographic primitive called *multi-key hierarchical identity-based signatures* (multi-key HIBS). The essence of this new primitive is as follows. We operate in the setting of HIBC, but we assume each user owns multiple identifiers and thus possesses a set of corresponding private or signing keys. These identifiers may be located at arbitrary positions in the hierarchy. When a user generates a signature on a message, he uses a subset of his private keys for signing. Informally, then, a multi-key HIBS scheme is used to produce a single signature on a selected message using a set of signing keys. We discuss two potential applications for multi-key HIBS next, and then explain the relationship between our new primitive and related cryptographic concepts.

Motivating Examples. Our first example is related to access control in open distributed environments, such as grid computing systems [15].

There has been a recent trend in research on access control in which user authentication and access control are achieved in a unified way using cryptographic techniques, for example, policy-based encryption [3,33], attribute-based encryption [20,28,29] and role signatures [13]. Particularly, the use of roles as identifiers (in the context of HIBC), or role identifiers and their associated signatures for access control in [13] has inspired the work presented in this paper.

Role-based access control (RBAC) [31] is well-known for being more scalable than access control based on user identities. Users are granted membership into roles based on their competencies and responsibilities in their organisation. The operations that a user is permitted to perform are based on the user's role. Role signatures, as proposed in [13], are designed to address the problems of inter-domain principal mapping and authentication of user credentials simultaneously.

It is observed in [13] that a HIBS scheme can be used naturally for RBAC within a distributed computing environment, such as a grid system, which has a hierarchical structure. A verification key can be defined by a role identifier defined within a hierarchical namespace. User authentication and access control is unified, and credential verification is rendered trivial. An authorisation service is required to only verify a single signature, produced from the signing key associated to the role identifier, to both confirm that the user is an authenticated member of an organisation and occupies a particular role within that organisation.

Suppose, for example, that Alice has been given the roles $r_1 =$ lecturer, $r_2 =$ professor and $r_3 =$ IEEE member. Let Alice be an employee of university X, which in turn is part of the Open University, OU. Then, a role identifier, defined in a hierarchical namespace, can be of the form "OU, X, r_i", where $1 \leq i \leq 3$. In order to abide by the principle of least privilege, Alice should be given the options of selecting which role(s) she wishes to use when accessing some resources.

Now assume that Alice wants to access some restricted digital documents stored in the Open Library using roles r_2 and r_3. In principle, Alice must then sign an access request using private keys S_2 and S_3 which correspond to role identifiers "OU, X, r_2" and "OU, X, r_3", respectively. Our multi-key HIBS primitive allows the access request to be signed using the pair of private keys S_2 and S_3, while verification of the signed request uses the pair of associated role identifiers.

Our second example is related to mobile ad hoc network (MANET) applications. Here, conserving computation and bandwidth are at a premium, so the use of identity-based techniques is attractive [24]. However, it is commonly assumed in the MANET setting that there are no trusted entities that can play the role of PKGs (or trusted authorities). Moreover, nodes may be compromised or unavailable. Therefore, it is desirable to distribute the functions of PKGs across multiple network nodes. In the most general setting, the distributed identity-based key infrastructure that results may even be hierarchical in nature. Hence a user in the network may receive multiple private keys from different nodes, yet wish to efficiently demonstrate possession of some or all of these keys when creating a signature. For example, this may be required as part of an identity-based key exchange protocol used to create a session key shared between two network nodes. In this situation, a multi-key HIBS scheme is just what is needed.

Related Concepts. Our concept of multi-key HIBS is closely related to the concepts of identity-based multi-signatures (IBMS) [6,18] and identity-based aggregate signatures (IBAS) [18].

In multi-signatures, a set of users all sign the same message and collectively produce a single signature. An IBMS scheme may be either non-interactive or interactive; this property is related to whether or not individual signatures can be combined to produce a multi-signature by an outside agency. An example of the former type of IBMS scheme can be found in [18], and of the latter in [6]. In an aggregate signature scheme [9], each user can sign a different message; subsequently a set of signatures can be turned into a single signature via an aggregation process which may be executed by any entity. The first efficient constructions of IBAS were obtained in [18], but under the restriction that signers coordinate some state before signing. A non-interactive IBMS scheme can be obtained from any IBAS scheme.

It is easy to see that the basic functionality needed in our motivating examples can be obtained by just using an IBMS scheme (of either type), and hence also by using an IBAS scheme. However, this assumes that hierarchical IBMS or IBAS schemes with the required degree of flexibility can be obtained; currently available schemes operate only in the basic IBC setting having a single level in the hierarchy. Moreover, the fact that all signing keys are possessed by a single party in the multi-key HIBS setting opens up the possibility of making significant efficiency gains (particularly in the signing algorithm) by using a scheme designed from scratch. Indeed, we show here that such gains can be realised in practice. This efficiency gain highlights one of the main differences between our new concept of multi-key HIBS and the existing notions of IBMS/IBAS.

A second main difference arises when we consider the strength of security model that we introduce here for multi-key HIBS. Our model allows the adversary to obtain all but one of the signing keys involved in its output forgery and gives the adversary full access to a signing oracle. This is comparable to the multi-signer security model introduced for interactive IBMS in [6], which seems to be the strongest security model for IBMS introduced to date. However, our model is much stronger than the model for non-interactive IBMS given in [18]: the restrictions on adversarial behaviour in [18] are such that no identity may be involved in two different queries to the signing oracle (on different messages). This seems very limited in comparison to our model.[1]

Yet a third difference arises from the hierarchical and flexible nature of our multi-key HIBS definition and its concrete instantiations: in contrast to existing definitions and realisations of IBMS/IBAS, we work in a fully hierarchical setting, and allow the identifiers involved to be located at arbitrary positions in the hierarchy. A special case of our instantiations yields an efficient multi-key IBS (one-level) scheme.

Our new primitive is also loosely related to the concept of IBTS [2]. Generally speaking, a (t, n)-IBTS scheme allows a set of t (the threshold) out of n parties to first compute individual signature shares; these shares are then combined into one single signature. However, IBTS schemes are interactive, and generally much less efficient than our multi-key HIBS schemes. Another related idea is Boneh and Franklin's technique [8] of distributing an identity-based master secret over multiple PKGs using Shamir secret sharing. Boneh and Franklin observed that

---

[1] Note that the paper [18] does not explicitly include a security model for IBMS, though it does contain a concrete IBMS scheme and claim of security for that scheme. However, one can infer the IBMS model intended by the authors of [18] by specialising their state-oriented IBAS model to the stateless IBMS setting.

the master secret in their IBE scheme can be distributed in a t-out-of-n fashion by giving each of the n PKGs one share of the master secret. A user can then construct his private key by obtaining shares of his private key from each of the t chosen PKGs and then combining these using the appropriate Lagrange coefficients. Our new primitive can be seen as being related to a special case of this idea in which $t = n$. However, we deal with signatures and not encryption, we work in the hierarchical setting and not the basic identity-based setting, and our multi-key HIBS schemes allow signatures to be constructed in a more flexible way from private keys corresponding to nodes at arbitrary positions in the hierarchy.

Our Results. We provide a formal definition of multi-key HIBS and an appropriate security model for this primitive. We then present a concrete and provably secure multi-key HIBS scheme. Our scheme is provably secure in the Random Oracle Model (ROM), assuming the hardness of the computational Diffie-Hellman problem in groups equipped with a pairing. We consider the strongest possible security setting, by allowing the adversary arbitrary access to signing and key extraction oracles (subject to obvious limitations needed to prevent the adversary from trivially winning the security games).

We then show how the complexity of the verification algorithm of our scheme can be reduced in special cases that depend on the relative positions of nodes associated with the signing keys used to generate a signature. We also demonstrate that, in situations where multi-key HIBS are applicable, our concrete scheme has a much more efficient signing procedure than the currently best available IBMS schemes [6,18]. Since these IBMS schemes are only one-level (i.e. non-hierarchical), we make the comparison in the one-level case, even though our concrete scheme is far more flexible.

In the next section, we provide definitions for our new multi-key HIBS primitive. In Section 3, we propose a concrete multi-key HIBS scheme. Its security analysis is given in Section 4. Section 5 explains how to optimise our multi-key HIBS scheme in specific cases. Then, in Section 6, we compare the performance of our concrete multi-key HIBS scheme with the best IBMS schemes available in the literature. In Section 7, we present our conclusions and some open problems suggested by our work.

2   Definitions

2.1   Pairings and Associated Problems

Our scheme makes use of pairings. Let $\mathbb{G}$ and $\mathbb{G}_T$ be two cyclic groups where $|\mathbb{G}| = |\mathbb{G}_T| = q$, a large prime. Then an admissible pairing $e : \mathbb{G} \times \mathbb{G} \to \mathbb{G}_T$ has the following properties:

  – *Bilinear*: Given $P, Q, R \in \mathbb{G}$, we have

$$e(P, Q + R) = e(P, Q) \cdot e(P, R) \text{ and } e(P + Q, R) = e(P, R) \cdot e(Q, R).$$

Hence, for any $a, b \in \mathbb{Z}_q^*$,

$$e(aP, bQ) = e(abP, Q) = e(P, abQ) = e(aP, Q)^b = e(P, Q)^{ab}.$$

- *Non-degenerate*: There exists $P \in \mathbb{G}$ such that $e(P, P) \neq 1$.
- *Computable*: If $P, Q \in \mathbb{G}$, then $e(P, Q)$ can be efficiently computed.

Further details on pairings and their implementations using elliptic curves and the Weil, Tate or related pairings can be found in [16]. Our scheme is easily adapted to cope with pairings $e : \mathbb{G}_1 \times \mathbb{G}_2 \to \mathbb{G}_T$, but we focus here on the simpler case for ease of presentation.

Let λ denote the security parameter. We say that a randomised algorithm $\mathcal{G}$ is a Bilinear Diffie-Hellman (BDH) parameter generator if: (i) $\mathcal{G}$ takes $\lambda > 0$ as input; (ii) $\mathcal{G}$ runs in time polynomial in λ; and (iii) $\mathcal{G}$ outputs the description of two groups $\mathbb{G}$ and $\mathbb{G}_T$ of the same prime order q and the description of an admissible pairing $e : \mathbb{G} \times \mathbb{G} \to \mathbb{G}_T$. The security of the schemes presented in this paper is based on the assumed hardness of the computational Diffie-Hellman (CDH) problem in groups $\mathbb{G}$ produced by such a generator. This problem is defined as follows:

Definition 1. *(CDH in a group equipped with a pairing) Given $\mathbb{G}$, $\mathbb{G}_T$ of the same prime order q, an admissible pairing $e : \mathbb{G} \times \mathbb{G} \to \mathbb{G}_T$, and $\langle P, aP, bP \rangle \in \mathbb{G}$ for some random $P \in \mathbb{G}$ and randomly chosen $a, b \in \mathbb{Z}_q^*$, the CDH problem in $\mathbb{G}$ is to compute $abP \in \mathbb{G}$.*

An algorithm $\mathcal{A}$ is said to have advantage ϵ in solving the CDH problem in $\mathbb{G}$ if

$$\Pr\left[\mathcal{A}(P, aP, bP) = abP \geq \epsilon\right]$$

where the probability is over the random choice of P in $\mathbb{G}$, the random scalars a and b in $\mathbb{Z}_q^*$, and the random bits used by $\mathcal{A}$. (Here we suppress the additional inputs to $\mathcal{A}$.)

2.2 Multi-key HIBS Scheme

Hierarchical identity-based cryptography involves nodes arranged in a tree structure, with each node having an identifier. The identifier of an entity is the concatenation of the node identifiers in the path from the root to the node associated with that entity. Assuming the root Private Key Generator (PKG) is located at level 0, then the identifier of an entity at level t is the concatenation of node identifiers $id_1, \ldots, id_t$ in which each $id_i \in \{0, 1\}^*$. We denote the concatenation of node identifiers $id_1, \ldots, id_t$ by identifier ID_t. The entity with identifier ID_t has an ancestor at level i with identifier $\text{ID}_i = id_1, \ldots, id_i$ for $1 \leq i < t$; this entity's parent is the node with identifier ID_{t-1} and its children are all the nodes with identifiers of the form $\text{ID}_{t+1} = id_1, \ldots, id_t, id_{t+1}$. We use (P_t, S_t) to represent the public/private key pair of the entity with identifier ID_t.

A multi-key hierarchical identity-based signature (multi-key HIBS) scheme can be regarded as an extended HIBS scheme. It produces signatures in the hierarchical identity-based setting using sets of signing keys. A multi-key HIBS scheme is specified by the following algorithms:

ROOT SETUP: This algorithm is performed by the root PKG. It generates the system parameters and a master secret on input a security parameter λ. The system parameters, which include a description of the message space $\mathcal{M}$ and the signature space $\mathcal{S}$, will be made publicly available to all entities (PKGs or users). However, the master secret is known only to the root PKG.

LOWER-LEVEL SETUP: All entities at lower levels must obtain the system parameters generated by the root PKG. This algorithm allows a lower-level PKG to establish a secret value to be used to issue private keys to its children.

EXTRACT: This algorithm is performed by a PKG (root or lower-level PKG) with identifier ID_t to compute a private key S_{t+1} for any of its children using the system parameters and its private key (and any other secret information).

SIGN: Given a set $\mathsf{SK} = \{S_{t_j}^j : 1 \leq j \leq n\}$ of signing (private) keys, a message $M \in \mathcal{M}$, and the system parameters, this algorithm outputs a signature $\sigma \in \mathcal{S}$. Here t_j denotes the level of the j-th signing key in the set SK.

VERIFY: Given a signature $\sigma \in \mathcal{S}$, a set $\mathsf{ID} = \{\mathsf{ID}_{t_j}^j : 1 \leq j \leq n\}$ of identifiers, a message $M \in \mathcal{M}$, and the system parameters, this algorithm outputs valid or invalid.

We have the obvious consistency requirement: if σ is output by SIGN on input a set SK of private keys and message M, then VERIFY outputs valid when given input σ, the set ID of identifiers corresponding to SK, and M.

We remark that the first three algorithms specified above are identical to those of a HIBS scheme. Moreover, a multi-key HIBS scheme, when used with a single signing key, is essentially just a normal HIBS scheme. We also remark that while we have chosen to work with sets of identifiers and private keys, the definitions (as well as the security model and concrete construction to follow) can easily be adapted to deal with ordered lists of identifiers and keys.

2.3   Security Model

The security model for a multi-key HIBS scheme is based on the following game between a challenger and an adversary that extends the normal HIBS security game [19]:

1. The challenger runs the ROOT SETUP algorithm of the multi-key HIBS scheme. The resulting system parameters are given to the adversary. The master secret, however, is kept secret by the challenger.
2. The adversary adaptively issues queries to the challenger. Each query can be one of the following:

- Extract: The adversary can ask for the private key associated with any identifier $ID_{t_j}^j = id_1^j, \ldots, id_{t_j}^j$. The challenger responds by running the EXTRACT algorithm to generate a private key $S_{t_j}^j$, which is then returned to the adversary.
- Sign: The adversary can ask for the signature associated with a set ID of identifiers on a message M of its choice. The challenger responds by running the SIGN algorithm using as input a set of signing keys SK corresponding to the set of identifiers ID, message M, and the system parameters. The resulting signature σ is returned to the adversary.

3. The adversary outputs a string σ^*, a set of target identifiers ID*, and a message M^*. The adversary wins the game if the following are all true:
   - VERIFY(σ^*, ID*, M^*) = valid;
   - There exists an identifier $ID' \in ID^*$ for which the adversary has not made an Extract query on ID' or any of its ancestors;
   - The adversary has not made a Sign query on input ID*, M^*.

The advantage of an adversary $\mathcal{A}$ in the above game is defined to be

$$\text{Adv}_{\mathcal{A}} = \Pr[\mathcal{A} \ wins]$$

where the probability is taken over all coin tosses made by the challenger and the adversary.

Note that the usual security model for HIBS is recovered in the case where all queries involve a single identifier.

3   Construction

We now present a concrete multi-key HIBS scheme, which is adapted from the Gentry-Silverberg HIBS scheme [19]. We will show in Section 5 how the cost of the verification algorithm of this scheme can be reduced in specific situations.

ROOT SETUP: The root PKG:
1. runs $\mathcal{G}$ on input λ to generate $\mathbb{G}$ and $\mathbb{G}_T$ of prime order q and an admissible pairing $e : \mathbb{G} \times \mathbb{G} \to \mathbb{G}_T$;
2. chooses a generator $P_0 \in \mathbb{G}$;
3. picks a random value $s_0 \in \mathbb{Z}_q^*$ and sets $Q_0 = s_0 P_0$;
4. selects cryptographic hash functions $H_1 : \{0,1\}^* \to \mathbb{G}$ and $H_2 : \{0,1\}^* \to \mathbb{G}$.

The root PKG's master secret is s_0 and the system parameters are $\langle \mathbb{G}, \mathbb{G}_T, e, q, P_0, Q_0, H_1, H_2 \rangle$. The message space is $\mathcal{M} = \{0,1\}^*$ and the signature space is $\mathcal{S} = \bigcup_{t \geq 0} \mathbb{G}^{t+1}$.

LOWER-LEVEL SETUP: A lower-level entity (lower-level PKG or user) at level $t \geq 1$ picks a random secret $s_t \in \mathbb{Z}_q^*$.

EXTRACT: For an entity with identifier $ID_t = id_1, \ldots, id_t$, the entity's parent:

1. computes $P_t = H_1(\mathrm{ID}_t) \in \mathbb{G}$;
2. sets $S_t = \sum_{i=1}^{t} s_{i-1} P_i = S_{t-1} + s_{t-1} P_t$;
3. defines $Q_i = s_i P_0$ for $1 \leq i \leq t - 1$.

The private key $\langle S_t, Q_1, \ldots, Q_{t-1} \rangle$ is given to the entity by its parent. Note that up to this point, our scheme is identical to the Gentry-Silverberg HIBS scheme.

SIGN: Given any $n \geq 1$ and a set $\mathsf{SK} = \{\langle S_{t_j}^j, Q_1^j, \ldots, Q_{t_j-1}^j \rangle : 1 \leq j \leq n\}$ of n private keys associated with a set $\mathsf{ID} = \{\mathrm{ID}_{t_j}^j : 1 \leq j \leq n\}$ of identifiers, and a message M, the signer:

1. chooses a secret value $s_\varphi \in \mathbb{Z}_q^*$;
2. computes $P_M = H_2(\mathrm{ID}_{t_1}^1, \ldots, \mathrm{ID}_{t_n}^n, M)$ (where we assume the identifiers are first placed in lexicographic order if necessary);
3. calculates

$$\varphi = \sum_{j=1}^{n} S_{t_j}^j + s_\varphi P_M \quad \text{and} \quad Q_\varphi = s_\varphi P_0.$$

The algorithm outputs the signature $\sigma = \langle \varphi, \mathsf{Q}, Q_\varphi \rangle$, where $\mathsf{Q} = \{Q_i^j : 1 \leq i \leq t_j - 1, 1 \leq j \leq n\}$.

VERIFY: Given $\sigma = \langle \varphi, \mathsf{Q}, Q_\varphi \rangle$, a set of identifiers $\mathsf{ID} = \{\mathrm{ID}_{t_1}^1, \ldots, \mathrm{ID}_{t_n}^n\}$ and a message M, the verifier:

1. computes $P_i^j = H_1(\mathrm{ID}_i^j)$ for $1 \leq i \leq t_j$ and $1 \leq j \leq n$;
2. computes $P_M = H_2(\mathrm{ID}_{t_1}^1, \ldots, \mathrm{ID}_{t_n}^n, M)$ (first arranging the identifiers lexicographically if they are not already in this order);
3. checks if $e(P_0, \varphi)$ is equal to

$$\left(\prod_{j=1}^{n} \prod_{i=1}^{t_j} e(Q_{i-1}^j, P_i^j) \right) \cdot e(Q_\varphi, P_M),$$

outputting `valid` if this equation holds, and `invalid` otherwise.

It is not hard to verify that this scheme is consistent. We remark that two different hash functions H_1 and H_2 are used in the construction to make the security proof easier to follow; a similar scheme using only one hash function is easily constructed. However, in such a scheme we need to take care to distinguish the different types of strings input to the hash. Even with two hash functions, we need to assume that any input to H_2 can be uniquely parsed as a list of identifiers concatenated with a (possibly empty) message. This prevents trivial re-encoding attacks against the scheme. This requirement can be met in a number of ways. For example, we can assume that there is a special separating symbol which marks the end of the identifier portion and which does not appear in any identifier, or we can encode each identity bit id as $0id$ and each message bit m as $1m$. We note that this parsing requirement is not explicitly specified for the HIBS scheme of [19], meaning that (trivially preventable) re-encoding attacks *are* possible against that scheme.

4   Security Analysis

In this section, we establish the security of the concrete multi-key HIBS scheme of Section 3. For simplicity of presentation, we focus here on the case in which all identifiers lie at level 1 in the hierarchy (in which case they have the root PKG as their parent). Thus our proof is actually for the multi-key IBS scheme that arises as a special case of our more general hierarchical scheme. We slightly extend the techniques of [8] to directly relate this scheme's security to the hardness of the CDH problem in groups $\mathbb{G}$ equipped with a pairing. After the proof, we sketch how our approach might be extended to handle the more general case by using additional techniques from [19].

Theorem 2. *Suppose that $\mathcal{A}$ is a forger against our multi-key IBS scheme that has success probability ϵ. Let hash functions H_1 and H_2 be modeled as random oracles. Then there is an algorithm $\mathcal{B}$ which solves the CDH problem in groups $\mathbb{G}$ equipped with a pairing, with advantage at least*

$$\epsilon/(\mathbf{e} \cdot q_{H_1} \cdot q_{H_2})$$

and which has running time $\mathcal{O}(time(\mathcal{A}))$. Here, q_{H_1} is the maximum number of H_1 queries made by $\mathcal{A}$ during its attack, q_{H_2} is the maximum number of H_2 queries made by $\mathcal{A}$, and $\mathbf{e}$ denotes the base of natural logarithms.

Proof: Algorithm $\mathcal{B}$ is given as input an admissible pairing $e : \mathbb{G} \times \mathbb{G} \to \mathbb{G}_T$, where $|\mathbb{G}| = |\mathbb{G}_T| = q$, and an instance $\langle P, aP, bP \rangle$ of the CDH problem in $\mathbb{G}$, generated by $\mathcal{G}$. It will interact with algorithm $\mathcal{A}$, as follows, in an attempt to compute abP.

Setup: Algorithm $\mathcal{B}$ sets the system parameters of the root PKG to be $\langle \mathbb{G}, \mathbb{G}_T, e, P_0 = P, Q_0 = aP, H_1, H_2 \rangle$, so that the master secret is the unknown value a. The system parameters are then forwarded to $\mathcal{A}$. Here, H_1 and H_2 are random oracles controlled by $\mathcal{B}$. Additionally, $\mathcal{B}$ randomly selects $c \in \{1, \ldots, q_{H_1}\}$, where q_{H_1} is the maximum number of queries to H_1 made by $\mathcal{A}$.

H_1 queries: $\mathcal{A}$ can query H_1 on any input ID $\in \{0,1\}^*$. In responding to these queries, $\mathcal{B}$ maintains a list Λ_{H_1} containing tuples of the form $\langle \text{ID}^i, r_i, R_i \rangle$. The list Λ_{H_1} is initially empty. If ID already appears in the list Λ_{H_1}, in position i say, then $\mathcal{B}$ responds with R_i. Otherwise, $\mathcal{B}$ responds as follows:

1. If this is the c-th distinct query to H_1, then set $\text{ID}' = \text{ID}$;
2. Select r at random from $\mathbb{Z}_q^*$;
3. If $\text{ID} = \text{ID}'$, then add $\langle \text{ID}, r, R = bP + rP \rangle$ to the list Λ_{H_1}.
4. Otherwise, add $\langle \text{ID}, r, R = rP \rangle$ to the list Λ_{H_1}.
5. Output R as the response to the H_1 query.

Notice that the output R is always chosen uniformly in $\mathbb{G}$, as required.

H_2 queries: $\mathcal{A}$ can query H_2 on any list of identifiers $\mathrm{ID}^1, \ldots, \mathrm{ID}^n$ and any message M. To respond to these queries, $\mathcal{B}$ maintains a list Λ_{H_2} of tuples $\langle (\mathrm{ID}^1_i, \ldots, \mathrm{ID}^n_i), M_i, z_i, f_i, d_i, T_i \rangle$. The list is initially empty, and $\mathcal{B}$ uses the list to reply consistently with values T_i to $\mathcal{A}$'s queries. $\mathcal{B}$ first queries H_1 on each identifier ID^j and sets $\mathrm{ID} = \{\mathrm{ID}^1, \ldots, \mathrm{ID}^n\}$. Then $\mathcal{B}$ responds as follows:

1. If $\mathrm{ID}' \in \mathrm{ID}$, then generate a coin $f \in \{0, 1\}$ where $\Pr[f = 0] = \delta$ for some δ to be determined later;
   (a) if $f = 0$, then set $d = 0$;
   (b) otherwise select d at random from $\mathbb{Z}^*_q$;
2. If $\mathrm{ID}' \notin \mathrm{ID}$, then set $f = 0$ and $d = 0$;
3. Set $T = zP - d(bP)$ where z is selected at random from $\mathbb{Z}^*_q$;
4. Record $\langle (\mathrm{ID}^1, \ldots, \mathrm{ID}^n), M, z, f, d, T \rangle$ in Λ_{H_2};
5. Return T to $\mathcal{A}$ as the output of H_2.

Again, the output T is always chosen uniformly in $\mathbb{G}$.

Extract queries: When algorithm $\mathcal{A}$ requests a private key associated with identifier ID, algorithm $\mathcal{B}$ responds as follows:

1. Recover the associated tuple $\langle \mathrm{ID}, r, R \rangle$ from Λ_{H_1} (first making a query to H_1 if necessary);
2. If $\mathrm{ID} = \mathrm{ID}'$, then $\mathcal{B}$ aborts;
3. Otherwise $\mathcal{B}$ returns $r(aP)$ to $\mathcal{A}$.

It is easy to see that $\mathcal{A}$ receives a valid private key for identifier ID, provided $\mathcal{B}$ does not abort.

Sign queries: Algorithm $\mathcal{A}$ can request a signature for any set of identifiers $\mathrm{ID} = \{\mathrm{ID}^1, \ldots, \mathrm{ID}^n\}$ and any message M. We can assume the list $\mathrm{ID}^1, \ldots, \mathrm{ID}^n$ is ordered lexicographically.

In responding to $\mathcal{A}$'s request, $\mathcal{B}$ performs the following steps:

1. Recover $\langle (\mathrm{ID}^1, \ldots, \mathrm{ID}^n), M, z, f, d, T \rangle$ from Λ_{H_2} (if $\mathcal{A}$ has not already queried H_2 on the appropriate input, then $\mathcal{B}$ first makes the relevant H_2 query himself);
2. If $\mathrm{ID}' \notin \mathrm{ID}$, then:
   (a) Select s_φ at random from $\mathbb{Z}^*_q$;
   (b) Run Extract on input ID^j to recover a private key S^j, for each $1 \leq j \leq n$;
   (c) Compute $\varphi = \sum_{j=1}^n S^j + s_\varphi T$ and $Q_\varphi = s_\varphi P$;
   (d) Return $\langle \varphi, Q_\varphi \rangle$ to $\mathcal{A}$.
3. If $\mathrm{ID}' \in \mathrm{ID}$ and $f = 0$, then abort;
4. Otherwise (when $\mathrm{ID}' \in \mathrm{ID}$ and $f = 1$, so that $T = zP - d(bP)$):
   (a) Let $\mathrm{ID}^{j'} = \mathrm{ID}'$, and obtain entry $\langle \mathrm{ID}', r', R' \rangle$ from Λ_{H_1};
   (b) Set $v = d^{-1} \bmod q$ and, for each $j \neq j'$, run Extract on input ID^j to recover a private key S^j.
   (c) Set

$$\varphi = \sum_{j=1, j \neq j'}^n S^j + r'(aP) + vz(aP), \quad Q_\varphi = vaP;$$

   (d) Return $\langle \varphi, Q_\varphi \rangle$ to $\mathcal{A}$.

We remark that in step 4 above, $\mathcal{B}$ is not able to compute the private key corresponding to identifier ID' (which should equal $abP + ar'P$), but is still able to compute a valid signature. This can be shown from the definitions of T and R', which were made with the aim of causing certain cancelations to occur in the signature computation: if $\mathcal{B}$'s output is to be a valid signature, we should have $\varphi = \sum_{j=1}^{n} S^j + s_{\varphi} H_2(\text{ID}^1, \ldots, \text{ID}^n, M)$ with $s_{\varphi} = va$. Indeed,

$$
\begin{aligned}
&\textstyle\sum_{j=1}^{n} S^j + s_{\varphi} H_2(\text{ID}^1, \ldots, \text{ID}^n, M) \\
&= \textstyle\sum_{j=1, j\neq j'}^{n} S^j + (abP + ar'P) + va(zP - d(bP)) \\
&= \textstyle\sum_{j=1, j\neq j'}^{n} S^j + r'(aP) + abP - vdabP + vz(aP) \\
&= \varphi
\end{aligned}
$$

Forgery: Eventually, algorithm $\mathcal{A}$ outputs a set ID^* of n target identifiers, a message M^* and a pair $\sigma^* = \langle \varphi^*, Q_{\varphi}^* \rangle$. If σ^* is to be a valid signature then it should satisfy the verification equation:

$$
e(P, \varphi^*) = e(Q_0, \sum_{j=1}^{n} H_1(\text{ID}^{*j})) \cdot e(Q_{\varphi}^*, P_{M*})
$$

where the identifiers in ID^* are, in lexicographic order, $\text{ID}^{*1}, \ldots, \text{ID}^{*n}$, and where

$$
P_{M*} = H_2(\text{ID}^{*1}, \ldots, \text{ID}^{*n}, M^*).
$$

This ends our description of $\mathcal{B}$'s interaction with $\mathcal{A}$. If $\mathcal{B}$ does not abort, then $\mathcal{A}$'s view is identical to that in a real attack. Hence the probability that $\mathcal{A}$ outputs a valid signature is at least ϵ, provided $\mathcal{B}$ does not abort.

Now let $\mathcal{E}_1$ be the event that $\mathcal{A}$ has queried H_1 on every identifier ID^{*j}. It is obvious from inspecting the above verification equation for $\mathcal{A}$'s output that if $\mathcal{E}_1$ does not occur, then $\mathcal{A}$'s success probability is negligible. Because ϵ is non-negligible (assuming $\mathcal{B}$ has not aborted), it follows that $\mathcal{E}_1$ does occur. Now the probability that both $\text{ID}' \in \text{ID}^*$ and that ID' is not the subject of an Extract query is at least $1/q_{H_1}$. This follows because we selected $c \in \{1, \ldots, q_{H_1}\}$ uniformly at random and set ID' to be the identifier in the c-th query to H_1, and because at least one identity appearing in ID^* is not the subject of an Extract query.

Let $\mathcal{E}_2$ be the event that $\mathcal{A}$ has made an H_2 query on input $\text{ID}^{*1}, \ldots, \text{ID}^{*n}$ and M^*. If $\mathcal{E}_2$ does not occur, then $\mathcal{A}$'s success probability is again negligible. For otherwise, P_{M*} is uniformly distributed over $\mathbb{G}$ and the probability that the above verification equation holds is $1/q$, which is negligible in λ, the security parameter. Because ϵ is non-negligible, it follows that $\mathcal{E}_2$ does occur, so there is an entry $\langle (\text{ID}^{*1}, \ldots, \text{ID}^{*n}), M^*, z^*, f^*, d^*, T^* \rangle$ in Λ_{H_2}.

We now explain how $\mathcal{B}$ analyzes $\mathcal{A}$'s output. If $f^* = 1$ or if $\text{ID}' \notin \text{ID}^*$, then $\mathcal{B}$ aborts. Otherwise, we can assume that $f^* = 0$ and $\text{ID}' = \text{ID}_t^{*j'}$ for some $\text{ID}_t^{*j'} \in \text{ID}^*$. Then, from the simulation of H_2, we have $d^* = 0$ and $P_{M*} = H_2(\text{ID}^{*1}, \ldots, \text{ID}^{*n}, M^*) = z^*P$. Then, by rearranging the verification equation for $\mathcal{A}$'s output, we obtain:

$$
\begin{aligned}
e(P, \varphi^*) &= e(Q_0, \textstyle\sum_{j=1}^{n} H_1(\mathrm{ID}^{*j})) \cdot e(Q_\varphi^*, P_{M*}) \\
&= e(aP, \textstyle\sum_{j=1}^{n} H_1(\mathrm{ID}^{*j})) \cdot e(Q_\varphi^*, z^*P) \\
&= e(P, \textstyle\sum_{j=1}^{n} aH_1(\mathrm{ID}^{*j})) \cdot e(P, z^*Q_\varphi^*) \\
&= e(P, \textstyle\sum_{j=1}^{n} aH_1(\mathrm{ID}^{*j}) + z^*Q_\varphi^*).
\end{aligned}
$$

From the non-degeneracy of the pairing, we may now deduce

$$
\varphi^* = \sum_{j=1}^{n} aH_1(\mathrm{ID}^{*j}) + z^*Q_\varphi^*.
$$

But, for $j \neq j'$, we have $aH_1(\mathrm{ID}^{*j}) = r_j(aP)$ for some value r_j available in list Λ_{H_1}, while $aH_1(\mathrm{ID}^{*j'}) = r'(aP) + abP$. It is now easy to see that $\mathcal{B}$ can recover abP from $\mathcal{A}$'s output and its knowledge of the r_j values and z^*, provided $\mathcal{A}$'s output is a valid signature. Hence, if $\mathcal{B}$ does not abort, then $\mathcal{B}$ can extract the value abP with probability at least ϵ.

So it remains to calculate the probability that $\mathcal{B}$ does not abort. $\mathcal{B}$ is forced to abort if $\mathcal{A}$ makes an Extract query on ID', if $\mathcal{A}$ makes a Sign query on a set ID containing ID' for which $f = 0$, or if the set ID^* does not contain ID', or if $f^* = 1$. Hence $\mathcal{B}$ does *not* abort if all of the following conditions are met:

- $\mathcal{A}$ does not make an Extract query on ID';
- $f = 1$ in every Sign query involving ID' made by $\mathcal{A}$;
- $\mathrm{ID}' \in \mathrm{ID}^*$ and $f^* = 0$.

We have already seen that the probability that $\mathcal{A}$ does not make an Extract query on ID' and $\mathrm{ID}' \in \mathrm{ID}^*$ is $1/q_{H_1}$. Setting $\delta = \Pr[f = 0] = 1/q_{H_2}$, where q_{H_2} denotes a bound on the number of queries made to H_2, it is then easy to see that $\mathcal{B}$'s success probability is at least

$$
\epsilon \cdot \frac{1}{q_{H_1}} \left(1 - \frac{1}{q_{H_2}}\right)^{q_S} \cdot \frac{1}{q_{H_2}}
$$

where q_S is the number of Sign queries involving ID' made by $\mathcal{A}$. In turn, since $q_S \leq q_{H_2}$ and, for large q_{H_2}, we have $(1 - 1/q_{H_2})^{q_{H_2}} \approx 1/\mathbf{e}$ where $\mathbf{e}$ is the base of natural logarithms, we obtain (neglecting negligible terms) that $\mathcal{B}$'s success probability is at least

$$
\epsilon/(\mathbf{e} \cdot q_{H_1} \cdot q_{H_2})
$$

as required. This completes the proof. □

We now sketch how this proof might be extended to cope with the more complicated situation of our multi-key HIBS (rather than IBS). The main idea is to borrow the simulation techniques used in proving the security of the Gentry-Silverberg HIBE scheme, [19, Lemma 2]. There, it is shown how to simulate H_1

and Extract queries for a HIBE scheme that has the same key generation procedures as our multi-key HIBS scheme, in such a way as to embed a value abP into the private key held by some proportion of the entities, whilst allowing Extract queries to be answered for all other entities (in the more challenging hierarchical setting). When combined with our approach to handling Sign queries, this simulation technique should yield a security reduction for our multi-key HIBS scheme. However, we stress that we have so far only obtained a security proof for some special cases in the hierarchical setting using the sketched techniques. Constructing a proof for the general case remains an open problem, and, unfortunately, as with the proofs in [19], we expect that any reduction obtained using this approach will not be very tight.

5   Reducing the Cost of Verification

In this section, we study how the complexity of the verification algorithm for our concrete multi-key HIBS scheme can be reduced in special cases.

The main situation we consider is where all the identifiers involved in signing are at the same level in the hierarchy and have a common parent at the level above. This includes as a special case the situation where all identifiers are at the first level in the hierarchy. In this case, the SIGN and VERIFY algorithms of multi-key HIBS scheme can be modified as follows.

SIGN: Given any $n \geq 1$, and a set SK of n signing keys corresponding to a set of identifiers $\{\mathrm{ID}_t^1, \ldots, \mathrm{ID}_t^n\}$ having a common parent, and a message M, the signer:
   1. chooses a secret value $s_\varphi \in \mathbb{Z}_q^*$;
   2. computes $P_M = H_2(\mathrm{ID}_t^1, \ldots, \mathrm{ID}_t^n, M)$ (where we assume the identifiers are first placed in lexicographic order if necessary);
   3. calculates

$$\varphi = \sum_{j=1}^{n} S_t^j + s_\varphi P_M \quad \text{and} \quad Q_\varphi = s_\varphi P_0.$$

The algorithm outputs the signature $\sigma = \langle \varphi, Q_1, \ldots, Q_{t-1}, Q_\varphi \rangle$. Here, we assume (because of the common parent) that each signing key in SK involves the same list $Q_1, \ldots, Q_{t-1}$ of Q-values.

VERIFY: Given $\sigma = \langle \varphi, Q_1, \ldots, Q_{t-1}, Q_\varphi \rangle$, a set of identifiers $\mathrm{ID} = \{\mathrm{ID}_t^1, \ldots, \mathrm{ID}_t^n\}$ having a common parent, and a message M, the verifier:
   1. computes $P_i = H_1(\mathrm{ID}_i)$ for $1 \leq i \leq t-1$ and $P_t^j = H_1(\mathrm{ID}_t^j)$ for $1 \leq j \leq n$;
   2. computes $P_M = H_2(\mathrm{ID}_t^1, \ldots, \mathrm{ID}_t^n, M)$ (first arranging the identifiers lexicographically if they are not already in this order);
   3. checks if $e(P_0, \varphi)$ is equal to

$$e\left(Q_{t-1}, \sum_{j=1}^{n} P_t^j\right) \cdot e(Q_\varphi, P_M) \cdot \left(\prod_{i=1}^{t-1} e(Q_{i-1}, P_i)\right)^n,$$

outputting valid if this equation holds, and invalid otherwise.

Note that the above verification algorithm requires only $t + 2$ pairing computations as compared to the $n \cdot t + 2$ that would be needed with the unoptimised verification algorithm of Section 3.

For $n = 1$, the above SIGN and VERIFY algorithms are essentially those of the Gentry-Silverberg HIBS scheme [19]. In the special case where we take $t = 1$ throughout in the above scheme, we obtain a multi-key IBS (i.e. one-level) scheme in which only 3 pairing computations are needed during verification.

In more general situations, we can obtain a more efficient verification algorithm whenever the nodes associated with the signing keys used to generate a signature have some common ancestors and common Q-values. For then certain pairing computations in the verification equation can be eliminated. Recall that this equation involves a term $\prod_{j=1}^{n} \prod_{i=1}^{t_j} e(Q_{i-1}^j, P_i^j)$. To illustrate the possibility of eliminating pairing computations, suppose for example that all n nodes have a single common ancestor at some level k (and therefore at all levels above level k as well). Then the n terms $e(Q_{i-1}^j, P_i^j)$, $1 \leq j \leq n$, are in fact equal for each value $i \leq k$, and so we can replace these n ostensibly different pairings at level i with a power of a single pairing, for each $i \leq k$. Similar remarks also apply when some (but not necessarily all) nodes share a common ancestor at some level in the hierarchy.

6   Efficiency Comparison

To gauge the efficiency gain offered by our multi-key HIBS primitive in comparison to IBMS schemes, we compare the multi-key IBS scheme that arises from our multi-key HIBS scheme in the one-level case (with an optimised verification equation) with the RSA-based Bellare-Neven IBMS scheme [6] and the pairing-based Gentry-Ramzan IBMS scheme [18]. The computational costs for these schemes are shown in Table 1.

From Table 1, it is evident that our multi-key IBS scheme is much more efficient than both the Bellare-Neven and the Gentry-Ramzan IBMS schemes in terms of signing cost at an equivalent security level. Suppose our scheme and the Gentry-Ramzan IBMS scheme are instantiated using pairings defined on an elliptic curve at the 80-bit security level, while the Bellare-Neven scheme is instantiated using a 1024-bit RSA modulus (also offering roughly 80 bits of security). Then the main signing cost in our scheme is 2 elliptic curve point multiplications, whereas the Gentry-Ramzan IBMS scheme needs $2n$. The signing cost of the Bellare-Neven scheme is much greater for reasonable values of n, and this is due to its interactive nature. This is as expected, since our scheme exploits the fact that signing keys can be "aggregated" before generating a signature.

On the other hand, the verification cost for the Gentry-Ramzan scheme and our scheme is the same. The verification cost for the Bellare-Neven scheme appears to be the least among the three schemes (since a modular exponentiation is expected to be faster than a pairing computation).

Our signatures have the same length as in the Gentry-Ramzan scheme and are generally shorter than signatures in the Bellare-Neven scheme, as they com-

Table 1. Computational costs for the Bellare-Neven IBMS scheme, the Gentry-Ramzan IBMS scheme and our multi-key IBS (one-level) scheme. Here, ADD denotes the number of elliptic curve point additions, eMUL the number of elliptic curve point multiplications, PAI the number of pairing computations, HASH the number of hash operations, mMUL the number of modular multiplications, and EXP the number of modular exponentiations.

	ADD	eMUL	PAI	HASH	mMUL	EXP
Bellare-Neven IBMS						
signing	-	-	-	$n(n+1)$	n^2+n-1	$2n$
verification	-	-	-	$n-1$	n	2
Gentry-Ramzan IBMS						
signing	$3n-2$	$2n$	0	n	-	-
verification	$n-1$	0	3	$n+1$	-	-
Multi-key IBS						
signing	n	2	0	1	-	-
verification	$n-1$	0	3	$n+1$	-	-

prise of only two group elements (at around 320 bits in total), instead of the 1184 bits quoted in [6]. Note that bandwidth savings can be more critical than computational efficiency gains in many contexts.

7   Conclusions and Open Problems

There exist practical applications which require users to demonstrate possession of more than a single private signing key. We developed a new primitive called multi-key hierarchical identity-based signatures (multi-key HIBS) which can achieve this more efficiently than with existing approaches. We showed that our multi-key HIBS primitive can make use of a security model that is at least as strong as, if not even stronger than, existing related security models for multi-signature schemes and aggregate signature schemes. We also showed how the new primitive could be efficiently instantiated in the Random Oracle Model.

Currently, our multi-key HIBS primitive is restricted to demonstrating the possession of all the private keys in a given set. This should be compared to threshold cryptographic schemes in which signing parties, for example, effectively demonstrate knowledge of a subset of size k of a set of private keys of size n. It will be interesting to generalise our multi-key HIBS concept to the threshold setting. This could involve multiple hierarchies with different roots, or potentially, a single hierarchy, but working with a threshold of private keys within that hierarchy.

Naturally, there exists a version of our multi-key HIBS primitive in the normal public-key setting. It seems easy to design a one-level non-identity-based multi-key scheme by adapting the BGLS aggregate signature scheme [9], for example. Nevertheless, it will be interesting to explore security models and concrete schemes that support hierarchical signing keys in that setting. It seems likely that it will be possible to adapt the Bellare-Neven IBMS scheme to produce a more efficient RSA-based multi-key IBS. However, finding an efficient, RSA-based, multi-key HIBS may be a more challenging task.

It may also be interesting to try to construct a multi-key HIBS scheme secure in the standard model, perhaps by adapting the work of Lu *et al.* [25].

References

1. Adida, B., Chau, D., Hohenberger, S., Rivest, R.L.: Lightweight signatures for email. In: De Prisco, R., Yung, M. (eds.) SCN 2006. LNCS, vol. 4116, Springer, Heidelberg (2006)
2. Baek, J., Zheng, Y.: Identity-based threshold signature scheme from the bilinear pairings. In: ITCC 2004. Proceedings of the International Conference on Information Technology: Coding and Computing, vol. 1, pp. 124–128. IEEE Computer Society Press, Los Alamitos (2004)
3. Bagga, W., Molva, R.: Policy-based cryptography and applications. In: Patrick, A.S., Yung, M. (eds.) FC 2005. LNCS, vol. 3570, pp. 72–87. Springer, Heidelberg (2005)
4. Barreto, P.S.L.M., Libert, B., McCullagh, N., Quisquater, J.: Efficient and provably-secure identity-based signatures and signcryption from Bilinear maps. In: Roy, B. (ed.) ASIACRYPT 2005. LNCS, vol. 3788, pp. 515–532. Springer, Heidelberg (2005)
5. Bellare, M., Namprempre, C., Neven, G.: Security proofs for identity-based identification and signature schemes. In: Cachin, C., Camenisch, J.L. (eds.) EUROCRYPT 2004. LNCS, vol. 3027, pp. 268–286. Springer, Heidelberg (2004)
6. Bellare, M., Neven, G.: Identity-based multi-signatures from RSA. In: Abe, M. (ed.) CT-RSA 2007. LNCS, vol. 4377, pp. 145–162. Springer, Heidelberg (2006)
7. Boneh, D., Boyen, X., Goh, E.: Hierarchical identity based encryption with constant size ciphertext. In: Cramer, R.J.F. (ed.) EUROCRYPT 2005. LNCS, vol. 3494, pp. 440–456. Springer, Heidelberg (2005)
8. Boneh, D., Franklin, M.: Identity-based encryption from the Weil pairing. In: Kilian, J. (ed.) CRYPTO 2001. LNCS, vol. 2139, pp. 213–229. Springer, Heidelberg (2001)
9. Boneh, D., Gentry, C., Lynn, B., Shacham, H.: Aggregate and verifiably encrypted signatures from bilinear maps. In: Biham, E. (ed.) EUROCRPYT 2003. LNCS, vol. 2656, pp. 416–432. Springer, Heidelberg (2003)
10. Boyen, X., Waters, B.: Anonymous hierarchical identity-based encryption (without random oracles). In: Dwork, C. (ed.) CRYPTO 2006. LNCS, vol. 4117, pp. 290–307. Springer, Heidelberg (2006)
11. Burnett, A., Byrne, F., Dowling, T., Duffy, A.: A biometric identity based signature scheme. International Journal of Network Security 5(3), 317–326 (2007)
12. Cha, J.C., Cheon, J.H.: An identity-based signature from Gap Diffie-Hellman groups. In: Desmedt, Y.G. (ed.) PKC 2003. LNCS, vol. 2567, pp. 18–30. Springer, Heidelberg (2002)
13. Crampton, J., Lim, H.W.: Role Signatures for Access Control in Grid Computing. Royal Holloway, University of London, Technical Report RHUL-MA-2007-2, (May 2007)

14. Crampton, J., Lim, H.W., Paterson, K.G., Price, G.: A certificate-free grid security infrastructure supporting password-based user authentication. In: Proceedings of the 6th Annual PKI R&D Workshop 2007. NIST Interagency Report (2007)
15. Foster, I., Kesselman, C., Tuecke, S.: The anatomy of the Grid: Enabling scalable virtual organizations. International Journal of High Performance Computing Applications 15(3), 200–222 (2001)
16. Galbraith, S.D., Paterson, K.G., Smart, N.P.: Pairings for Cryptographers. Cryptology ePrint Archive, Report 2006/165 (May 2006), http://eprint.iacr.org/2006/165
17. Galindo, D., Herranz, J., Kiltz, E.: On the generic construction of identity-based signatures with additional properties. In: Lai, X., Chen, K. (eds.) ASIACRYPT 2006. LNCS, vol. 4284, pp. 178–193. Springer, Heidelberg (2006)
18. Gentry, C., Ramzan, Z.: Identity-based aggregate signatures. In: Yung, M., Dodis, Y., Kiayias, A., Malkin, T.G. (eds.) PKC 2006. LNCS, vol. 3958, pp. 257–273. Springer, Heidelberg (2006)
19. Gentry, C., Silverberg, A.: Hierarchical ID-based cryptography. In: Zheng, Y. (ed.) ASIACRYPT 2002. LNCS, vol. 2501, pp. 548–566. Springer, Heidelberg (2002)
20. Goyal, V., Pandey, O., Sahai, A., Waters, B.: Attribute-based encryption for fine-grained access control of encrypted data. In: Wright, R.N., di Vimercati, S.D.C., Shmatikov, V. (eds.) CCS 2006. Proceedings of the 13th ACM Computer and Communications Security Conference, pp. 89–98. ACM Press, New York (2006)
21. Hess, F.: Efficient identity based signature schemes based on pairings. In: Nyberg, K., Heys, H. (eds.) SAC 2002. LNCS, vol. 2593, pp. 310–324. Springer, Heidelberg (2003)
22. Horwitz, J., Lynn, B.: Towards hierarchical identity-based encryption. In: Knudsen, L.R. (ed.) EUROCRYPT 2002. LNCS, vol. 2332, pp. 466–481. Springer, Heidelberg (2002)
23. Joux, A.: A one round protocol for tripartite Diffie-Hellman. In: Bosma, W. (ed.) ANTS-IV. LNCS, vol. 1838, pp. 385–394. Springer, Heidelberg (2000)
24. Khalili, A., Katz, J., Arbaugh, W.A.: Toward secure key distribution in truly ad-hoc networks. In: SAINT 2003. Proceedings of the 2003 Symposium on Applications and the Internet Workshops, pp. 342–346. IEEE Computer Society Press, Los Alamitos (2003)
25. Lu, S., Ostrovsky, R., Sahai, A., Shacham, H., Waters, B.: Sequential aggregate signatures and multisignatures without random oracles. In: Vaudenay, S. (ed.) EUROCRYPT 2006. LNCS, vol. 4004, pp. 465–485. Springer, Heidelberg (2006)
26. Paterson, K.G.: ID-based signatures from pairings on elliptic curves. Electronics Letters 38(18), 1025–1026 (2002)
27. Paterson, K.G., Schuldt, J.C.N.: Efficient identity-based signatures secure in the standard model. In: Batten, L.M., Safavi-Naini, R. (eds.) ACISP 2006. LNCS, vol. 4058, pp. 207–222. Springer, Heidelberg (2006)
28. Pirretti, M., Traynor, P., McDaniel, P., Waters, B.: Secure attribute-based systems. In: Wright, R.N., di Vimercati, S.D.C., Shmatikov, V. (eds.) CCS 2006. Proceedings of the 13th ACM Computer and Communications Security Conference, pp. 99–112. ACM Press, New York (2006)
29. Sahai, A., Waters, B.: Fuzzy identity-based encryption. In: Cramer, R. (ed.) EUROCRYPT 2005. LNCS, vol. 3494, pp. 457–473. Springer, Heidelberg (2005)

30. Sakai, R., Ohgishi, K., Kasahara, M.: Cryptosystems based on pairing. In: SCIS 2000. Proceedings of the 2000 Symposium on Cryptography and Information Security (January 2000)
31. Sandhu, R.S., Coyne, E.J., Feinstein, H.L., Youman, C.E.: Role-based access control models. IEEE Computer 29(2), 38–47 (1996)
32. Shamir, A.: Identity-based cryptosystems and signature schemes. In: Blakely, G.R., Chaum, D. (eds.) CRYPTO 1984. LNCS, vol. 196, pp. 47–53. Springer, Heidelberg (1985)
33. Smart, N.P.: Access control using pairing based cryptography. In: Joye, M. (ed.) CT-RSA 2003. LNCS, vol. 2612, pp. 111–121. Springer, Heidelberg (2003)

Verifier-Key-Flexible Universal Designated-Verifier Signatures

Raylin Tso, Juan Manuel Gonzàlez Nieto[♯], Takeshi Okamoto[†], Colin Boyd[♯], and Eiji Okamoto[‡]

Department of Risk Engineering, Graduate School of Systems and Information Engineering, University of Tsukuba, 1-1-1 Tennodai, Tsukuba, Ibaraki, 305-8573, Japan

[♯] Information Security Institute, Queensland University of Technology, GPO BOX 2434, Brisbane Q 4001, Australia

{raylin, †ken, ‡okamoto}@risk.tsukuba.ac.jp, juanma@isrc.qut.edu.au, c.boyd@qut.edu.au

Abstract. Universal Designated-Verifier Signatures (UDVS) are proposed to protect the privacy of a signature holder. Since UDVS schemes reduce to standard signatures when no verifier designation is performed, from the perspective of a signer, it is natural to ask if a UDVS can be constructed from widely used *standardized*-signatures so that the existing public key infrastructures for these schemes can be used without modification. Additionally, if designated-verifiers already have their own private/public key-pairs (which may be of a different type from the signer's), then, for the convenience of designated-verifiers, it is also natural to ask if designated-verifiers can use their own private keys to verify designated signatures instead of using a *new* key compatible with the UDVS system. In this paper, we address these problems and propose a new UDVS scheme. In our scheme, the signature is generated by a signer using DSA/ECDSA, and the designated-signature can be verified using the original private key (RSA-based or DL-based) of the designated-verifier instead of using a new key. We call this new property *verifier-key-flexible*. The security of the scheme is proved in the random oracle model.

Keywords: ECDSA, random oracle, universal designated-verifier signature, verifier-key-flexible.

1 Introduction

There are many ways for a signer to protect his own privacy in the electronic world. One such example is an undeniable signature introduced by Chaum and van Antwerpen [7]; this is a kind of digital signature which has the appealing property of the signature not being able to be verified unless there is an interaction with the signer; the signature cannot be denied if the signer has actually generated the signature. Some other examples are: designated confirmer signatures [6], limited verifier signatures [2], and designated verifier signatures [10].

S.D. Galbraith (Eds.): Cryptography and Coding 2007, LNCS 4887, pp. 403–421, 2007.

In particular, a designated verifier signature, introduced by Jakobsson *et al.*, is designed to allow a signer to prove the validity of a signature to a specific verifier in such a way that the verifier can only check the validity of the signature, but he cannot transfer this conviction to a third party. On the other hand, if the document to be signed is a certificate for a user, for example, a diploma of a graduate, a health insurance card, or a transcript of some academic certificates, then privacy issues concerning the holder of a signature instead of the signer need to be taken into consideration. To protect a signature holder's privacy, Steinfeld *et al.* in 2003 firstly defined and proposed an extension of designated-verifier signatures called universal designated-verifier signatures (UDVS) [14].

A UDVS allows *any* holder of a signature to designate the signature to any desired designated-verifier such that the designated-verifier can only verify the correctness of the signature but cannot convince any third party of this fact. This property is useful for preventing the abuse or dissemination of personal information contained in the certificate, thus protecting the privacy of the certificate holder. An appealing feature of a UDVS scheme is the convenience for signers (e.g. the certificate issuer CA) who sign using a standard digital signature. Following Steinfeld *et al.*'s pioneer work, many new UDVS schemes and improvements have been proposed (eg. [3,15,18,19]).

Motivation: It was pointed out by Baek *et al.* [3] that one inconvenience of previously known UDVSs is that they require the designated-verifier to create a public key using the signer's public key parameters. In addition, the public key must be certified via the *Verifier Key-Registration Protocol* in order to ensure that the resulting public key is compatible with the setting that the signer provided as well as to ensure that the designated-verifier knows the private key corresponding to his public key. But, in some situations, to enforce the protocol for conducting such a task (i.e., verifier key generation and key registration) to be a sub-protocol of UDVSs may be unrealistic. For, example, when proving knowledge of a signature obtained from the original signer is only in the designator's interest, the verifier may not be willing to go through such a key-setup process. In addition, if a verifier already has a private/public key pair (which may be of a different type to the signer's), he may not be willing to generate a *new* key just for the purpose of verifying a designated-signature. This is because that the unrealistic key setup involving management of Public Key Infrastructure (PKI) may incur significant cost from the view of the verifier.

Baek *et al.*'s scheme [3] solves this problem by employing an interactive proof between the holder of a signature and the designated-verifier. However, from the perspective of the designated-verifier, a question that directly arises from this model is whether the designated-verification processes can be done non-interactively and using existing public keys instead of new keys.

Except for the schemes proposed by Steinfeld *et al.* [15], which are based on Schnorr/RSA signatures, most previously known UDVSs are constructed from bilinear pairings[1]. However, since UDVS schemes reduce to standard signatures

---

[1] For more details about bilinear pairings, see for example [4,5].

when no verifier designation is performed, from the perspective of a signer it is natural to ask if a UDVS can be constructed from some widely used *standardized* signatures such as DSA [11] or ECDSA [16]. This will allow UDVS functionality to be added to (widely used) standardized signatures and the existing public key infrastructure (PKI) for these signatures can be used without modification.

Our Contributions: We propose a new UDVS scheme which can solve the above mentioned problems at the same time. The new scheme, called Verifier-Key-Flexible Universal Designated-Verifier Signature (VKF-UDVS), achieves both non-interactiveness and general public keys for the designated-signatures verification. Our scheme also allows a signer to sign a message by DSA/ECDSA which is one of the most widely used standardized signature algorithms. We will give concrete security proof of our scheme in the random oracle model.

Paper Organization: The rest of this paper is organized as follows: Section 2 presents the notation and computational assumptions used in the paper. In Section 3, we define the notion of a VKF-UDVS and the security requirements. Section 4 describes the proposed schemes and Section 5 presents the corresponding security analysis. Section 6 compares the performance of the new schemes with other existing ones and Section 7 is the conclusion of this paper.

2   Preliminaries

This section gives some notations and security assumptions required for our construction.

2.1   Notations

By $\{0,1\}^*$, we mean the set of all finite binary strings. Let A be a probabilistic Turin machine (PPTM) running in polynomial time, and let x be an input for A. We use $a \leftarrow A(x)$ to denote the assignment to a of a random element from the output of A on input x according to the probability distribution induced by the internal random choices of A. For a finite set X, we use $x \leftarrow_R X$ to denote the assignment to x of a random element from X chosen uniformly at random. We use $y \in X$ to denote that y is an element of the finite set X.

2.2   Security Assumptions

Definition 1. Discrete Logarithm (DL) Problem: Let p and q be two large primes such that $q|(p-1)$. Let $g \in Z_p^*$ of order q and h a randomly picked element from the subgroup $\langle g \rangle$. When given g and h, the DL problem is to find an element $x \in Z_q$ such that $g^x \equiv h \bmod p$.

Definition 2. Elliptic Curve Discrete Logarithm (ECDL) Problem: The DL problem in the elliptic curve setting is defined as follows:
Let $E(F_q)$ denote an elliptic curve E defined over a finite field F_q. Given a point $P \in E(F_q)$ of order n, and a point $Q = lP$ where $0 \le l \le n - 1$, determine l.

The DL problem as well as ECDL problem are believed to be difficult.

3   Verifier-Key-Flexible Universal Designated-Verifier Signature (VKF-UDVS)

In this section, we formally define the notion of VKF-UDVS and the security requirements behind the scheme.

3.1   Formal Model

VKF-UDVS schemes involve three entities: a signer, a designator (signature holder) and a designated-verifier. The signer uses his private key to sign a message and transmits a signature to the designator together with the message. The designator, after verifying the correctness of the signature, creates a transformed signature (i.e. a designated-verifier signature) for a designated-verifier. The designated-verifier verifies the validity of the transformed signature in an non-interactive way using the public key of the original signer and his own private/public key pair.

We formally define the model of our VKF-UDVS scheme as shown below.

Definition 3. Regardless of the verifier-key-generation process which is assumed to be done outside of our scheme. A VKF-UDVS scheme consists of the following polynomial-time algorithms:

- **Signer Key Generation:** A probabilistic algorithm $SigKeyGen$ which takes a security parameter 1^λ as input, and outputs a private/public key-pair (sk_s, pk_s) for the signer.
- **Signature Generation:** A probabilistic algorithm $Sign$ which takes a signer's private key sk_s and a message m as input, and outputs a standard signature σ on the message m. For ease of description, we call this signature the public verifiable signature (PV-signature).
- **Verification:** A deterministic algorithm $Verify$ which takes a signer's public key pk_s, a PV-signature σ and a message m as input, and outputs 1 or 0 for accepting or rejecting the signature σ, respectively.
- **Designation:** A probabilistic algorithm DS which is used when the designated-verifier has his own (RSA-based or any DL-based) private/public key-pair (sk_v, pk_v). DS takes a signer's public key pk_s, a designated-verifier's public key pk_v and a message m as input, and outputs a designated-verifier signature $\tilde{\sigma}$. We call $\tilde{\sigma}$ the DV-signature.
- **Designated Verification:** A deterministic algorithm $DVeri$ which takes a signer's public key pk_s, a designated-verifier's public key pk_v, a DV-signature $\tilde{\sigma}$ and a message m as input, the algorithm outputs 1 or 0 for accepting or rejecting the signature σ, respectively.
- **Verifier Key-Registration:** a protocol between a "Key Registration Authority" (KRA) and a "Verifier" (VER) who wishes to register a verifier's public key. The algorithms KRA and VER interact by sending messages alternately form one to another. At the end of the protocol, KRA outputs a pair $(pk_v, Auth)$, where pk_v is a verifier's public key and $Auth \in \{Acc, Rej\}$ is a key-registration authorization decision.

The purpose of the Verifier Key-Registration Protocol is to ensure that the verifier knows the private key which corresponds to his public key; this enforces the non-transferability privacy property. In VKF-UDVS, we omit this protocol and simply assume that designated-verifiers' key-pairs are generated by trusted KGCs. The validity of the keys are guaranteed by the trusted KGCs or by the certificates.

3.2 Security Requirements

We describe the security requirements of a UDVS scheme, unforgeability and non-transferability, in this section.

Unforgeability. In a UDVS scheme, there are two types of unforgeability properties to consider: PV-signature unforgeability (PV-unforgeability) and DV-signature unforgeability (DV-unforgeability).

PV-unforgeability is just the usual Existential Unforgeability notion under Adaptive Chosen Message Attack (EUF-ACMA) [8] for the standard PV-signature scheme induced by the UDVS scheme, which is defined as follows:

Definition 4. A signature scheme is said to be EUF-ACMA secure, if for any polynomial-time adversary $\mathcal{F}$, the advantage defined by

$$Adv_{\mathcal{F}}^{EUF-ACMA} \triangleq Pr\left[Verify(pk, m^*, \sigma^*) = 1 \, \middle| \, \begin{array}{l} (sk, pk) \leftarrow KeyGen(1^k) \\ (m^*, \sigma^*) \leftarrow \mathcal{F}^{\mathcal{OS}}(pk) \end{array} \right]$$

is less than $\frac{1}{\lambda^c}$ for sufficiently large λ and some constant c. Here $\mathcal{OS}$ is the signing oracle that $\mathcal{F}$ can access. The probability is taken over the coin tosses of the algorithms, of the oracles, and of the forger.

PV-unforgeability prevents attacks intended to impersonate the signer. In contrast, DV-unforgeability prevents attacks to fool the designated-verifier, possibly mounted by a dishonest designator. As defined by Steinfeld *et al.* [14], DV-unforgeability means that it is difficult for an attacker to forge a DV-signature $\widetilde{\sigma}^*$ by the signer on a new message m^*, such that the pair $(m^*, \widetilde{\sigma}^*)$ passes the DV-verification test with respect to a designated-verifier's and signer's public key.

   We define the DV-unforgeability of our VKF-UDVS via the following game. This game is executed between a challenger $\mathcal{C}$ and an adaptively chosen message adversary $\mathcal{F}$.

Definition 5. Let $UDVS = (SigKeyGen, Sign, Verify, DS, DVeri)$ be a VKF-UDVS scheme. Let $\mathcal{G}$ be a key-generation algorithm outside of our scheme and the private/public key-pair for the designated-verifier DV is generated as $(sk_v, pk_v) \leftarrow \mathcal{G}(1^k)$.

- **Setup:** The challenger $\mathcal{C}$ runs $SigKeyGen$ to generate the signer's private/ public key-pair (sk_s, pk_s). $\mathcal{C}$ gives the public keys pk_s and pk_v to $\mathcal{F}$ and allows $\mathcal{F}$ to run.

- **Query:** $\mathcal{F}$ can adaptively issues q_h times random oracle queries ($\mathcal{H}$ Query), q_s times signing queries ($Sign$ Query), and q_d times designation queries (DS Query):
  - $\mathcal{H}$ **Query:** For each random oracle query on a message $m \in \{0,1\}^*$, $\mathcal{C}$ responds with a random element $\mathcal{H}(m) \in \{0, \cdots, p-1\}$. $(m, \mathcal{H}(m))$ is then added to a List $\mathcal{H}_{List}$ in order to avoid a collision. $\mathcal{H}_{List}$ is assumed to be initially empty.
  - $Sign$ **Query:** For any signing query asked by $\mathcal{F}$ on a message m under the public key pk_s, $\mathcal{C}$ runs the $Sign$ algorithm and returns the output σ to $\mathcal{F}$. $\mathcal{C}$ then adds the message/signature (m, σ) to a List $Sign_{List}$ which is assumed to be initially empty.
  - DS **Query:** For any DS query asked by $\mathcal{F}$ on a message m under the public keys (pk_s, pk_v), $\mathcal{C}$ first checks if $(m, \sigma) \in Sign_{List}$ or not. If not, then $\mathcal{C}$ first runs the $Sign$ algorithm to obtain the PV-signature as $\sigma \leftarrow Sign(sk_s, m)$ and adds (m, σ) to the $Sign_{List}$. $\mathcal{C}$ then runs DS algorithm to obtain the DV-signature as $\widetilde{\sigma} \leftarrow DV(pk_s, pk_v, \sigma, m)$. $\mathcal{C}$ returns $\widetilde{\sigma}$ to $\mathcal{F}$ as the reply and add $(m, \widetilde{\sigma})$ to a List, DV_{List}, which is assumed to be initially empty.
- **Forge:** $\mathcal{F}$ outputs a forgery $(m^*, \widetilde{\sigma}^*)$.

We say $\mathcal{F}$ wins the game if: $DVeri(pk_s, pk_v, \widetilde{\sigma}^*, m^*) = 1$, and $m^* \notin Sign_{List}$.

Definition 6. A VKF-UDVS scheme provides DV-unforgeability against adaptive chosen message attack if, for any PPT forging algorithm $\mathcal{F}$ that plays the above game, $\mathcal{F}$ wins the above game with probability at most $\frac{1}{\lambda^c}$ for sufficiently large λ and some constant c. The probability is taken over the coin flips the algorithms, of the oracles, and of the forger.

Non-transferability. Informally, non-transferability of UDVS means that only the designated verifier can be convinced by the UDVS, and, from using the DV-signature and the message m, he cannot produce evidence to convince a third party that the message was signed by the signer, even if he reveals his private key.

Definition 7. Let $UDVS = (SigKeyGen, Sign, Verify, DS, DVeri)$ be a VKF-UDVS scheme and $\mathcal{G}$ be a key-generation algorithm outside of our scheme, as defined in Definition 5. In addition, a new algorithm $\overline{DS}$ is required in this simulation. $\overline{DS}$ is a probabilistic algorithm which, on input a signer's public key pk_s, a designated-verifier's private key sk_v and a message m, outputs a (simulated) designated verifier signature $\overline{\sigma_{DS}} \leftarrow \overline{DS}(pk_s, sk_v, m)$.

$\mathcal{C}$ gives all the public keys pk_v and pk_s to $\mathcal{A}$ and allows $\mathcal{A}$ to run.

- **Phase 1:** At any time, $\mathcal{A}$ can ask for a $Sign$ Query, and a DS Query as defined in Definition 5.
- **Challenge:** The above queries can be executed in polynomially many number of times. After enough executions, $\mathcal{A}$ submits (m^*, pk_s, pk_v^*) to the challenger $\mathcal{C}$ as the challenge with the constraints that the signing query on

(m^*, pk_s) has not been asked and the DS query on (m^*, pk_s, pk_v^*) has not been asked during Phase 1. As response for the challenge, C flips a fair coin and obtains a bit $b \in \{0, 1\}$. if $b = 0$, S runs the DS algorithm and returns a DV-signature $\tilde{\sigma}$ to A. Otherwise, runs $\overline{DS}$ and returns $\overline{\sigma_{DS}}$ to A.

- **Phase** 2: Phase 1 is repeated in polynomial number of times with the same constraints as those in the Challenge Phase.
- **Guess:** Finally, A outputs a bit $b' \in \{0, 1\}$.

We say the adversary A wins the game if $b = b'$.

Definition 8. A VKF-UDVS scheme provides unconditionally non- transferability if, for any adversary A playing the above game, A wins the game with probability at most $1/2 + \frac{1}{\lambda^c}$ for sufficiently large λ and some constant c.. The probability is taken over the coin flips of the algorithms and of the oracles.

4   Proposed VKF-UDVS

In our scheme, a designated-verifier is denoted by DV_{RSA} if he has a RSA-based private/public key-pair and by DV_{DL} if he has a DL-based private/public key-pair. For DV_{RSA}, his private key is $\bar{d}$ and public key is $(\bar{e}, \bar{N}, \mathcal{H}_{RSA})$ where $\bar{N}$ is a product of two large prime numbers, $\bar{e}\bar{d} \equiv 1 \mod \phi(\bar{N})$ and $\mathcal{H}_{RSA} : \{0,1\}^* \to \mathbb{Z}_{\bar{N}}$ is a one-way hash function. For DV_{DL}, his public key is $(\bar{p}, \bar{q}, \bar{g}, \bar{y}, \mathcal{H}_{DL})$ and private key is $\bar{x} \in \mathbb{Z}_{\bar{q}}$, where $\bar{p}, \bar{q}$ are prime, $\bar{q}|(\bar{p}-1)$, $\bar{g} \in \mathbb{Z}_{\bar{p}}^*$ of order $\bar{q}$ and $\bar{y} = \bar{g}^{\bar{x}} \mod \bar{p}$. $\mathcal{H}_{DL} : \{0,1\}^* \to \mathbb{Z}_{\bar{q}}$ is a one-way hash function. We assume that all of these are done outside of our scheme.

Signer Key Generation: (Identical to ECDSA). A signer launches an ECDSA-Setup Algorithm $ECDSASet$ on input a security parameter 1^λ and sets up the following parameters:

- q: a large prime greater than 2^{160}.
- $E(F_q)$: an elliptic curve defined over the finite field F_q.
- G: a point on $E(F_q)$ such that the ECDL problem in the subgroup $\langle G \rangle$ generated by G is infeasible.
- p: the prime order of G.
- $\mathcal{H} : \{0,1\}^* \to \{0, \cdots, p-1\}$: a one way hash function.
- $d_s \leftarrow_R \{1, \cdots, p-1\}$.
- $Q_s = d_s G$.

The signer sets the public key as $pk = (q, E(F_q), G, \langle G \rangle, p, \mathcal{H}, Q_s)$ and the private key as d_s.

Signature Generation: (Identical to ECDSA). To sign a message $m \in \{0,1\}^*$, the signer with private key d_s does the following steps:

1. Pick $k \in \{0, \cdots, p-1\}$ randomly.
2. Compute $kG = (x, y)$ and $r = x \mod p$, if $r = 0$ then go to step 1.
3. Compute $s = (\mathcal{H}(m) + d_s r)/k \mod p$, if $s = 0$ then go to step 1.

The signature on m is $\sigma = (r, s)$. The signer then sends σ to a receiver (i.e., designator) $\mathcal{R}$.

Verification: To verify σ, $\mathcal{R}$ does the following steps:

- $e_1 = \mathcal{H}(m)/s \bmod p$ and $e_2 = r/s \bmod p$.
- $\zeta = (x, y) = e_1 G + e_2 Q_s$.
- Accept σ if and only if $x \bmod p = r$.

Designation: To designate the signature σ, $\mathcal{R}$ first picks a random number $h \in \{1, \cdots, p-1\}$, he then computes $t = s/h \bmod p$ and $T = h\zeta$ where ζ is generated at the verification phase. Then, if the designated-verifier is DV_{RSA}, $\mathcal{R}$ (using both the public keys of the signer and the designer verifier) carries out the following steps:

- Pick $\alpha \in \{0, \cdots, p-1\}$ randomly, and compute $a_1 = \alpha\zeta$ and $c_2 = \mathcal{H}_{RSA}(m, T, \bar{e}, a_1)$.
- Pick $\mu_2 \in \mathbb{Z}_{\bar{N}}$ and compute $a_2 = c_2 + \mu_2^{\bar{e}} \bmod \bar{N}$ and $c_1 = \mathcal{H}(m, T, \bar{e}, a_2)$.
- Compute $\mu_1 = \alpha - hc_1 \bmod p$.

If the designated-verifier is DV_{DL}, $\mathcal{R}$ first computes α and a_1 in the same way as in the above steps. Then, using both the public keys of the signer and the designer verifier, $\mathcal{R}$ carries out the following steps:

- Compute $c_2 = \mathcal{H}_{DL}(m, T, \bar{y}, a_1)$.
- Pick $\mu_2 \in \mathbb{Z}_{\bar{q}}$ and compute $a_2 = \bar{g}^{\mu_2} \bar{y}^{c_2} \bmod \bar{p}$ and $c_1 = \mathcal{H}(m, T, \bar{y}, a_2)$.
- Compute $\mu_1 = \alpha - hc_1 \bmod p$.

For DV_{RSA} or DV_{DL}, the designated signature on m is $\tilde{\sigma} = (\zeta, t, c_1, \mu_1, \mu_2)$.

Designated Verification: A designated-verifier first computes $r = x \bmod p$ from $\zeta = (x, y)$, then, using t to compute $e_1' = \mathcal{H}(m)/t \bmod p$, $e_2' = r/t \bmod p$, and $T = e_1' G + e_2' Q_s$. Furthermore,

- for DV_{RSA}, he computes $\hat{a}_1 = \mu_1 \zeta + c_1 T$, $\hat{c}_2 = \mathcal{H}_{RSA}(m, T, \bar{e}, \hat{a}_1)$, $\hat{a}_2 = \hat{c}_2 + \mu_2^{\bar{e}} \bmod \bar{N}$, $\hat{c}_1 = \mathcal{H}(m, T, \bar{e}, \hat{a}_2)$.
- for DV_{DL}, he computes $\hat{a}_1 = \mu_1 \zeta + c_1 T$, $\hat{c}_2 = \mathcal{H}_{DL}(m, T, \bar{e}, \hat{a}_1)$ in the same way as in the above steps. Then he computes $\hat{a}_2 = \bar{g}^{\mu_2} \bar{y}^{\hat{c}_2} \bmod \bar{p}$ and $\hat{c}_1 = \mathcal{H}(m, T, \bar{y}, \hat{a}_2)$.

A designated-verifier DV_{RSA} or DV_{DL} accepts the signature $\tilde{\sigma}$ if and only if $\hat{c}_1 = c_1 \in \tilde{\sigma}$.

Discussion. In the above scheme, the DV-signature is $\tilde{\sigma} = (\zeta, t, c_1, \mu_1, \mu_2)$ where ζ is a point on the elliptic curve $E(F_q)$ defined over F_q and the others (except μ_2) are elements in $\{0, \cdots, p-1\}$. If the designated-verifier DV_i has a DL-based key-pair (key_{DL} for short), then $\mu_2 \in \mathbb{Z}_{\bar{q}}$. Each of them is about 160 bit-length so the total size of the DV-signature is about 800-bit in this case. If DV_i has a RSA-based key-pair (key_{RSA}), then $\mu_2 \in \mathbb{Z}_{\bar{N}}$ with $|\bar{N}| = 1024$; so the size of a DV-signature is about 1664-bit (i.e., $1024 + 160 \times 4$).

Since ECDSA is the elliptic analogue of DSA, the above technique can also be used on DSA so as to add a UDVS functionality as an optional feature for DSA users. But, when using DSA as a basic scheme, ζ in a DV-signature will have size of 1024-bit since $\zeta \in \mathbb{Z}_p$ with $|p| = 1024$. So the size of a DV-signature will become 1664-bit if key_{DL} and 2528-bit if key_{RSA}, which is about twice the size of the (ECDSA-based) DV-signature.

Steinfeld *et al.* [15] proposed UDVSs based on Schnorr signatures [13] and RSA signatures. Although the security assumption of Schnorr signatures and ECDSA are both based on the DL-problem, in practice, one may prefer to use an ECDSA as a basic scheme instead of using a Schnorr signature. This is because ECDSA is widely used and has been accepted as an ISO standard (ISO 14888-3), an ANSI standard (ANSI X9.62), an IEEE standard (IEEE P1363) and a FIPS standard (FIPS 186-2). In addition, because an ECDSA is more complicated than a Schnorr signature, it is more difficult to construct a UDVS based on ECDSA instead of on Schnorr signatures. We emphasize that our idea is also possible to use on Schnorr signatures but the idea in [15] cannot be used to extend a ECDSA signature into a UDVS scheme.

Notice that the above protocol for Designated Verification is a protocol for proving knowledge of h, which is the solution of the ECDL problem of T to the basis ζ. Since $t = s/h$ and $r = x \bmod p$, knowing h means that the designator $\mathcal{R}$ possesses a valid ECDSA signature (r, s). Therefore, $\mathcal{R}$ can convince a designated-verifier if and only if the transformed signature passes the designated verification protocol. This kind of knowledge-proof protocols supporting both DL-based and RSA-based key users are first proposed by Abe et al. [1]. The security is also proved in their paper in a random oracle model.

5   Security Analysis

The correctness of the PV-signature and the DV-signature is straightforward. PV-unforgeability is based on the unforgeability of the ECDSA. Up to now, the provable security results are still not applicable to the widely standardized ECDSA, unless the schemes are modified so that the signer hashes both the message and the random group element generated, rather than just the message. Although the security of ECDSA has not been proven yet, it is widely used in practice and is believed to be EUF-ACMA secure (see Definition 4). Until now, no significant security flaws on ECDSA are known. Consequently, in this section, we only consider the security concerning the DV-signature (since the PV-signature of our scheme is a ECDSA signature).

Before considering the unforgeability of the DV-signature, we first consider the non-transferability of the DV-signature.

Theorem 1. The DV-signature of the proposed scheme provides unconditional non-transferability.

Proof: To prove unconditional non-transferability, we first show how a simulation algorithm $\overline{DS}$ can simulate an output of a valid DV-signature on the input (pk_s, sk_v, m) without the knowledge of the signer's private key. In the following simulation, by $(sk_v, pk_v) \in KEY_{DL}$, we mean the designated-verifier's key-pair is a DL-based key and by $(sk_v, pk_v) \in KEY_{RSA}$, we mean the key-pair is a RSA-based key. As described in Section 4, if $(sk_v, pk_v) \in KEY_{DL}$, then $pk_v = (\bar{p}, \bar{q}, \bar{g}, \bar{y}, \mathcal{H}_{DL})$ and $sk_v = \bar{x} \in \mathbb{Z}_{\bar{q}}$, where $\bar{p}, \bar{q}$ are prime, $\bar{q}|(\bar{p}-1)$, $\bar{g} \in \mathbb{Z}_{\bar{p}}^*$ of order $\bar{q}$ and $\bar{y} = \bar{g}^{\bar{x}} \bmod \bar{p}$. If $(sk_v, pk_v) \in KEY_{RSA}$, then $sk_v = \bar{d}$ and $pk_v = (\bar{e}, \bar{N}, \mathcal{H}_{RSA})$, where $\bar{N}$ is a product of two equal-length prime numbers and $\bar{e}\bar{d} \equiv 1 \bmod \phi(\bar{N})$.

Case (1). $(sk_v, pk_v) \in KEY_{DL}$: To simulate a DV-signature on a message m signed under the signer's private key $sk_s = d_s$ corresponding to the public key $pk_s = (q, E(F_q), G, \langle G \rangle, p, \mathcal{H}, Q_s)$ and the designated-verifier's public key $pk_v = (\bar{p}, \bar{q}, \bar{g}, \bar{y}, \mathcal{H}_{DL})$ corresponding to the private key $sk_v = \bar{x}$, $\overline{DS}$ does the following steps:

- Pick $\zeta \in \langle G \rangle$ and $t \in \{1, \cdots, p-1\}$ randomly.
- Using ζ, t, compute T ($\in \langle G \rangle$). This is done in the same way described in the Designated Verification Phase of the proposed scheme.
- Select $\bar{k} \in \mathbb{Z}_{\bar{q}}$ randomly and compute $a_2 = \bar{g}^{\bar{k}} \bmod \bar{p}$ and $c_1 = \mathcal{H}(m, T, \bar{y}, a_2)$.
- Pick $\mu_1 \in \{0, \cdots, p-1\}$ randomly, compute $a_1 = \mu_1 \zeta + c_1 T$ and $c_2 = \mathcal{H}_{DL}(m, T, \bar{y}, a_1)$.
- Compute $\mu_2 = \bar{k} - \bar{x}c_2 \bmod \bar{q}$.

The simulated DV-signature is $(\zeta, t, c_1, \mu_1, \mu_2)$.

Case (2). $(sk_v, pk_v) \in KEY_{RSA}$: To simulate a DV-signature on a message m signed under the signer's private key $sk_s = d_s$ corresponding to the public key $pk_s = (q, E(F_q), G, \langle G \rangle, p, \mathcal{H}, Q_s)$ and the designated-verifier's public key $pk_v = (\bar{e}, \bar{N}, \mathcal{H}_{RSA})$ corresponding to the private key $sk_v = \bar{d}$, $\overline{DS}$ does the following steps:

- Pick $\zeta \in \langle G \rangle$ and $t \in \{1, \cdots, p-1\}$ randomly.
- Using ζ, t, compute T ($\in \langle G \rangle$).
- Select $a_2 \in \mathbb{Z}_{\bar{N}}$ randomly and compute $c_1 = \mathcal{H}(m, T, \bar{y}, a_2)$.
- Pick $\mu_1 \in \{0, \cdots, p-1\}$ randomly, compute $a_1 = \mu_1 \zeta + c_1 T$ and $c_2 = \mathcal{H}_{RSA}(m, T, \bar{y}, a_1)$.
- Compute $\mu_2 = (a_2 - c_2)^{\bar{d}} \bmod \bar{N}$.

The simulated DV-signature is $(\zeta, t, c_1, \mu_1, \mu_2)$.
By a straightforward computation the same as that in the Designated Verification Phase of the proposed scheme, one can easily conclude that $(\zeta, t, c_1, \mu_1, \mu_2)$ is a valid DV-signature.

Now, we prove that a DV-signature is simulated by $\overline{DS}$ with an indistinguishable probability distribution. Both cases of $(sk_v, pk_v) \in KEY_{DL}$ and $(sk_v, pk_v) \in KEY_{RSA}$ can be analyzed in the same way, so we only consider the case of $(sk_v, pk_v) \in KEY_{DL}$.

Lemma 1. Given a VKF-UDVS scheme, $UDVS = (SigKeyGen, Sign, Verify, DS, DVeri)$, and a simulator $\overline{DS}$, the distributions of a DV-signature generated by DS and a DV-signature simulated by $\overline{DS}$ are the same.

Proof: Assume $\tilde{\sigma} = (\zeta, t, c_1, \mu_1, \mu_2)$ is generated by DS (for detail of signature generation, see Section 4) and $\sigma' = (\zeta', t', c_1', \mu_1', \mu_2')$ is generated by $\overline{DS}$ on the same message m and under the same signer with key-pair (pk_s, sk_s) and designated-verifier with key-pair (pk_v, sk_v), we show the distributions of these two signatures are the same.

First, assume $\mathcal{L} = \{\sigma_1, \cdots, \sigma_l\}$ be a set of all DV-signatures on the message m under the signer with key-pair (pk_s, sk_s) and the designated-verifier with key-pair (pk_v, sk_v), we randomly pick a sequence $\sigma_i = (a, b, c, d, e) \in \mathcal{L}$ where $a \in \langle G \rangle, b \in \{1, \cdots, p-1\}, c \in \{0, \cdots, p-1\}, d \in \{0, \cdots, p-1\}$ and $e \in \mathbb{Z}_{\bar{q}}$. Then we compute the probability of appearance of this sequence following each distribution of probabilities:

$$\mathrm{Pr}_{\tilde{\sigma}}\left[(\zeta, t, c_1, \mu_1, \mu_2) = (a, b, c, d, e) \right]$$

$$= \mathrm{Pr}\left[\begin{array}{ll} \zeta = a, & \zeta = e_1 G + e_2 Q_s \in \langle G \rangle, \\ t = b, & t = s/h \bmod p \in \{1, \cdots, p-1\}, \\ & (h \leftarrow_R \{1, \cdots, p-1\}) \\ \mu_2 = e, & \mu_2 \leftarrow_R \mathbb{Z}_{\bar{q}}, \\ c_1 = c, & c_1 = \mathcal{H}(,, a_2) \\ \mu_1 = d. & \mu_1 = \alpha - h c_1 \bmod p \end{array} \right] = \frac{1}{q \cdot (p-1) \cdot \bar{q} \cdot p}.$$

$$\mathrm{Pr}_{\sigma'}\left[(\zeta', t', c_1', \mu_1', \mu_2') = (a, b, c, d, e) \right]$$

$$= \mathrm{Pr}\left[\begin{array}{ll} \zeta' = a, & \zeta' \leftarrow_R \langle G \rangle, \\ t' = b, & t' \leftarrow_R \{1, \cdots, p-1\}, \\ c_1' = c, & c_1' = \mathcal{H}(,, a_2), \\ & (a_2 = \bar{g}^{\bar{k}}; \bar{k} \leftarrow_R \mathbb{Z}_{\bar{q}}) \\ \mu_1' = d, & \mu_1' \leftarrow_R \{0, \cdots, p-1\} \\ \mu_2' = e. & \mu_2' = \bar{k} - \bar{x} c_2 \bmod \bar{q} \end{array} \right] = \frac{1}{q \cdot (p-1) \cdot \bar{q} \cdot p}.$$

Note that μ_1 has no freedom when ζ, t, μ_2 and c_1 fixed, and μ_2' has no freedom when ζ', t', c_1' and μ_1' fixed. Also note that because

$$c_1 = \mathcal{H}(,, a_2); a_2 = \bar{g}^{\mu_2} \bar{y}^{c_2}; c2 = \mathcal{H}_{DL}(,, a_1); a_1 = \alpha \zeta; \alpha \leftarrow_R \{0, \cdots, p-1\},$$

so, when μ_2 fixed, there are only p possible choices of c_1. The case of $(sk_v, pk_v) \in KEY_{RSA}$ can be analyzed in the same way. Therefore, whether a DV-signature is generated by DS or by $\overline{DS}$ is probabilistically indistinguishable. $\square$

Combining the above lemma with the proof of Theorem 1, we know that any DV-signature can be generated by a signer or a designated-verifier with equal probability. Therefore, we conclude that the proposed scheme is unconditional non-transferable. This concludes the proof of Theorem 1. $\square$

Theorem 2. The DV-signature of the proposed VKF-UDVS scheme is EUF-ACMA secure under the assumptions that the ECDSA is EUF-ACMA secure and the DL/ECDL Problem is hard.

Proof: Since both cases when a designated-verifier is DV_{DL} or DL_{RSA} can be proved in a similar way, we only consider the case when the designated-verifier is DV_{DL} with a DL-based key-pair. Suppose there exists an adversary $\mathcal{B}$ which can break the EUF-ACMA security of the DV-signature of the proposed scheme, then, using $\mathcal{B}$ as a black-box, there exists another adversary $\mathcal{A}$ which can either break the EUF-ACMA security of ECDSA or break the discrete logarithm problem.

First, a challenger $\mathcal{C}$ runs the signer-key-generation algorithm *SignKeyGen* to generate a signer's private/public key-pair $(sk_s, pk_s) = (d_s, Q_s)$ of an ECDSA system. In addition, $\mathcal{C}$ generates $(\bar{p}, \bar{q}, \bar{g}, \bar{y})$ where $\bar{p}, \bar{q}$ are two large primes such that $\bar{q}|(\bar{p}-1)$, $\bar{g} \in \mathbb{Z}_{\bar{p}}^*$ is an element in $\mathbb{Z}_{\bar{p}}^*$ of order $\bar{q}$, and $\bar{y}$ is an random element picked from $\langle \bar{g} \rangle$. Note that $(\bar{p}, \bar{q}, \bar{g}, \bar{y})$ are independent of (sk_s, pk_s), the key-pair of an ECDSA system. $\mathcal{C}$ then gives the public key pk_s of an ECDSA system and the discrete logarithm problem $(\bar{p}, \bar{q}, \bar{g}, \bar{y})$ to A.

$\mathcal{A}$ is possible to access to the random oracle $\mathcal{H}$ and the signing oracle $\mathcal{SQ}$ of the ECDSA system. The purpose of $\mathcal{A}$ is either to forge an ECDSA signature corresponding to the public key pk_s or to find $\log_{\bar{g}}^{\bar{y}}$, the solution to the DL-problem of $\bar{y}$ to the basis $\bar{g}$. Note that both challenges (i.e., to forge an ECDSA signature and to break the DL-problem) are independent and solving one challenge does not help $\mathcal{A}$ to solve the other challenge. We say $\mathcal{A}$ succeeds if it can solve either one of the challenges.

In order to solve the challenges given by $\mathcal{C}$, $\mathcal{A}$ utilizes $\mathcal{B}$ as a black-box. To get the black-box $\mathcal{B}$ run properly, $\mathcal{A}$ simulates the environments of the proposed VKF-UDVS scheme and the random oracles that corresponds to the hash functions $\mathcal{H}$, $\mathcal{H}_{DL}$, the signing oracle $\mathcal{SQ}$ and the designation oracle $\mathcal{DSQ}$ which can be accessed by $\mathcal{B}$.

- **Environment Setting:** $\mathcal{A}$ sets $pk_v = (\bar{p}, \bar{q}, \bar{g}, \bar{y}, \mathcal{H}_{DL})$ as the designated-verifier's public key, where $\mathcal{H}_{DL}$ is a hash function simulated by $\mathcal{A}$. $\mathcal{A}$ then gives the public key pk_v together with the signer's public key pk_s to $\mathcal{B}$ and allows $\mathcal{B}$ to run.
- **Query:** $\mathcal{B}$ is allowed to ask the following queries in polynomial number of times:
  - $\mathcal{H}_{DL}$ **Query:** For each $\mathcal{H}_{DL}$ Query on input of the form $(m_i, T_i, \bar{y}, a_{(1,i)})$, $\mathcal{A}$ picks a random number $c_i \in \mathbb{Z}_{\bar{q}}^*$ and responds with c_i to $\mathcal{B}$ as the answer. To avoid collision, that is, to make sure that each different query (with different input) has different answer, $\mathcal{A}$ records $(c_i, m_i, T_i, \bar{y}, a_{(1,i)})$ to a $\mathcal{H}_{DL}$-List which is initially empty.
  - $\mathcal{H}$ **Query:** According to the proposal, the $\mathcal{H}$ oracle can be queried in two types: (1) to query on input of a message m_i (2) to query on input of the form $(m_i, T_i, \bar{y}, a_{(2,i)})$. To avoid confusion, we may consider these two queries are different. We split the $\mathcal{H}$ query into two parts: $\mathcal{H}_1$ query on input m_i, and $\mathcal{H}_2$ query on input $(m_i, T_i, \bar{y}, a_{(2,i)})$. For each $\mathcal{H}_1$ query on a message $m_i \in \{0,1\}^*$, $\mathcal{A}$ just queries to random oracle $\mathcal{HQ}$ of itself and forwards the answer (denoted by $\mathcal{H}(m_i)$) form $\mathcal{HQ}$ to $\mathcal{B}$. For each $\mathcal{H}_2$ query on input of the form $(m_i, T_i, \bar{y}, a_{(2,i)})$, $\mathcal{A}$ replies in the same

way as that when making $\mathcal{H}_1$ query (i.e., ask $\mathcal{HQ}$ of itself and forwards the answer (denoted by $c_{(1,i)}$) from $\mathcal{HQ}$ to $\mathcal{B}$). Although there is no difference between these two queries, we split them in two parts in order for our security proof. In addition, since the input on $\mathcal{H}_1$ query (which is any binary string) and the input on $\mathcal{H}_2$ query (which is a sequence $(m_i, T_i, \bar{y}, a_{(2,i)})$ with $T_i, a_{(2,i)} \in \langle G \rangle$, and $\bar{y} \in \mathbb{Z}_q^*$), we may further assume that $\mathcal{A}$ always knows whether it is an $\mathcal{H}_1$ query or an $\mathcal{H}_2$ query. $\mathcal{A}$ then records $(m_i, \mathcal{H}(m_i))$ to the $\mathcal{H}_1$-List and $(m_i, T_i, \bar{y}, a_{(2,i)}, c_{(1,i)})$ to the $\mathcal{H}_2$-List which are initially empty.

- *Sign* **Query:** For each signing query on a message m_i asked by $\mathcal{B}$, $\mathcal{A}$ asks the signing oracle $\mathcal{SQ}$ of itself and responds the answer σ_i to $\mathcal{B}$. (m_i, σ_i) is then added to a *Sign*-List which is initial empty.
- **Designated (DS) Query:** For any DS query asked by $\mathcal{B}$ on a message m_i, $\mathcal{A}$ first checks its *Sign*-List. If $(m_i, \sigma_i) \notin Sign$-List, then $\mathcal{A}$ first asks the signing oracle $\mathcal{SQ}$ of itself and records the answer (m_i, σ_i) to the *Sign*-List. $\mathcal{A}$ then runs a DS algorithm to obtain the DV-signature as $\tilde{\sigma}_i \leftarrow (pk_s, pk_v, \sigma_i)$. For detail computation of the DS algorithm, see Section 3.1 and also the designation phase in Section 4. $\tilde{\sigma}_i$ is then the answer to the DS-query.

- **Forge the DV-signature:** At the end of the simulation, $\mathcal{B}$ outputs a forged DV-signature $\tilde{\sigma}^* = (\zeta^*, t^*, c_1^*, \mu_1^*, \mu_2^*)$ on m^*. Assume $\mathcal{B}$ wins the game and the forged DV-signature is valid (i.e., $DVeri(pk_s, pk_v, \tilde{\sigma}^*, m^*) = 1$, and $m^* \notin Sign$-List).

We need the following lemma.

Lemma 2. In a DV-signature $(\zeta, t, c_1, \mu_1, \mu_2)$, the sequence (c_1, μ_1, μ_2) is a proof of knowledge of $\log_\zeta^T$ or the knowledge of the designated-verifier's private key, where T is publicly computable via ζ, t and the message m according to the designated verification protocol described in Section 4. In addition, the protocol of the knowledge proof is EUF-ACMA secure, in the random oracle model.

The above mentioned knowledge proof protocol is proposed and the EUF-ACMA security of the protocol is proved in the random oracle model by Abe *et al.* in [1]. Denote Λ be the event that $\mathcal{B}$ wins the game and outputs a valid forgery $\tilde{\sigma}^* = (\zeta^*, t^*, c_1^*, \mu_1^*, \mu_2^*)$ of the DV-signature. The event Λ can be split in 3 disjoint sub-cases:

Λ_1: $\mathcal{B}$ wins the game without knowing both the knowledge of $h^* = \log_{\zeta^*}^{T^*}$ and the designated-verifier's private key, at the end of the game.
Λ_2: $\mathcal{B}$ wins the game with the knowledge of $h^* = \log_{\zeta^*}^{T^*}$ at the end of the game.
Λ_3: $\mathcal{B}$ wins the game with the knowledge of the designated verifier's private key sk_v at the end of the game.

According to Lemma 2, we know that $\mathcal{B}$ cannot win the game via event Λ_1 since it contradicts the EUF-ACMA unforgeability of the scheme in [1]. In other words, if $\tilde{\sigma}^* = (\zeta^*, t^*, c_1^*, \mu_1^*, \mu_2^*)$ is a successful forgery on m^* in our scheme while $\mathcal{B}$ does not know the private key sk_v or $h^* = \log_{\zeta^*}^{T^*}$, then, $\mathcal{B}$ succeeded in forging a knowledge proof $(c_1^*, \mu_1^*, \mu_2^*)$ of the knowledge proof protocol in [1] without knowing the knowledge $\log_{\zeta^*}^{T^*}$ or sk_v. Here T^* can be computed publicly according to the designated verification protocol described in Section 4. If event Λ_1 happened, then, using the same security proof in [1], one can further use the adversary $\mathcal{B}$ to solve a discrete logarithm problem.

If event Λ_2 happened, then we show how to extract h^* from $\mathcal{B}$. Using h^*, $\mathcal{A}$ can generate a valid forgery of ECDSA. The proof uses the forking technique [12] which involves running the attacker $\mathcal{B}$ for solving our scheme twice, answering its i^*-th $\mathcal{H}_2$ query differently in the two runs to obtain two distinct solutions $(c_{(1,i^*)}, \mu_{(1,i^*)})$ and $(c'_{(1,i^*)}, \mu'_{(1,i^*)})$, from which the solution $h^* = \frac{\mu_{(1,i^*)} - \mu'_{(1,i^*)}}{c'_{(1,i^*)} - c_{(1,i^*)}}$ can be recovered.

Let Θ, Ω be the random tapes given to the simulator $\mathcal{A}$ and the adversary $\mathcal{B}$, respectively, such that $\mathcal{B}$ outputs a forged DV-signature. Notice that the success probability of $\mathcal{B}$ is taken over the space defined by Θ, Ω and the random oracles. At the first run, the simulator acts in exactly the same manner as that was described at the beginning of this proof. At the end of this run, $\mathcal{B}$ outputs a successful forgery $\tilde{\sigma}^* = (\zeta^*, t^*, c_1^*, \mu_1^*, \mu_2^*)$ on a message m^*.

Note: The method of computing c_1^* is equivalent to querying $(m^*, T^*, \bar{y}, a_2^*)$ to $\mathcal{H}_2$ (where a_2^* is an element in $\mathbb{Z}_{\bar{p}}$). Due to the ideal randomness of the hash function $\mathcal{H}$ (with regard to the $\mathcal{H}_2$ queries), with probability at least $1 - 1/p$, there exists a $\mathcal{H}_2$ query on input $(m^*, T^*, \bar{y}, a_2^*)$ if $\tilde{\sigma}^*$ is a successful forgery. Assume this query occurs at the i^*-th $\mathcal{H}_2$ query.

At the second run, with the same random tapes Θ, Ω given to the simulator $\mathcal{A}$ and the adversary $\mathcal{B}$, this run is almost the same as the first run except the simulation of the $\mathcal{H}_2$ oracle. This time, for any j-th $\mathcal{H}_2$ query with $j < i^*$, $\mathcal{A}$ responds to $\mathcal{B}$ with the same value as that at the first run. In other words, $\mathcal{A}$ queries the random oracle $\mathcal{HQ}$ of itself and forwards the answer to $\mathcal{B}$ (Actually, $\mathcal{A}$ is not necessary to query $\mathcal{HQ}$ again since all the queries made by $\mathcal{B}$ as well as the answers from $\mathcal{HQ}$ in the first run has already been recorded in the $\mathcal{H}_2$-List, $\mathcal{A}$ can just check the List and make a response). However, for any j-th $\mathcal{H}_2$ query with $j \geq i^*$, $\mathcal{A}$ picks a random number $c'_{(1,j)} \in \{0, \cdots, p-1\}$ and responds with $c'_{(1,j)}$. There is no change to the other oracles comparing to the first run. Finally, at the end of the second run, $\mathcal{B}$ outputs its forgery $\tilde{\sigma}' = (\zeta', t', c'_1, \mu'_1, \mu'_2)$ on a message m'.

Assume $\mathcal{B}$ can query at most $q_{\mathcal{H}_2}$ times to the $\mathcal{H}_2$ oracle. For $i \in \{1, \cdots, q_{\mathcal{H}_2}\}$, we call a run of $\mathcal{B}$ i-successful if $\mathcal{B}$ succeeds and $i^* = i$. Note that if both runs of $\mathcal{B}$ are i-successful for some i with regard to the $\mathcal{H}_2$ query, then, since the view of $\mathcal{B}$ in both runs is the same up to the i-th $\mathcal{H}_2$ response, $(m^*, T^*, \bar{y}, a_2^*)$ (which is the input of the i-th $\mathcal{H}_2$ query in the first run) must be equal to $(m', T', \bar{y}, a_2')$ (which is the input of the i-th $\mathcal{H}_2$ query in the second run). We first show that when $m^* = m'$, $T^* = T'$ and $a_2^* = a_2'$, then $t^* = t'$ in both runs. Otherwise, $\mathcal{A}$ can find the signing key of the ECDSA signer, which means $\mathcal{A}$ can totally break the ECDSA (this is assumed to be impossible).

According to the protocol, we have

$$T^* = e_1^*G + e_2^*Q_s = \mathcal{H}(m^*)/t^*G + r^*/t^*Q_s \qquad \text{and} \qquad (1)$$

$$T' = e_1'G + e_2'Q_s = \mathcal{H}(m')/t'G + r'/t'Q_s. \qquad (2)$$

Since $T^* = T'$, combine equations (1) and (2), we have

$$(t'\mathcal{H}(m^*) - t^*\mathcal{H}(m'))G = (t^*r' - t'r^*)Q_s. \qquad (3)$$

If $t^* \neq t'$, then $t'\mathcal{H}(m^*) - t^*\mathcal{H}(m') \neq 0$ since $\mathcal{H}(m^*) = \mathcal{H}(m')$. This implies that $t^*r' - t'r^* \neq 0$. Therefore, in this case, $\mathcal{A}$ can find the signing key, $sk_s = \log_G^{Q_s}$, of the ECDSA signer:

$$sk_s = \log_G^{Q_s} = \frac{t'\mathcal{H}(m^*) - t^*\mathcal{H}(m')}{t^*r' - t'r^*}.$$

Since it is impossible if ECDSA is EUF-ACMA secure, it must be the case that $t^* = t'$. In addition, $r^* = r'$ in this case.

Next, we show that $\zeta^* = \zeta'$ in both runs. Since $r^* = x^* \mod p$ with $(x^*, y^*) = \zeta^*$ in the first run and $r' = x' \mod p$ with $(x', y') = \zeta'$ in the second run, if $r^* = r'$, then $\zeta^* = \pm\zeta'$. On the other hand, since we give the same random tapes Θ, Ω to $\mathcal{A}$ and $\mathcal{B}$, respectively, in both runs, with overwhelming probability, we are in the case of $\zeta^* = \zeta'$ [2] if $r^* = r'$.

Up to the present, we have proved that if both runs of $\mathcal{B}$ are i-successful with a valid forgery $(\zeta^*, t^*, c_1^*, \mu_1^*, \mu_2^*)$ in the end of the first run and a valid forgery $(\zeta', t', c_1', \mu_1', \mu_2')$ in the end of the second run, then, $\zeta^* = \zeta'$, $t^* = t'$, $T^* = T'$ and $a_2^* = a_2'$. Therefore, $\log_{\zeta^*}^{T^*} = h^* = \log_{\zeta'}^{T'}$ which means that both runs have the same h^*. We then show that $\mu_2^* = \mu_2'$ and $c_2^* = c_2'$, otherwise, $\mathcal{A}$ can solve the DL-problem of $\log_{\bar{g}}^{\bar{y}}$ given by the challenger.

According to the protocol, $a_2^* = \bar{g}^{\mu_2^*}\bar{y}^{c_2^*} \mod \bar{p}$ and $a_2' = \bar{g}^{\mu_2'}\bar{y}^{c_2'} \mod \bar{p}$. Since $a_2^* = a_2'$, we have $\bar{g}^{\mu_2^*}\bar{y}^{c_2^*} = \bar{g}^{\mu_2'}\bar{y}^{c_2'}$ so

$$\log_{\bar{g}}^{\bar{y}} = \frac{\mu_2^* - \mu_2'}{c_2' - c_2^*},$$

since each μ_2 and c_2 can be computed publicly from the forged signatures, this is the solution to the DL-problem given by the challenger. Therefore, if $\mathcal{A}$ does not succeed in this phase, then it must be the case that $\mu_2^* = \mu_2'$ and $c_2^* = c_2'$.

Note that according to the protocol, $c_2^* = c_2'$ implies $a_1^* = a_1'$. In addition, since $\zeta^* = \zeta'$ and $a_1^* = a_1'$, so $\alpha^* = \alpha'$. Also note that $c_1^* \neq c_1'$ with probability $1 - 1/p$ since c_1' is randomly picked from $\{0, \cdots, p-1\}$. Consequently, $\mu_1^* \neq \mu_1'$ according to the equation: $\mu_1 = \alpha - hc_1$.

---

[2] In the worst case when $\zeta^* = -\zeta'$, then $h^* = -h'$. In this case, it is also possible to prove the security of the scheme. The technique is exactly the same as that when $\zeta^* = \zeta'$ so the detail is omitted.

Since h^* is the same in both runs but $c_1^* \neq c_1'$ and $\mu_1^* \neq \mu_1'$, we can extract h^* as $h^* = \frac{\mu_1^* - \mu_1'}{c_1^* - c_1'}$.

If $\mathcal{A}$ can extract h^*, then, with the knowledge h^* and $(\zeta = (X^*, Y^*), t)$ form the forged signature σ^*, $\mathcal{A}$ outputs its forgery of ECDSA signature as (r^*, s^*) where $r^* = X^* \bmod p$ and $s^* = h^* t^* \bmod p$. It is easy to see that this is a valid ECDSA signature.

Since ECDSA is believed to be EUF-ACMA secure, we found a contradiction.

It remains to estimate the probability of the event $\mathcal{S}^*$ that both runs of $\mathcal{B}$ are i-successful with regard to the $\mathcal{H}_2$ query for some $i \in \{1, \cdots, q_{\mathcal{H}_s}\}$. To do this, we split $\mathcal{S}^*$ into $q_{\mathcal{H}_2}$ distinct subevents $\mathcal{S}_i^*$ according the value of i and bound each one. For each i, let Γ_i denote the outcome space for the random variable $\alpha_i = (\Theta, \Omega, pk_s, pk_v, c_{(1,1)}, \cdots, c_{(1,i-1)})$ consisting of the view of $\mathcal{B}$ up to the i-th query to $\mathcal{H}_2$, and let Υ_i denote the outcome space for the independent random variable $\beta_i = (c_{(1,i)}, \cdots, c_{(1,q_{\mathcal{H}_2})})$ consisting of the view of $\mathcal{B}$ after the i-th query to $\mathcal{H}_2$. We need the following lemma.

Lemma 3. (The Splitting Lemma):[12] Let $\mathcal{S} \subset \Gamma \times \Upsilon$ such that $Pr[(\alpha, \beta) \in \mathcal{S}] \geq \varepsilon$. For any $\lambda < \varepsilon$, define

$$\varphi = \left\{ (\alpha, \beta) \in \Gamma \times \Upsilon \mid \Pr_{\beta' \in \Upsilon}[(\alpha, \beta') \in \mathcal{S}] \geq \varepsilon - \lambda \right.,$$

then the following statements hold:

(i) $Pr[\varphi] \geq \lambda$.
(ii) $\forall (\alpha, \beta) \in \varphi$, $\Pr_{\beta' \in \Upsilon}[(\alpha, \beta') \in \mathcal{S}] \geq \varepsilon - \lambda$.

Define $\mathcal{S}_i$ be the event that a run of $\mathcal{B}$ is i-successful. Then, $\mathcal{S}_i$ is a subset of $\Gamma_i \times \Upsilon_i$ with probability $p_i \triangleq Pr[(\alpha_i, \beta_i) \in \mathcal{S}_i]$. Applying the Splitting Lemma and set $\lambda \leftarrow p_i/2$, we know that there exists a subevent φ_i of $\mathcal{S}_i$ such that $\Pr[(\alpha_i, \beta_i) \in \varphi_i] \geq p_i/2$ (according to (i)), and for each $(\alpha, \beta) \in \varphi_i$, the probability that $(\alpha, \beta') \in \mathcal{S}_i$ over a random choice of β' in φ_i is also at least $p_i/2$ (according to (ii). Therefore, the probability that the outcome (α, β) of the first run of $\mathcal{B}$ in our algorithm is in φ_i is at least $p_i/2$, and, for each of those outcomes, the probability over the random choice of $\beta' = (c_{(1,i)}', \cdots, c_{(1,q_{\mathcal{H}_2})}')$ that the second run outcome (α, β') is in $\mathcal{S}_i$ is at least $p_i - p_i/2 = p_i/2$. Since $c_1' = c_{(1,i)}'$ is uniformly chosen in $\{0, \cdots, p-1\}$, with probability $1/p$ it will collide with $c_{(1,i)} = c_{(1,i)}'$. Consequently, we have that $(\alpha, \beta) \in \varphi_i$, $(\alpha, \beta') \in \mathcal{S}_i$ and $c_{(1,i)} = c_{(1,i)}'$ with probability at least $p_i/2 \cdot (p_i/2 - 1/p)$ which implies that both runs are i-successful and $c_{(1,i)} = c_{(1,i)}'$. That is, the event $\mathcal{S}_i^*$ occurs.

Since p_i is the probability that a run of $\mathcal{B}$ is i-successful, define $Adv_{\mathcal{B}}^{UDVS}$ be the probability of event Λ_2 that $\mathcal{B}$ breaks the unforgeability of our scheme and gains the knowledge of $h^* = \log_{\zeta^*}^{T^*}$. We have

$$Adv_{\mathcal{B}}^{UDVS} = \Sigma_{i=1}^{q_{\mathcal{H}_s}} p_i \quad \text{and}$$

$$Pr[\mathcal{S}^*] = \Sigma_{i=1}^{q_{\mathcal{H}_s}} Pr[\mathcal{S}_i^*] = \Sigma_{i=1}^{q_{\mathcal{H}_s}} p_i/2 \cdot (p_i/2 - 1/p) = \Sigma_{i=1}^{q_{\mathcal{H}_s}} (p_i^2/4 - p_i/(2p))$$
$$\geq 1/(4q_{\mathcal{H}_2}) \cdot (\Sigma_{i=1}^{q_{\mathcal{H}_s}} p_i)^2 - \Sigma_{i=1}^{q_{\mathcal{H}_s}} p_i/(2p)$$
$$= 1/(4q_{\mathcal{H}_2}) \cdot (Adv_{\mathcal{B}}^{UDVS})^2 - Adv_{\mathcal{B}}^{UDVS}/(2p).$$

$Pr[\mathcal{S}^*]$ is the success probability on $\mathcal{A}$ to extract h^* where the inequality above comes from the Cauchy-Schwartz inequality. Note that $Adv_{\mathcal{B}}^{UDVS}/(2p)$ can be neglect since P is large. This ends the proof of event Λ_2.

If event Λ_3 happened, using the security proof similar to that for proving event Λ_2 (this time, changes the output of the $\mathcal{H}_{DL}$ oracle at the second run, instead of the $\mathcal{H}_2$ oracle), $\mathcal{A}$ can extract the private key sk_v corresponding to the public key pk_v, which is equal to $\log_{\bar{g}}^{\bar{y}}$ and is the answer of the DL-problem given by $\mathcal{C}$. Since the technique is the same as that for proofing the event Λ_2, we omit the detail of this proof.   □

6   Performance Comparison

In this section, we compare our VKF-UDVS scheme with most of the existing UDVS schemes in terms of the extended signature, the verifier key flexibility (Key-Flex), the computation cost for a designator and the computation cost for a designated-verifier. By $1E_{\langle G \rangle}$, we mean 1 elliptic curve multiplication on the group $\langle G \rangle$. $1Ex_p$ means 1 exponentiation on the group $\mathbb{Z}_p^*$ and $1P$ means 1 pairing computation. Usually, $1P$ is about 10 times more expensive than $1E_{(.)}$. All of these schemes are proved in the random oracle model, except ZFI05 [18] and Verg06$_2$ [17] which are proved in the standard model.

Table 1. Performance Comparison

Scheme	Extended Sign.	Key-Flex	DV-Sig Length	Cost (Designator)		Cost (DVerifier)
				PV-verify	Designation	DV-verify
Ours-DL	**ECDSA**	**Yes**	$5 \cdot 160$ $(\approx 0.8kb)$	$2E_{\langle G \rangle}$	$2E_{\langle G \rangle} + 2Ex_{\bar{p}}$	$4E_{\langle G \rangle} + 2Ex_{\bar{p}}$
Ours-RSA	**ECDSA**	**Yes**	$1024 + 160 \cdot 4$ $(\approx 1.6kb)$	$2E_{\langle G \rangle}$	$2E_{\langle G \rangle} + 1Ex_{\bar{N}}$	$4E_{\langle G \rangle} + 1Ex_{\bar{N}}$
BNS05 [3]	BLS	N/A	$160 + |Proof|$ $(\approx 0.6kb)$	$2P$	$1E_{G_1} + 2E_{G_2}$	$2P + 2E_{G_2}$
SPWP03 [14]	BLS	No	1024 $(\approx 1.0kb)$	$2P$	$1P$	$1P + 1E_{G_1}$
Verg06$_1$ [17]	BLS	No	$2 \cdot 160$ $(\approx 0.3kb)$	$2P$	$1E_{G_1} + 1E_{G_2}$	$2P + 1E_{G_1}$
SWP04$_1$ [15]	Schnorr	No	$2 \cdot 1024$ $(\approx 2.0kb)$	$2Ex_p$	$1Ex_p$	$2Ex_p$
SWP04$_2$ [15]	Schnorr	No	$1024 + 3 \cdot 160$ $(\approx 1.5kb)$	$2Ex_p$	$1Ex_p + TH$	$3Ex_p + TH$
SWP04$_3$ [15]	RSA	No	$1024 + l_F + l_J +$ $\lceil l_J / \log_2^e \rceil \cdot 1024$ $(\approx 11.6kb)$	$1Ex_N$	$2(\lceil l_J / \log_2^e \rceil$ $+1)Ex. + TH$	$(\lceil l_J / \log_2^e \rceil$ $+1)Ex. + TH$
ZFI05 [18]	Variant BB	No	$1024 + 2 \cdot 160$ $(\approx 1.3kb)$	$1P + 2E_{G_2}$	$1P + 2E_{G_2}$	$2P + 2E_{G_2}$
Verg06$_2$ [17]	BB	No	$3 \cdot 160$ $(\approx 0.5kb)$	$1P + 2E_{G_2}$	$1E_{G_1} + 2E_{G_2}$	$2P + 3E_{G_2}$

In [15], TH denotes the cost of evaluating the trapdoor hash function F_{pk}. l_J is the bit length of J.

We can see in Table 1 that our scheme is the only scheme providing verifier-key-flexibility. The verifier in BNS05 does not need to have a key-pair. However this scheme is interactive, requiring a 3-move communication for the knowledge proof between designator and designated verifier. Our scheme is quite efficient compared to the other schemes, especially to those schemes based on pairing-based signatures (i.e., BLS and BB signature). The SWP04 schemes are based on Schnorr/RSA signatures and are also very efficient, but the length of the DV-signature is comparatively large, especially their RSA-based scheme ($11.6kb$). In our scheme, when the designated-verifier has a DL-based private/public key-pair, then the size of a DV-signature of our scheme is just $0.8kb$. Also notice that our scheme is the only scheme based on widely used standardized signatures.

7   Conclusion

This paper describes an improvement on previous UDVS schemes. We propose a new UDVS scheme which allows more flexibility in the verifier keys and allows using some widely used signature standards such as DSA or ECDSA as the original signature. In our scheme, the verifier keys can be either RSA-based or DL-based with system settings different from that of signer's. This is the first work on UDVS to achieve both non-interactiveness and general public keys for the designated-signature verification.

Acknowledgement. The authors thank the anonymous referees for the constructive comments.

References

1. Abe, M., Ohkubo, M., Suzuki, K.: 1-out-of-n signatures from a variety of keys. In: Zheng, Y. (ed.) ASIACRYPT 2002. LNCS, vol. 2501, pp. 415–432. Springer, Heidelberg (2002)
2. Araki, S., Uehara, S., Imamura, K.: The limited verifier signature and its application. IEICE Trans. Fundamentals E82-A(1), 63–68 (1999)
3. Baek, J., Safavi-Naini, R., Susilo, W.: Universal designated verifier signature proof (or how to efficiently prove knowledge of a signature). In: Roy, B. (ed.) ASIACRYPT 2005. LNCS, vol. 3788, pp. 644–661. Springer, Heidelberg (2005)
4. Boneh, D., Franklin, M.: Identity-based encryption from the Weil pairing. In: Kilian, J. (ed.) CRYPTO 2001. LNCS, vol. 2139, pp. 213–229. Springer, Heidelberg (2001)
5. Boneh, D., Lynn, B., Shacham, H.: Short signatures from the Weil pairing. In: Boyd, C. (ed.) ASIACRYPT 2001. LNCS, vol. 2248, pp. 514–532. Springer, Heidelberg (2001)
6. Chaum, D.: Designated confirmer signatures. In: De Santis, A. (ed.) EUROCRYPT 1994. LNCS, vol. 950, pp. 86–91. Springer, Heidelberg (1995)
7. Chaum, D., van Antwerpen, H.: Undeniable signatures. In: Brassard, G. (ed.) CRYPTO 1989. LNCS, vol. 435, pp. 212–216. Springer, Heidelberg (1990)

8. Goldwasser, S., Micali, S., Rivest, R.: A digital signature scheme secure against adaptively chosen message atttacks. SIAM Journal on Computing 17(2), 281–308 (1988)
9. Huan, X., Susilo, W., Mi, Y., Wu, W.: Universal designated verifier signature without delegatability. In: Ning, P., Qing, S., Li, N. (eds.) ICICS 2006. LNCS, vol. 4307, pp. 479–498. Springer, Heidelberg (2006)
10. Jakkobsson, M., Sako, K., Impagliazzo, T.: Designated verifier proofs and their applications. In: Maurer, U.M. (ed.) EUROCRYPT 1996. LNCS, vol. 1070, pp. 143–154. Springer, Heidelberg (1996)
11. NIST (National Institute for Standard and Technology), Digital Signature Standard (DSS), FIPS PUB, vol. 186 (1994)
12. Pointcheval, D., Stern, J.: Security argrments for digital signatures and blind signatures. Journal of Cryptology 13(3), 361–396 (2000)
13. Schnorr, C.P.: Efficient signature generation by smart cards. Journal of Cryptology 4(3), 161–174 (1991)
14. Steinfeld, R., Bull, L., Wang, H., Pieprzyk, J.: Universal designated-verifier signatures. In: Laih, C.-S. (ed.) ASIACRYPT 2003. LNCS, vol. 2894, pp. 523–542. Springer, Heidelberg (2003)
15. Steinfeld, R., Wang, H., Pieprzyk, J.: Efficient extension of standard Schnorr/RSA signatures into universal designated-verifier signatures. In: Bao, F., Deng, R., Zhou, J. (eds.) PKC 2004. LNCS, vol. 2947, pp. 86–100. Springer, Heidelberg (2004)
16. Vanstone, S.: Responses to NIST's proposal. Communications of the ACM 35, 50–52 (1992)
17. Vergnaud, D.: New extensions of pairing-based signatures into universal designated verifier signatures. In: Bugliesi, M., Preneel, B., Sassone, V., Wegener, I. (eds.) ICALP 2006. LNCS, vol. 4052, pp. 58–69. Springer, Heidelberg (2006)
18. Zhang, R., Furukawa, J., Imai, H.: Sort signature and universal designated verifier signature without random oracles. In: Ioannidis, J., Keromytis, A.D., Yung, M. (eds.) ACNS 2005. LNCS, vol. 3531, pp. 483–498. Springer, Heidelberg (2005)
19. Zhang, F., Susilo, W., Mu, Y., Chen, X.: Identity-based universal designated verifier signatures. In: Enokido, T., Yan, L., Xiao, B., Kim, D., Dai, Y., Yang, L.T. (eds.) EUC 2005. LNCS, vol. 3823, pp. 825–834. Springer, Heidelberg (2005)

Author Index

Lecture Notes in Computer Science

Sublibrary 4: Security and Cryptology

Vol. 4296: M.S. Rhee, B. Lee (Eds.), Information Security and Cryptology – ICISC 2006. XIII, 358 pages. 2006.

Vol. 4284: X. Lai, K. Chen (Eds.), Advances in Cryptology – ASIACRYPT 2006. XIV, 468 pages. 2006.

Vol. 4283: Y.Q. Shi, B. Jeon (Eds.), Digital Watermarking. XII, 474 pages. 2006.

Vol. 4266: H. Yoshiura, K. Sakurai, K. Rannenberg, Y. Murayama, S.-i. Kawamura (Eds.), Advances in Information and Computer Security. XIII, 438 pages. 2006.

Vol. 4258: G. Danezis, P. Golle (Eds.), Privacy Enhancing Technologies. VIII, 431 pages. 2006.

Vol. 4249: L. Goubin, M. Matsui (Eds.), Cryptographic Hardware and Embedded Systems - CHES 2006. XII, 462 pages. 2006.

Vol. 4237: H. Leitold, E.P. Markatos (Eds.), Communications and Multimedia Security. XII, 253 pages. 2006.

Vol. 4236: L. Breveglieri, I. Koren, D. Naccache, J.-P. Seifert (Eds.), Fault Diagnosis and Tolerance in Cryptography. XIII, 253 pages. 2006.

Vol. 4219: D. Zamboni, C. Krügel (Eds.), Recent Advances in Intrusion Detection. XII, 331 pages. 2006.

Vol. 4189: D. Gollmann, J. Meier, A. Sabelfeld (Eds.), Computer Security – ESORICS 2006. XI, 548 pages. 2006.

Vol. 4176: S.K. Katsikas, J. López, M. Backes, S. Gritzalis, B. Preneel (Eds.), Information Security. XIV, 548 pages. 2006.

Vol. 4117: C. Dwork (Ed.), Advances in Cryptology - CRYPTO 2006. XIII, 621 pages. 2006.

Vol. 4116: R. De Prisco, M. Yung (Eds.), Security and Cryptography for Networks. XI, 366 pages. 2006.

Vol. 4107: G. Di Crescenzo, A. Rubin (Eds.), Financial Cryptography and Data Security. XI, 327 pages. 2006.

Vol. 4083: S. Fischer-Hübner, S. Furnell, C. Lambrinoudakis (Eds.), Trust and Privacy in Digital Business. XIII, 243 pages. 2006.

Vol. 4064: R. Büschkes, P. Laskov (Eds.), Detection of Intrusions and Malware & Vulnerability Assessment. X, 195 pages. 2006.

Vol. 4058: L.M. Batten, R. Safavi-Naini (Eds.), Information Security and Privacy. XII, 446 pages. 2006.

Vol. 4047: M.J.B. Robshaw (Ed.), Fast Software Encryption. XI, 434 pages. 2006.

Vol. 4043: A.S. Atzeni, A. Lioy (Eds.), Public Key Infrastructure. XI, 261 pages. 2006.

Vol. 4004: S. Vaudenay (Ed.), Advances in Cryptology - EUROCRYPT 2006. XIV, 613 pages. 2006.

Vol. 3995: G. Müller (Ed.), Emerging Trends in Information and Communication Security. XX, 524 pages. 2006.

Vol. 3989: J. Zhou, M. Yung, F. Bao (Eds.), Applied Cryptography and Network Security. XIV, 488 pages. 2006.

Vol. 3969: Ø. Ytrehus (Ed.), Coding and Cryptography. XI, 443 pages. 2006.

Vol. 3958: M. Yung, Y. Dodis, A. Kiayias, T.G. Malkin (Eds.), Public Key Cryptography - PKC 2006. XIV, 543 pages. 2006.

Vol. 3957: B. Christianson, B. Crispo, J.A. Malcolm, M. Roe (Eds.), Security Protocols. IX, 325 pages. 2006.

Vol. 3956: G. Barthe, B. Grégoire, M. Huisman, J.-L. Lanet (Eds.), Construction and Analysis of Safe, Secure, and Interoperable Smart Devices. IX, 175 pages. 2006.

Vol. 3935: D.H. Won, S. Kim (Eds.), Information Security and Cryptology - ICISC 2005. XIV, 458 pages. 2006.

Vol. 3934: J.A. Clark, R.F. Paige, F.A.C. Polack, P.J. Brooke (Eds.), Security in Pervasive Computing. X, 243 pages. 2006.

Vol. 3928: J. Domingo-Ferrer, J. Posegga, D. Schreckling (Eds.), Smart Card Research and Advanced Applications. XI, 359 pages. 2006.

Vol. 3919: R. Safavi-Naini, M. Yung (Eds.), Digital Rights Management. XI, 357 pages. 2006.

Vol. 3903: K. Chen, R. Deng, X. Lai, J. Zhou (Eds.), Information Security Practice and Experience. XIV, 392 pages. 2006.

Vol. 3897: B. Preneel, S. Tavares (Eds.), Selected Areas in Cryptography. XI, 371 pages. 2006.

Vol. 3876: S. Halevi, T. Rabin (Eds.), Theory of Cryptography. XI, 617 pages. 2006.

Vol. 3866: T. Dimitrakos, F. Martinelli, P.Y.A. Ryan, S. Schneider (Eds.), Formal Aspects in Security and Trust. X, 259 pages. 2006.

Vol. 3860: D. Pointcheval (Ed.), Topics in Cryptology – CT-RSA 2006. XI, 365 pages. 2006.

Vol. 3858: A. Valdes, D. Zamboni (Eds.), Recent Advances in Intrusion Detection. X, 351 pages. 2006.

Vol. 3856: G. Danezis, D. Martin (Eds.), Privacy Enhancing Technologies. VIII, 273 pages. 2006.

Vol. 3786: J.-S. Song, T. Kwon, M. Yung (Eds.), Information Security Applications. XI, 378 pages. 2006.

Vol. 3108: H. Wang, J. Pieprzyk, V. Varadharajan (Eds.), Information Security and Privacy. XII, 494 pages. 2004.

Vol. 2951: M. Naor (Ed.), Theory of Cryptography. XI, 523 pages. 2004.

Vol. 2742: R.N. Wright (Ed.), Financial Cryptography. VIII, 321 pages. 2003.